Discrete Optimization Algorithms

with Pascal Programs

Discrete Optimization Algorithms

with Pascal Programs

MACIEJ M. SYSŁO

Universität Bonn
West Germany

NARSINGH DEO

Washington State University
Pullman, Washington

JANUSZ S. KOWALIK

Washington State University
Pullman, Washington

Prentice-Hall, Inc.
Englewood Cliffs, NJ 07632

Library of Congress Cataloging in Publication Data

SYSŁO, MACIEJ M.
    Discrete optimization algorithms.

    Includes index.
      1. Mathematical optimization—Data processing.
2. PASCAL (Computer program language) I. Deo, Narsingh,
1936– . II. Kowalik, Janusz S. III. Title.
QA402.5.S94 1983   519.7   83-10983
ISBN 0-13-215509-5

Editorial / production supervision and
interior design: *Aliza Greenblatt*

Cover design: *Edsal Enterprises*

Manufacturing buyer: *Gordon Osbourne*

Printed in the United States of America

10 9 8 7 6 5 4 3 2 1

ISBN 0-13-215509-5

PRENTICE-HALL INTERNATIONAL, INC., *London*
PRENTICE-HALL OF AUSTRALIA PTY. LIMITED, *Sydney*
EDITORA PRENTICE-HALL DO BRASIL, LTDA., *Rio de Janeiro*
PRENTICE-HALL CANADA INC., *Toronto*
PRENTICE-HALL OF INDIA PRIVATE LIMITED, *New Delhi*
PRENTICE-HALL OF JAPAN, INC., *Tokyo*
PRENTICE-HALL OF SOUTHEAST ASIA PTE. LTD., *Singapore*
WHITEHALL BOOKS LIMITED, *Wellington, New Zealand*

TO OUR CHILDREN

Tomasz

Alok, Sandhya, Jyoti

Olaf

Contents

PREFACE   xi

1 LINEAR AND INTEGER PROGRAMMING   1

1.1 Linear Programming   2

1.1.1 Construction of the Revised Simplex Method   3 1.1.2 Numerical Stability   18 1.1.3 Large LP Problems   21
1.1.4 Convergence and Time Complexity   21
1.1.5 A Polynomial-Time Algorithm   23
1.1.6 Problems Equivalent to Linear Programming   26

Problems   28

References and Remarks   35

1.2 Dual Linear Programming Method   37

Problems   50

References and Remarks   53

1.3 Transportation Problem   54

1.3.1 Maximum Flow in a Network   56
1.3.2 Maximum Flow Solution Method of the Transportation Problem   60

Problems   71

References and Remarks   76

1.4   Integer Programming   77

   1.4.1   Algorithms for Integer Programming   80
1.4.2   Construction of the Gomory All-Integer Dual
Method   82   1.4.3   Gomory's All-Integer Dual
Algorithm   86

Problems   96

References and Remarks   99

1.5   Zero-One Integer Programming   100

   1.5.1   Implicit Enumeration   102   1.5.2   The Balas
Zero-One Additive Algorithm   104

Problems   114

References and Remarks   115

2   PACKING AND COVERING   117

2.1   The Knapsack Problem   118

   2.1.1   Problem Formulations and
Applications   118   2.1.2   Lower and Upper
Bounds   123   2.1.3   Reduction Algorithm   126
2.1.4   Approximation Algorithms   135   2.1.5
Exact Methods   142   2.1.6   Computational Results
162

Problems   165

References and Remarks   171

2.2   Covering Problems   176

   2.2.1   Formulations and Applications   176
2.2.2   Reduction Algorithms   179   2.2.3   Implicit
Enumeration Method for the Set-Partitioning
Problem   194   2.2.4   Computational Results   210

Problems   211

References and Remarks   217

3   OPTIMIZATION ON NETWORKS   221

3.1   Computer Representation of a Network   223

3.2   Paths and Trees   226

3.3   Shortest Path Problems   227

   *3.3.1   Single-Source Paths, Nonnegative
   Weights   228   3.3.2   Single-Source Paths, Arbitrary
   Weights   235   3.3.3   Shortest Paths between All
   Pairs of Nodes   242
   3.3.4   Comparative Performances of Shortest-Path
   Algorithms   246*

   Problems   247

   References and Remarks   249

3.4   Minimum Spanning Tree Problem   253

   *3.4.1   Kruskal's Algorithm   253   3.4.2   Prim's
   Algorithm   259   3.4.3   Comparative Performance
   of MST Algorithm   264*

   Problems   265

   References and Remarks   267

3.5   Maximum Flow Problem   269

   Problems   296

   References and Remarks   298

3.6   Minimum-Cost Flow Problem   301

   Problems   314

   References and Remarks   317

3.7   Maximum-Cardinality Matching   320

Problems   337

References and Remarks   340

3.8   Traveling Salesman Problem   343

3.8.1   A Branch-and-Bound Algorithm   346
3.8.2   Approximate Algorithms   361 3.8.3   Local
Search Heuristics   368 3.8.4   Modifications   380

Problems   382

References and Remarks   386

4   COLORING AND SCHEDULING   393

4.1   Graph Coloring   394

4.1.1   Definitions and Basic Properties   395
4.1.2   Independent Set Approach   399
4.1.3   Approximate Sequential Algorithms   404
4.1.4   Backtracking Sequential Algorithm   424
4.1.5   Computational Results   433

Problems   438

References and Remarks   442

4.2   Scheduling Problems   445

4.2.1   Network Scheduling   446 4.2.2   Resource
Constrained Network Scheduling   454
4.2.3   Flow Shop and Job Shop Scheduling   474
4.2.4   General Scheduling Problem   484
4.2.5   Single-Machine Scheduling   489
4.2.6   Parallel Machine Scheduling   496
4.2.7   Parallel Machine Scheduling with
Precedence Constraints   503

Problems   517

References and Remarks   527

INDEX   535

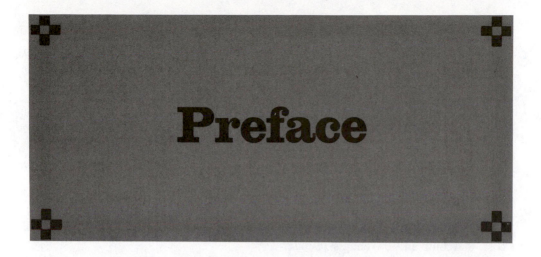

Preface

The field of discrete optimization, which includes linear and integer programming, covering, knapsack, graph-theoretic problems, network analysis, and scheduling, is well established and rich in publications. Yet there are very few books offering a collection of ready-to-use computer programs together with their derivation and performance characteristics. The objective of this book is to fill this gap, at least partly. Our intention has been to write a book in a style similar to the one by Forsythe, Malcolm and Moler on numerical analysis, entitled "Computer Methods for Mathematical Computations," Prentice-Hall, 1977.

We expect the reader to have some familiarity with design, analysis and use of computer algorithms, and programming in Pascal. We have selected Pascal as the implementation language since it is widely used for instruction, description and publication of algorithms. In addition, almost every major computer installation has a Pascal compiler. An elementary background in discrete structures, combinatorics, graph theory, linear and integer programming and matrix algebra is also assumed. Ideally, the reader should have access to a computer to execute programs using the Pascal procedures given in this book. Given these prerequisites the book is self-contained and does not have to be read or used in the order corresponding to the chapter numbering. Each chapter is self-contained.

The book can be used in at least a couple of ways. Firstly, as a *supporting textbook* in discrete optimization courses, such as: discrete structures, optimization, design and analysis of algorithms, and concrete complexity of algorithms.

Supplemented by the readily available reference material the book is appropriate for upper level undergraduate students or graduate students of Computer Science, Mathematics, Electrical Engineering, Industrial Engineering or Operations Research to study algorithmic aspects of discrete optimization and to find a frequently missing link between mathematical description of computer algorithms and quality computer implementations.

Secondly, it can serve as a *software handbook* containing 26 programs which execute the most common algorithms in each topic area. Such a source of codes should be appreciated by academic users and industrial practitioners who need to solve numerically their scientific and applied problems. These procedures and functions can be used directly to solve the problems they have been designed for, or as building blocks for larger computer programs.

The first chapter treats general linear programming, integer linear programming and 0-1 programming problems, and the problem of single-commodity transportation.

Chapter 2 presents special cases of the 0-1 integer programming which have wide applications to packing and covering. In particular, we discuss the knapsack problem, the set partitioning and covering problems.

In Chapter 3, several fundamental network optimization problems are covered. Among them are: shortest path, minimum spanning tree, max-flow and min-flow cost, matching and traveling salesman.

The last chapter comprises coloring and scheduling problems.

Each section covering a particular class of problems starts with the problem definition and applications, followed by a description of the solution methods. The sections on computer implementation contain a user's guide, a sample run, and a Pascal source code. We have used standard Pascal, as defined in Jensen, Wirth's Pascal User Manual and Report, Springer-Verlag, 1978. All codes have been executed on the Amdahl 470 V/6 or V/8 computers at Washington State University. Careful attention has been given to correctness and efficiency of the Pascal procedures. All programs have been extensively tested. In some cases, timing results are provided to give the reader some feeling of how long it would take to solve small- or medium-size problems using the algorithm implemented. Often, timing comparisons of different codes (methods) working on the same input are included.

Although the authors consider the book as their joint project the prime responsibilities for its content and style are as follows: Maciej M. Sysło, Chapters 2 and 4; Narsingh Deo, Chapter 3; and Janusz S. Kowalik, Chapter 1.

We are indebted to many people for their assistance and help. We would particularly like to mention M. S. Krishnamoorthy, M. Kubale and M. A. Langston who shared with us their comments on the manuscript; F. M. Prabhu, M. J. Quinn, W. A. Slough, Y. B. Yoo who prepared the Pascal programs for Chapter 3, Karen Deo for editing assistance and G. L. Crooker, J. A. Frigon, G. L. Rowland, H. Sledziewska, and S. L. Sperry who typed the book.

We thank the Department of Transportation and the National Science Foundation for the funding which helped to develop the book, and the Washington State University for the use of computing facilities.

K. Karlstrom of Prentice-Hall offered us his encouragement and valuable advice.

Maciej M. Sysło
Narsingh Deo
Janusz S. Kowalik

Discrete Optimization Algorithms

with Pascal Programs

1

Linear and Integer Programming

Mathematical programming has both theoretical and practical importance, and constitutes the basis of many computational algorithms for optimization problems. We begin Chapter 1 with the linear programming problem in its general form, where the objective function to be optimized and the constraints imposed on the problem variables are linear. This problem is combinatorial in nature; that is, an optimum solution, if it exists, can be found among a finite set of candidate solutions defined by the linear constraints. In the worst cases the linear programming problems are hard to solve, but in an overwhelming majority of practical cases they are solved very efficiently using the celebrated simplex method. We present two versions of the simplex method; the primal and the dual methods. In addition, a brief outline of a polynomial-time algorithm which has been recently invented by Soviet mathematicians is given.

Some network problems can be formulated as linear programming problems and solved by specialized methods. In Section 1.3 we present one of these problems; the max-flow problem and the associated primal–dual algorithm of Ford and Fulkerson. An interesting feature of this problem is that given integer input data defining the max-flow problem, its optimal solution, obtained for instance by using the simplex method, is also integer. This is not true for a general integer linear programming problem, where special methods have to be employed to obtain an optimal integer solution. One of them, the all-integer Gomory method is described in Section 1.4.

The last section of the chapter presents the Balas method for zero–one integer programming problems. We have selected the methods of Gomory and Balas to illustrate two very different approaches to integer linear programming. The method of Gomory is based on the cutting plane concept, and the method of Balas is based on a very general algorithmic approach called implicit enumeration.

1.1.   LINEAR PROGRAMMING

A linear programming problem is designed to identify a set of nonnegative variables minimizing a linear objective function subject to a set of linear constraints. A *standard form* of the *linear program* (LP) is:

minimize
$$z = \mathbf{c}^T \mathbf{x} \tag{1-1}$$

subject to
$$A\mathbf{x} = \mathbf{b} \tag{1-2}$$

$$\mathbf{x} \geqslant 0 \tag{1-3}$$

where A is a given matrix of order $m \times n$, $m \leqslant n$, $\mathbf{c}$ is an n-row cost vector, $\mathbf{b}$ is an m vector, and $\mathbf{x}$ is an unknown vector of n components. The superscript T denotes vector transposition.

Many practical problems of economics, operations research, decision theory, and engineering can be formulated as LP models. This unquestionably accounts for the enormous popularity of linear programming and its phenomenal growth in the last 30 years. Three other factors also contributed to this growth: (1) elegant mathematics based on the theory of linear algebraic inequalities and equations; (2) rapid development of digital computers, which are necessary tools in the application of linear programming; and (3) linear programming approximations are used in the solution of nearly all mathematical programming problems.

George Dantzig, Leonid Kantorovich, John von Neumann, and Allan Tucker are among many fine mathematicians who have developed theory, computational methodology and uses of linear programming. Dantzig [1963] in Chapter 3 of his comprehensive book on the subject of linear programming and extensions gives an interesting description of the origins and influences of linear programming. Since the late 1940s hundreds of books and thousands of papers on linear programming theory and applications have been written. The reader will have no difficulty in finding suitable sources of information to supplement our concise presentation on the subject of LP.

Before considering linear programming in its standard form [(1-1)–(1-3)], we indicate how various other formulations of linear programs can be converted to standard form.

An inequality constraint $a_{11}x_1 + a_{12}x_2 + \ldots + a_{1n}x_n \leqslant b_1$ can be converted to an equality by introducing a positive variable (*slack variable*)$x_{n+1} \geqslant 0$ as follows:

$$a_{11}x_1 + a_{12}x_2 + \ldots + a_{1n}x_n + x_{n+1} = b_1$$

If the inequality in the original constraint is reversed, the variable x_{n+1} is subtracted in the corresponding equation.

If a linear program is given in standard form except that one or more variables are free (not required to be nonnegative), the problem can be transformed to standard form by replacing unrestricted x_i by $x_i = \bar{x}_i - \hat{x}_i$, where both $\bar{x}_i$ and $\hat{x}_i$ are nonnegative.

If the objective function is maximized, we change the maximization to the minimization by multiplying the objective function by -1.

1.1.1. CONSTRUCTION OF THE REVISED SIMPLEX METHOD

Consider a linear programming problem in its standard form [(1-1)–(1-3)], and assume that the rank of A is m. If B is any nonsingular $m \times m$ submatrix of A, and N is the remaining submatrix of A, we can write (1-2) as

$$[B, N]\begin{bmatrix} \mathbf{x}_B \\ \mathbf{x}_N \end{bmatrix} = \mathbf{b} \tag{1-4}$$

where $\mathbf{x}_B$ and $\mathbf{x}_N$ have m and $n - m$ components, respectively. It is assumed for convenience that B consists of the first m columns of A.

If $\mathbf{x}_N = 0$, the solution to (1-4) is said to be a *basic solution* $\mathbf{x}_B = B^{-1}\mathbf{b}$ and the nonsingular matrix B is the *basis*. If, in addition, $\mathbf{x}_B \geqslant 0$, we say that $\mathbf{x}_B$ is a *basic feasible solution*.

The following theorem establishes the important role of basic feasible solutions.

Theorem 1-1

If there is a feasible solution [satisfying constraints (1-2) and (1-3)], there is a basic feasible solution. Furthermore, if there is an optimal solution minimizing z, there is an optimal basic solution. ∎

We refer the reader to Luenberger [1973] for a proof. This fundamental theorem is crucial to developing the *simplex algorithm*, the most powerful solution method for the general linear programming problems. There exist several distinct versions of this method and many numerical implementations. We will describe the primal simplex method in its revised version, which is the most common LP solver.

First, observe that by partitioning the cost vector $\mathbf{c}$ into two sets of elements related to $\mathbf{x}_B$ and $\mathbf{x}_N$ we can express the objective function z as

$$z = \mathbf{c}_B^T \mathbf{x}_B + \mathbf{c}_N^T \mathbf{x}_N$$

and using

$$\mathbf{x}_B = B^{-1}\mathbf{b} - B^{-1}N\mathbf{x}_N \tag{1-5}$$

we obtain

$$z = \mathbf{c}_B^T B^{-1}\mathbf{b} + \left(\mathbf{c}_N^T - \mathbf{c}_B^T B^{-1}N\right)\mathbf{x}_N$$
$$= z_0 + \mathbf{p}^T \mathbf{x}_N \tag{1-6}$$

where

$$z_0 = \mathbf{c}_B^T B^{-1}\mathbf{b} \tag{1-7}$$

and

$$\mathbf{p}^T = \mathbf{c}_N^T - \mathbf{c}_B^T B^{-1}N \tag{1-8}$$

The value of the objective function z can be improved in the next iteration if we find a negative component of $\mathbf{p}$ and introduce the corresponding nonbasic variable to the basic set. The vector $\mathbf{p}$ can be conveniently computed in two steps:

$$B^T\lambda = \mathbf{c}_B$$

and

$$\mathbf{p}^T = \mathbf{c}_N^T - \lambda^T N$$

where λ is called the vector of *simplex multipliers* and $\mathbf{p}$ the *relative cost vector*. If

$$p_k = c_{Nk} - \lambda^T\mathbf{a}_k < 0$$

for some $m + 1 \leqslant k \leqslant n$, then $\mathbf{a}_k$ is a nonbasic column of A which can enter the basis B in the next iteration and the function value will be reduced. It remains to find the column of B that will be dropped from the basis. Calling the current feasible basic solution $\mathbf{x}_0$, we require that

$$\mathbf{x}_B = \mathbf{x}_0 - B^{-1}\mathbf{a}_k x_{Nk} \geqslant 0 \qquad (1\text{-}9)$$

or

$$\mathbf{x}_B = \mathbf{x}_0 - \mathbf{y}x_{Nk} \geqslant 0$$

where $\mathbf{y}$ is obtained by solving

$$B\mathbf{y} = \mathbf{a}_k$$

Inequality (1-9) must be satisfied in order to maintain feasibility of the next solution $\mathbf{x}_B$. The column leaving the basis can now be found by examining x_{0i}/y_i for $y_i > 0$, $i = 1, 2, \ldots, m$. If

$$\theta = \frac{x_{0l}}{y_l} = \min_{\substack{1 \leqslant i \leqslant m \\ y_i > 0}} \left[\frac{x_{0i}}{y_i} \right]$$

then the variable x_{Bl} becomes zero and is moved to the nonbasic set. Accordingly, the column $\mathbf{a}_l$ joins the nonbasic matrix N.

The Revised Simplex Algorithm

An iteration of the revised simplex algorithm proceeds as follows:

Step 1.   Given is the basis B such that

$$\mathbf{x}_B = B^{-1}\mathbf{b} \geqslant 0$$

Step 2.   Solve

$$B^T\lambda = \mathbf{c}_B$$

for the vector of simplex multipliers λ.

Step 3. Select a column $\mathbf{a}_k$ of N such that

$$p_k = c_{Nk} - \boldsymbol{\lambda}^T \mathbf{a}_k < 0$$

We may, for example, select the $\mathbf{a}_k$ which gives the largest negative value of p_k. If

$$\mathbf{p}^T = \mathbf{c}_N^T - \boldsymbol{\lambda}^T N \geqslant 0$$

stop; the current solution is optimal.

Step 4. Solve for $\mathbf{y}$ the system of equations

$$B\mathbf{y} = \mathbf{a}_k$$

Step 5. Find

$$\theta = \frac{x_{0l}}{y_l} = \min_{\substack{1 \leqslant i \leqslant m \\ y_i > 0}} \left[\frac{x_{0i}}{y_i} \right]$$

If none of the y_i's are positive, then the set of solutions to $A\mathbf{x} = \mathbf{b}, \mathbf{x} \geqslant 0$ is unbounded and z can be made an arbitrarily large negative number. Terminate the computation.

Step 6. Update the basic solution using (1-9):

$$\bar{x}_i = x_i - \theta y_i, \qquad i \neq k$$
$$\bar{x}_k = \theta$$

where $\bar{x}_k$ is the new basic variable.

Step 7. Update the basis B and repeat from step 2.

Step 1 assumes that an initial feasible basic solution is available. In order to find such a solution to $A\mathbf{x} = \mathbf{b}, \mathbf{x} \geqslant 0$, consider an auxiliary minimization problem

$$\min \sum_{i=1}^{m} \hat{y}_i \qquad (1\text{-}10)$$

subject to

$$A\mathbf{x} + I\hat{\mathbf{y}} = \mathbf{b}$$
$$\mathbf{x} \geqslant 0$$
$$\hat{\mathbf{y}} \geqslant 0$$

where $\hat{\mathbf{y}}$ is a vector of *artificial variables*. If we can find an optimal feasible solution to (1-10) such that

$$\sum_{i=1}^{m} \hat{y}_i = 0$$

then we have also obtained a basis yielding solution x_B. If (1-10) has minimum value greater than 0, there is no feasible solution to $Ax = b, x \geqslant 0$. Problem (1-10) is easy to solve using the same simplex method since it has an obvious initial feasible solution $x = 0, \hat{y} = b$, for $B = I$. This *two-phase method* is implemented in this chapter to solve general linear programming problems. Phase I is used to find a feasible solution to $Ax = b, x \geqslant 0$, or to determine that no feasible solution exists. Phase II uses the basic feasible solution generated in phase I and solves problem (1-1)–(1-3).

The main computational effort in the revised simplex method is related to solving repeatedly the systems of equations $B^T \lambda = c_B$ and $By = a_k$, where B's differ only by one column between any two subsequent iterations. A simple way to solve these equations is to compute B^{-1}, and calculate λ and y by the matrix-vector multiplications. Updating of B^{-1} between two subsequent iterations can be accomplished by *pivot operations* applied to B^{-1}. If B^{-1} is the current matrix inverse and $\hat{B}^{-1}$ is the new updated inverse, the following relationship holds (see Problem 1-4):

$$\hat{B}^{-1} = EB^{-1}$$

where E is an elementary matrix of the form

$$E = \begin{bmatrix} 1 & & & & \eta_1 & & \\ & \ddots & 0 & & \vdots & 0 & \\ & & 1 & & \vdots & & \\ & & & \ddots & & & \\ & 0 & & & \eta_l & & \\ & & & & \vdots & \ddots & \\ & & & & \eta_m & & 1 \end{bmatrix}$$

where

$$\eta^T = \left(-\frac{y_1}{y_l}, \ldots, -\frac{y_{l-1}}{y_l}, \frac{1}{y_l}, -\frac{y_{l+1}}{y_l}, \ldots, -\frac{y_m}{y_l} \right)$$

We now have the option of updating the elements of B^{-1} explicitly or developing the *product form of the inverse* (PFI). The first approach is practical as long as m^2 locations of the random access computer memory can be used to store B^{-1}. For large-scale problems that require sequential storage media such as disks or tapes the PFI method offers substantial advantage. In this method we compute a sequence of the E matrices and represent B^{-1} after the pth iteration by the product

$$B^{-1} = E_p \ldots E_3 E_2 E_1 \tag{1-11}$$

We have assumed in (1-11) that the original basis was the unit matrix I. This form of the basis inverse is very convenient for implementation with sequential memory devices. The vector λ^T is calculated by the formula

$$\lambda^T = \left(\left(c_B^T E_p \right) E_{p-1} \right) \ldots E_1$$

and to calculate $\mathbf{y}$ we use

$$\mathbf{y} = E_p \ldots \left(E_2 (E_1 \mathbf{a}_k) \right)$$

Each iteration of the simplex method adds one more vector $\boldsymbol{\eta}$. In practice, only nonzero elements of E_j are stored and, in general, this requires less storage than the explicit matrix B^{-1}. Thus the PFI method has been used successfully for large problems with sparse matrices.

However, as the simplex iterations continue, the PFI representation of B^{-1} requires storing new elementary matrices. To limit the increasing storage requirements and preserve numerical accuracy it is necessary to halt the iterative process periodically and reinvert the matrix B using the columns of A. The reinversion step is then followed by the recomputation of $\mathbf{x}_B$ from the equation $\mathbf{x}_B = B^{-1}\mathbf{b}$, and a sequence of iterative steps until a subsequent reinversion is executed.

Example 1-1

To illustrate the simplex method we will perform one iteration of the revised simplex algorithm using the following example problem.

minimize $\qquad z = -0.5x_1 - 0.4x_2$

subject to

$$x_1 + 2x_2 + x_3 \qquad\qquad = 24$$

$$1.5x_1 + x_2 \qquad + x_4 \qquad = 18$$

$$x_1 \qquad\qquad\qquad + x_5 = 11$$

$$x_1, x_2, \ldots, x_5 \geq 0$$

We have the following vectors and matrices defined:

$$A = \begin{bmatrix} 1 & 2 & 1 & 0 & 0 \\ 1.5 & 1 & 0 & 1 & 0 \\ 1 & 0 & 0 & 0 & 1 \end{bmatrix}, \quad \mathbf{b} = \begin{bmatrix} 24 \\ 18 \\ 11 \end{bmatrix}, \quad \mathbf{c} = \begin{bmatrix} -0.5 \\ -0.4 \\ 0 \\ 0 \\ 0 \end{bmatrix}$$

To start the computation it is convenient to choose

$$\mathbf{x}_B = \begin{bmatrix} x_3 \\ x_4 \\ x_5 \end{bmatrix}, \qquad B = B^{-1} = I, \qquad \mathbf{c}_B = \begin{bmatrix} 0 \\ 0 \\ 0 \end{bmatrix}$$

which gives

$$\mathbf{x}_B = B^{-1}\mathbf{b} = \begin{bmatrix} 24 \\ 18 \\ 11 \end{bmatrix}$$

We calculate λ and $\mathbf{p}$:

$$\lambda^T = \mathbf{c}_B^T B^{-1} = (0,0,0)$$
$$\mathbf{p}^T = \mathbf{c}_N^T - \lambda^T N = (-0.5, -0.4)$$

and select x_1 to enter the basis. Solving for $\mathbf{y}$ the system of equations $B\mathbf{y} = \mathbf{a}_1$ gives

$$\mathbf{y} = \begin{bmatrix} 1 \\ 1.5 \\ 1 \end{bmatrix}$$

The value of θ is

$$\theta = \min\{24/1, 18/1.5, 11/1\} = 11$$

and x_5 leaves the basis. The new basis inverse and basic solution are

$$\hat{B}^{-1} = EB^{-1} = \begin{bmatrix} 1 & 0 & -1 \\ 0 & 1 & -1.5 \\ 0 & 0 & 1 \end{bmatrix}\begin{bmatrix} 1 & 0 & 0 \\ 0 & 1 & 0 \\ 0 & 0 & 1 \end{bmatrix} = \begin{bmatrix} 1 & 0 & -1 \\ 0 & 1 & -1.5 \\ 0 & 0 & 1 \end{bmatrix}$$

$$\hat{\mathbf{x}}_B = \hat{B}^{-1}\mathbf{b} = \begin{bmatrix} 13 \\ 1.5 \\ 11 \end{bmatrix}$$

The first iteration is completed.   ∎

Implemented Algorithm

The PSIMPLEX procedure is based on the revised simplex method. We will describe the computational procedure fully, but some details relating this implementation to our outline of the revised simplex method given in the preceding section are omitted. The reader is asked to determine this relationship as an exercise or consult Gass [1975].

We consider the general linear programming problem in its standard form and assume that $\mathbf{b} \geqslant 0$. For the revised simplex procedure we write this problem as follows:

maximize x_{n+m+1} (1-12)

subject to $a_{11}x_1 + a_{12}x_2 + \ldots + a_{1n}x_n + x_{n+1} \qquad\qquad = b_1$

$\cdots\cdots\cdots\cdots\cdots\cdots\cdots\cdots\cdots\cdots\cdots\cdots\cdots\cdots$

$\qquad a_{m1}x_1 + a_{m2}x_2 + \ldots + a_{mn}x_n + x_{n+m} \qquad\qquad = b_m$

$\qquad a_{m+1,1}x_1 + a_{m+1,2}x_2 + \ldots + a_{m+1,n}x_n + x_{n+m+1} \quad = 0$

$\qquad a_{m+2,1}x_1 + a_{m+2,2}x_2 + \ldots + a_{m+2,n}x_n + x_{n+m+2} \quad = b_{m+2}$

$$x_i \geqslant 0, \quad i = 1, \ldots, n + m$$

where

$$a_{m+1,j} = c_j, \qquad a_{m+2,j} = -\sum_{i=1}^{m} a_{ij}, \qquad j = 1, 2, \ldots, n$$

$$b_{m+2} = -\sum_{i=1}^{m} b_i,$$

We use phase I to determine feasibility and phase II to determine the optimum solution. The last equation in (1-12) is redundant but is needed in phase I of the algorithm. The nonnegative artificial variables, x_{n+i}, $i = 1, \ldots, m$, have been added to the structural variables $x_1, \ldots, x_n$, to create a simple basis matrix $B = I$. Since

$$x_{n+m+2} = -(x_{n+1} + \ldots + x_{n+m})$$

the variable x_{n+m+2} is the negative sum of the artificial variables. Clearly, we must have $x_{n+m+2} \leqslant 0$.

The revised problem has $m + 2$ equations in $n + m + 2$ variables. The basic feasible solution, if it exists, has $m + 2$ variables from the set $\{x_1, \ldots, x_n, x_{n+m+1}, x_{n+m+2}\}$, with $-x_{n+m+1}$ representing the optimal objective function value and $x_{n+m+2} = 0$.

In phase I we maximize x_{n+m+2} subject to the constraints of (1-12). If max $x_{n+m+2} = 0$, we begin phase II with the objective function

$$x_{n+m+1} = -(c_1 x_1 + \ldots + c_n x_n)$$

while keeping $x_{n+m+2} = 0$. We define the matrix $\bar{A}$ as

$$\bar{A} = \left[\begin{array}{c} A \\ \hline \begin{array}{ccc} a_{m+1,1} & \cdots & a_{m+1,n} \\ a_{m+2,1} & \cdots & a_{m+2,n} \end{array} \end{array} \right]$$

which is A augmented by two last rows. Row $m + 2$ is used to compute the relative cost vector $\mathbf{p}$ in phase I. Row $m + 1$ has the same role in phase II. We also introduce an $(m + 2) \times (m + 2)$ matrix which initially is

$$U = \left[\begin{array}{c|c} I_{m \times m} & 0 \\ \hline 0 & I_{2 \times 2} \end{array} \right]$$

The first m rows and columns of U will contain the inverse of B. The last two rows of U will be used to determine the vector entering the basis (row $m + 2$ in phase I and row $m + 1$ in phase II).

Using this notation and initial values $x_{n+i} = b_i$, $i = 1, 2, \ldots, m$, the computational steps in phase I and phase II are as follows:

Phase 1

Step 1.   If $x_{m+n+2} < 0$, calculate

$$\delta_j = \text{row}_{m+2}(U) \cdot \text{col}_j(\bar{A})$$

$$= \sum_{p=1}^{m+2} u_{m+2, p} a_{pj}, \qquad j = 1, 2, \ldots, n$$

and continue. If $x_{m+n+2} = 0$, go to phase II, step 1.

Comment: In phase I the objective function to be maximized is x_{m+n+2}, and all cost coefficients except $c_{m+n+2} = 1$ are zero. The values of δ_j are the components of the relative cost vector $\mathbf{p}$ defined by (1-8).

Step 2.   If all $\delta_j \geq 0$, then x_{n+m+2} is at its maximum and no feasible solution to the original LP problem exists. If at least one $\delta_j < 0$, then the variable to be introduced into the basic set is x_k such that

$$\delta_k = \min_{\substack{1 \leq j \leq n \\ \delta_j < 0}} \delta_j$$

Step 3.   Compute

$$y_i = \text{row}_i(U) \cdot \text{col}_k(\bar{A}) = \sum_{p=1}^{m+2} u_{ip} a_{pk}, \qquad i = 1, 2, \ldots, m + 2$$

Step 4.   Calculate

$$\min_{\substack{1 \leq i \leq m \\ y_i > 0}} \left[\frac{x_i}{y_i} \right] = \frac{x_l}{y_l} = \theta$$

If $y_i \leq 0$ for all $i = 1, 2, \ldots, m$, there is no feasible solution. Otherwise, the variable x_l is eliminated from the basic set.

Step 5.   Calculate the new values of the variables in the basic solution

$$\bar{x}_k = \theta$$
$$\bar{x}_i = x_i - \theta y_i \qquad (i \neq k)$$

for $i = 1, 2, \ldots, m + 2$, and

$$\bar{u}_{lj} = \frac{u_{lj}}{y_l}$$

$$\bar{u}_{ij} = u_{ij} - y_i \frac{u_{lj}}{y_l} \qquad (i \neq l)$$

for $i = 1, 2, \ldots, m + 2$, $j = 1, 2, \ldots, m$. Return to step 1.

Comment: The columns $m + 1$ and $m + 2$ of U do not change. The phase I iterations are continued until $x_{n+m+2} = 0$ or it is determined that no feasible solution exists. In the former case we go to phase II.

Phase II

Step 1.   Here we maintain $x_{m+n+2} = 0$. Compute

$$\gamma_j = \text{row}_{m+1}(U) \cdot \text{col}_j(\bar{A})$$
$$= \sum_{p=1}^{m+2} u_{m+1,p} a_{pj}, \qquad j = 1, 2, \ldots, n$$

Step 2.   Compute

$$\gamma_k = \min_{\substack{1 \leqslant j \leqslant n \\ \gamma_j < 0}} \gamma_j$$

The variable x_k is selected to enter the basic set. If all $\gamma_j \geqslant 0$, then x_{n+m+1} is at its maximum value and the original problem is solved.

Step 3.   Calculate

$$y_i = \sum_{p=1}^{m+2} u_{ip} a_{pk}, \qquad i = 1, 2, \ldots, m + 2$$

Step 4.   Find

$$\min_{\substack{1 \leqslant i \leqslant m \\ y_i > 0}} \left[\frac{x_i}{y_i} \right] = \frac{x_l}{y_l} = \theta$$

If all $y_i \leqslant 0$, the objective function x_{n+m+1} can be made arbitrarily large. The computation is terminated. Otherwise, proceed to step 5.

Step 5.   Calculate

$$\bar{x}_k = \theta$$

$$\bar{x}_i = x_i - \theta y_i \qquad (i \neq k)$$

for $i = 1, 2, \ldots, m + 1$ and

$$\bar{u}_{lj} = \frac{u_{lj}}{y_l}$$

$$\bar{u}_{ij} = u_{ij} - y_i \frac{u_{lj}}{y_l} \qquad (i \neq l)$$

for $i = 1, 2, \ldots, m + 1, j = 1, 2, \ldots, m$.

Return to step 1.

Comment: Steps 2 to 5 of phase I and phase II are very similar. The problems in phase I and phase II have different objective functions and termination conditions. The PSIMPLEX procedure automatically introduces the artificial variables x_{n+1} to x_{n+m}. In more refined versions of the revised simplex method it is possible to use other unit vectors present in the original constraints $A\mathbf{x} = \mathbf{b}$, if their cost coefficients are zero. If it happens that we have m of them, phase I can be omitted. A further improvement related to the technique of finding an initial feasible solution is known as crashing. *Crashing* is an attempt to introduce into the basis as many structural variables as possible. The computation is initialized with a unit basis created by introducing slacks to all inequality constraints and artificials to all equality constraints. Then every structural variable x_i is tested in turn and is made basic, without creating new infeasible variables, provided that this removes an artificial variable from the basis, has a positive reduced cost, or reduces the number of feasible variables.

In our computer implementation we test all nonbasic variables as candidates for entering the basis. Since this is very time consuming for large problems, a technique known as a multiple column selection can be applied to accelerate the function improvement in each iteration step. In this technique, we select $k > 1$ nonbasic columns of A with the largest negative reduced cost, calculate the related vectors $\mathbf{y}$, and perform the ratio tests of step 4, phase II. The variable x_i that causes the greatest reduction in the objective function is accepted as basic. The basis is undated and we seek a new variable that can be moved to the basic set.

Another technique is to select $k > 1$ nonbasic columns with the largest negative reduced cost and form a restricted problem involving the current basis and these k columns. In this suboptimization problem we allow interchange of variables between the basic and nonbasic sets.

In general, the large computer codes do not compute the entire vector of reduced costs to find the variable x_k which corresponds to the largest negative component to the reduced cost vector. Instead, the elements γ_j are calculated one by one and the first nonbasic variable with a sufficiently negative reduced cost is made basic. The time of each iteration is substantially reduced, but we may have to perform more iterations to solve the problem.

Computer Implementation of the Revised Simplex Algorithm

The PSIMPLEX procedure is a Pascal version of the ALGOL 60 procedure called *revsimplex* published in Kucharczyk and Sysło [1975]. It solves the LP problems in standard form:

$$\text{minimize} \qquad \mathbf{c}^T \mathbf{x}$$
$$\text{subject to} \qquad A\mathbf{x} = \mathbf{b}$$
$$\mathbf{x} \geqslant 0$$

where **b** is nonnegative. The detailed description of the algorithm is given in the preceding section.

Global Constants

M   number of constraints, m
N   number of variables $x_1, x_2, \ldots, x_n$
M2   integer whose value is $m + 2$

Data Types

```
TYPE   ARRM2M2 = ARRAY[1..M2,1..M2] OF REAL;
       ARRM2N  = ARRAY[1..M2,1..N] OF REAL;
       ARRM2   = ARRAY[1..M2] OF REAL;
       ARRN    = ARRAY[1..N] OF REAL;
       ARRM    = ARRAY[1..M] OF INTEGER;
```

Procedure Parameters

```
PROCEDURE PSIMPLEX(
     M,N                  :INTEGER;
     EPS                  :REAL;
     VAR A                :ARRM2N;
     VAR B,X              :ARRM2;
     VAR C                :ARRN;
     VAR W                :ARRM;
     VAR F                :REAL;
     VAR NOFEAS,NOSOL  :BOOLEAN);
```

Input

M	number of constraints, m
N	number of variables $x_1, \ldots, x_n$
EPS	small real number such that if for any real number a, $|a| <$ EPS, then $a := 0.0$; we have used $10^{-16} \leqslant$ EPS $\leqslant 10^{-4}$
A[1..M,1..N]	array of the left-hand-side coefficients of $A\mathbf{x} = \mathbf{b}$; additionally the rows M + 1 and M + 2 of this array are used in the procedure to compute the relative cost vector
B[1..M]	array of $\mathbf{b}$; additionally the elements B[M + 1] and B[M + 2] are used for intermediate computations
C[1...N]	cost array $\mathbf{c}$

Output

NOFEAS	Boolean variable equal to TRUE if there is no feasible solution, and FALSE, otherwise
NOSOL	Boolean variable equal to TRUE if the objective function is unbounded, and FALSE, otherwise
W[1..M]	array identifying the optimal basic variables
X[1..M2]	array whose first M elements give the optimal basic solution (X[I] is the value of the optimal variable numbered W[I])
F	optimal value of the objective function

Comment: The program calling PSIMPLEX should first check if NOFEAS = TRUE, then if NOSOL = TRUE. A finite optimal solution exists only if both NOFEAS and NOSOL are FALSE.

Example

Consider the following problem (Gass [1975]):

maximize

$$x_1 + 2x_2 + 3x_3 - x_4$$

subject to

$$x_1 + 2x_2 + 3x_3 \qquad = 15$$

$$2x_1 + x_2 + 5x_3 \qquad = 20$$

$$x_1 + 2x_2 + x_3 + x_4 = 10$$

$$x_i \geqslant 0, \quad i = 1, 2, 3, 4$$

For this problem the input is

$$N = 4, \quad M = 3, \quad M2 = 5 \text{ (global constants)}$$

$$A[1..3, 1..4] = \begin{bmatrix} 1 & 2 & 3 & 0 \\ 2 & 1 & 5 & 0 \\ 1 & 2 & 1 & 1 \end{bmatrix}$$

$$B[1..3] = [15, 20, 10]$$

$$C[1..4] = [-1, -2, -3, 1]$$

The solution obtained is

$$\text{NOFEAS} = \text{NOSOL} = \text{FALSE}$$

I	W[I]	X[I]
1	1	2.5
2	2	2.5
3	3	2.5

and $F_{MIN} = -15$. Hence, the optimal function value for the original problem is $F_{MAX} = 15$.

Pascal Procedure PSIMPLEX

```
PROCEDURE PSIMPLEX(
    M,N                 :INTEGER;
    EPS                 :REAL;
    VAR A               :ARRM2N;
    VAR B,X             :ARRM2;
    VAR C               :ARRN;
    VAR W               :ARRM;
    VAR F               :REAL;
    VAR NOFEAS,NOSOL    :BOOLEAN);


VAR  I,J,K,L,P,Q   :INTEGER;
     D,R,S         :REAL;
     U             :ARRM2M2;
     Y             :ARRM2;
     EX,PHASE,STOP :BOOLEAN;
BEGIN
    NOFEAS:=FALSE;
    NOSOL :=FALSE;
    P:=M+2;
    Q:=M+2;
    PHASE:=TRUE;
    K:=M+1;
    FOR J:=1 TO N DO
```

```
BEGIN
    A[K,J]:=C[J];
    S:=0.0;
    FOR I:=1 TO M DO S:=S-A[I,J];
    A[P,J]:=S
END;
S:=0.0;
FOR I:=1 TO M DO
BEGIN
    W[I]:=N+I;
    R:=B[I];
    X[I]:=R;
    S:=S-R
END;
X[K]:=0.0;
X[P]:=S;
FOR I:=1 TO P DO
BEGIN
    FOR J:=1 TO P DO U[I,J]:=0.0;
    U[I,I]:=1.0
END;
STOP:=FALSE;
REPEAT    (*  UNTIL STOP - PHASE 1  *)
    IF (X[P] >= -EPS) AND PHASE THEN
    BEGIN
        PHASE:=FALSE;
        Q:=M+1
    END;
    D:=0.0;         (*  PHASE 2  *)
    FOR J:=1 TO N DO
    BEGIN
        S:=0.0;
        FOR I:=1 TO P DO
            S:=S+U[Q,I]*A[I,J];
        IF D > S THEN BEGIN D:=S; K:=J END
    END;
    IF D > -EPS THEN
    BEGIN
        STOP:=TRUE;
        IF PHASE THEN NOFEAS:=TRUE
        ELSE F:=-X[Q]
    END
    ELSE
    BEGIN
        FOR I:=1 TO Q DO
        BEGIN
            S:=0.0;
            FOR J:=1 TO P DO
                S:=S+U[I,J]*A[J,K];
            Y[I]:=S
        END;
        EX:=TRUE;
        FOR I:=1 TO M DO
```

```
            IF Y[I] >= EPS THEN
            BEGIN
                S:=X[I]/Y[I];
                IF EX OR (S < D) THEN
                BEGIN D:=S; L:=I END;
                EX:=FALSE
            END;
        IF EX THEN
        BEGIN NOSOL:=TRUE;  STOP:=TRUE END
        ELSE
        BEGIN
            W[L]:=K;
            S:=1.0/Y[L];
            FOR J:=1 TO M DO U[L,J]:=U[L,J]*S;
            IF L = 1 THEN I:=2 ELSE I:=1;
            REPEAT
                S:=Y[I];
                X[I]:=X[I]-D*S;
                FOR J:=1 TO M DO
                    U[I,J]:=U[I,J]-U[L,J]*S;
                IF I = L-1 THEN I:=I+2 ELSE I:=I+1
            UNTIL I > Q;
            X[L]:=D
        END  (*  NOT EX  *)
    END  (*  D <= -EPS  *)
UNTIL STOP
END; (* PSIMPLEX *)
```

1.1.2.   NUMERICAL STABILITY

It is clear that in the process of updating B^{-1} we have no control of the pivots y_l and if some of them happen to be small, the inverse matrix may be computed inaccurately. Numerically safer versions of the revised simplex method are based on matrix factorizations. Two types of factorizations have been considered for this purpose:

1.   Triangular factorization LU, and

2.   Orthogonal factorization LQ,

where L is lower triangular, U is upper triangular, and Q is orthogonal.

A simplex method based on triangular factorization has been developed by Bartels and Golub [1969]. In the reinversion process the basis matrix B is triangularized by the Gaussian eliminations, that is,

$$B_{j+1} = M_j B_j, \qquad j = 1, 2, \ldots, m - 1$$

where $B_1 = B$ and $B_m = U$. After $m - 1$ eliminations we have

$$U = M_{m-1} M_{m-2} \cdots M_1 B$$

where the M_j are of the form

$$M_j = \begin{bmatrix} 1 & & & \\ & \ddots & & \\ & m_{ij} & \ddots & \\ & \vdots & & 1 \end{bmatrix}$$

and $|m_{ij}| \leqslant 1$ if row interchanges are used. Calling

$$\prod_{j=m-1}^{1} M_j = L^{-1}$$

we have

$$B = LU = M_1^{-1} M_2^{-1} \dots M_{m-1}^{-1} U$$

with

$$M_j^{-1} = \begin{bmatrix} 1 & & & \\ & \ddots & & \\ & -m_{ij} & \ddots & \\ & \vdots & & 1 \end{bmatrix}$$

If B is the current basis

$$B = [\mathbf{a}_1, \dots, \mathbf{a}_l, \dots, \mathbf{a}_m]$$

then the new basis B is formed by dropping $\mathbf{a}_l$, shifting columns $\mathbf{a}_{l+1}, \dots, \mathbf{a}_m$ to the left, and inserting $\mathbf{a}_k$ as the last column; that is,

$$\hat{B} = [\mathbf{a}_1, \dots, \mathbf{a}_{l-1}, \mathbf{a}_{l+1}, \dots, \mathbf{a}_m, \mathbf{a}_k]$$

We now have

$$H = L^{-1}\hat{B} = [\mathbf{u}_1, \dots, \mathbf{u}_{l-1}, \mathbf{u}_{l+1}, \dots, \mathbf{u}_m, L^{-1}\mathbf{a}_k]$$

where $\mathbf{u}_1, \dots, \mathbf{u}_m$ are the columns of U. The matrix H is an upper Hessenberg matrix of the form

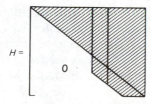

and can be triangularized by a sequence of Gaussian eliminations

$$\hat{U} = \hat{M}_{m-1} \ldots \hat{M}_{l+1} \hat{M}_l H$$

where $\hat{U}$ is triangular and

$$\hat{M}_j = \begin{bmatrix} 1 & & \\ & \ddots & \\ & \hat{m}_{j+1,j} & \ddots \\ & & & 1 \end{bmatrix}$$

If the elimination is accompanied by row interchanges (in each step there are two rows to choose from), then we have a numerically stable updating process where $|\hat{m}_{j+1,j}| \leqslant 1$.

The new updated basis is

$$B = LH = L\hat{M}_l^{-1}\hat{M}_{l+1}^{-1} \ldots \hat{M}_{m-1}^{-1}\hat{U}$$

$$= \hat{L}\hat{U}$$

The factors of L^{-1} can be stored in product form, as in the PFI method. The elements of U, however, have to be accessed individually and are usually stored in linked-list form. This form of storage is very convenient for inserting or deleting elements of U without shifting around the remaining nonzeros.

The LU method is numerically safe but in its pure form it ignores the problem of the proliferation of nonzeros in the process of reinversion and updating. The overall density of nonzero elements in many large-scale LP problems is well below 1%. It is, therefore, important that the process of computing B^{-1} or its equivalent gives sparse representation of the inverse. Reid [1971] has suggested a minor modification to the Bartels–Golub method, which allows a wider choice of pivot elements. The modification is to allow $|m_{ij}| \leqslant \gamma$, $\gamma > 1$, and choose a pivot that locally minimizes the increase of nonzeros in the factors of B. Thus compromise between sparsity preservation and numerical stability can be accomplished. However, flexibility of using the LU factorization in minimizing proliferation of nonzero elements is limited by the fact that the LU factorization does not exist for all row and column permutations of B, and for some it may be poorly determined numerically. Therefore, we do not have full freedom in choosing permutations solely on the basis of sparsity considerations.

An attractive alternative to the LU decomposition in the simplex algorithm is the orthogonal LQ factorization, well known for its favorable numerical stability properties. This factorization also has other nice properties which can be exploited to maintain sparsity of the L factor which is needed in the simplex iteration. Readers interested in this subject are referred to the studies by Gill and Murray [1970] and Saunders [1972].

1.1.3.   LARGE LP PROBLEMS

So far we have assumed that the matrix of constraint coefficients A had no special structure. In many applications, particularly, in large-system problems, the matrix A has definite structure features; for example, the nonzero elements of A may appear in diagonal blocks, except for relatively few rows or columns. Problems with such structures must be solved by specialized algorithms which take advantage of these features. Not only can significant gains in computing costs be achieved by these tailored algorithms, but in many instances very large problems cannot be solved by general LP solvers due to excessive storage and time requirements.

One of the first significant methods designed to solve large-scale problems with a block diagonal structure has been developed by Dantzig and Wolfe in 1960. Since this time many techniques handling special structures have been explored and applied in practice. Some of the most important algorithms for optimizing large linear and nonlinear systems are discussed by Lasdon [1970].

1.1.4.   CONVERGENCE AND TIME COMPLEXITY

The simplex method is based on the principle that the optimal value of a linear program, if it exists, is always attained at a basic solution. Assuming that all basic feasible solutions are nondegenerate, the simplex method produces optimal solution in a finite number of iterations since the number of possible bases is finite and no one is repeated more than once. The linear program is said to be nondegenerate if $B^{-1}\mathbf{b} > 0$. This is equivalent to saying that $\mathbf{b}$ cannot be expressed as a linear combination of any set of $m - 1$ or less columns of A. If the LP problem is nondegenerate, every basic feasible solution has a unique basis associated with it. In the degenerate case we may encounter a series of iterations generating a series of bases $B_i, B_{i+1}, \ldots, B_p$ which all correspond to the same basic feasible solution and the same objective function value. It may also happen that $B_i = B_p$, and the simplex method would cycle indefinitely. Cycling is not a serious problem and virtually never occurs in practical implementations. Most of the LP computer codes do not have any rules resolving degeneracy. However, cycling is a real problem in some integer linear problems solved by extensions of the simplex method and it has to be prevented in the integer programming codes.

To summarize the above, we state the following theorem.

Theorem 1-2

Starting with a feasible basis, the simplex method finds, after a finite number of iterations, one of the following:

1.   An optimal solution, or
2.   A feasible basis which allows us to determine that the LP problem is unbounded ∎

An important practical question arises: How many iterations are needed to solve a given LP problem with n variables and m constants? It is difficult to estimate precisely the number of iterations necessary to produce and optimal solution. For a problem with n variables and m constraints, there are at most

$$\binom{n}{m} = \frac{n!}{m!(n-m)!}$$

basic solutions. This is, however, an extremely poor bound. The simplex method does not examine all basic solutions whose number is very high for any medium-size problem. In fact, it examines a very small subset of the basic feasible solutions. Experience with many thousands of real-life problems show that the number of iterations varies between m and $3m$. Some claim that for a randomly chosen problem with fixed m, the number of iterations is proportional to n. A reasonably safe estimate for the number of iterations is $2(n+m)$.

The maximum possible number of basic feasible solutions is another matter. In one pivot step, the simplex algorithm moves from one extreme point of the set of feasible solutions to an adjacent extreme point. The objective function value decreases monotonically as the path progresses. This path of basic feasible solutions is called an *isotonic path* with respect to the objective function $z = \mathbf{c}^T \mathbf{x}$. The length of the path is defined to be l, the number of points in it, excluding the initial point. Klee and Minty [1972] have answered the following question: Given the LP problem: $\min\{\mathbf{c}^T\mathbf{x} : A\mathbf{x} \geqslant \mathbf{b}, \mathbf{x} \geqslant 0\}$ where A is $m \times n$, what is the maximum length of an isotonic path? The simplex method can be made to follow such a path by appropriately selecting basic variables in the iterative steps. They have constructed an example showing that the maximum-length path might be as high as $2^k - 1$, where

$$k = \frac{m}{2} = n$$

(see Problem 1-11). This number is not bounded by any polynomial in m and n. Thus it is possible to design a variant of the simplex method which would follow an isotonic path and this algorithm would not be polynomially bounded. However, if any standard simplex algorithm (such as the version developed in this chapter) is applied to the Klee–Minty example, only one iteration is needed to solve it. (Why?) Clausen [1980] presented an LP problem with n rows and $2n$ columns, which when solved by our algorithm would require 2^{n-1} iterations (see Problem 1-14).

Thus from the rigorous theoretical viewpoint the simplex method has the serious deficiency of not being polynomially bounded with respect to its time complexity. Instead, its bound is exponential, and this usually characterizes algorithms whose use is limited to relatively small problems. Hence a challenging question is: Why does the simplex method exhibit such a good performance in the vast majority of practical problems of any size?

1.1.5. A POLYNOMIAL-TIME ALGORITHM

A fundamentally new approach to linear programming has been suggested by Khachian [1979, 1980], who exploited some earlier ideas of N. Z. Shor, and investigated solvability of the linear inequality problem $A\mathbf{x} < \mathbf{b}$, with integer coefficients but rational $\mathbf{x}$. His departure point for constructing the new algorithm is the observation that if the feasible region is not empty, there exists a set S in the feasible region, of some given volume, and such that S is within a finite distance from the origin of the coordinate system.

The Khachian polynomial-time method determines whether a system of linear strict inequalities $A\mathbf{x} < \mathbf{b}$ (where A is $m \times n$) is satisfiable. It consists of the following steps:

Step 1.   Initialization. Set $\mathbf{x}_k = 0$, $A_k = 2^L I$ and $k = 0$, where I is a unit matrix of size $n \times n$.

Step 2.   If $\mathbf{x}_k$ is feasible (i.e., $A\mathbf{x}_k < \mathbf{b}$), terminate the computation. If not and $k = 6n^2 L$, also terminate since there is no solution. Otherwise, continue.

Step 3.   Select any inequality in $A\mathbf{x} < \mathbf{b}$ that is violated at $\mathbf{x}_k$ (e.g., $\mathbf{a}_i^T \mathbf{x}_k \geq b_i$) and set

$$\mathbf{x}_{k+1} = \mathbf{x}_k - \frac{1}{n+1} \frac{A_k \mathbf{a}_i}{\sqrt{\mathbf{a}_i^T A_k \mathbf{a}_i}}$$

and

$$A_{k+1} = \frac{n^2}{n^2 - 1} \left(A_k - \frac{2}{n+1} \frac{(A_k \mathbf{a}_i)(A_k \mathbf{a}_i)^T}{\mathbf{a}_i^T A_k \mathbf{a}_i} \right)$$

Set $k := k + 1$ and return to step 2.

The value of L depends on the number of coefficients in A and $\mathbf{b}$ and their size. Specifically,

$$L = \sum_{i,j=1}^{m,n} \log_2 \left(|a_{ij}| + 1 \right) + \sum_{i=1}^{m} \log_2 (|b_i| + 1) + \log_2 (mn) + 1.$$

Thus L is the total number of bits required to store all the coefficients in $A\mathbf{x} < \mathbf{b}$ in binary words of a computer. Assuming that all coefficients a_{ij} and b_i are approximately of the same magnitude, say 2^p, we get

$$L > 2 + \log_2 mn + mp + mnp > mp(n+1)$$

The polynomial-time property of Khachian's method follows from the fact that in the worst case the algorithm is halted after $K = 6n^2 L$ steps and L is polynomial in m and n.

The geometrical basis of the algorithm is the creation of a sequence of ellipsoids

$$(\mathbf{x} - \mathbf{x}_k)^T A_k^{-1}(\mathbf{x} - \mathbf{x}_k) \leq 1$$

each of decreasing volume and containing a feasible solution, if one exists. The algorithm starts with a sphere centered at $\mathbf{x}_0 = 0$ and with the radius 2^L sufficiently large to contain some set S of feasible points. If the center of ellipsoid $\mathbf{x}_k$ is not feasible, then a hyperplane parallel to a violated constraint going through the center is used to cut the ellipsoid in half. This half ellipsoid is used to create the next ellipsoid, which envelopes this half of the current ellipsoid. The new ellipsoid has a volume smaller than the preceding one (see Fig. 1-1). It passes through $\mathbf{z}_1$ and $\mathbf{z}_2$, has its center on the solution side of the chord c, and is tangential to the old ellipsoid at the point T. The new ellipsoid is the one with the smallest volume that has the aforementioned properties. It can be shown that if the algorithm does not find a solution to $A\mathbf{x} < \mathbf{b}$ after $k = 6n^2L$ steps, we can conclude that the feasible set is empty.

This result can be extended to show that the LP problems can, in theory, be solved in polynomial time.

Every primal LP problem

minimize $\mathbf{c}^T\mathbf{x}$

subject to $A\mathbf{x} \geq \mathbf{b}$

                                                  $\mathbf{x} \geq 0$

has an associated dual problem

maximize $\mathbf{b}^T\mathbf{y}$

subject to $A^T\mathbf{y} \leq \mathbf{c}$

                                                  $\mathbf{y} \geq 0$

such that $\min \mathbf{c}^T\mathbf{x} = \max \mathbf{b}^T\mathbf{y}$ if and only if the primal problem has a finite optimum (see Section 1.2, where duality is explained). It follows that the linear system

$$\mathbf{c}^T\mathbf{x} \geq \mathbf{b}^T\mathbf{y}$$
$$A\mathbf{x} \geq \mathbf{b}$$
$$A^T\mathbf{y} \leq \mathbf{c}$$
$$\mathbf{x}, \mathbf{y} \geq 0$$

has a solution if and only if the primal problem has a finite minimum. If we can find a solution $\mathbf{x}^*, \mathbf{y}^*$ to this linear system, then $\mathbf{x}^*$ is an optimal solution to the primal

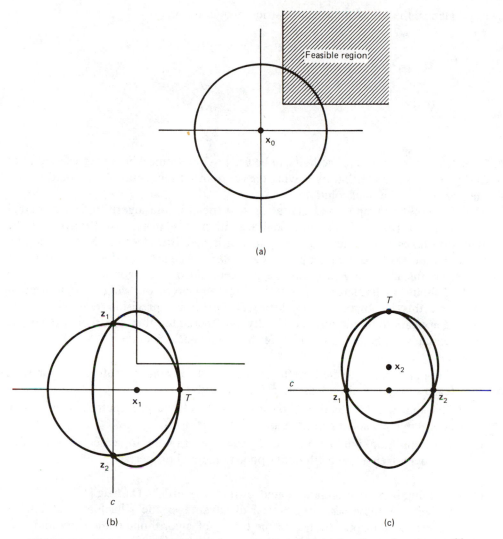

Figure 1-1 *Geometry of Khachian's method*: (*a*) *Initial sphere for the problem* $x_1 > 1, x_2 > 1$; (*b*) *First ellipsoid*; (*c*) *Second ellipsoid*.

LP problem. We can transform this system of weak inequalities to a system of strict equalities using the following lemma (see Gács and Lovász [1979]):

Lemma

The system of linear inequalities $\mathbf{a}_i^T \mathbf{x} \leqslant b_i$, $i = 1, 2, \ldots, m$, has a solution if and only if the system of linear strict inequalities $\mathbf{a}_i^T \mathbf{x} < b_i + 2^{-L}$, $i = 1, 2, \ldots, m$, has a solution. ∎

Hence, the transformed set of strict inequalities is

$$\mathbf{b}^T\mathbf{y} - \mathbf{c}^T\mathbf{x} < 2^{-L}$$
$$\mathbf{b} - A\mathbf{x} < 2^{-L}$$
$$A^T\mathbf{y} - \mathbf{c} < 2^{-L}$$
$$-\mathbf{x} < 2^{-L}$$
$$-\mathbf{y} < 2^{-L}$$

Now Khachian's ellipsoid method can be used to determine whether $\mathbf{x}^*$ exists and to find it if it does. Note that by solving the set of strict inequalities, we would obtain the primal and the dual solutions.

It should be emphasized that at present the ellipsoid algorithm in its original form is not a practical computational algorithm. The worst-case behavior of the algorithm takes place when a given system of inequalities does not have a solution, and the number of required iterations is $K = 6n^2L$. The numerical values of L would be astronomical for any medium-size problem. Also, the rigorous version of the method should be implemented with floating-point precision using $O(nL)$-bit numbers, and this is obviously unrealistic. The excessive precision required by the method appears to be the prime difficulty, but there are other nontrivial drawbacks of Khachian's method. A partial list of these drawbacks includes:

1. The points $\mathbf{x}_k$ generated by the method tend to wander erratically around the feasible region until a feasible point is found.
2. The matrices A_k tend to converge to a fully dense matrix. This is not acceptable for solving large sparse LP problems.
3. The condition number of the A_k matrices rapidly increases to very high numbers. Hence the method has poor numerical properties.

Thus "solving in a polynomially bounded time says little" (Dantzig [1979]).

In spite of these serious practical disadvantages the Khachian method has solved a very significant problem in the theory of algorithmic complexity and may be useful for other optimization problems. It has also generated new approaches to the LP and LP-related problems.

An extensive bibliography for the ellipsoid algorithm can be found in Wolfe [1980].

1.1.6. PROBLEMS EQUIVALENT TO LINEAR PROGRAMMING

It is of interest to identify a class of problems which are equivalent to LP in the sense that they are equally hard to solve. More specifically, an LP-equivalent problem can be reduced in polynomial time to LP. This notion of equivalence allows us to consider any LP-equivalent problem as a linear programming problem and apply the simplex method to solve it. It can be shown that many mathematical problems are LP-equivalent. Some of the more interesting cases include the following five problems.

Linear inequalities

Given is an integer $m \times n$ matrix A and integer m-vector $\mathbf{b}$, find if there exists a rational vector $\mathbf{x}$ such that $A\mathbf{x} \leqslant \mathbf{b}$. This problem can be cast in a geometric form.

The geometric form of the linear inequalities problem is: Given closed halfspaces $H_i = \{\mathbf{x}: \mathbf{a}_i^T\mathbf{x} \leqslant b_i\}$, $i = 1, 2, \ldots, n$, determine if the intersection $H_1 \cap H_2 \cap \ldots \cap H_n$ is nonempty. A *closed halfspace* is a set of points on or on one side of a hyperplane $\mathbf{a}_i^T\mathbf{x} \leqslant b_i$.

Note that in this problem we are only trying to determine if a solution exists rather than seeking a solution. The reader is asked to show that the linear inequality and LP problems are equivalent.

Relevancy

In formulating the LP problems we may be interested to know if a particular set of constraints is redundant. A set is *redundant* if there is a constraint that is always satisfied when all other constraints are satisfied. This leads to the problem of relevancy: Given a set of constraints $\mathbf{a}_i^T\mathbf{x} \leqslant b_i$, $i = 1, 2, \ldots, m$, determine if satisfying the last $m - 1$ constraints is equivalent to satisfying all m constraints. It is interesting to note that a similar problem—Given a set of linear constraints $\mathbf{a}_i^T\mathbf{x} \leqslant b_i$, $i = 1, 2, \ldots, m$, and an integer k, determine some set of k constraints equivalent to satisfying all the constraints—is a much more difficult problem, not LP-equivalent. (See Problem 1-16 for further comments.)

Simplified linear programming

Given an integer $m \times n$ matrix A, determine if there is a rational vector $\mathbf{x}$ such that $A\mathbf{x} = 0$, $\mathbf{x} \neq 0$, $\mathbf{x} \geqslant 0$. This problem is similar to the one we solve in phase I of the simplex method.

Extreme point

Given a set of points in the Euclidean space E^n, the *convex hull* is the smallest convex set containing these points.

A point $\mathbf{Q}$ is *extreme* with respect to the points $\mathbf{P}_1, \mathbf{P}_2, \ldots, \mathbf{P}_n$ if and only if $\mathbf{Q}$ is exterior to the convex hull of $\mathbf{P}_1, \ldots, \mathbf{P}_n$. The extreme point problem is: Given a set of points $\mathbf{P}_0, \mathbf{P}_1, \ldots, \mathbf{P}_n$ in E^n, determine if $\mathbf{P}_0$ is extreme with respect to $\mathbf{P}_1, \mathbf{P}_2, \ldots, \mathbf{P}_n$.

It is well known that $\mathbf{P}_0$ is interior to the convex hull of $\mathbf{P}_1, \ldots, \mathbf{P}_n$ if and only if

$$\sum_{i=1}^{m} x_i \mathbf{P}_i = \mathbf{P}_0, \quad x_i \geqslant 0 \quad \text{and} \quad \sum_{i=1}^{n} x_i = 1$$

and this problem can be solved by the simplex method.

Set–set separation

Two point sets are said to be *separable* if and only if there is a hyperplane H such that all points of one set lie on one side of H and all points of the other set lie on the other side of H.

The set–set separation problem is: Given points $\langle \mathbf{P}_1, \mathbf{P}_2, \ldots, \mathbf{P}_n \rangle$ and $\langle \mathbf{Q}_1, \mathbf{Q}_2, \ldots, \mathbf{Q}_n \rangle$, determine if the sets $\langle \mathbf{P}_1, \mathbf{P}_2, \ldots, \mathbf{P}_n \rangle$ and $\langle \mathbf{Q}_1, \mathbf{Q}_2, \ldots, \mathbf{Q}_n \rangle$ are separable.

More on the LP-equivalent problems and related proofs can be found in Dobkin and Reiss [1980].

PROBLEMS

1.1.   **Pivoting.** Developing the simplex method we have assumed that the original linear system $A\mathbf{x} = \mathbf{b}$ can be transformed to the *basic system*

$$\mathbf{x}_B + B^{-1}N\mathbf{x}_N = B^{-1}\mathbf{b}$$

which is equivalent to $A\mathbf{x} = \mathbf{b}$; that is, it has the same solutions. This transformation can be accomplished by the pivot operations. A *pivot operation* consists of the following steps:

1. Select a *pivot* $a_{lk} \neq 0$ in the system being transformed.
2. Replace equation l by the equation l divided by a_{lk}.
3. For $i = 1, 2, \ldots, m, i \neq l$, replace equation i by the sum of equation i and the replaced equation l multiplied by $-a_{ik}$.

The transformation from $A\mathbf{x} = \mathbf{b}$ to the basic system is accomplished by a sequence of m pivot operations.

(a) Verify that a pivot operation on a linear system of equations produces an equivalent system.

(b) Transform by pivoting operations the system

$$2x_1 + x_2 + 2x_3 = 4$$
$$3x_1 + 3x_2 + x_3 = 3$$

to an equivalent system of the form

$$x_1 \quad + ux_3 = v$$
$$x_2 + zx_3 = t$$

(c) Show that this can be accomplished by multiplying the original system by the inverse of

$$B = \begin{bmatrix} 2 & 1 \\ 3 & 3 \end{bmatrix}$$

1-2. Can such a transformation always be accomplished without interchanging rows or columns? Why? Give an example.

1-3. Given is a basic system of equations

$$x_i + \sum_{j=m+1}^{n} \bar{a}_{ij} x_j = \bar{b}_i, \qquad i = 1, 2, \ldots, m$$

We want to replace the basic variable x_l by a nonbasic variable x_k. This can be done by pivoting on $\bar{a}_{lk} \neq 0$.

(a) Show that if we perform this pivoting and set the nonbasic variables to zero, we obtain

$$x_i = \bar{b}_i - \frac{\bar{b}_l}{\bar{a}_{lk}} \bar{a}_{ik}, \qquad i = 1, \ldots, m; \quad i \neq k$$

$$x_k = \frac{\bar{b}_l}{\bar{a}_{lk}}$$

1-4. Assume that B and $\hat{B}$ are two basic matrices that differ only by one column vector, that is,

$$B = [\mathbf{a}_1, \mathbf{a}_2, \ldots, \mathbf{a}_l, \ldots, \mathbf{a}_m]$$
$$\hat{B} = [\mathbf{a}_1, \mathbf{a}_2, \ldots, \mathbf{a}_k, \ldots, \mathbf{a}_m]$$

Let

$$B^{-1}\mathbf{a}_k = (y_1, y_2, \ldots, y_l, \ldots, y_m)^T$$

Show that

$$\hat{B}^{-1} = EB^{-1}$$

where

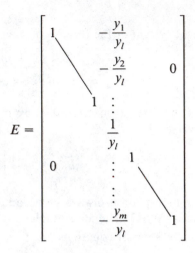

$$E = $$

1-5. Verify that pivoting on $\bar{a}_{lk}$ in Problem 1-3 is equivalent to substituting the column $\bar{a}_k$ for the column $\bar{a}_l$ in the basis.

1-6. Consider the *tableau simplex method*. Assume that for the standard linear programming problem we arrange the initial tableau as follows:

$$T = \left[\begin{array}{c|c} A & \mathbf{b} \\ \hline \mathbf{c}^T & 0 \end{array}\right] = \left[\begin{array}{c|c|c} B & N & \mathbf{b} \\ \hline \mathbf{c}_B^T & \mathbf{c}_N^T & 0 \end{array}\right]$$

Show that if we use the first m columns of A as the basis and pivot on $b_{11}, b_{22}, \ldots, b_{mm}$, we get

$$\hat{T} = \left[\begin{array}{c|c|c} I & B^{-1}N & B^{-1}\mathbf{b} \\ \hline \mathbf{0}^T & \mathbf{c}_N^T - \mathbf{c}_B^T B^{-1}N & -\mathbf{c}_B^T B^{-1}\mathbf{b} \end{array}\right]$$

This tableau contains all information required in the simplex method iterations: the basic solution $\mathbf{x}_B = B^{-1}\mathbf{b}$, the relative cost coefficients $\mathbf{p}^T = \mathbf{c}_N^T - \mathbf{c}_B^T B^{-1}N$, the objective function value $\mathbf{c}_B^T B^{-1}\mathbf{b}$, and the vectors $\mathbf{y}_j = B^{-1}\text{col}_j(N)$. This problem nicely relates the tableau simplex method to the revised simplex method described earlier. The matrix form of the simplex tableau shown has been given by Luenberger [1973]. (*Hint:* The pivoting operations transform the matrix

$$\begin{bmatrix} B \\ \mathbf{c}_B^T \end{bmatrix} \quad \text{to} \quad \begin{bmatrix} I \\ 0^T \end{bmatrix}$$

Construct an $(m + 1) \times (m + 1)$ matrix R such that

$$R\begin{bmatrix} B \\ \mathbf{c}_B^T \end{bmatrix} = \begin{bmatrix} I \\ 0^T \end{bmatrix}$$

and compute $\hat{T} = RT$.)

1-7. **The diet problem.** Assume that we can compose our diet of n different foods. The ith food sells at a price c_i per unit and each unit of it contains a_{ji} units of the jth nutrient. It is required that our diet contain at least b_j, $j = 1, \ldots, m$, units of the jth nutrient per unit of time under consideration. Formulate the linear program determining the least expensive diet that satisfies the minimum nutritional requirements.

1-8. **Nonuniqueness.** Consider the standard linear programming problem. Show that if $\mathbf{x}_1$ and $\mathbf{x}_2$ are two optimal solutions to this problem, then every convex combination of $\mathbf{x}_1$ and $\mathbf{x}_2$, $\mathbf{x} = \lambda\mathbf{x}_1 + (1 - \lambda)\mathbf{x}_2$, $0 \leqslant \lambda \leqslant 1$, is also an optimal solution.

1-9. If P_1 and P_2 are row and column permutation matrices, then the orthogonal factorization of the basis B,

$$P_1BP_2 = LQ$$

where L is lower triangular and Q is orthogonal, exists for all matrices P_1 and P_2. Show that the factor L does not depend on the column permutation matrix. This property of the orthogonal factorization can be used to preserve sparsity in the L matrix.

1-10. In the process of simplex iteration we need to solve two kinds of systems of linear equations:

$$(1) \quad B^T\lambda = \mathbf{c}_B$$
$$(2) \quad B\mathbf{y} = \mathbf{a}_k$$

Suppose that B has been factorized as follows: $B = LQ$, where L is lower triangular and Q is orthogonal. Demonstrate that (1) and (2) can be solved without Q.

1-11. (Klee and Minty [1972])   Consider the linear program in two variables (i.e., $n = 2$):

minimize $\qquad\qquad\qquad\qquad -x_2$

subject to $\qquad\qquad\qquad\qquad A\mathbf{x} \geqslant \mathbf{b}$

where

$$A = \begin{bmatrix} 1 & 0 \\ -1 & 0 \\ -\varepsilon & 1 \\ -\varepsilon & -1 \end{bmatrix}, \quad b = \begin{bmatrix} 0 \\ -1 \\ 0 \\ -1 \end{bmatrix}$$

and $0 < \varepsilon < 1/2$.

(a) Draw the set of feasible solutions and assume that the simplex method starts at $(0,0)$ and proceeds to the solution along the longer path. Verify that the length of this isotonic path is $2^n - 1$.

(b) Construct a similar problem with $n = 3$.

1-12. How many iterations would be required to solve Problem 1-11 if we use the standard simplex method developed in this section? Why?

1-13. Given the following LP problem:

minimize $\qquad\qquad 10x_1 + x_2$

subject to $\qquad\qquad x_1 \qquad\leqslant 1$

$$20x_1 + x_2 \leqslant 100$$

$$x_1, x_2 \geqslant 0$$

How many iterations does the simplex method take to find the solution of this problem? Use our Pascal PSIMPLEX procedure to verify your answer.

1-14. (Clausen [1980]) The following problem requires $2^n - 1$, $(n = 3)$ simplex iterations:

maximize $\quad \frac{1}{16}x_1 + \frac{1}{20}x_2 + \frac{1}{25}x_3$

subject to $\quad x_1 \qquad\qquad\quad + s_1 \qquad\qquad = 1$

$$\frac{5}{2}x_1 + x_2 \qquad\qquad + s_2 \qquad = 5$$

$$\frac{25}{8}x_1 + \frac{5}{2}x_2 + x_3 \qquad\qquad + s_3 = 25$$

$$x_1, x_2, x_3, s_1, s_2, s_3 \geqslant 0$$

Draw the polytope determined by $x_1, x_2, x_3 \geqslant 0$ and the three inequalities obtained by leaving out the slack variables, and verify that the number of iterations of the PSIMPLEX procedure is 7. It is interesting to note that this problem can be solved easily in $n = 3$ steps by a special version of the LP algorithm (Charnes et al. [1980]).

1-15. One test of the accuracy and robustness of the LP algorithms is the problem

(Brocklehurst and Dennis [1979])

minimize $\qquad \qquad \qquad \qquad \mathbf{c}^T\mathbf{x}$

subject to $\qquad \qquad \qquad \qquad A\mathbf{x} = \mathbf{b}$

$$\mathbf{x} \geqslant 0$$

where

$$a_{ij} = \frac{1}{i+j-1}, \qquad i, j = 1, 2, \ldots, n$$

$$c_i = b_i = \sum_{j=1}^{n} \frac{1}{i+j-1}, \qquad i = 1, 2, \ldots, n$$

The exact solution to this problem is $x_i = 1$, $i = 1, 2, \ldots, n$.

The matrix A involved in the test, called the Hilbert matrix, creates a very ill conditioned problem even for moderate values of n. By this we mean that the output results can be very sensitive to disturbances in the input data. We tested the PSIMPLEX procedure running the Hilbert LP problem on an Amdahl 470 V/8. We used double-precision arithmetic and set EPS $= 10^{-16}$. The following number of decimal digits were correct at solutions obtained for different values of n.

n	Correct Digits
6	$\geqslant 7$
7	$\geqslant 7$
8	$\geqslant 7$
9	6
10	4
11	1

Use a computer and determine the largest Hilbert problem you can solve with at least two-digit accuracy.

1-16. The relevancy problem described in Section 1.1.6 can be reduced to the following one:

Given a set of halfspaces $H_2 H_3, \ldots, H_m$, determine if a hyperplane $\mathbf{a}_1^T\mathbf{x} \leqslant b_1$ intersects $H_2 \cap H_3 \cap \ldots \cap H_m$.

Show that this problem is LP-equivalent.

1-17. Using the PSIMPLEX procedure, solve the set of linear equations $A\mathbf{x} = \mathbf{b}$, where A and $\mathbf{b}$ are defined as in Problem 1-15. Try different values of n.

1-18.   Convert the following problem to a LP program in standard form:

minimize $\qquad\qquad\qquad \sum_{i=1}^{n} |x_i|$

subject to $\qquad\qquad\qquad A\mathbf{x} \leqslant \mathbf{b}$

1-19.   Given is an overdetermined system of linear equations, $\mathbf{a}_i^T\mathbf{x} = b_i$, $i = 1, 2, \ldots, m$, $\mathbf{x}^T = (x_1, x_2, \ldots, x_n)$, $m > n$. Show that the following problem can be formulated as an LP problem:

$$\min_{\mathbf{x}} \max_{1 \leqslant i \leqslant m} |b_i - \mathbf{a}_i^T\mathbf{x}|$$

1-20.   Under the same assumptions as in Problem 1-19, show that the problem

$$\min_{\mathbf{x}} \sum_{i=1}^{m} |b_i - \mathbf{a}_i^T\mathbf{x}|$$

can also be formulated as an LP problem.

1-21.   Given is a set of linear algebraic equations

$$\sum_{j=1}^{n} a_{ij}x_j = b_i, \qquad i = 1, 2, \ldots, n$$

Explain why the solution can be obtained by solving the following LP problem:

minimize $\qquad\qquad\qquad y$

subject to $\qquad \sum_{j=1}^{n} a_{ij}z_j - \left(\sum_{j=1}^{n} a_{ij} \right) y = b_i, \qquad i = 1, 2, \ldots, n$

$$z_j \geqslant 0, \quad y \geqslant 0$$

and calculating x_j from

$$x_j = z_j - y, \qquad j = 1, 2, \ldots, n$$

1-22.   Using the PSIMPLEX procedure, solve the following set of equations:

$$
\begin{aligned}
x_1 + x_2 - 3x_3 - x_4 + 3x_5 &= 1 \\
2x_1 - 2x_2 + x_3 - 5x_4 + 2x_5 &= 50 \\
x_1 - x_2 - x_3 \qquad\quad + x_5 &= 0 \\
4x_1 + 3x_2 + x_3 + x_4 - x_5 &= 2 \\
-3x_1 + 4x_2 - 2x_3 + 2x_4 \qquad\;\; &= -32
\end{aligned}
$$

REFERENCES AND REMARKS

In Section 1.1 we refer to the following sources of information:

BARTELS, R. H., and G. H. GOLUB, The Simplex Method of Linear Programming Using LU Decomposition, *Comm. ACM* 12(1969), 266–268

BROCKLEHURST, E. R., and K. DENNIS, A Comparison of Four FORTRAN Routines for Dense Linear Programs, National Physical Laboratory Report NACS 19/79, Teddington, England, Oct. 1979

CHARNES, A., W. W. COOPER, S. DUFFUAA, and M. KRESS, Complexity and Computability of Solutions to Linear Programming Systems, *Internat. J. Comput. Inform. Sci.* 9(1980), 483–506

CLAUSEN, J., A Tutorial Note on the Complexity of the Simplex Algorithm, in J. Krarup and S. Walukiewicz (eds.), *Proceedings of DAPS-79*, Institute of Datalogy, University of Copenhagen, Copenhagen, Apr. 1980

DANTZIG, G. B., *Linear Programming and Extensions*, Princeton University Press, Princeton, N.J., 1963

DANTZIG, G. B., Comments on Khachian's Algorithm for Linear Programming, Technical Report SOL 79-22, Dept. of OR, Stanford University, Stanford, Calif., 1979

DOBKIN, D. P., and S. P. REISS, The Complexity of Linear Programs, *Theoret. Comput. Sci.* 11(1980), 1–18

GÁCS, P., and L. LOVÁSZ, Khachiyan's Algorithm for Linear Programming, *Math. Programming Stud.* 14(1981), 61–68

GASS, S. I., *Linear Programming—Methods and Applications*, 4th ed., McGraw-Hill, New York, 1975

GILL, P. E., and W. MURRAY, A Numerically Stable Form of the Simplex Method, Technical Report No. Math. 87, National Physical Laboratory, Teddington, England, 1970

KHACHIAN, L. G., A Polynomial Algorithm in Linear Programming, *Dokl. Akad. Nauk SSSR*, *Nova Seria* 244:5(1979), 1093–1096; translated as *Soviet Math. Dokl.* 20(1979), 191–194

KHACHIAN, L. G., A Polynomial Algorithm in Linear Programming, *Ž. Vyčisl. Math. i Math. Fiz.* 20(1980), 51–68

KLEE, V., and G. J. MINTY, How Good Is the Simplex Method? in O. Shisha (ed.), *Inequalities III*, Academic Press, New York, 1972, pp. 159–175

KUCHARCZYK, J., and M. M. SYSŁO, *Optimization Algorithms in ALGOL 60*, Państwowe Wydawnictwo Naukowe, Warsaw, 1975 (in Polish)

LASDON, L. S., *Optimization Theory for Large Systems*, Macmillan, New York, 1970

LUENBERGER, D. G., *Introduction to Linear and Nonlinear Programming*, Addison-Wesley, Reading, Mass., 1973

REID, J. K., A Note on the Stability of Gaussian Elimination, *J. Inst. Math. Appl.* 8(1971), 374–375

SAUNDERS, M. A., Product Form of the Choleski Factorization for Large-Scale Linear Programs, STAN-CS-72-301, Stanford University, Stanford, Calif., 1972

WOLFE, P., A Bibliography for the Ellipsoid Algorithm, Research Report RC 8237, Math. Sci. Dept., IBM Thomas J. Watson Research Center, Yorktown Heights, N.Y., 1980

The reader who is interested in practical applications and software aspects of linear programming is referred to

BEALE, E. M. L., *Mathematical Programming in Practice*, Wiley, New York, 1968

The book is intended primarily for practitioners and students of operational research. It deals with methods of organizing real problems so that they can be solved numerically by computer codes. The reader should also consult

GREENBERG, H. J. (ed.), *Design and Implementation of Optimization Software*, NATO Advanced Study Institute Series, Series E: Applied Science No. 28, 1978

This text consists of tutorials that focus on optimization software.

Sparseness in linear programming has recently received the attention of many mathematicians and users. We suggest, as an introduction, the following sources of information on this subject:

REID, J. K. (ed.), *Large Sparse Sets of Linear Equations*, Academic Press, New York, 1971

GILL, P. E., and W. MURRAY (eds.), *Numerical Methods for Constrained Optimization*, Academic Press, New York, 1974, Chapter 4

The book

DAELLENBACH, H. G., and E. J. BELL, *User's Guide to Linear Programming*, Prentice-Hall, Englewood Cliffs N.J., 1970

provides a good discussion of large linear programming problems in industries and existing computer codes.

One of the most popular LP computer codes is an IBM package MPSX. A good introductory description of this program's use is contained in the book

RANDOLPH, P. H., and H. D. MEEKS, *Applied Linear Optimization*, Grid, Inc., Columbus, Ohio, 1978

The reader interested in applications of LP to various problems of numerical analysis, such as approximations, curve fitting, linear equations, differential equations, and others, should consult

RABINOWITZ, P., Applications of Linear Programming to Numerical Analysis, *SIAM Rev.* 10(1968), 121–159

The standard simplex algorithm chooses in each iteration the nonbasic variable that gives the best reduced cost. Goldfarb and Reid [1977] tried algorithms selecting edges of the polytope of feasible solutions on which the objective function decreases most rapidly with respect to distance in the space of all the variables. A worthwhile overall gain over the standard simplex method was shown.

GOLDFARB, D., and J. K. REID, A Practicable Steepest-Edge Simplex Algorithm, *Math. Programming* 12(1977), 361–371

A detailed exposition of the ellipsoid algorithm can be found in

PAPADIMITRIOU, C. H., and K. STEIGLITZ, *Combinatorial Optimization*: *Algorithms and Complexity*, Prentice-Hall, Englewood Cliffs, N.J., 1982, Chapter 8.

Some results related to the simplex algorithm average performance have been obtained by S. Smale and reported in *Science*, 217(1982), 39.

1.2. DUAL LINEAR PROGRAMMING METHOD

An important concept in linear programming is that of duality. It leads to better theoretical understanding of linear programming and to new computational methods which in some cases are more efficient than the primal simplex method considered in Section 1.1. Since the duality relationship is symmetric for problems expressed in terms of inequalities, we first consider the *primal problem* in the form

minimize $\qquad\qquad\qquad\qquad \mathbf{c}^T\mathbf{x}$

subject to $\qquad\qquad\qquad\qquad A\mathbf{x} \geqslant \mathbf{b}$ $\qquad\qquad\qquad$ (1-13)

$\qquad\qquad\qquad\qquad\qquad \mathbf{x} \geqslant 0$

The primal simplex method solves this problem by generating a sequence of feasible solutions which improve the value of the objective function as the computation progresses from the ith iteration to the $(i + 1)$st iteration. It terminates if no further improvement of the objective function is possible. In this section an algorithm is developed that generates infeasible solutions $\mathbf{x}_B$ and terminates once a feasible solution is reached. This computational procedure is based on some properties of the *dual problem* associated with (1-13), which is

maximize $\qquad\qquad\qquad\qquad \lambda^T\mathbf{b}$

subject to $\qquad\qquad\qquad\qquad \lambda^T A \leqslant \mathbf{c}^T$

$\qquad\qquad\qquad\qquad\qquad \lambda \geqslant 0$

where λ is the dual variable vector. It is simply to verify that if we slightly modify the dual problem so that it conforms with the primal problem formulation, then its dual is the original primal problem.

If a linear programming problem is in the standard form,

minimize $\qquad\qquad\qquad\qquad \mathbf{c}^T\mathbf{x} \qquad\qquad\qquad\qquad\qquad$ (1-14)

subject to $\qquad\qquad\qquad\qquad A\mathbf{x} = \mathbf{b}$

$$\mathbf{x} \geqslant 0$$

then its dual is

maximize $\qquad\qquad\qquad\qquad \lambda^T\mathbf{b} \qquad\qquad\qquad\qquad\qquad$ (1-15)

subject to $\qquad\qquad\qquad\qquad \lambda^T A \leqslant \mathbf{c}^T$

with no restriction that λ must be nonnegative (see Problem 1-23).

In general, every linear programming problem has its dual, which can be established by following certain transformation rules. In this section our attention is restricted to the second pair of linear problems.

Given such a pair of problems, and the vectors $\mathbf{x}$ and λ which are feasible for (1-14) and (1-15), respectively, we can easily establish that $\mathbf{c}^T\mathbf{x} \geqslant \lambda^T\mathbf{b}$. This is true since

$$\lambda^T\mathbf{b} = \lambda^T A\mathbf{x} \leqslant \mathbf{c}^T\mathbf{x}$$

and $\mathbf{x} \geqslant 0$. We conclude that if $(\lambda^*)^T\mathbf{b} = \mathbf{c}^T\mathbf{x}^*$ and $A\mathbf{x}^* = \mathbf{b}$, $A^T\lambda^* \leqslant \mathbf{c}$, then λ^* and $\mathbf{x}^*$ are optimal solutions for their respective problems. The following duality theorem relates precisely the primal and the dual solutions $\mathbf{x}^*$ and λ^*.

Theorem 1-3 (Duality Theorem)

If either (1-14) or (1-15) has a finite optimal solution, so does the other problem. The corresponding values of the objective functions are equal. Furthermore, if either problem has an unbounded solution, the other problem is infeasible.   ∎

The reader interested in a proof is referred to Luenberger [1973]. Using this fundamental result we can show that once the primal problem is solved, the solution to the dual problem is also available.

Assume that $\mathbf{x}_B = B^{-1}\mathbf{b}$ is feasible and optimal for (1-14). Then the relative cost vector is nonnegative, $\mathbf{p}^T = \mathbf{c}_N^T - \mathbf{c}_B^T B^{-1}N \geqslant 0$, and we have

$$\mathbf{c}_B^T B^{-1}N \leqslant \mathbf{c}_N^T$$

It turns out that the vector of simplex multipliers

$$\lambda^T = \mathbf{c}_B^T B^{-1}$$

is an optimal solution to the dual problem. This vector is dual feasible since

$$\lambda^T A = \lambda^T(B, N) = \left(c_B^T, c_B^T B^{-1} N\right)$$
$$\leqslant \left(c_B^T, c_N^T\right) = c^T$$

and it produces the value of the dual objective function equal to the optimal value of the primal function,

$$\lambda^T b = c_B^T B^{-1} b = c_B^T x_B$$

We conclude that if primal problem (1-14) has an optimal solution $x_B = B^{-1}b$, then dual problem (1-15) has the optimal solution $\lambda^T = c_B^T B^{-1}$, which can be readily calculated since the basis matrix inverse is available.

Construction of a Dual Simplex Method

To develop a dual simplex method we need to discuss an additional relationship between the optimal pair of x and λ. This relationship is provided by the complementary slackness theorem.

Theorem 1-4 (Complementary Slackness Theorem)

If x and λ are feasible for their respective problems (1-14) and (1-15), then a necessary and sufficient condition for their optimality is that for all i:

1. $x_i > 0$ implies that $\lambda^T \text{col}_i(A) = c_i$.
2. $\lambda^T \text{col}_i(A) < c_i$ implies that $x_i = 0$. ∎

Clearly, for such a pair of x and λ we have

$$(\lambda^T A - c^T)x = 0 \tag{1-16}$$

and

$$\lambda^T b = c^T x$$

Observe that (1-16) is enforced in the primal simplex method, that is,

$$(\lambda^T A - c^T)x = (\lambda^T B - c_B^T)x_B + (\lambda^T N - c_N^T)x_N = 0$$

since $\lambda^T B - c_B^T = 0$ and $x_N = 0$. However, λ is infeasible, except at the optimal solution, where

$$\lambda^T N \leqslant c_N^T$$

This relationship, combined with

$$\lambda^T B = c_B^T$$

gives

$$\lambda^T A \leqslant c^T$$

meaning that the dual feasibility is satisfied for the optimal simplex multipliers λ.

We can now develop a new solution algorithm called the *dual simplex method*, which also enforces (1-16) but in a different way. In this version of the simplex method the relationship $\lambda^T A \leqslant c^T$ is always satisfied, but x_B is not feasible (may have negative components) except for the optimal solution. Such a method is useful if we have a pair of x and λ such that λ is dual feasible but x is not primal feasible. For example, we may have solved an LP problem, and then constructed a new problem with somewhat modified vector b. The simplex multipliers are feasible for the dual problem, but the new vector x_B may not be nonnegative. We may also have a problem for which it is easy to generate a dual feasible solution. For instance, let us consider problem (1-13) with $c \geqslant 0$ and $b \not\leqslant 0$. Using nonnegative slack variables, we obtain an equivalent problem:

minimize $\qquad\qquad\qquad c^T x$

subject to $\qquad\qquad -Ax + I\bar{x} = -b$

$$x \geqslant 0, \quad \bar{x} \geqslant 0$$

Let us choose $B = I$ and $x_B = \bar{x} = -b \not\geqslant 0$. For this base the vector $\lambda^T = c_B^T B^{-1} = 0$ is feasible for the dual problem. In this case we say that the corresponding basic solution to the primal problem, $x_B = B^{-1}b$, is *dual feasible*.

Note that if

$$p^T = c_N^T - c_B^T B^{-1} N = c_N^T - \lambda^T N \geqslant 0$$

then

$$\lambda^T A = (\lambda^T B, \lambda^T N) \leqslant (c_B^T, c_N^T) = c^T$$

That is, the primal basic solution is dual feasible if the vector of relative costs is nonnegative.

In the discussion that follows we assume for convenience that the basis matrix B consists of the first m columns of A. Since equation (1-16) is enforced in the method we develop, we have

$$\lambda^T \text{col}_j(A) = c_j \qquad j = 1, 2, \ldots, m$$

and (1-17)

$$\lambda^T \text{col}_j(A) < c_j, \qquad j = m + 1, \ldots, n$$

In one iteration of the dual simplex method we find a new vector $\hat{\lambda}$ such that one of the equalities moves to the set of inequalities, and vice versa, and the objective function $\lambda^T \mathbf{b}$ increases. This exchange implies that one column of the basis matrix B is replaced by a new column corresponding to a variable that becomes basic in this iteration.

Suppose that the equation number l, $1 \leqslant l \leqslant m$, becomes an inequality. Then perturbing $\hat{\lambda}$ as follows:

$$\hat{\lambda}^T = \lambda^T - \varepsilon \, \text{row}_l(B^{-1})$$

where $\varepsilon > 0$, we get

$$\hat{\lambda}^T \text{col}_j(A) = \lambda^T \text{col}_j(A) - \varepsilon \, \text{row}_l(B^{-1})\text{col}_j(A)$$

If we define $z_j = \lambda^T \text{col}_j(A)$ and recall that

$$B^{-1}\text{col}_j(A) = \mathbf{y}_j$$

we get

$$\text{row}_l(B^{-1})\text{col}_j(A) = y_{lj}, \qquad j = m + 1, \ldots, n$$

For the new vector $\hat{\lambda}$ relationships (1-17) change to

$$\hat{\lambda}^T \text{col}_j(A) = c_j, \qquad j = 1, 2, \ldots, m; \quad j \neq l$$

$$\hat{\lambda}^T \text{col}_l(A) = c_l - \varepsilon \, \text{row}_l(B^{-1})\text{col}_l(A)$$

$$= c_l - \varepsilon \qquad\qquad (1-18)$$

$$\hat{\lambda}^T \text{col}_j(A) = z_j - \varepsilon y_{lj}, \qquad j = m + 1, \ldots, n \qquad (1-19)$$

If $\varepsilon > 0$, then (1-18) gives

$$\hat{\lambda}^T \text{col}_l(A) < c_l$$

which is the result we desired. At the same time we choose ε so that one of the inequalities in (1-17) becomes the equality. This requires that

$$c_j = z_j - \varepsilon y_{lj}$$

for some index j. The value of ε is determined from

$$\varepsilon = \frac{z_k - c_k}{y_{lk}} = \min\left[\frac{z_j - c_j}{y_{lj}}\right]$$

over all $y_{lj} < 0$, $j = m + 1, m + 2, \ldots, n$. For this ε the kth inequality becomes an equation and the kth variable moves to the basic set.

The modified value of the objective function is

$$\hat{\boldsymbol{\lambda}}^T \mathbf{b} = \boldsymbol{\lambda}^T \mathbf{b} - \varepsilon \, \text{row}_l(B^{-1}) \mathbf{b}$$
$$= \boldsymbol{\lambda}^T \mathbf{b} - \varepsilon \, \text{component}_l(\mathbf{x}_B)$$

If $\mathbf{x}_B$ is feasible (i.e., $\mathbf{x}_B \geqslant 0$), the dual objective function cannot be improved and this implies that the vector $\mathbf{x}_B$ is optimal. If at least one component of $\mathbf{x}_B$ is negative, we can repeat the iteration just described.

Now we can state the dual simplex algorithm originally given by Lemke [1954].

Dual Simplex Algorithm

Given is a basic solution $\mathbf{x}_B = B^{-1}\mathbf{b}$ which is dual feasible; that is, the related vector of simplex multipliers $\boldsymbol{\lambda}^T = \mathbf{c}_B^T B^{-1}$ satisfies the relationship $\mathbf{c}_N^T - \boldsymbol{\lambda}^T N \geqslant 0$.

Step 1.   If $\mathbf{x}_B \geqslant 0$, then $\mathbf{x}_B$ is optimal and the computation is terminated. Otherwise, select a negative component of $\mathbf{x}_B$, say the lth element of $\mathbf{x}_B$.

Comment: The variable x_l will be removed from the basic set, and the lth column of A will be removed from the basis B.

Step 2.   Calculate

$$\text{row}_l(B^{-1})\text{col}_j(A) = y_{lj}$$

for $j = m + 1, m + 2, \ldots, n$. If all $y_{lj} \geqslant 0$, then $\hat{\boldsymbol{\lambda}}^T = \boldsymbol{\lambda}^T - \varepsilon \, \text{row}_l(B^{-1})$ is feasible for an arbitrarily large value of $\varepsilon > 0$, and the dual problem has no finite solution. This follows by considering (1-19). Otherwise, calculate for all $j = m + 1, m + 2, \ldots, n$ such that $y_{lj} < 0$,

$$\varepsilon = \min \left[\frac{z_j - c_j}{y_{lj}} \right] = \frac{z_k - c_k}{y_{lk}}$$

where

$$z_j = \boldsymbol{\lambda}^T \text{col}_j(A)$$
$$= \mathbf{c}_B^T B^{-1} \text{col}_j(A)$$

Comment: The column $\text{col}_k(A)$ replaces the $\text{col}_l(A)$ in the basis.

Step 3.   Compute the new vector $\boldsymbol{\lambda}^T$,

$$\boldsymbol{\lambda}^T := \boldsymbol{\lambda}^T - \varepsilon \, \text{row}_l(B^{-1})$$

Step 4. Update the inverse of the basis matrix B^{-1}, and calculate the new solution vector

$$\mathbf{x}_B = B^{-1}\mathbf{b}$$

Step 5. Repeat from step 1.

> *Comment:*   The new dual variable λ is also feasible. It can be computed as shown in step 3 or using the relationship $\lambda^T = \mathbf{c}_B^T B^{-1}$ after the basis inverse is updated. The value of the dual objective function $\lambda^T \mathbf{b}$ is improved (increased) in each subsequent iteration. The dual simplex procedure terminates after a finite number of iterations with $\mathbf{x}_B \geqslant 0$. It is important to understand that the dual simplex method solves the primal problem by maintaining optimality condition while working toward feasibility. In terms of the dual problem, it maintains feasibility while working toward optimality.

Implemented Algorithm

The implemented dual simplex method solves the following problem:

minimize (or maximize)

$$a_{00} = \sum_{j=1}^{n} a_{0j}x_j$$

subject to

$$\sum_{j=1}^{n-m} a_{ij}x_j + x_{n-m+i} = a_{i0}, \qquad i = 1, 2, \ldots, m$$

$$x_j \geqslant 0, \qquad j = 1, 2, \ldots, n$$

where all $a_{0j} \geqslant 0$ for minimization, and $a_{0j} \leqslant 0$ for maximization.

In the matrix form we have

minimize (or maximize) $z = \mathbf{c}^T\mathbf{x} + \hat{\mathbf{c}}^T\hat{\mathbf{x}}$

subject to $A\mathbf{x} + I\hat{\mathbf{x}} = \mathbf{b}$ (1-20)

$$\mathbf{x}, \hat{\mathbf{x}} \geqslant 0$$

For this problem the primal basic solution $\hat{\mathbf{x}} = \mathbf{b}$ is dual feasible if

$$\mathbf{p}^T = \mathbf{c}^T - (\hat{\mathbf{c}})^T A \geqslant 0$$

First, the procedure determines if $\hat{\mathbf{x}}$ is dual feasible. It also determines a set H of indices corresponding to the basic variables, and a set J corresponding to the nonbasic variables.

One iteration of the method consists of the following steps:

Step 1. If for all $i \in H$, $a_{i0} \geqslant 0$, then the basic primal solution is primally feasible and the computation is terminated. Otherwise, determine H_1, which is a subset of H such that

$$H_1 = \{i : a_{i0} < 0\}$$

Step 2.   If there is such $i \in H_1$ that for every $j \in J$ we have $a_{ij} \geq 0$, the problem has no finite solution. Otherwise, proceed to step 3.

Step 3.   Find indices l and k such that

a.
$$a_{l0} = \min\{a_{i0} : i \in H_1\}$$

b.
$$\left|\frac{a_{0k}}{a_{lk}}\right| = \min\left\{\left|\frac{a_{0j}}{a_{lj}}\right| : j \in J, a_{lj} < 0\right\}$$

Comment: The absolute quotients are used in step 3b because all the values $a_{0j} = c_j - z_j$ are nonnegative. To find the appropriate pivot element we compute the ratios $(z_j - c_j)/a_{lj} > 0$. The values of the a_{lj}'s correspond to the y_{lj}'s used in the preceding section.

Step 4.   Use the pivot element a_{lk} to update the tableau A by performing the pivoting operation. Update the sets H and J as follows:

$$H := H - \{l\} + \{k\}, \qquad J := J - \{k\} + \{l\}$$

Step 5.   Repeat from step 1.

Computer Implementation
of the Dual Simplex Algorithm

The DSIMPLEX procedure is a Pascal version of the ALGOL 60 procedure called *dusimplex* published by Kucharczyk and Sysło [1975].

To apply the DSIMPLEX procedure, the linear programming problem must have the form shown in (1-20). This implies that there is available a basic solution which is dual feasible. Such a situation may arise if, for example, an optimal solution $\mathbf{x}_B$ to some LP problem has been calculated and then the right-hand side vector $\mathbf{b}$ of constraints is modified, say to $\mathbf{b} + \delta\mathbf{b}$. In this case we recompute the basic solution

$$\mathbf{x}_B = B_{\text{opt}}^{-1}(\mathbf{b} + \delta\mathbf{b})$$

If $\mathbf{x}_B \geq 0$, this vector is a new optimal solution and no recomputation is required. If $\mathbf{x}_B \ngeq 0$, we have a dual feasible but primal infeasible solution. It is then advantageous to use the dual simplex method to find an optimal solution.

Global Constants

M	number of constraints, m
N	number of variables $x_1, \ldots, x_n$
NMINM	integer whose value is $n - m$

Data Types

```
TYPE   ARRMN  = ARRAY[0..M,0..N] OF REAL;
       ARRN   = ARRAY[0..N] OF INTEGER;
       ARRMIN = ARRAY[1..NMINM] OF INTEGER;
```

Procedure Parameters

```
PROCEDURE DSIMPLEX(
    M,N,FOPT              :INTEGER;
    VAR A                 :ARRMN;
    VAR U                 :ARRN;
    EPS,INF               :REAL;
    VAR NOFEAS,NOSOL    :BOOLEAN);
```

Input

M   number of constraints, m

N   number of variables $x_1, x_2, \ldots, x_n$

A[0..M,0..N]   array containing A, **b**, **c**, and $\hat{\mathbf{c}}$ which define LP problem (1-20) and are assigned to the elements of A as follows (the other elements of A are not required as input):

$$A[0..M,0..N] = \begin{array}{|c|c|c|} \hline & a_{01} \ldots a_{0,\,n-m} & a_{0,\,n-m+1} \cdots a_{0n} \\ \hline a_{10} & a_{11} \ldots a_{1,\,n-m} & \\ \vdots & \vdots \quad A \quad \vdots & \\ a_{m0} & a_{m1} \ldots a_{m,\,n-m} & \\ \hline \end{array}$$

where

$$\mathbf{c}^T = \left(a_{01}, \ldots, a_{0,\,n-m} \right)$$
$$\hat{\mathbf{c}}^T = \left(a_{0,\,n-m+1}, \ldots, a_{0n} \right)$$
$$\mathbf{b}^T = \left(a_{10}, \ldots, a_{m0} \right)$$

and

$$A = \left[a_{ij} \right], \qquad i = 1, \ldots, m; \quad j = 1, \ldots, n - m$$

The remaining input data are:

FOPT   flag; FOPT = -1 if a_{00} is minimized, FOPT = 1, if a_{00} is maximized

EPS   small real number such that if for any number a, $|a| <$ EPS, then $a := 0.0$; we have used EPS = 10^{-4}

INF   maximal real number available in the floating-point number system that is used

Output

NOFEAS   Boolean variable equal to TRUE if the initial solution $\hat{\mathbf{x}} = \mathbf{b}$ is not dually feasible and FALSE–otherwise

NOSOL   Boolean variable equal to TRUE if the problem has no finite solution and FALSE–otherwise

A[0,0]   optimal value of the objective function

U[1..M]   array identifying the optimal basic variables

A[1..M,0]   optimal solution array; (A[I,0] is the value of the optimal variable numbered U[I])

Comment: The program calling DSIMPLEX should first check if NOFEAS = TRUE, then if NOSOL = TRUE. A finite optimal solution exists only if both NOFEAS and NOSOL are FALSE.

Example

Consider the following problem (Dantzig [1963], p. 243]):

$$
\begin{array}{llllllll}
\text{minimize} & x_1 & +3x_2 & +2x_3 \\
\text{subject to} & 4x_1 & -5x_2 & +7x_3 & +x_4 & & & = & 8 \\
& -2x_1 & +4x_2 & -2x_3 & & +x_5 & & = & -2 \\
& x_1 & -3x_2 & +2x_3 & & & +x_6 & = & 2
\end{array}
$$

$$x_i \geqslant 0, \quad i = 1, 2, \dots, 6$$

In this problem we have

$$M = 3, \quad N = 6, \quad \text{NMINM} = 3 \text{ (global constants)}$$

$$
A[0..3, 0..6] = \begin{bmatrix}
- & 1 & 3 & 2 & 0 & 0 & 0 \\
8 & 4 & -5 & 7 & - & - & - \\
-2 & -2 & 4 & -2 & - & - & - \\
2 & 1 & -3 & 2 & - & - & -
\end{bmatrix}
$$

The solution obtained is

$$\text{NOFEAS} = \text{NOSOL} = \text{FALSE}$$
$$A[0,0] = 1$$

I	U[I]	A[I,0]
1	1	1.0
2	4	4.0
3	6	1.0

Pascal Procedure DSIMPLEX

```
PROCEDURE DSIMPLEX(
    M,N,FOPT              :INTEGER;
    VAR A                 :ARRMN;
    VAR U                 :ARRN;
    EPS,INF               :REAL;
    VAR NOFEAS,NOSOL      :BOOLEAN);



VAR    I,J,K,K1,K2,K3,K4,L,W :INTEGER;
       MIN,XM,XS             :REAL;
       B,STOP                :BOOLEAN;
       Z,Z1                  :ARRMIN;
BEGIN
   NOFEAS:=FALSE;
   NOSOL:=FALSE;
   K4:=N-M;
   A[0,0]:=0.0;
   FOR I:=0 TO K4 DO
   BEGIN
      XS:=0.0;
      FOR J:=1 TO M DO
         XS:=XS+A[J,I]*A[0,K4+J];
      A[0,I]:=XS-A[0,I]
   END;
   FOR I:=K4+1 TO N DO
      FOR J:=1 TO M DO
         IF I = K4+J THEN
            A[J,I]:=1.0
         ELSE A[J,I]:=0.0;
   I:=0;
   WHILE (NOT NOFEAS) AND (I < K4) DO
   BEGIN
      I:=I+1;
      XS:=A[0,I];
      NOFEAS:=(ABS(XS) > EPS) AND (XS*FOPT < 0);
      IF NOT NOFEAS THEN U[M+I]:=I
   END;
   IF NOT NOFEAS THEN
   BEGIN
      FOR I:=1 TO M DO  U[I]:=K4+I ;
      STOP:=FALSE;
      REPEAT   (*  UNTIL STOP  *)
         MIN:=0.0;
         B:=TRUE;
         I:=0;
         REPEAT  (*  UNTIL STOP OR I >= M  *)
            I:=I+1;
            J:=M;
            XS:=A[I,0];
            IF XS < -EPS THEN
```

```
            BEGIN
                STOP:=TRUE;
                WHILE (J < N) AND STOP DO
                BEGIN
                    J:=J+1;
                    W:=U[J];
                    STOP:=A[I,W] >= -EPS
                END;
                IF STOP THEN NOSOL:=TRUE
                ELSE
                BEGIN
                    B:=FALSE;
                    IF XS-MIN < -EPS THEN
                    BEGIN
                        MIN:=XS;
                        L:=I
                    END
                END  (*  NOT STOP  *)
            END  (*  XS <- EPS  *)
        UNTIL STOP OR (I >= M);
        IF NOT STOP THEN
        BEGIN
            IF B THEN
            BEGIN NOSOL:=FALSE;  STOP:=TRUE END
            ELSE
            BEGIN
                MIN:=INF;
                FOR J:=1 TO K4 DO Z1[J]:=M+J;
                FOR I:=0 TO M DO
                    IF (I <> 1) AND (NOT B) THEN
                    BEGIN
                        K:=0;
                        FOR J:=1 TO K4 DO Z[J]:=Z1[J];
                        K3:=1;
                        FOR J:=M+1 TO N DO
                            IF J = Z[K3] THEN
                            BEGIN
                                K3:=K3+1;
                                W:=U[J];
                                XS:=A[L,W];
                                IF XS < -EPS THEN
                                BEGIN
                                    XS:=ABS(A[I,W]/XS);
                                    XM:=XS-MIN;
                                    IF ABS(XM) < EPS THEN
                                    BEGIN
                                        K:=K+1;
                                        Z1[K]:=J;
                                        B:=FALSE
                                    END
```

```
                        ELSE
                            IF XM < 0.0 THEN
                            BEGIN
                                MIN:=XS;
                                K1:=J;
                                K2:=W;
                                Z1[1]:=1;
                                K:=1;
                                FOR W:=2 TO K4 DO
                                    Z1[W]:= 0;
                                B:=TRUE
                            END
                        END (* XS < -EPS *)
                    END (* J = Z[K3] *)
                END (* I <> 1 AND (NOT B) *);
                MIN:=1.0/A[L,K2];
                U[K1]:=U[L];
                IF L = 0 THEN I:=1 ELSE I:=0;
                REPEAT
                    XS:=A[I,K2]*MIN;
                    A[I,0]:=A[I,0]-A[L,0]*XS;
                    FOR J:=M+1 TO N DO
                    BEGIN
                        W:=U[J];
                        A[I,W]:=A[I,W]-A[L,W]*XS
                    END;
                    IF I = L-1 THEN I:=I+2 ELSE I:=I+1
                UNTIL I > M;
                FOR J:=M+1 TO N DO
                BEGIN
                    W:=U[J];
                    A[L,W]:=A[L,W]*MIN
                END;
                A[L,0]:=A[L,0]*MIN;
                FOR I:=0 TO M DO
                    IF I = 1 THEN
                        A[I,K2]:=1.0
                    ELSE A[I,K2]:=0.0;
                U[L]:=K2
            END  (*  NOT B  *)
        END  (*  NOT STOP  *)
    UNTIL STOP
 END  (*  NOT NOFEAS  *)
END;   (*  END DSIMPLEX  *)
```

PROBLEMS

1-23. We define duality through the pair of problems as follows:

Primal	Dual
min $\mathbf{c}^T\mathbf{x}$	max $\mathbf{b}^T\lambda$
s.t. $A\mathbf{x} \geqslant \mathbf{b}$	s.t. $\lambda^T A \leqslant \mathbf{c}^T$
$\mathbf{x} \geqslant 0$	$\lambda \geqslant 0$

The dual of any other LP problem can be found by converting the problem to the primal as shown above. Verify that a linear program in the standard form (1-14) has the corresponding dual (1-15). (*Hint:* Replace the equality constraints by two opposing inequalities.)

1-24. Consider the primal problem

$$\text{minimize} \qquad \mathbf{c}^T\mathbf{x}$$
$$\text{subject to} \qquad A\mathbf{x} = \mathbf{b}$$

Find the dual of this problem.

1-25. Given is the LP problem

$$\text{minimize} \qquad \mathbf{c}^T\mathbf{x} - \mathbf{b}^T\lambda$$
$$\text{subject to} \qquad A\mathbf{x} \geqslant \mathbf{b}$$
$$-A^T\lambda \geqslant -\mathbf{c}$$
$$\mathbf{x} \geqslant 0, \quad \lambda \geqslant 0$$

where A, $\mathbf{b}$, and $\mathbf{c}$ are matrices of order $m \times n$, $m \times 1$, and $n \times 1$, respectively. Prove that this problem is either infeasible or has a solution vector $(\mathbf{x}^T, \lambda^T)$ such that $\mathbf{c}^T\mathbf{x} - \mathbf{b}^T\lambda = 0$.

1-26. Using the results of Problem 1-25, show that the LP problem

$$\text{minimize} \qquad \mathbf{c}^T\mathbf{x}$$
$$\text{subject to} \qquad A\mathbf{x} \geqslant \mathbf{b}$$
$$\mathbf{x} \geqslant 0$$

can be reduced to the following existence problem: Given the vector $\mathbf{w}^T = (w_1, w_2, \ldots, w_r)$ and the $r \times r$ matrix M, find $\mathbf{z}$ and $\mathbf{q}$ such that

$$M\mathbf{z} + \mathbf{q} = \mathbf{w}, \qquad \mathbf{w}^T\mathbf{z} = 0$$
$$\mathbf{z} \geqslant 0, \qquad \mathbf{q} \geqslant 0$$

1-27. The feasibility and unboundedness in the primal and dual problems can be summarized as follows:

		Primal	
		Feasible Solutions Exist	No Feasible Solutions
Dual	Feasible Solutions Exist	$z_D \leqslant z_P$ $\max z_D = \min z_P$	$\max z_D = +\infty$
	No Feasible Solutions	$\min z_P = -\infty$	Possible

where z_P and z_D are the primal and the dual objective functions, respectively. Construct a pair of primal and dual problems such that neither has feasible solutions.

1-28. Determine if the following system of inequalities has any solution using the dual simplex algorithm:

$$5x_1 + 4x_2 - 7x_3 \leqslant 1$$
$$- x_1 + 2x_2 - x_3 \leqslant -4$$
$$-3x_1 - 2x_2 + 4x_3 \leqslant 3$$
$$x_1, x_2, x_3 \geqslant 0$$

1-29. (Murty [1976]) Show that the following primal problem is identical to its dual

minimize $x_1 + x_2 + x_3$

subject to $-x_2 + x_3 \geqslant -1$

$$x_1 \qquad - x_3 \geqslant -1$$

$$-x_1 + x_2 \qquad \geqslant -1$$

$$x_1, x_2, x_3 \geqslant 0$$

1-30. Assume that A is a square matrix. Obtain sufficient conditions on **c**, A, and **b** under which the LP problem

minimize $\mathbf{c}^T \mathbf{x}$

subject to $\qquad\qquad\qquad\qquad\qquad A\mathbf{x} \leqslant \mathbf{b}$

$$\mathbf{x} \geqslant 0$$

and its dual are identical.

1-31. Given the following LP problem:

minimize $\qquad\qquad x_1 + 4x_2 \qquad\qquad + 3x_4 + x_5 + x_6$

subject to $\qquad -x_1 + 2x_2 + x_3 - x_4 + x_5 \qquad\qquad = -3$

$$2x_1 + x_2 - 4x_3 - x_4 \qquad + x_6 = -2$$

$$x_i \geqslant 0, \quad i = 1,\ldots, 4$$

Verify that the primal solution $x_5 = -3$, $x_6 = -2$ is dual feasible.

1-32. Consider the example

minimize $\qquad\qquad\qquad\qquad 2x_3 + x_4$

subject to $\qquad\qquad x_1 \quad + 3x_3 + x_4 = 1$

$$x_2 - x_3 + 5x_4 = 2$$

$$x_i \geqslant 0, \quad i = 1,\ldots, 4$$

The primal solution $x_1 = 1$, $x_2 = 2$ is optimal. Now we add the constraint $x_1 + x_2 \leqslant 2$, which is not satisfied by the optimal solution. Show how to obtain the standard form of this problem with a basic solution primally infeasible but dual feasible.

1-33. **Sensitivity.** Assume that we have an optimal solution $\mathbf{x}_B = B^{-1}\mathbf{b}$ to the LP problem: minimize $\mathbf{c}^T\mathbf{x}$ subject to $A\mathbf{x} = \mathbf{b}$ and $\mathbf{x} \geqslant 0$. Now we have slightly perturbed the vector $\mathbf{b}$ by $\delta\mathbf{b}$, but in such a way that $B^{-1}(\mathbf{b} + \delta\mathbf{b}) \geqslant 0$, so that B is still an optimal basis. Show that the vector of simplex multipliers λ (dual solution) gives the sensitivity of the optimal objective function $\mathbf{c}^T\mathbf{x}$ with respect to small perturbations of the vector $\mathbf{b}$.

1-34. Explain why the diet problem described in Section 1.1 can be solved easily by the dual simplex method.

1-35. **(a)** Using the PSIMPLEX procedure, solve

maximize $\qquad\qquad\qquad 2x_1 + 3x_2 + x_3$

subject to $\qquad\qquad\qquad x_1 + x_2 + x_3 \leqslant 3$

$$x_1 + 4x_2 + 7x_3 \leqslant 9$$

$$x_i \geqslant 0, \quad i = 1, 2, 3$$

(b) Using the result of part (a) and the DSIMPLEX procedure, solve this problem with the right-hand side of the first equation changed to 15.

1-36. Assume that we have solved the standard linear programming problem and consider adding new constraints of the form

$$Dx + Ix_s = \hat{b}$$

where x_s is a vector of slack variables with zero cost coefficients. Assume that for the optimal solution x_B, $Dx_B > \hat{b}$ and show how the dual simplex method can be used to solve the new modified problem.

1-37. Given is the following LP problem:

maximize $x_1 + 2x_2 + 3x_3 + \ldots + nx_n$

subject to $x_1 + x_2 + x_3 + \ldots + x_n \leqslant 1$

$x_2 + x_3 + \ldots + x_n \leqslant 1$

$\ldots\ldots\ldots$

$\ldots\ldots$

$x_n \leqslant 1$

$x_i \geqslant 0$ for all i

Consider its dual problem and determine the optimal solution without any numerical computation.

REFERENCES AND REMARKS

In Section 1.2 we have referred to the following sources:

KUCHARCZYK, J., and M. M. SYSŁO, *Optimization Algorithms in ALGOL 60*, Państwowe Wydawnictwo Naukowe, Warsaw, 1975 (in Polish)

LEMKE, C. E., The Dual Method for Solving the Linear Programming Problem, *Naval Res. Logist. Quart.* 1(1954), 36–47

LUENBERGER, D. G., *Introduction to Linear and Nonlinear Programming*, Addison-Wesley, Reading, Mass., 1973

MURTY, K. G., *Linear and Combinatorial Programming*, Wiley, New York, 1976

Versions of the simplex method have been developed that enable one to combine the primal and the dual algorithm to solve LP problems. Two of these techniques are the *composite-simplex* method described in

DANTZIG, G. B., Composite Simplex–Dual Simplex Algorithm I, Rand Report RM-1274, Rand Corporation, Santa Monica, Calif., 1954

ORCHARD - HAYS, W., A Composite Simplex Algorithm II, Rand Report Rm-1275, Rand Corporation, Santa Monica, Calif., 1954

and the *primal–dual* method of

DANTZIG, G. B., L. R. FORD, JR., and D. R. FULKERSON, A Primal–Dual Algorithm, in H. U. KUHN and A. W. TUCKER (eds.), *Linear Inequalities and Related Systems*, Annals of Mathematics 38, Princeton University Press, Princeton, N.J., 1956, pp. 171–181

The latter method works simultaneously on the primal and dual problem, and is useful in solving network flow problems.

The book

SIMONNARD, M., *Linear Programming*, Prentice-Hall, Englewood Cliffs, N.J., 1966

contains a very good discussion of duality.

Also there are very illuminating economic interpretations of the simplex multipliers, and we refer the reader to

DANTZIG, G. B., *Linear Programming and Extensions*, Princeton University Press, Princeton, N.J., 1963

for explanations.

1.3. TRANSPORTATION PROBLEM

There exists a large group of specialized linear programming problems which have network or network-related formulations. They include such mathematical problems as transshipment, transportation, personnel assignment, computer networks, plant location, and multicommodity flow. Problems of this nature cover a wide spectrum of scientific and industrial applications: for example, cash flow, airline scheduling, shipping, water resource optimization, personnel planning, machine loading, and telecommunications. In theory, these models can be solved using general linear programming codes, but this would frequently require excessive amounts of computer time and considerably more storage than special-purpose linear programming and network methods.

In this section we consider the *transportation problem*

minimize
$$z = \sum_{i=1}^{m} \sum_{j=1}^{n} c_{ij} x_{ij} \qquad (1\text{-}21)$$

subject to
$$\sum_{j=1}^{n} x_{ij} \leqslant a_i, \qquad i = 1, 2, \ldots, m \tag{1-22}$$

$$\sum_{i=1}^{m} x_{ij} \geqslant b_j, \qquad j = 1, 2, \ldots, n \tag{1-23}$$

$$x_{ij} \geqslant 0, \qquad i = 1, 2, \ldots, m; \quad j = 1, 2, \ldots, n$$

where a_i, b_j, and c_{ij} are all nonnegative integers.

The transportation problem has a simple network interpretation. Assume that we have a *directed network* (also called a *weighted digraph*) defined by a set of nodes V and a set of directed edges E. In the transportation problem the network is *bipartite* and *complete*; that is, all its nodes can be arranged in two groups, the *supply nodes* numbered $i = 1, 2, \ldots, m$ and the *demand nodes* numbered $j = 1, 2, \ldots, n$, and every supply node has n outgoing edges to all demand nodes (see Fig. 1-2). For each edge there is a shipping cost per unit of the transported commodity, c_{ij}. The problem is to determine shipping amounts x_{ij}'s which minimize the total transportation cost z. The first m inequalities apply to the supply nodes; the next n inequalities apply to the demand nodes. The transportation problem has a feasible solution if

$$\sum_{i=1}^{m} a_i \geqslant \sum_{j=1}^{n} b_j$$

and the demand constraints hold as strict equalities at the optimal solution, that is,

$$\sum_{i=1}^{m} x_{ij} = b_j$$

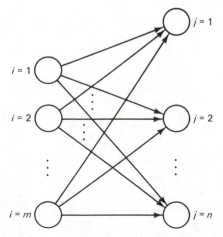

Figure 1-2 *Depiction of the transportation network.*

for all j (see Problem 1-42). Moreover, if

$$\sum_{i=1}^{m} a_i = \sum_{j=1}^{n} b_j$$

then every feasible solution satisfies all the inequality constraints as equalities.

Without loss of generality, we can use equality constraints in (1-22) and (1-23), since we can always introduce a fictitious $(n + 1)$st demand node with demand

$$b_{n+1} = \sum_{i=1}^{m} a_i - \sum_{j=1}^{n} b_j$$

and costs

$$c_{i, n+1} = 0, \qquad i = 1, 2, \ldots, m$$

The transportation problem is an instance of the LP problem and can be solved by a specialized primal simplex method. There is also an alternative approach known as the primal–dual method, based on iterative use of the maximum-flow algorithm. It exploits the necessary and sufficient linear programming conditions for optimality. It is this type of method we present and implement.

First, we describe an algorithm for finding a maximum flow in a network, then show a maximum-flow formulation of the transportation problem, and finally give the primal–dual algorithm due to Ford and Fulkerson [1962].

1.3.1. MAXIMUM FLOW IN A NETWORK

Consider the network $N = (V, E)$, where two nodes are specified as the *source node s* and the *sink node t*. Each edge in the network is assigned a nonnegative integer k_{ij}, called *capacity* of an edge.

A *flow* in the network N is an assignment of a real number x_{ij} (*edge flow*) to each edge, such that:

1. $0 \leqslant x_{ij} \leqslant k_{ij}$, for all the edges $(i, j) \in E$.
2.   For each node j, other than s and t, the *flow conservation* equation is satisfied,

$$\sum_{i} x_{ij} - \sum_{l} x_{jl} = 0$$

where i is related to the set of edges incoming to the node j, and l to the set of edges outgoing from the node j.

The *flow value* $f(t)$ is defined as the sum of the edge flows into the sink.

The *maximum-flow problem* (*max-flow*) is to find the values of x_{ij} for all the edges such that $f(t)$ is maximum. To find a maximum flow a technique of augmenting path can be used. An *augmenting path* is a sequence of pairwise adjacent edges from s to t, which allows us to increase the value of flow. If the (i, j)th edge orientation coincides with the direction of the path, then in order to push more flow through it, x_{ij} must be less than k_{ij}. If the (i, j)th edge points in the opposite direction, then in order to push some additional flow through it, we must reduce its flow, and $x_{ij} > 0$ is required.

Labeling Algorithm

To find an augmenting path from s to t, a labeling procedure is used. The labeling algorithm assumes that there exists an initial flow in the network. We may have, for instance, all $x_{ij} = 0$. Then labels of the form (j, ε) or $(j, -\varepsilon)$, where ε is a positive number or infinity are assigned to each node, beginning with the source s. If it is possible to label the sink t, a change of the flow from s to t is made and the labeling is repeated. If it is impossible to label the sink, the flow is optimal. The procedure uses two routines, A and B. During each step of routine A, a node is in one of three states: unlabeled and unscanned, labeled and unscanned, or labeled and scanned. In the beginning all nodes are unlabeled and unscanned.

Routine A (Labeling)

Step 1.   Label the source with $(-, \infty)$.

Step 2.   For any labeled and unscanned node j with label $(i, \varepsilon_j$ (or $-\varepsilon_j))$, scan it by examining all unlabeled nodes l, adjacent to j.

    a.   If (j, l) is an edge and $x_{jl} < k_{jl}$, then label node l with (j, ε_l), where $\varepsilon_l = \min\{\varepsilon_j, k_{jl} - x_{jl}\}$.

    b.   If (l, j) is an edge and $x_{lj} > 0$, then label node l with $(j, -\varepsilon_l)$, where $\varepsilon_l = \min\{\varepsilon_j, x_{lj}\}$.

Step 3.   The nodes l are now labeled and unscanned, and node j is labeled and scanned. Repeat step 2 until either the sink t is labeled or it is impossible to label the sink. In the first case we have a *breakthrough* and routine B is initiated. In the second case we have a *nonbreakthrough* and the algorithm is terminated, the flow is optimal.

Routine B (Flow Change)

The sink t has been labeled with (l, ε_t). Therefore, the network with the current flow admits an augmenting path from s to t which can increase the flow value by ε_t, and l is the second last node on this path. Hence set $x_{lt} \leftarrow x_{lt} + \varepsilon_t$. Now look at the node l labeled $(j, \varepsilon_l$ (or $-\varepsilon_l))$. If the second label is ε_l then l has been labeled from j by using the edge (j, l), therefore set $x_{jl} \leftarrow x_{jl} + \varepsilon_t$. Otherwise, the edge (l, j) has been used, and set $x_{lj} \leftarrow x_{lj} - \varepsilon_t$. Continue the flow change indicated by the first element of the labels until the node s is reached. Discard the labels and return to routine A.

Example 1-2

Consider the numerical example shown in Fig. 1-3. The numbers in the square brackets are k_{ij} and x_{ij}. Initial edge flows are zero. According to our algorithm the following steps can be executed:

> Label the source $(-, \infty)$.
> Label nodes 1 and 2 with $(s, 5)$ and $(s, 4)$, respectively.
> Label node 4 with $(2, 4)$.
> Label nodes 3 and 6 with $(4, 1)$ and $(4, 2)$, respectively.
> Label node 5 with $(3, 1)$.
> Label the sink with $(5, 1)$.

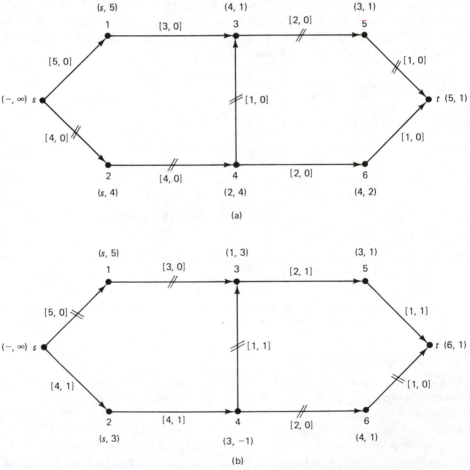

Figure 1-3   *Labeling algorithm:* (*a*) *Initial flow and first labeling;* (*b*) *Second labeling.*

There is a breakthrough and the augmenting path is indicated by $\nparallel$. The first labeling produces the flow of 1 through the edges $(s, 2)$, $(2, 4)$, $(4, 3)$, $(3, 5)$, and $(5, t)$. The second labeling is shown in Fig. 1-3(b). The third labeling is attempted but no breakthrough occurs.

The algorithm terminates and produces the final solution: $x_{s2} = x_{s1} = x_{13} = x_{24} = x_{35} = x_{46} = x_{5t} = x_{6t} = 1$, and the flow value is 2. ∎

Optimality and Finiteness
of the Labeling Algorithm

Assuming that the algorithm terminates, the last labeling does not reach the sink node t. Let S be the set of nodes labeled in the last labeling attempt and $\bar{S}$ the set of unlabeled nodes. If an edge (i, j) is directed from S to $\bar{S}$, it must be saturated, that is, $x_{ij} = k_{ij}$; otherwise, j would have been labeled when i was scanned. Also, all edges (j, i) from $\bar{S}$ to S must have zero flow; otherwise, j would have been labeled when i was scanned. Observe, that s belongs to S and t belongs to $\bar{S}$. It is fairly obvious that the flow value is not greater than the sum of capacities of any set of edges (called a *cut*) which contains at least one edge of every path from s to t. Hence the flow value $f(t)$ is optimal and equals the sum of capacities of the edges between S and $\bar{S}$. We state this as the theorem known as the min-cut max-flow theorem.

Theorem 1-5 (Ford–Fulkerson)

When the labeling algorithm terminates, the flow value $f(t)$ is optimal and equal to the capacity of the minimum cut. ∎

The question of whether the algorithm always terminates also needs to be considered. To see that it does if all initial edge flows and capacities are integer, we need to make two observations. First, the algorithm adds and subtracts only and does not introduce fractional flows. Second, if t is labeled the flow value is increased by at least one unit. Since the flow value is bounded from above (e.g., by $\Sigma_i k_{it}$, which is finite), the labeling algorithm must terminate.

However, unless we better define the labeling process (process A) the algorithm can be inefficient in some pathological cases. Modify the capacities of the network in Fig. 1-3 as shown in Fig. 1-4 and assume that M is a very large number. If the labeling algorithm starts with $f(t) = 0$ and alternatively uses the same augmenting paths as shown in Fig. 1-3, it will require $2M$ iterations of routines A and B to find the optimum flow value $f(t) = 2M$. Here the number of iterations depends on the problem capacities.

Edmonds and Karp [1972] corrected this deficiency and showed that if the labeling procedure always uses the augmenting paths as short as possible, its time complexity is $O(nm^2)$, in an n-node m-edge network. Had we used the shortest augmenting path in the example shown in Fig. 1-4 we would have used routines A and B only twice.

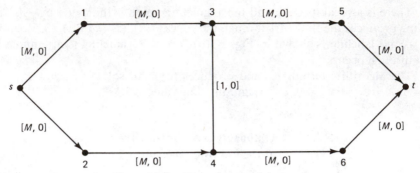

Figure 1-4 *Worst-case performance.*

Other maximum-flow algorithms have been developed which do not use the concept of flow augmenting paths. A recent such algorithm by Malhotra, Kumar, and Maheshwari is described in Section 3.3.

1.3.2. MAXIMUM-FLOW SOLUTION METHOD OF THE TRANSPORTATION PROBLEM

If we rewrite (1-22) as

$$- \sum_{j=1}^{n} x_{ij} \geq -a_i$$

and assign the dual variables u_i to (1-22) and v_j to (1-23), the dual linear problem of the transportation problem (1-21)–(1-23) is

maximize
$$- \sum_{i=1}^{m} u_i a_i + \sum_{j=1}^{n} v_j b_j \qquad (1\text{-}24)$$

subject to
$$- u_i + v_j \leq c_{ij} \qquad (1\text{-}25)$$
$$u_i, v_j \geq 0$$

for $i = 1, 2, \ldots, m$ and $j = 1, 2, \ldots, n$.

Suppose that the optimal solution to the primal problem, $\hat{x}_{ij}$, and to the dual problem, $\hat{u}_i$ and $\hat{v}_j$, are known. Then from the complementary slackness condition (Theorem 1-4) we get

$$\left(c_{ij} - \left(-\hat{u}_i + \hat{v}_j \right) \right) \hat{x}_{ij} = 0 \qquad (1\text{-}26)$$

Usually, we do not know the optimal solution to the dual problem, but it is easy to find an arbitrary feasible dual solution satisfying (1-25). In the primal–dual algorithm for the transportation problem we try to satisfy the primal constraints and the complementary slackness condition by seeking a solution to the following problem:

maximize
$$z = \sum_{i=1}^{m} \sum_{j=1}^{n} x_{ij}$$

subject to
$$\sum_{j=1}^{n} x_{ij} \leqslant a_i, \qquad i = 1, \ldots, m \tag{1-27}$$

$$\sum_{i=1}^{m} x_{ij} \leqslant b_j, \qquad j = 1, \ldots, n$$

$$x_{ij} = \begin{cases} 0, & \text{if } -u_i + v_j < c_{ij} \\ \text{nonnegative}, & \text{otherwise} \end{cases}$$

This is a maximum-flow problem that can be solved by the maximum-flow algorithm.

Figure 1-5 shows the network diagram of the maximum-flow formulation of the transportation problem, given by (1-27). In this flow problem the flows x_{ij} must be zero if $-u_i + v_j < c_{ij}$. This condition can be enforced by removing the corresponding edges (s_i, t_j) from the network. The remaining edges (s_i, t_j) have infinite capacities. The capacities of edges from s to s_i are a_i, and from t_j to t are b_j.

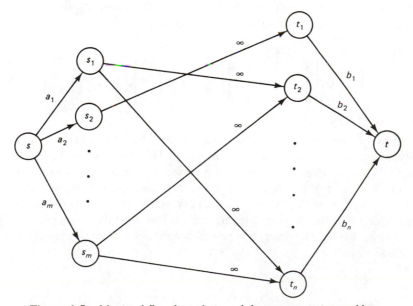

Figure 1-5 *Maximal flow formulation of the transportation problem.*

Note that the first two constraints in (1-27) are inequalities. For an arbitrary choice of the dual solution $\mathbf{u}, \mathbf{v}$ it might be impossible to satisfy the primal inequality constraints of the transportation problem [(1-21)–(1-23)]. If this happens, we calculate a new set of feasible dual variables and solve (1-27), again obtaining a new increased set of edge flows. Eventually, the primal constraints of the transportation problem are satisfied, and the algorithm terminates.

Hence the method for solving the transportation problem works as follows:

> **begin**
>   calculate an initial dual solution;
>   **while** primal constraints
>
> $$\sum_{j=1}^{n} x_{ij} \leq a_i, \quad \sum_{i=1}^{m} x_{ij} \geq b_j \text{ are not satisfied for all } i \text{ and } j \text{ do}$$
>
>   **begin**
>   **repeat**
>   use labeling routine A and flow change routine B
>   **until**   nonbreakthrough occurs in routine A, that is, until the current
>   maximum flow problem is solved;
>   change dual variables, thus obtain a new maximum flow problem of
>   the form (1-27)
>   **end**
> **end**

Example 1-3

To illustrate the primal–dual method for solving the transportation problem, consider a numerical example defined by the following arrays:

$$[c_{ij}] = \begin{bmatrix} 3 & 7 & 3 & 4 \\ 5 & 7 & 2 & 6 \\ 8 & 13 & 9 & 3 \end{bmatrix}, \quad \mathbf{a} = \begin{bmatrix} 15 \\ 30 \\ 55 \end{bmatrix}, \quad \mathbf{b} = \begin{bmatrix} 30 \\ 10 \\ 15 \\ 45 \end{bmatrix}$$

The values of the dual variables must satisfy the inequalities $-u_i + v_j \leq c_{ij}$. Initially, we can set $u_i = 0$ and $v_j = \min_i c_{ij}$. This gives $\mathbf{u}^T = (0,0,0)$ and $\mathbf{v}^T = (3,7,2,4)$. With this choice of $\mathbf{u}$ and $\mathbf{v}$ the nonzero flows x_{ij} are possible only in the edges: (s_1, t_1), (s_1, t_2), (s_1, t_4), (s_2, t_2), (s_2, t_3), and (s_3, t_4), since for these edge indices $c_{ij} + u_i - v_j = 0$. The initial network is shown in Fig. 1-6(a). Using the labeling algorithm for solving the maximum-flow problem, we get the solution shown in Fig. 1-6(b). However, the primal solution x_{ij} does not satisfy the primal constraints

$$\sum_{i=1}^{m} x_{ij} \geq b_j, \quad j = 1, 2, \ldots, n \qquad \blacksquare$$

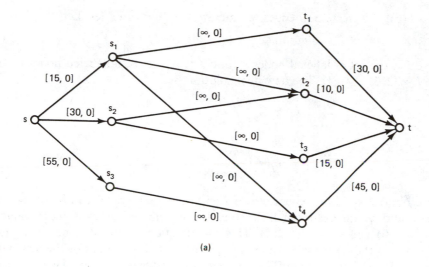

(a)

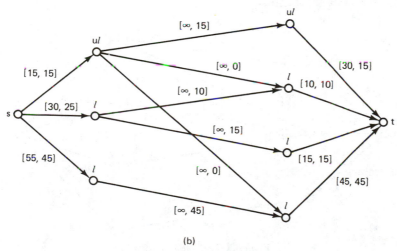

(b)

Figure 1-6 *First network.*

There is no difficulty in satisfying (1-22) as long as our network has the capacities $k_{s,\,s_i} = a_i$. Now we change the dual variables in such a way that the new dual variables are feasible and the current edge flows are not disturbed. To increase the flow value it is necessary to create new edges, particularly from the labeled nodes s_i to the unlabeled nodes t_j in the last labeling iteration of the maximum-flow algorithm.

Clearly, in the last labeling process of Example 1-3 the nodes s_1 and t_1 were not labeled. These are unlabeled nodes (ul), and the remaining nodes are labeled (l) [see

Fig. 1-6(b)]. To create new edges, we modify the dual variables. Let

$$\delta = \min\{c_{ij} + u_i - v_j : i \in I, \, j \in \bar{J}\} \qquad (1\text{-}28)$$

where I is the set of labeled nodes s_i, and $\bar{J}$ is the set of unlabeled nodes t_j. In our case $I = \{2, 3\}$, $\bar{J} = \{1\}$. The new dual solution is

$$\bar{u}_i = \begin{cases} u_i, & i \in I \\ u_i + \delta, & i \in \bar{I} \end{cases}$$

$$\bar{v}_j = \begin{cases} v_j, & j \in J \\ v_j + \delta, & j \in \bar{J} \end{cases}$$

where $\bar{I}$ and J are complement sets of I and $\bar{J}$, respectively. It can be easily verified that $\bar{u}$ and $\bar{v}$ are again dual feasible. In our numerical case these values are $\bar{u}^T = (2, 0, 0)$ and $\bar{v}^T = (5, 7, 2, 4)$. This modification creates one new edge (s_2, t_1) and removes the edges (s_1, t_2) and (s_1, t_4) with edge flow zero. For the new network shown in Fig. 1-7 the current flow is not optimal, since we can now label the node t_1. The labeling algorithm can be used again and the process continues until the demand constraints are satisfied.

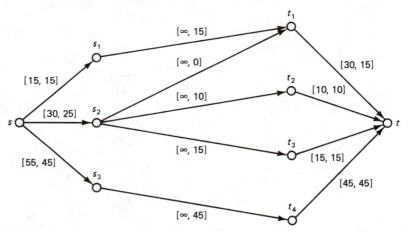

Figure 1-7 *Second network.*

In this algorithm we alternate between changing the primal and the dual solutions; hence the algorithm can be classified as a *primal–dual* method. In its computer implementation the algorithm manipulates the transportation table of the variables x_{ij} with the $m \times n$ cells. Cells of the table for which $-u_i + v_j = c_{ij}$ are called *admissible*. Other cells are *inadmissible*.

The computation progresses by a sequence of labeling steps (routine A). We use labels (l, ε_i) for rows and (l, δ_j) for columns. If a breakthrough is accomplished, the flow is increased by routine B. If a nonbreakthrough results, the dual variables u_i, v_j are modified by routine C. The algorithm terminates when a flow is obtained that satisfies the primal demand constraints.

The Ford–Fulkerson Transportation Algorithm

Step 1. *Initialization of dual variables.* Set

$$u_i = 0 \quad \text{and} \quad v_j = \min_{1 \leqslant i \leqslant m} c_{ij}, \qquad i = 1, 2, \ldots, m; \quad j = 1, 2, \ldots, n$$

and

$$x_{ij} = 0, \qquad i = 1, 2, \ldots, m; \quad j = 1, 2, \ldots, n$$

Routine A (Labeling)

Step 2. Assign labels

$$\left(0, \; \varepsilon_i = a_i - \sum_{j=1}^{n} x_{ij} \right)$$

to all rows i for which

$$\sum_{j=1}^{n} x_{ij} < a_i$$

Step 3. Select a labeled and unscanned row i and scan it for all unlabeled columns j such that the cell (i, j) is admissible; label the column $(i, \delta_j = \varepsilon_i)$. Repeat until the labeled rows have all been scanned.

Step 4. Select a labeled and unscanned column j and scan it for all unlabeled rows i such that $x_{ij} > 0$; label such rows with $(j, \varepsilon_i = \min\{x_{ij}, \delta_j\})$. Repeat until previously labeled columns have all been scanned.

Step 5. Repeat steps 3 and 4 until either a column is labeled for which

$$\sum_{i=1}^{m} x_{ij} < b_j$$

(go to step 6, breakthrough is accomplished), or no more labels can be assigned (nonbreakthrough, and go to step 8).

Routine B (Flow Change)

Step 6. There is a column k with (l, δ_k) such that

$$\sum_{i=1}^{m} x_{ik} < b_k$$

Alternately add and subtract from x_{ij} the value

$$\varepsilon = \min\left\{ \delta_k, b_k - \sum_{i=1}^{m} x_{ik} \right\}$$

along the path starting at element x_{lk} and indicated by the first elements of the labels. This process is finished when we reach a label whose first element is zero.

Step 7. If all column demands have been satisfied, the algorithm terminates. Otherwise, discard the old labels and return to routine A.

Routine C (Dual Variable Change)

Step 8. The labeling process resulted in nonbreakthrough. Let I and J be the index set of labeled rows and columns, respectively.

Step 9. Calculate new feasible dual variables as follows:

$$\delta = \min\{c_{ij} + u_i - v_j : i \in I, j \in \bar{J}\}$$

and

$$\bar{u}_i = \begin{cases} u_i, & i \in I \\ u_i + \delta, & i \in \bar{I} \end{cases}$$

$$\bar{v}_j = \begin{cases} v_j, & j \in J \\ v_j + \delta, & j \in \bar{J} \end{cases}$$

where $\bar{I}$ and $\bar{J}$ are complement sets of I and J with respect to the sets $\{1, 2, \ldots, m\}$ and $\{1, 2, \ldots, n\}$, respectively.

Step 10. Return to step 2.

Finiteness and Optimality of the Ford–Fulkerson Algorithm

It is possible to establish that the labeling algorithm solving (1-21)–(1-23) terminates by the following argument. Each occurrence of breakthrough increases the flow value by at least one unit, and the number of such occurrences is bounded above by the total demand $\sum_j b_j$. Each nonbreakthrough increases the dual objective function by at least one unit, and this function value is bounded above by $\sum_i \sum_j c_{ij} x_{ij}$ for any feasible $X = [x_{ij}]$. Furthermore, upon termination of the computation the primal and dual solutions are optimal, since both solutions satisfy the primal and dual constraints, and the complementary slackness condition.

Computer Implementation of the Transportation Algorithm

The TRANSPORT procedure is a Pascal version of the ALGOL 60 procedure called *trHFF* published by Kucharczyk and Sysło [1975]. It solves the transportation problem (1-21)–(1-23), where

$$\sum_{i=1}^{m} a_i \geqslant \sum_{j=1}^{n} b_j$$

The primal–dual algorithm of Ford and Fulkerson [1962] described above is used.

Global Constants

M   number of rows in the transportation tableau (supply nodes)
N   number of columns in the transportation tableau (demand nodes)

Data Types

```
TYPE   ARRMN = ARRAY[1..M,1..N] OF INTEGER;
       ARRM  = ARRAY[1..M] OF INTEGER;
       ARRN  = ARRAY[1..N] OF INTEGER;
```

Procedure Parameters

```
PROCEDURE TRANSPORT(
    M,N,INF  :INTEGER;
    VAR A    :ARRM;
    VAR B    :ARRN;
    VAR C,X  :ARRMN;
    VAR KO   :INTEGER);
```

Input

M	numbers of rows in the transportation tableau (supply nodes)
N	numbers of columns in the transportation tableau (demand nodes)
A[1..M]	array of supplies
B[1..N]	array of demands
C[1..M,1..N]	array of unit transportation costs
INF	very large integer number

Output

X[1..M,1..N]	array of optimal shipping
KO	optimal shipping cost

Example

Consider a transportation problem with the following input data (Murty [1976]):

$$M = 4, \quad N = 6, \quad INF = 10E6$$
$$A[1..4] = [30, 10, 45, 30]$$
$$B[1..6] = [25, 20, 6, 7, 22, 35]$$
$$C[1..4,1..6] = \begin{bmatrix} 30 & 11 & 5 & 35 & 8 & 29 \\ 2 & 5 & 2 & 5 & 1 & 9 \\ 35 & 20 & 6 & 40 & 8 & 33 \\ 19 & 2 & 4 & 30 & 10 & 25 \end{bmatrix}$$

The optimal solution obtained is

$$X[1..4,1..6] = \begin{bmatrix} 0 & 12 & 0 & 0 & 0 & 18 \\ 3 & 0 & 0 & 7 & 0 & 0 \\ 0 & 0 & 6 & 0 & 22 & 17 \\ 22 & 8 & 0 & 0 & 0 & 0 \end{bmatrix}$$
$$KO = 1902$$

Pascal Procedure TRANSPORT

```
PROCEDURE TRANSPORT(
  M,N,INF  :INTEGER;
  VAR  A    :ARRM;
  VAR  B    :ARRN;
  VAR  C,X  :ARRMN;
  VAR  KO   :INTEGER);


VAR I,J,SF,R,RA   :INTEGER;
    LAB,LAB1,LAB2:BOOLEAN;
    U,W,EPS       :ARRM;
    V,K,DEL       :ARRN;
BEGIN   (*  INITIALIZATION OF DUAL VARIABLES U AND V  *)
  FOR I:=1 TO M DO U[I]:=0;
  FOR J:=1 TO N DO
  BEGIN
    R:=INF;
    FOR I:=1 TO M DO
    BEGIN
      X[I,J]:=0;
      SF:=C[I,J];
      IF SF < R THEN R:=SF
    END ;
    V[J]:=R
  END ;
```

```
LAB1:=FALSE;
REPEAT  (* UNTIL LAB1. LAB1 IS TRUE IF THE OPTIMAL SOLUTION  *)
        (* HAS BEEN FOUND AND FALSE OTHERWISE *)
   (*   INITIALIZATION OF ROW AND COLUMN LABELS  *)
   FOR I:=1 TO M DO
   BEGIN
      W[I]:=0;
      EPS[I]:=A[I]
   END;
   FOR J:=1 TO N DO
   BEGIN
      K[J]:=0;
      DEL[J]:=0
   END;
   REPEAT  (*  UNTIL LAB  *)
      (*  PROCESS OF LABELING ROWS AND COLUMNS  *)
      LAB:=TRUE; (* LAB BECOMES FALSE WHEN A COLUMN IS LABELED *)
      LAB2:=TRUE;      (*  LAB2 IS FALSE IF BREAKTHROUGH  *)
      I:=0;            (*  APPEARS, AND TRUE OTHERWISE  *)
      REPEAT  (*  UNTIL I = M OR LAB2 = FALSE  *)
         I:=I+1;
         SF:=EPS[I];  EPS[I]:=-SF;
         IF SF > 0 THEN
         BEGIN    (*  ROW I BECOMES LABELED  *)
            RA:=U[I];
            J:=0;
            REPEAT   (*  UNTIL J = N OR LAB2 = FALSE  *)
               J:=J+1;
               IF (DEL[J] = 0) AND (V[J]-RA = C[I,J]) THEN
               BEGIN          (*  ELEMENT I,J IS ADMISSIBLE  *)
                  K[J]:=I;    (*  COLUMN J CAN BE LABELED  *)
                  DEL[J]:=SF;
                  LAB:=FALSE;
                  IF B[J] > 0 THEN
                  BEGIN (*  BREAKTHROUGH  *)
                     LAB:=TRUE;  LAB2:=FALSE;
                     SF:=ABS(DEL[J]);
                     R:=B[J];
                     IF R < SF THEN SF:=R;
                     B[J]:=R-SF;
                     REPEAT
                        I:=K[J];
                        X[I,J]:=X[I,J]+SF;
                        J:=W[I];
                        IF J <> 0 THEN
                           X[I,J]:=X[I,J]-SF
                     UNTIL J = 0;
                     A[I]:=A[I]-SF;
                     J:=0;
                     REPEAT
                        J:=J+1;
                        LAB1:=B[J] <= 0
                     UNTIL (J = N) OR NOT LAB1;
                     IF LAB1 THEN
```

```
                        BEGIN  (* OPTIMAL SOLUTION HAS BEEN FOUND *)
                            SF:=0;
                            FOR I:=1 TO M DO
                                FOR J:=1 TO N DO
                                BEGIN
                                    R:=X[I,J];
                                    IF R > 0 THEN SF:=SF+R*C[I,J]
                                END ;
                            KO:=SF
                        END  (* LAB1  *)
                    END  (* BREAKTHROUGH  *)
                END  (*  LABELING COLUMN J  *)
            UNTIL (J = N) OR NOT LAB2
        END   (*  SF > 0  *)
    UNTIL (I = M) OR NOT LAB2;
    IF NOT LAB THEN
    BEGIN  (*  LABELING ROWS FROM COLUMNS  *)
        LAB:=TRUE;
        FOR J:=1 TO N DO
        BEGIN
            SF:=DEL[J];
            IF SF > 0 THEN
            BEGIN
                FOR I:=1 TO M DO
                    IF EPS[I] = 0 THEN
                    BEGIN
                        R:=X[I,J];
                        IF R > 0 THEN
                        BEGIN
                            W[I]:=J;
                            IF R <= SF THEN EPS[I]:=R
                            ELSE EPS[I]:=SF;
                            LAB:=FALSE;
                        END
                    END(* I , EPS[I] = 0 *);
                DEL[J]:=-SF
            END (*  SF > 0 *);
        END  (* J *)
    END  (*  NOT LAB  *)
    UNTIL LAB; (*  END OF LABELING  *)
    IF LAB2 THEN
    BEGIN  (*  MODIFYING DUAL VARIABLES  *)
        R:=INF;
        FOR I:=1 TO M DO
            IF EPS[I] <> 0 THEN
            BEGIN
                RA:=U[I];
                FOR J:=1 TO N DO
                    IF DEL[J] = 0 THEN
                    BEGIN
                        SF:=C[I,J]+RA- V[J];
                        IF R > SF THEN R:=SF
```

```
              END
         END (* I , EPS[I] <> 0 *);
      FOR I:=1 TO M DO
         IF EPS[I] = 0 THEN U[I]:=U[I]+R;
      FOR J:=1 TO N DO
         IF DEL[J] = 0 THEN V[J]:=V[J]+R;
   END  (*  LAB2  *)
 UNTIL LAB1
END; (*END TRANSPORT*)
```

PROBLEMS

1-38. The max-flow problem as a linear programming problem.

   (a) Formulate the max-flow problem as a standard linear programming problem. Identify its constraint matrix.

   (b) Formulate the max-flow problem as an LP problem with the following constraint matrix D. Let edges be numbered $e_1, e_2, \ldots, e_m$ and $P_1, P_2, \ldots, P_k$ be an enumeration of all paths from s to t. Form the edge-path incidence matrix $D = [d_{ij}]$ defined as follows:

$$d_{ij} = \begin{cases} 1, & \text{if } e_i \text{ is in } P_j \\ 0, & \text{otherwise} \end{cases}$$

   for $i = 1, 2, \ldots, m$; $j = 1, 2, \ldots, k$.

1-39. Show that every maximum flow problem for a network with m edges has an optimal flow that can be decomposed into the sum of flows along no more than m paths. [*Hint:* Make use of the formulation identified in Problem 1-38(b).]

1-40. Consider the following network:

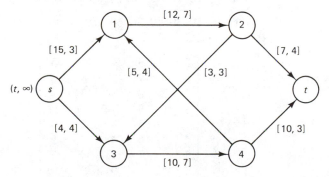

   where the numbers in the square brackets are k_{ij} and x_{ij}. The initial total flow is 7. Use the labeling algorithm and show that the network after the

termination of the algorithm is

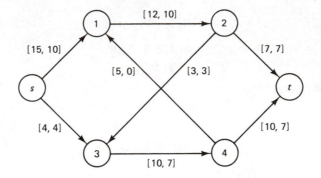

1-41. Continue until termination the numerical calculations of Example 1-3.

1-42. Consider transportation problem (1-21)–(1-23) and assume that

$$\sum_{i=1}^{m} a_i \geq \sum_{j=1}^{n} b_j$$

Show that for any optimal solution $[x_{ij}^*]$ it is always true that

$$\sum_{i=1}^{m} x_{ij}^* = b_j, \qquad j = 1, 2, \ldots, n$$

1-43. Show that a feasible solution to the transportation problem (1-21)–(1-23) always exists.

1-44. Using the TRANSPORT procedure, solve a transportation problem with the following input data:

$$\mathbf{a}^T = (20, 10, 40, 75, 15, 60, 20, 30)$$
$$\mathbf{b}^T = (60, 30, 60, 30, 60, 30)$$

$$[c_{ij}] = \begin{bmatrix} 5 & 3 & 6 & 4 & 7 & 5 \\ 6 & 4 & 7 & 5 & 8 & 6 \\ 7 & 2 & 8 & 3 & 9 & 4 \\ 8 & 9 & 9 & 10 & 10 & 11 \\ * & * & 5 & 3 & 6 & 4 \\ * & * & 6 & 4 & 7 & 5 \\ * & * & 7 & 2 & 8 & 3 \\ * & * & 8 & 9 & 9 & 10 \end{bmatrix}$$

The stars indicate an impossible shipping.

1-45.   The transportation problem in the matrix form is

minimize $\qquad\qquad\qquad z = \mathbf{c}^T\mathbf{x}$

subject to $\qquad\qquad\qquad A\mathbf{x} = \begin{bmatrix} \mathbf{a} \\ \mathbf{b} \end{bmatrix}$ $\qquad\qquad$ (1-29)

$$\mathbf{x} \geqslant 0$$

where the cost coefficients and variables are assumed to form vectors

$$\mathbf{c}^T = \left(c_{11}, c_{12}, \ldots, c_{1n}, c_{21}, \ldots, c_{mn} \right)$$
$$\mathbf{x}^T = \left(x_{11}, x_{12}, \ldots, x_{1n}, x_{21}, \ldots, x_{mn} \right)$$

and

$$A = \begin{bmatrix} \mathbf{1}^T & \mathbf{0}^T & \cdots & \mathbf{0}^T \\ \mathbf{0}^T & \mathbf{1}^T & \cdots & \mathbf{0}^T \\ \vdots & \vdots & & \vdots \\ \mathbf{0}^T & \mathbf{0}^T & \cdots & \mathbf{1}^T \\ I & I & \cdots & I \end{bmatrix}$$

where $\mathbf{1}^T = (1, 1, \ldots, 1) = $ a row vector with n components
$\qquad\mathbf{0}^T = $ a zero-vector with n components
$\qquad I = $ an $n \times n$ unit matrix

(a) Show that the constraint matrix A is not full rank.

(b) Prove that the matrix A is of rank $m + n - 1$, and that if we remove any one row from A, the remaining $m + n - 1$ rows are linearly independent. (*Hint:* Consider the matrix M formed from A by taking columns $n, 2n, \ldots, mn, 1, 2, \ldots, n - 1$ and rows $1, \ldots, m + n - 1$.)

1-46.   Demonstrate that if $\bar{A}$ is formed by choosing any $m + n - 1$ linearly independent columns of A in (1-29), then if any single row of $\bar{A}$ is crossed out, the remaining matrix of order $m + n - 1$ is nonsingular. The basis matrices B of linear programming problem (1-29) can be obtained this way.

1-47.   Let B be a nonsingular square submatrix of $\bar{A}$ of order $m + n - 1$ ($\bar{A}$ has been defined in Problem 1-46). Prove that every matrix B (a basis for the transportation problem) is triangular.

1-48.   A triangular system of linear equations can be easily solved by substitution. Show that the values of all basic variables in the transportation problem are integer if all a_i, $i = 1, 2, \ldots, m$, and b_j, $j = 1, 2, \ldots, n$, are positive integers. *Remark.* The optimum solution to the transportation problem with all a_i's and b_j's positive integers is an integer solution, and it can be obtained by the standard simplex method. Thus the transportation problem is a special case of an integer programming problem that is easy to solve.

1-49.   The dual problem of (1-21)–(1-23) is presented in (1-24) and (1-25). The reduced costs can be obtained from

$$p_{ij} = c_{ij} + u_i - v_j$$

(for $i = 1, 2, \ldots, m$ and $j = 1, 2, \ldots, n$) which is a special case of the equation

$$\mathbf{p}_j = \mathbf{c}_j - \lambda^T \text{col}_j(A), \qquad j = 1, 2, \ldots, n$$

discussed in Section 1.1 and follows from the special structure of the matrix A. From the complementarity condition for optimality $p_{ij} x_{ij} = 0$ we get

$$c_{ij} = -u_i + v_j \qquad \text{if } x_{ij} > 0 \tag{1-30}$$

Since we have $n + m - 1$ basic variables x_{ij} and $n + m$ variables u_i, v_j, there is one degree of freedom and one dual variable can be chosen arbitrarily (e.g., $u_1 = 0$). Thus the u_i and v_j values are indeterminate up to an additive constant. Furthermore, given the optimal primal solution, equations (1-30) can be easily solved since their matrix of coefficients is triangular.

Find the optimal dual solution for the transportation problem solved in the Computer Implementation of the transportation algorithm.

1-50.   The numerical problem solved in the Computer Implementation displays an interesting feature. If a_2 is changed to 20 and b_2 is changed to 30, the optimal solution is

$$[x_{ij}] = \begin{bmatrix} 0 & 12 & 0 & 0 & 0 & 18 \\ 13 & 0 & 0 & 7 & 0 & 0 \\ 0 & 0 & 6 & 0 & 22 & 17 \\ 12 & 18 & 0 & 0 & 0 & 0 \end{bmatrix}$$

and its cost is 1752. It is natural to expect that the total cost of this modified problem would be higher since more goods are transported. Explain this paradox using the optimum dual solution obtained in Problem 1-49.

1-51.   **Assignment problem.** Suppose that we have to assign n persons to n jobs, and we can determine the cost of assigning the ith person to the jth job, c_{ij}. The *assignment problem* is

minimize $\displaystyle\sum_{i, j = 1}^{n} c_{ij} x_{ij}$

subject to $\displaystyle\sum_{i=1}^{n} x_{ij} = 1, \qquad j = 1, 2, \ldots, n$

$\displaystyle\sum_{j=1}^{n} x_{ij} = 1, \qquad i = 1, 2, \ldots, n$

$x_{ij} = 0 \text{ or } 1, \qquad i = 1, 2, \ldots, n; \quad j = 1, 2, \ldots, n$

The first two constraints imply that each person can be assigned to only one job, and that each job can be assigned to only one person. It is quite obvious that the assignment problem is a special case of the transportation problem.

(a) Show that it is also true that any transportation problem can be formulated as an assignment problem.

(b) Let π be the set of all permutations of the sequence $(1, 2, 3, \ldots, n)$ and let $p = (p(1), p(2), \ldots, p(n))$ denote a permutation. Show that the assignment problem is equivalent to

$$\text{minimize} \qquad \sum_{i=1}^{n} c_{i,\,p(i)}$$

where minimum is taken over all p in π.

1-52. **Machine-assignment problem.** The machine-assignment problem is to find nonnegative x_{ij} such that

$$\sum_{j=1}^{n} d_{ij} x_{ij} \leqslant a_i, \qquad i = 1, 2, \ldots, m$$

$$\sum_{i=1}^{m} x_{ij} = b_j, \qquad j = 1, 2, \ldots, n$$

and minimize the cost

$$z = \sum_{i=1}^{m} \sum_{j=1}^{n} c_{ij} x_{ij}$$

where we use the following notation:

m number of machines

n number of products

d_{ij} time required to process one unit of product j on machine i

a_i time available on machine i

b_j number of units of product j which must be completed

c_{ij} cost of processing one unit of product j on machine i

In this problem the basic solution need not have integral values, and certain classes of the machine-assignment problem must be solved using a general integer LP code. Show that if the ratio of the times required to process one unit of a given product on any two different machines is the same for all machines, then the machine-assignment problem reduces to the standard transportation problem.

1-53. Formulate the *time transportation problem*, where we want to minimize the longest transportation time subject to the usual supply and demand constraints. Given are an $m \times n$ array of the transportation times related to all possible edges (s_i, t_j), a supply vector **a**, and a demand vector **b**.

1-54. Formulate a special assignment problem, called the *bottleneck assignment problem*, in which we want to minimize the maximum cost of assigning the ith person to the jth job.

REFERENCES AND REMARKS

The references used in this section are as follows:

EDMONDS, J., and R. M. KARP, Theoretical Improvements in Algorithmic Efficiency for Network Flow Problems, *J. ACM* 19(1972), 248–264

FORD, L. R., JR., and D. R. FULKERSON, *Flows in Networks*, Princeton University Press, Princeton, N.J., 1962

KUCHARCZYK, J., and M. M. SYSŁO, *Optimization Algorithms in ALGOL 60*, Państwowe Wydawnictwo Naukowe, Warsaw, 1975 (in Polish)

MURTY, K. G., *Linear and Combinatorial Programming*, Wiley, New York, 1976

Large-scale network and network-related problems can be found in

BRADLEY, G. H., G. G. BROWN, and G. W. GRAVES, Design and Implementation of Large Scale Primal Transshipment Algorithms, *Management Sci.* 24(1977), 1–34

KLINGMAN, D., Capsule View of Future Developments on Large Scale Network and Network-Related Problems, Research Report CS 238, Center for Cybernetic Studies, University of Texas, Austin, 1975

The paper by Bradley et al. explains the unified mathematical framework of linear programming as applied to networks and presents data structures crucial for computer implementations.

The reader interested in important data structures and computer science techniques applied to optimization problems should consult

FOX, B. L., Data Structures and Computer Science Techniques in Operations Research, *Oper. Res.* 26(1978), 686–717

The paper by Fox discusses details of the primal simplex method specialized to solve the transportation problems. It also describes the pioneering approach of Glover and Klingman to data structures used in their implementation of the primal simplex for large-scale problems.

A book concerned with combinatorial optimization problems which can be formulated in terms of networks and matroids is

LAWLER, E. L., *Combinatorial Optimization: Networks and Matroids*, Holt, Rinehart and Winston, New York, 1976

An attempt to integrate many central constructs and results in mathematical programming has been made by Shapiro in his book

SHAPIRO, J. F., *Mathematical Programming: Structures and Algorithms*, Wiley-Interscience, New York, 1979

Tomizawa has proposed an improvement over the classic primal–dual algorithm by Ford and Fulkerson

TOMIZAWA, N., On Some Techniques Useful for Solution of Transportation Network Problems, *Networks* 1(1971), 173–194

A revised version of Tomizawa's algorithm for solving the linear assignment problem is described in

DORHOUT, B., Experiments with Some Algorithms for the Linear Assignment Problem, Report BW 39/77, Stichtig Mathematisch Centrum, Amsterdam, 1977

More on linear and bottleneck assignment problems can be found in

BURKARD, R. E., and U. DERIGS, *Assignment and Matching Problems: Solution Methods with FORTRAN Programs*, Springer-Verlag, New York, 1980

The time transportation problem has been considered in

HAMMER, P. L., Time Minimizing Transportation Problems, *Naval Res. Logist. Quart.* 16(1969), 345–357

SZWARC, W., Some Remarks on the Time Transportation Problem, *Naval Res. Logist. Quart.* 18(1971), 473–486

1.4.   INTEGER PROGRAMMING

In many practical applications of linear programming the optimal solution variables are constrained to be integer. Depending on the nature of a particular optimization integer problem we may be able to solve it using a standard continuous linear programming algorithm, round the integer-constrained variables, and obtain an acceptable integer solution. If, for instance, the continuous LP method produces rather large numerical values, then rounding them up or down to the nearest integer value may be perfectly acceptable for our practical purposes. On the other hand, if we consider a problem of scheduling airplanes servicing several routes and the continuous optimal solution (the number of flights) is a vector consisting of numbers such as $0.3, 1.4, 2.7, \ldots$, then the rounding may result in values far from an optimal integer solution or in solutions that are infeasible. The rounding technique in such applications is not applicable and we have to resort to an integer linear programming method. A list of real-life problems requiring integer algorithms includes engineering design, facilities location, resource-task scheduling, and loading and airline crew scheduling, all of which appear in industry, science, and government environments.

Some Applications

There are other important reasons for integer programming. Consider an optimization problem with a piecewise linear constraint or objective function in one variable shown in Fig. 1-8. Any given value of y between y_1 and y_n can be expressed as a convex combination of y_i and y_{i+1}, $y = \lambda_i y_i + \lambda_{i+1} y_{i+1}$, where $\lambda_i + \lambda_{i+1} = 1$, $\lambda_i, \lambda_{i+1} \geq 0$. Similarly, $f(y) = \lambda_i f_i + \lambda_{i+1} f_{i+1}$, where f_k stands for $f(y_k)$. Using integer variables, we can express $f(y)$ in the entire interval $[y_1, y_n]$ as follows:

$$f(y) = \sum_{i=1}^{n} \lambda_i f_i$$

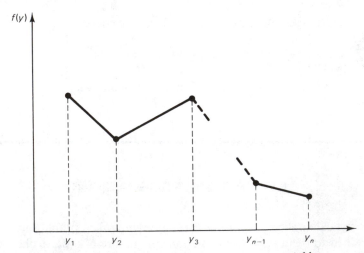

Figure 1-8   *Piecewise linear function in one variable.*

where

$$\sum_{i=1}^{n} \lambda_i y_i = y, \qquad \sum_{i=1}^{n} \lambda_i = 1$$

$$\lambda_i \geqslant 0, \qquad\qquad i = 1, 2, \ldots, n$$

$$\lambda_1 \leqslant x_1$$

$$\lambda_i \leqslant x_{i-1} + x_i, \qquad i = 2, \ldots, n - 1$$

$$\lambda_n \leqslant x_{n-1}$$

$$\sum_{i=1}^{n-1} x_i = 1$$

$$x_i = 0 \text{ or } 1, \qquad\qquad i = 1, 2, \ldots, n - 1$$

Only one value of x_i can be 1. This implies that only λ_i and λ_{i+1} can be nonzero and $\lambda_i + \lambda_{i+1} = 1$.

Now the optimization problem involves the zero–one integer variables $x_1, \ldots,$ x_{n-1}, which define the interval containing y, and the continuous variables $\lambda_1, \ldots, \lambda_n$ determining precisely the value of y. Unfortunately, for problems with many piecewise linear functions the number of discrete variables and associated constraints may be excessive for numerical computation. Note that this technique can be used to approximate nonlinear functions by piecewise linear functions and transform problems with nonlinear objective functions to integer linear programming problems.

Consider *fixed-charge* problems where the objective function includes a startup cost:

$$f(x) = \begin{cases} ax + b, & \text{if } x > 0 \\ 0, & \text{if } x = 0 \end{cases}$$

Assuming that $b > 0$ and $f(x)$ is to be minimized, $f(x)$ can be reformulated as follows:

$$f(x) = ax + bz$$

$$x \geqslant 0$$

$$z = 0 \text{ or } 1$$

$$x(1 - z) = 0$$

where z is an indicator of whether or not activity is undertaken. The last constraint guarantees that if $x > 0$, then $z = 1$. Minimizing $f(x)$ will ensure that $z = 0$ when $x = 0$.

Integer variables may also be helpful in solving mathematical programming problems with nonconvex feasible regions. Suppose that the feasible region is defined as follows:

$$x_1 + x_2 \leqslant b$$
$$x_1 \geqslant 1 \quad \text{or} \quad x_2 \geqslant 1$$
$$x_1, x_2 \geqslant 0$$

Using two zero–one variables z_1 and z_2 we can define an equivalent feasible region as

$$x_1 + x_2 \leqslant b$$
$$x_1 - 1 + (1 - z_1) \geqslant 0$$
$$x_2 - 1 + (1 - z_2) \geqslant 0$$
$$z_1 + z_2 = 1, \qquad z_1, z_2 = 0 \text{ or } 1$$

We refer the reader to Dantzig [1960] and Salkin [1975] for more examples of the uses and applications of integer programming.

1.4.1. ALGORITHMS FOR INTEGER PROGRAMMING

We have seen in Section 1.3 that transportation and network flow problems can be solved as linear programming problems. Such problems are special cases in which solutions generated by the simplex method are automatically integer valued. In these problems the pivots used in the course of performing the simplex method are either $+1$ or -1. Consequently, no arithmetic divisions are used and if the initial problem is defined in terms of integers, its solution will be integer. In general, however, integer programs cannot be solved by the simplex method and special computational techniques must be used. The principal approaches for solving integer linear programming problems include:

1. Cutting planes
2. Enumeration
3. Decomposition (partitioning)
4. Group theory

We briefly characterize these methods and comment on their computational properties.

Cutting Plane Technique

The main idea is to use a *cut*, which is a derived constraint with the property of cutting off part of the set of feasible solutions while not excluding any integer solution. In these methods we construct cut constraints and use the simplex method iteratively until an optimal integer solution is found. Within this general framework two types of algorithms are possible: dual cutting plane and primal cutting plane algorithms. The first category maintains dual feasible linear programs and uses an extension of the dual simplex method. One of the main contributors in this area is Ralph Gomory, who developed three general, but somewhat different methods:

1. The dual all-integer integer programming method, where the pivots used are always -1's and the dual simplex tableau is maintained integer during the computational process. This method can be regarded as being a direct extension of the dual simplex method and is presented in detail in this section.

2. The dual fractional method, which utilizes the dual simplex method and allows fractional numbers in computations.

3. The extension of the above for the mixed integer programs, where some variables are continuous and some integer.

The *dual cutting plane* methods do not produce a feasible solution until the optimal solution is determined. An alternative approach has been developed where cutting plane methods maintain primal feasibility and produce feasible integer solutions. These methods can be categorized as *primal cutting plane* methods.

Enumerative Techniques

An obvious approach for solving an integer or mixed integer program is to enumerate all possible solution candidates and find the optimal solution. Clearly, such an explicit enumeration is not practical for any nontrivial problem of a reasonable size. What, however, can be done is to examine only some sets of integer solutions which can produce optimal solutions. By looking at some integrality, nonnegativity, and other constraint requirements we can often eliminate these sets of integer solutions, which do not have to be examined, so they are *implicitly enumerated*. Within this general methodology the implicit enumeration methods use some heuristic rules and their efficiency depends mainly on the elimination criteria used in the process of implicit enumeration.

We describe this approach in Section 1.5, devoted to the solution of the zero–one programming problems.

Decomposition Algorithms

Mixed integer programming problems can be solved by special decomposition algorithms, where we handle a series of related integer and continuous linear programming problems. The original problem is represented as a set of independent LP problems, except for some linking variables which may be integer variables. Then a problem is constructed which involves only the linking variables. The decomposition approach has been used to solve some special problems whose subproblems reduce to transportation problems and can be efficiently solved. Until now, the partitioning approach has not resulted in an efficient computational procedure capable of solving general integer programming problems.

Group-Theoretic Methods

This approach has been suggested and explored by Gomory [1965], who discovered that by relaxing the nonnegativity, but not integrality constraints on certain variables, any integer program can be represented as an optimization problem defined on a group. If such a problem is solved and yields a nonnegative solution, this is equivalent to solving the original integer program. The reader interested in group-theoretic methods is referred to Salkin [1975].

In spite of the very extensive research effort that has been made to develop efficient algorithms for linear integer problems, we cannot expect that every mixed or pure integer problem can be solved by presently available methods in an acceptable length of computer time. This is particularly true for problems that do not exhibit any special structure and whose number of variables and/or constraints exceeds roughly 100. What is worse, there are some relatively small problems that defy some or most of the available methods. It should be added that there exists a considerable gap between the massive research material accumulated in the area of integer programming and the rather modest amount of reported experimental work.

For the purpose of introducing the most important and distinct approaches to integer programming we have selected the Gomory all-integer methods and the implicit enumeration method of Balas for the zero–one programming problem (Section 1.5) as the two methods which are described in detail, and programmed. Also, in Chapter 2 we present packing and covering problems which are special instances of the all-integer programming problem.

1.4.2.   CONSTRUCTION OF THE GOMORY ALL-INTEGER DUAL METHOD

The Gomory all-integer dual method is a direct extension of the dual simplex method discussed in Section 1.2. The main difference is that the pivot row in the all-integer method is generated at each iteration and the pivot value is -1. This ensures that the dual simplex tableau remains all-integer. The intent of the algorithm is to reduce the feasible region to one whose optimum vertex is all-integer. Since the method is dual, primal solutions are infeasible until the last, optimal solution is obtained.

Lexicographic Dual Method

As indicated previously, there is no real danger of cycling in the continuous simplex method. However, in the integer versions of the dual simplex method cycling can easily occur due to massive degeneracy and special measures must be taken to avoid infinite looping. To see how cycling can be prevented we will consider a lexico-graphic form of the dual simplex method. In the cutting plane literature, the dual simplex tableau is recorded in a special manner shown in Table 1-1. This table corresponds to the integer problem written as follows:

minimize x_0 (1-31)

subject to $x_i = p_{i0} + \sum_{j \in J} p_{ij}(-x_j)$

$x_i \geq 0$ and integer, $i = 0, 1, \ldots, n$

where J is the set of nonbasic variable indices. In Table 1-1 we have assumed, for convenience, that the first m variables are basic. In the vector notation we have

minimize x_0

subject to $\mathbf{x} = \mathbf{p}_0 + \sum_{j \in J} \mathbf{p}_j(-x_j)$

$x_i \geq 0$ and integer, $i = 0, 1, 2, \ldots, n$

Table 1-1. Explicit Dual Simplex Tableau

		variable entering the basis ↓			
Variable	Constant	$-x_{m+1}$	$\cdots$	$-x_k$	$\cdots$	$-x_n$
x_0	p_{00}	$p_{0,m+1}$	$\cdots$	$p_{0,k}$	$\cdots$	$p_{0,n}$
x_1	p_{10}	$p_{1,m+1}$	$\cdots$	$p_{1,k}$	$\cdots$	$p_{1,n}$
$\vdots$	$\vdots$	$\vdots$		$\vdots$		$\vdots$
x_l	p_{l0}	$p_{l,m+1}$	$\cdots$	Pivot $\boxed{p_{l,k}}$	$\cdots$	$p_{l,n}$
$\vdots$	$\vdots$	$\vdots$		$\vdots$		$\vdots$
x_m	p_{m0}	$p_{m,m+1}$	$\cdots$	$p_{m,k}$	$\cdots$	$p_{m,n}$
x_{m+1}	0	-1				
$\vdots$	$\vdots$			0		
x_n	0			0		-1

variable leaving the basis →

basic variables

nonbasic variables

A vector $\mathbf{v} \neq 0$ is *lexicographically positive (negative)* if its first nonzero component is positive (negative). To denote lexicographically positive (negative) vector we use the notation $\mathbf{v} \overset{l}{>} 0 \left(\mathbf{v} \overset{l}{<} 0 \right)$.

A vector $\mathbf{v}$ is lexicographically greater (smaller) than a vector $\mathbf{w}$ if $\overset{l}{>} \left(\overset{l}{<} \right) 0$ and a sequence of vectors $\mathbf{v}^t$, $t = 1, 2, \ldots$, is lexicographically decreasing (increasing) if $\mathbf{v}^t - \mathbf{v}^{t+1} \overset{l}{>} \left(\overset{l}{<} \right) 0$.

The lexicographic dual algorithm replaces (1-31) by the following formulation: find the lexicographic minimum $\mathbf{x}$

subject to
$$\mathbf{x} \geq 0$$
$$\mathbf{x} = \mathbf{p}_0 + \sum_{j \in J} \mathbf{p}_j(-x_j)$$

The variable x_l leaving the basis is determined in the same way as that of the ordinary dual method:

$$p_{l0} = \min_{1 \leq i \leq m} \{ p_{i0} : p_{i0} < 0 \}$$

The variable x_k entering the basis is found using the entry test

$$\frac{1}{p_{lk}} \mathbf{p}_k = \operatorname*{lexmax}_{m+1 \leq j \leq n} \left\{ \frac{1}{p_{lj}} \mathbf{p}_j : p_{lj} < 0 \right\}$$

where lexmax means lexicographically maximum. The pivoting operation on p_{lk} transforms the tableau (Table 1-1) into a new tableau with the columns $\bar{\mathbf{p}}_j$,

$$\bar{\mathbf{p}}_j = \mathbf{p}_j - \frac{p_{lj}}{p_{lk}} \mathbf{p}_k, \qquad j \neq k$$

$$\bar{\mathbf{p}}_k = \frac{1}{p_{lk}} \mathbf{p}_k$$

$$(1\text{-}32)$$

If the initial tableau is dual feasible in the lexicographic sense (i.e., the vectors $\mathbf{p}_j$, $j = m + 1, \ldots, n$, are lexicographically negative), formulas (1-32) guarantee that the new column vectors $\bar{\mathbf{p}}_j$ are also lexicographically negative. It can also be shown that the basic solution $\mathbf{p}_0$ is strictly increasing (in the lexicographic sense) at each iteration, and no basis can be repeated, hence there can be no cycling.

General Outline of the Gomory All-Integer Dual Method

The Gomory method is started with an all-integer tableau and a lexicographic dual feasible solution; that is, the vectors $\mathbf{p}_j$, $j = m + 1, \ldots, n$, are lexicographically negative for a problem of minimization and positive for a problem of maximization.

The method proceeds as follows:

Step 1. Select a row with $p_{r0} < 0$, $r \neq 0$. This is a cut-generating row. If none exists, the current solution is optimal.

Step 2. Find the pivot column $\mathbf{p}_k$ which is lexicographically largest among those having $p_{rj} < 0$. If none exists, there is no integer feasible solution.

Step 3. Derive a cut inequality from the row r which is not satisfied at the current primal solution. The new row is appended at the bottom of the tableau and it is a pivot row which has a -1 coefficient used as pivot in the current iteration.

Step 4. Perform one dual simplex pivoting operation.

Step 5. Delete the appended row which has become trivial $[\bar{x} = -(-\bar{x})]$, and return to step 1.

We show now how a cut inequality is obtained.

The All-Integer Method Cut

Assume that the rth row of the dual tableau is selected as the cut-generating row,

$$x_r = p_{r0} + \sum_{j \in J} p_{rj}(-x_j)$$

Define $\lfloor y \rfloor$ as the largest integer less or equal y, and λ as a positive number. Any number p satisfies the following relationship:

$$p = \left\lfloor \frac{p}{\lambda} \right\rfloor \lambda + R \tag{1-33}$$

where $0 \leqslant R < \lambda$. Applying (1-33) to the generating row, we get

$$\sum_{j \in J} R_j x_j + R_r x_r = R_0 + \lambda \left\{ \left\lfloor \frac{p_{r0}}{\lambda} \right\rfloor + \sum_{j \in J} \left\lfloor \frac{p_{rj}}{\lambda} \right\rfloor (-x_j) + \left\lfloor \frac{1}{\lambda} \right\rfloor (-x_r) \right\}$$

$$= R_0 + \lambda \bar{x} \tag{1-34}$$

For any nonnegative solution satisfying (1-34) the value of

$$\bar{x} = \left\lfloor \frac{p_{r0}}{\lambda} \right\rfloor + \sum_{j \in J} \left\lfloor \frac{p_{rj}}{\lambda} \right\rfloor (-x_j) + \left\lfloor \frac{1}{\lambda} \right\rfloor (-x_r) \tag{1-35}$$

must be integer since all coefficients of (1-35) are integer. It is also true that $\bar{x} \geqslant 0$, since $0 \leqslant R_0 < \lambda$, and if $\bar{x}$ is a negative integer, then $R_0 + \lambda \bar{x} < 0$. But the last inequality is not possible since the left-hand side of (1-34) is nonnegative. We conclude that $\bar{x} \geqslant 0$.

If we choose $\lambda \geqslant 1$, the inequality $\bar{x} \geqslant 0$ simplifies to

$$\left\lfloor \frac{p_{r0}}{\lambda} \right\rfloor + \sum_{j \in J} \left\lfloor \frac{p_{rj}}{\lambda} \right\rfloor (-x_j) \geqslant 0 \qquad (1\text{-}36)$$

Setting $\pi_{rj} = \lfloor p_{rj}/\lambda \rfloor$, the all-integer Gomory cut is defined as follows

$$\bar{x} = \pi_{r0} + \sum_{j \in J} \pi_{rj}(-x_j) \geqslant 0 \qquad (1\text{-}37)$$

Constraint (1-37) is the pivot row and $\bar{x}$ is a new nonnegative slack variable. It remains to determine the value of λ.

Determination of λ

The value of λ is determined in three steps:

Step 1. Find the pivot column k,

$$\mathbf{p}_k = \operatorname*{lexmax}_{j \in \bar{J}} \mathbf{p}_j,$$

where $\bar{J} = \{ j : p_{rj} < 0, j \neq 0 \}$ and the rth row is generating a cut.

Step 2. Determine the largest integer e_j such that

$$-\frac{1}{e_j} \mathbf{p}_j \overset{l}{\geqslant} -\mathbf{p}_k, \qquad j \in \bar{J}, \ j \neq k$$

Let $e_k = 1$.

Step 3. Take $\lambda = \max_{j \in \bar{J}} \{\lambda_j = -p_{rj}/e_j\}$.

Deriving (1-36), we assumed that $\lambda \geqslant 1$. This inequality holds since $\lambda \geqslant \lambda_k = -p_{rk} \geqslant 1$. If $\lambda = 1$, then the generating row is the pivot row and there is no new constraint. By its derivation λ need not be an integer. Choosing this value of λ we obtain a -1 pivot and a pivot row which results in the greatest lexicographic increase of the column $\mathbf{p}_0$. The latter property of the pivot row increases the rate of convergence. The reader interested in detailed justification of this choice of λ is referred to Salkin [1975, p. 105].

1.4.3. GOMORY'S ALL-INTEGER DUAL ALGORITHM

It is assumed that the columns of $\mathbf{p}_j$, $j = m + 1, \ldots, n$, are lexicographically dual feasible.

Step 1. If $\mathbf{p}_0 \geqslant 0$, then the primal solution is feasible and optimal, and the computation is terminated. Otherwise, select a cut generating row with $p_{r0} < 0$.

Step 2. Identify the pivot column $\mathbf{p}_k$, $m + 1 \leqslant k \leqslant n$, to be the maximum lexico-graphically among those having $p_{rj} < 0$. If all $p_{rj} \geqslant 0$, then there is no feasible integer solution; stop. Otherwise, continue.

Step 3. For every $j \in \bar{J}$, $j \neq k$, find the largest integer number e_j such that $-(1/e_j)\mathbf{p}_j \geqslant -\mathbf{p}_k$, and set $e_k = 1$. Calculate $\lambda = \max_{j \in \bar{J}}\{-p_{rj}/e_j\}$.

Step 4. Add the constraint

$$\bar{x} = \pi_{r0} + \sum_{j \in J} \pi_{rj}(-x_j) \geqslant 0$$

to the dual tableau [see (1-37)].

Step 5. Perform one dual simplex pivoting operation using $\pi_{rk} = -1$ as a pivot.

Step 6. Drop the added constraint which has become trivial and return to step 1.

In step 1 we can choose the minimal $p_{i0} < 0$ to determine the cut-generating row. This does not guarantee convergence but yields in practice an efficient algorithm.

Theoretically, we should choose a rule which guarantees that if component p_{i0} becomes and remains negative, row i will be selected as a cut-generating row at some iteration. This can be accomplished in several ways. For example, we may:

1. Select a cut row randomly.
2. Select a cut row by some cyclic process.
3. Always select the first eligible row with $p_{i0} < 0$.

Example 1-4

To illustrate the algorithm, consider the following integer linear programming problem:

minimize
$$x_0 = 3x_1 + 5x_2$$

subject to
$$x_3 = -5 + x_1 + 4x_2 \geqslant 0$$

$$x_4 = -7 + 3x_1 + 2x_2 \geqslant 0$$

$$x_1, x_2, \ldots, x_4 \geqslant 0 \text{ and integer}$$

The explicit dual tableau for this problem is

	Variable	Constant	$\downarrow k$ $-x_1$	$-x_2$
	x_0	0	-3	-5
	x_3	-5	-1	-4
$\overset{r}{\rightarrow}$	x_4	-7	-3	-2
			Pivot	
	x_5	-3	-1	-1

The trivial constraints with only one entry -1 are omitted. The generating row is the row with x_4, and the pivot column is $k = 1$. The values of e_j are $e_1 = 1$ and $e_2 = 1$, and $\lambda = \max\{2/1, 3/1\} = 3$. The additional constraint is

$$x_5 = \lfloor -\tfrac{7}{3} \rfloor + \lfloor -\tfrac{3}{3} \rfloor(-x_1) + \lfloor -\tfrac{2}{3} \rfloor(-x_2) \geqslant 0$$

or

$$x_5 = -3 - 1(-x_1) - 1(-x_2) \geqslant 0$$

Pivoting on $p_{51} = -1$ yields the new tableau

Variable	Constant	$-x_5$	$-x_2$
x_0	0	-3	-2
x_1	3	-1	1
x_3	-2	-1	-3
x_4	2	-3	1

This iteration is shown graphically on Fig. 1-9. The reader is asked to continue the iterations until the optimal solution is found. ∎

Computational Performance of the Method

Even though we can construct a cutting plane algorithm which converges in a finite number of iterations, this finite number can be very large. The cutting plane algorithms are not polynomial-time algorithms. The Gomory cutting plane methods are most successful in solving large set partitioning and set covering problems (see Problem 1-58 and Section 2.2.4). They are less effective in solving general integer programming problems. However, they are useful in some hybrid approaches where

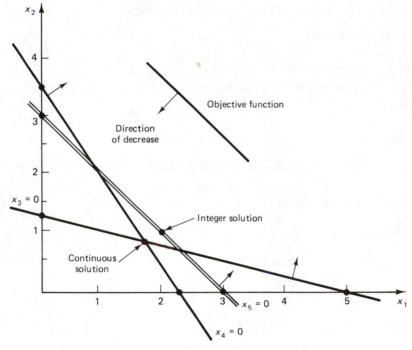

Figure 1-9 *First cut constraint.*

the cutting plane methods are used in conjunction with branch-and-bound algorithms.

In general, the cutting plane methods tend to be unpredictable in terms of computer run time. This has been attributed to the fact that frequently the cutting plane methods make substantial initial progress in the first few iterations and then slow down dramatically. On the other hand, they could be tried first because if they converge at all, the solution is obtained quickly. If this first attempt is not successful, we may try implicit enumerative codes.

Computer Implementation of the Gomory Algorithm

The ALLINTEGER procedure is a Pascal version of the ALGOL 60 procedure called *Gomory* published in Kucharczyk and Sysło [1975]. It solves the following integer LP problem:

minimize
$$\sum_{j=1}^{n} a_{0j} x_j$$

subject to
$$\sum_{j=1}^{n} a_{ij} x_j \leq a_{i, n+1}, \qquad i = 1, 2, \ldots, m - n$$

where $x_j \geqslant 0$ ($j = 1, 2, \ldots, n$) are required to be integer.

Global constants

N   number of variables $x_1, \ldots, x_n$

M n + the number of constraints

N1 $n + 1$

Data types

TYPE   ARRMN1 = ARRAY[0..M,1..N1] OF INTEGER;

Procedure Parameters

```
PROCEDURE ALLINTEGER(
     M,N          :INTEGER;
     VAR  A       :ARRMN1;
     VAR  COUNT  :INTEGER;
     VAR  NOFEAS :BOOLEAN);
```

Input

N   number of variables $x_1, x_2, \ldots, x_n$

M n + the number of constraints

A[0..M,1..N1]   array defining the integer LP problem

The array A has the following form:

$a_{01} \cdots\cdots\cdots\cdots\cdots\cdots a_{0n}$	$a_{0, n+1}$	
$a_{11} \cdots\cdots\cdots\cdots\cdots\cdots a_{1n}$	$a_{1, n+1}$	
$a_{m-n, 1} \cdots\cdots\cdots\cdots\cdots a_{m-n, n}$	$a_{m-n, n+1}$	
-1 $\cdots$ 0	$a_{m-n+1, n+1}$	
0 -1	$a_{m, n+1}$	

The values of $a_{0, n+1}$ and $a_{m-n+1, n+1}, \ldots, a_{m, n+1}$ should be initially set to 0.

It is assumed that the first n columns of A are lexicographically positive, since in our input we use the variables $x_j, j = 1, 2, \ldots, n$, rather than $-x_j, j = 1, 2, \ldots, n$. Also note that the $(n + 1)$st column contains the array of constants $a_{i, n+1}$, $i = 1, 2, \ldots, m - n$.

Output

A[0,N + 1]	optimal function value multiplied by -1
A[M − N + J,N + 1]	(J = 1, 2, ..., N) optimal solution vector
COUNT	number of dual simplex iterations performed
NOFEAS	Boolean variable equal to TRUE if there is no feasible solution, FALSE if there is

Example (Haldi [1964])

minimize $\qquad 13x_1 + 15x_2 + 14x_3 + 11x_4$

subject to

$$- 4x_1 - 5x_2 - 3x_3 - 6x_4 \leqslant - 96$$

$$-20x_1 - 21x_2 - 17x_3 - 12x_4 \leqslant -200$$

$$-11x_1 - 12x_2 - 12x_3 - 7x_4 \leqslant -101$$

$$x_1, x_2, x_3, x_4 \geqslant 0 \text{ and integer}$$

In this problem we have

$$N = 4, \quad M = 7,$$

A[0..7,1..5] =

13	15	14	11	0
− 4	−5	−3	−6	−96
−20	−21	−17	−12	−200
−11	−12	−12	−7	−101
− 1				0
	−1		0	0
		−1		0
0			−1	0

The solution obtained is

$$-A[0,5] = 187$$
$$x_1 = A[4,5] = 0$$
$$x_2 = A[5,5] = 0$$
$$x_3 = A[6,5] = 0$$
$$x_4 = A[7,5] = 17$$
$$\text{COUNT} = 17, \quad \text{NOFEAS} = \text{FALSE}$$

Pascal Procedure ALLINTEGER

```pascal
PROCEDURE ALLINTEGER(
   M,N            :INTEGER;
   VAR    A       :ARRMN1;
   VAR    COUNT   :INTEGER;
   VAR    NOFEAS  :BOOLEAN);


VAR    C,DENOM,I,J,K,L,NP,NUM,R,R1,S,T :INTEGER;
       B,ITER                          :BOOLEAN;

FUNCTION EUCLID(U,V :INTEGER):INTEGER;
VAR    W :INTEGER;
BEGIN
   W:=U DIV V;
   IF W*V > U THEN W:=W-1;
   IF (W+1)*V <= U THEN W:=W+1;
   EUCLID:=W
END; (* END OF EUCLID *)

BEGIN
   COUNT:=0;
   NP:=N+1;
   REPEAT
      COUNT:=COUNT+1;
      R:=0;
      REPEAT
         R:=R+1;
         ITER:=A[R,NP] < 0
      UNTIL ITER OR (R = M);
      IF ITER THEN
      BEGIN
         K:=0;
         REPEAT
            K:=K+1;
            ITER:=A[R,K] < 0
         UNTIL ITER OR (K = N);
         NOFEAS:= NOT ITER;
         IF ITER THEN
         BEGIN
            L:=K;
            FOR J:=K+1 TO N DO
               IF A[R,J] < 0 THEN
               BEGIN
                  I:=-1;
                  REPEAT
                     I:=I+1;
                     S:=A[I,J]-A[I,L]
                  UNTIL S <> 0;
                  IF S < 0 THEN L:=J
               END;
```

```
            S:=0;
            WHILE A[S,L]=0 DO S:=S+1;
            NUM:=-A[R,L];
            DENOM:=1;
            FOR J:=1 TO N DO
                IF (A[R,J] < 0) AND (J <> L) THEN
                BEGIN
                    I:=S-1;
                    B:=TRUE;
                    WHILE B AND (I >=0) DO
                    BEGIN
                        B:=A[I,J] = 0;
                        I:=I-1
                    END;
                    IF B THEN
                    BEGIN
                        I:=A[S,J];
                        R1:=A[S,L];
                        T:=EUCLID(I,R1);
                        IF (T*R1 = I) AND (T > 1) THEN
                        BEGIN
                            I:=S;
                            REPEAT
                                I:=I+1;
                                R1:=T*A[I,L]-A[I,J]
                            UNTIL R1 <> 0;
                            IF R1 > 0 THEN T:=T-1
                        END;
                        C:=-A[R,J];
                        IF C*DENOM > T*NUM THEN
                        BEGIN
                            NUM:=C;
                            DENOM:=T
                        END
                    END  (* B *)
                END;  (*  J  *)
            FOR J:=1 TO NP DO
                IF J <> L THEN
                BEGIN
                    C:=EUCLID(A[R,J]*DENOM,NUM);
                    IF C <> 0 THEN
                        FOR I:=0 TO M DO
                            A[I,J]:=A[I,J]+C*A[I,L]
                END
        END  (* ITER: A[R,K] < 0  *)
    END  (*  ITER: A[R,NP] < 0  *)
    UNTIL NOT ITER OR NOFEAS
END; (* END OF ALLINTEGER *)
```

Code Application and Some Computational Results

The reader should use the ALLINTEGER procedure with caution. The computational experience with cutting plane methods is mixed, at best. Some small and medium-size problems are easily solved by the plane cutting methods, whereas similar problems cannot be solved in any reasonable period of computer time. Table 1-2 illustrates the performance of the ALLINTEGER procedure on several standard test problems. Note the dramatic difference between the execution times for problems IBM4 and IBM5, which differ only in the right-hand side vector. The system design problem given in Table 1-3 was not solved after 350,000 iterations. The same problem was solved in less than 8 seconds on the Amdahl 470 V/6 using the 0–1 BALAS procedure (Section 1.5).

To illustrate the sensitivity of some integer problems to relatively minor changes in the problem data, we have solved the following allocation problem (Trauth and Woolsey [1969]):

minimize $\quad 20x_1 + 18x_2 + 17x_3 + 15x_4 + 15x_5 + 10x_6 + 5x_7 + 3x_8 + x_9 + x_{10}$

subject to $\quad -30x_1 - 25x_2 - 20x_3 - 18x_4 - 17x_5 - 11x_6 - 5x_7 - 2x_8$

$$-x_9 - x_{10} \leqslant -b$$

$$x_i \leqslant 1 \text{ and integer}, \quad i = 1, 2, \ldots, 10$$

Table 1-4 shows the solution for five different values of b.

Table 1-2.   Numerical Tests

Problem	Source	Size $(m - n), n$	Execution Time (msec)[a]	Number of Iterations, COUNT
IBM1	Haldi [1964, p. 12][b]	7, 7	7	10
IBM2	Haldi [1964, p. 12]	7, 7	13	17
IBM3	Haldi [1964, p. 13]	3, 4	4	17
IBM4	Haldi [1964, p. 13]	15, 15	43	18
IBM5	Haldi [1964, p. 13]	15, 15	5928	2570
System Design	Plane and McMillan [1971, pp. 157–159]	9, 14	No solution obtained after 15 minutes of computation time (350,000 iterations)	

[a] Amdahl 470 V/6.
[b] Haldi's problems can also be found in Garfinkel and Nemhauser [1972].

Table 1-3. Coefficients of the System Design Problem

i						j									Right-hand Side
	1	2	3	4	5	6	7	8	9	10	11	12	13	14	
0	50	60	40	20	60	75	75	90	3	4	4	4	10		
1					−10	−13	−14	−16							$\leqslant -40$
2	−10	−14	−9	−8	7	12	15	10							$\leqslant 0$
3	20	28	18	16											$\leqslant 0$
4	2								−10	−13	−14	−22			$\leqslant 0$
5						1			−1				−100	4	$\leqslant 4$
6							1							−3	$\leqslant 0$
7	50	75	60	50	600	600	800	750	25	17	20	26			$\leqslant 2300$
8													1		$\leqslant 1$
9														1	$\leqslant 1$

Table 1-4. Sensitivity Test

b	f_{min}	x_1	x_2	x_3	x_4	x_5	x_6	x_7	x_8	x_9	x_{10}	Time (msec)[a]	Number of Iterations, COUNT
95	72	1	1	1	1	0	0	0	0	1	1	22	14
70	53	1	1	0	1	0	0	0	0	0	0	29	19
65	48	1	1	0	0	0	1	0	0	0	0	15	8
60	43	1	1	0	0	0	0	1	0	0	0	25	12
55	38	1	1	0	0	0	0	0	0	0	0	10	5

[a]Amdahl 470 V/6.

PROBLEMS

1-55. Perform one more iteration for the problem of Example 1-4 and show that the optimal solution is $x_1 = 2$, $x_2 = x_3 = x_4 = 1$. Draw the second cut constraint. The first cut constraint is shown in Fig. 1-9.

1-56. Permutation of constraints can effect the number of iterations of the all-integer Gomory method. Use the ALLINTEGER procedure to solve the following problems:

minimize $x_1 + x_2 + x_3$

subject to $x_1 \qquad + 2x_3 \geqslant 4$

$\qquad\qquad\qquad x_2 + x_3 \geqslant 15$

$$x_1, x_2, x_3 \geqslant 0 \text{ and integer}$$

Permute the constraints and solve the problem again. Compare the values of the variable COUNT.

1-57. Show that the dual simplex method presented in Section 1.2 and the lexicographic dual method perform identical computations except in cases of dual degeneracy.

1-58. The *set-covering* problem is

minimize $c^T x$

subject to $Ax \geqslant e$

$$x_j = 0 \text{ or } 1, \quad j = 1, 2, \ldots, n$$

where A is an $m \times n$ matrix whose elements are 0 or 1, c is a positive cost vector, $e^T = (1, 1, \ldots, 1)$. Use ALLINTEGER to solve a set-covering problem with the following data:

$$\mathbf{c}^T = (1, 1, \ldots, 1)$$

$$A = \begin{bmatrix} 1 & 1 & 1 & 0 & 0 & 0 & 0 & 0 \\ 1 & 0 & 0 & 1 & 1 & 0 & 0 & 0 \\ 0 & 0 & 0 & 1 & 0 & 1 & 1 & 0 \\ 0 & 1 & 0 & 0 & 0 & 1 & 0 & 1 \\ 0 & 0 & 1 & 0 & 1 & 0 & 1 & 1 \end{bmatrix}$$

Explain why the constraints $x_j = 0$ or 1 do not have to be entered explicitly (i.e., they are satisfied automatically). See more on this problem in Section 2.2.

1-59. A cutting plane is defined to be a hyperplane that strictly separates the optimal solution of the continuous linear programming problem from the set of integer feasible solutions of this linear problem. If we call the continuous optimal solution $\hat{x}$ and assume that $\hat{x}$ is noninteger, then $\sum a_j x_j \geqslant b_i$ is a cutting plane if $\sum a_j \hat{x}_j < b_i$ and $\sum a_j x_j \geqslant b_i$ for all feasible integer solutions. Consider numerical Problem 1-55 and show that the constraint $-3 + x_1 + x_2 \geqslant 0$ is a cutting plane.

1-60. All-integer primal algorithm. The primal integer programming algorithms are modifications of the primal simplex method. A simple all-integer primal algorithm is presented here. Assume that the integer linear problem has all coefficients integer and, is in the following form:

maximize
$$x_0 = a_{00} + \sum_{j=1}^{n} a_{0j}(-x_j)$$

subject to
$$x_{n+i} = a_{i0} + \sum_{j=1}^{n} a_{ij}(-x_j), \qquad i = 1, 2, \ldots, m$$

$$x_j \geqslant 0 \text{ and integer,} \qquad j = 1, 2, \ldots, n,$$

where $x_1, \ldots, x_n$ are nonbasic and $x_{n+1}, \ldots, x_{n+m}$ are basic variables. If the ith constraint is selected as the cut-generating row, then the new cut constraint is

$$s = \sum_{j=1}^{n} \left\lfloor \frac{a_{ij}}{\lambda} \right\rfloor (-x_j) + \left\lfloor \frac{a_{i0}}{\lambda} \right\rfloor \geqslant 0 \qquad (1\text{-}38)$$

The algorithm consists of the following steps:

Step 1. Generate a basic feasible solution as in the standard simplex method.

Step 2. If $a_{0j} \geqslant 0$, $j = 1, 2, \ldots, n$, the current solution is optimal. Otherwise, find the pivot column k using

$$a_{0k} = \min_{1 \leqslant j \leqslant n} \{a_{0j} : a_{0j} < 0\}$$

Step 3.　Find the rth row using

$$\frac{a_{r0}}{a_{rk}} = \min_{a_{ik}>0}\left[\frac{a_{i0}}{a_{ik}}\right]$$

breaking ties arbitrarily, if necessary. If all $a_{ik} \leq 0$, the optimal solution is unbounded.

Step 4.　Form a cut according to (1-38) using the rth row and $\lambda = a_{rk}$. Append this row, perform a primal simplex pivot step (the pivot is 1), and return to step 2.

Using the outlined primal method, solve the following numerical problem given in Young [1965]:

maximize　　　　　　　$x_0 = \quad 3x_1 + x_2$

subject to　　　　　　$x_3 = 6 - 2x_1 - 3x_2$

$$x_4 = 3 - 2x_1 + 3x_2$$

$$x_1, \ldots, x_4 \geq 0 \text{ and integer}$$

The first iteration is presented.

	1	$-x_1$	$-x_2$			1	$-s_1$	$-x_2$
x_0	0	-3	-1		x_0	3	3	-7
x_1	0	-1	0		x_1	1	1	-2
x_2	0	0	-1		x_2	0	0	-1
x_3	6	2	3		x_3	4	-2	7
x_4	3	2	-3		x_4	1	-2	1
s_1	1	Pivot ①	-2					

Before pivoting　　　　　　　　　　　　After pivoting

The entering variable is x_1 since $-3 < -1$. The fourth row generates the cut since $3/2 < 6/2$. The cut equation is

$$s_1 = \lfloor \tfrac{2}{2} \rfloor(-x_1) + \lfloor -\tfrac{3}{2} \rfloor(-x_2) + \lfloor \tfrac{3}{2} \rfloor \geq 0$$

or

$$-x_1 + 2x_2 + 1 \geq 0$$

With the slack variable s_1 the new constraint

$$s_1 = 1 - x_1 + 2x_2$$

is appended. Pivoting on ① gives the new tableau.

The algorithm presented does not necessarily converge due to cycling. It has been used, however, as a basis for developing a convergent algorithm (see Salkin [1975, p. 130]).

REFERENCES AND REMARKS

In Section 1.4 references have been made to the following sources:

DANTZIG, G. B., On the Significance of Solving Linear Programming Problems with Some Integer Variables, *Econometrica* 28(1960), 30–44

GARFINKEL, R. S., and G. L. NEMHAUSER, *Integer Programming*, Wiley, New York, 1972

GOMORY, R., On the Relation between Integer and Non-integer Solutions to Linear Programs, *Proc. Nat. Acad. Sci., USA* 53(1965), 260–265

HALDI, J., 25 Integer Programming Test Problems, Working Paper No. 43, Stanford University, Stanford, Calif., 1964

KUCHARCZYK, J., and M. M. SYSŁO, *Optimization Algorithms in ALGOL 60*, Państwowe Wydawnictwo Naukowe, Warsaw, 1975 (in Polish)

PLANE, D. R., and C. McMILLAN, *Discrete Optimization, Integer Programming, and Network Analysis for Management Decisions*, Prentice-Hall, Englewood Cliffs, N.J., 1971

SALKIN, H. M., *Integer Programming*, Addison-Wesley, Reading, Mass., 1975

TRAUTH, C. A., JR., and R. E. WOOLSEY, Integer Linear Programming: A Study in Computational Efficiency, *Management Sci.* 15(1969), 481–493

YOUNG, R., A Primal (All-Integer) Integer Programming Algorithm, *J. Res. Nat. Bur. Standards* 69B(3)(1965), 213–249

The cutting plane algorithm presented in this chapter was developed by Ralph Gomory in his paper

GOMORY, R., All-Integer Integer Programming, IBM Research Center, Report RC-189, 1960

Decomposition and partitioning algorithms for linear and mixed-variable problems are well explained in

LASDON, L., *Optimization Theory for Large Systems*, Macmillan, New York, 1970

A book that presents the state-of-the-art review of combinatorial optimization including branch and bound methods and theory of cutting planes is

CHRISTOFIDES, N., A. MINGOZZI, P. TOTH, and C. SANDI (eds.), *Combinatorial Optimization*, Wiley, New York, 1979

Most of the existing commercial codes are directed at large linear programs with few integer variables. They are usually enumerative algorithms that can solve mixed integer or pure integer programming problems. A description of the available commercial and academic integer and mixed integer programming codes can be found in

Johnson, E. L., and S. Powell, Integer Programming Codes, in H. J. Greenberg (ed.), *Design and Implementation of Optimization Software*, NATO Advanced Study Institute Series, Sijthoff & Noordhoff, The Netherlands, 1978

Kuester, J. L., and J. H. Mize, *Optimization Techniques with FORTRAN*, McGraw-Hill, New York, 1973

Land, A., and S. Powell, Computer Codes for Problems in Integer Programming, *Ann. Discrete Math.* 5(1979), 221–269

Land, A., and S. Powell, *FORTRAN Codes for Mathematical Programming*, Wiley, New York, 1973

The reader interested in the practical aspects of solving integer linear programming problems should consult

Woolsey, R. E. D., How to Do Integer Programming in the Real World, in Salkin [1975]

Woolsey, R. E. D., Some Practical Aspects of Solving Integer Programming Problems, in S. Zionts (ed.), *Linear and Integer Programming*, Prentice-Hall, Englewood Cliffs, N.J., 1974

The computational complexity of integer programming is considered in

Papadimitriou, C. H., On the Complexity of Integer Programming, *J. ACM* 28(1981), 765–768

A variation of the Gomory cutting plane method to solve large set-covering problems is given in

Martin, G., An Accelerated Euclidean Algorithm for Integer Linear Programming, in R. Graves and P. Wolfe (eds.), *Recent Advances in Mathematical Programming*, McGraw-Hill, New York, 1963

1.5.   ZERO–ONE INTEGER PROGRAMMING

Many linear programming problems have the special feature that all the variables are restricted to only two values: zero and one. In these combinatorial problems with the binary variables we frequently consider decisions in which just two values are needed: for example, we either assign a job to a particular machine or we do not, we undertake certain activity or we do not, and so on.

A general form of the binary (zero–one) linear programming problem is

$$\text{minimize} \qquad z = \sum_{j=1}^{n} c_j x_j \qquad \qquad \text{(1-39)}$$

subject to
$$\sum_{j=1}^{n} a_{ij}x_j \leqslant b_i, \qquad i = 1, 2, \ldots, m$$

$$x_j = 0 \text{ or } 1, \qquad j = 1, 2, \ldots, n$$

It is worth mentioning that some integer problems can be reduced to binary programming problems. Such reduction is practical if the number of integer variables is not large and their maximum values are relatively small. If the bounds on an integer variable y are $0 \leqslant y \leqslant 2^p$, then y can be uniquely represented as

$$y = \sum_{i=0}^{p} 2^i x_i \tag{1-40}$$

where

$$x_i = 0 \text{ or } 1, \qquad i = 0, 1, 2, \ldots, p$$

Thus the integer variable y is replaced by the $p + 1$ binary variables x_i.

Optimization problems that involve binary variables include knapsack, capital budgeting, assembly line balancing, matching, set covering, and facility location. Some of these problems are formulated and analyzed in other sections of this book. In this section we are concerned with the general zero–one problem which does not have any special structure. Problems with distinct structural features are usually solved by specialized algorithms.

To illustrate how the binary variables come about in optimization, we present the *facilities location problem*. Assume that there are m facility locations which produce a single commodity for n customers each with a demand for d_j units, $j = 1, 2, \ldots, n$. If a particular facility is built, it has a fixed cost $c_i \geqslant 0$ and the production capacity $p_i > 0$. A cost of shipping a unit commodity from facility i to customer j is a_{ij}. We wish to find facility locations such that the total cost is minimized and the demands d_j are satisfied. If y_{ij} is the amount shipped from facility i to customer j, and $x_i = 1$ if facility i is built and $x_i = 0$ otherwise, the facility location problem is

minimize
$$\sum_{i=1}^{m} \sum_{j=1}^{n} a_{ij}y_{ij} + \sum_{i=1}^{m} c_i x_i \tag{1-41}$$

subject to
$$\sum_{i=1}^{m} y_{ij} = d_j, \qquad j = 1, 2, \ldots, n \tag{1-42}$$

$$\sum_{j=1}^{n} y_{ij} \leqslant p_i x_i, \qquad i = 1, 2, \ldots, m \tag{1-43}$$

where $y_{ij} \geqslant 0$ are continuous and x_i are binary variables.

The optimization problems of this structure can be solved using partitioning algorithms mentioned in Section 1.4. However, we will not pursue this subject any further but return to the general zero–one programming problem.

1.5.1.   IMPLICIT ENUMERATION

Consider the zero–one integer problem as given by (1-39). It is convenient to assume that the vector of prices $\mathbf{c}$ is nonnegative. Any zero–one problem can be converted to a problem with nonnegative costs by substituting $1 - \bar{x}_j$ for x_j if $c_j < 0$. To solve this problem, search algorithms are used which enumerate either explicitly or implicitly all 2^n possible zero–one vectors $\mathbf{x}$. In such procedures the vast majority of solutions are enumerated implicitly. The enumerative procedure can be illustrated by a search tree composed of nodes and branches. A node corresponds to a zero–one candidate solution $\mathbf{x}$. Two nodes connected by a branch differ in the state of one variable. Each variable can be in one of three states: *fixed at 1, fixed at 0*, or *free*. A new node is defined by fixing a variable to 1 (forward step), and a node is revisited by fixing a variable to 0 (backtrack step). Figure 1-10 illustrates a search tree.

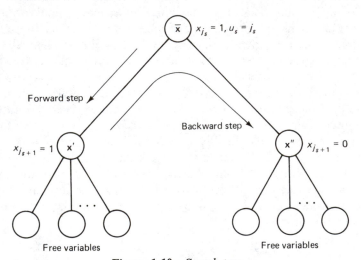

Figure 1-10   *Search tree.*

To explain how the search tree is traversed, let us assume that the current *partial solution* under consideration is given by the vector $\mathbf{u}^T = (u_1, u_2, \ldots, u_s, 0, 0, \ldots, 0)$ with entries interpreted as follows:

1. $u_k = j_k$, $1 \leqslant k \leqslant s$, means that x_{j_k} has been fixed to be 1 in accordance with certain rules that will be presented. Its complement ($x_{j_k} = 0$) has not yet been considered.

2. $u_k = -j_k$, $1 \leqslant k \leqslant s$, means that $x_{j_k} = 1$ or 0, and that the complements of these values have already been considered.

3. $u_k = 0$ for $k > s$.

We assume that we have an *incumbent solution* $\hat{x}$, which is the best feasible solution found so far. Finally, for the purpose of our discussion let us assume that $x_{j_s} = 1$, $u_s = j_s$ (i.e., the value of $x_{j_s} = 0$ has not been yet considered) and that at the node $\bar{x}$ the partial solution $\bar{x}^T = (x_{j_1}, x_{j_2}, \ldots, x_{j_s}, 0, \ldots, 0)$ is infeasible (Fig. 1-10).

Suppose now that we use a test which indicates that by adding another variable $x_{j_{s+1}}$ we can reduce the infeasibility and improve the objective function value. We move forward and add $u_{s+1} = j_{s+1}$ to the vector $\mathbf{u}$ (i.e., set $x_{j_{s+1}} = 1$). At the new node x' we find that x' is feasible and it produces the objective function value better than the incumbent solution. We replace the incumbent solution by x' and update the function value accordingly. All *completions* in this partial solution with $x_{j_{s+1}} = 1$ have been enumerated since no feasible solution can be better by fixing any new free variable at 1 (remember that $\mathbf{c} \geqslant 0$). We say that this partial solution has been *fathomed*; that is, all completions of the partial solution have been implicitly enumerated. Thus we can consider all completions of $\bar{x}$ with $x_{j_{s+1}} = 0$. We backtrack to the node x''. Suppose that there is no attractive completion for this node; that is, no completion of this partial solution can produce an optimal solution. At this point, we have enumerated all possible completions of $\bar{x}$ with $x_{j_{s+1}} = 1$ and $x_{j_{s+1}} = 0$. We are ready to backtrack again and consider $x_{j_s} = 0$. The element u_{j_s} of the vector $\mathbf{u}$ is now set to $-j_s$ (we assumed previously that $x_{j_s} = 0$ had not been considered). If, on the other hand, $u_s = -j_s$ and $x_{j_s} = 0$ (i.e., $x_{j_s} = 1$ was considered previously), we find the rightmost positive element of $\mathbf{u}$, change its sign to negative, and examine the new partial completions. In this case we backtrack more than one branch of the search tree.

To summarize, the basic enumeration approach is as follows:

1.   Fix a free variable $x_{j_{s+1}}$ at value 1.

2.   Resolve the subproblem in the remaining free variables.

3.   Fix $x_{j_{s+1}}$ at value 0 and repeat the process for the subproblem with $x_{j_{s+1}} = 0$.

To accomplish successfully the tree traversal we may have to move backward (backtrack) several branches of the tree. Resolving the current subproblems in the remaining free variables means enumerating explicitly or implicitly each of the completions of the particular partial solution.

Implicit Enumeration Criteria

1.   Objective function and constraint improvement.   The purpose of this test is to find out if it is possible to improve the objective function value and reduce constraint infeasibilities. We create the set T of variables such that

$$T = \{\, j : x_j \text{ is free}, z + c_j < \hat{z},$$

$$a_{ij} < 0 \text{ for such } i \text{ that}$$

$$y_i = b_i - \sum_{j \in J} a_{ij} x_j < 0\}$$

where J is a set of the current partial solution indices, $z = \sum_{j \in J} c_j$, and $\hat{z}$ is the best function value found so far. If T is empty, then a backward step (backtrack) is justified.

2.   *Infeasibility test.*   If $T \neq \varnothing$ we may be able to identify an index i such that $y_i < 0$ and $y_i - \sum_{j \in T} \min(0, a_{ij}) < 0$. Thus even if all variables in T are 1, the ith constraint will remain infeasible. In such cases, backtrack is also justified, since there is no feasible continuation.

3.   *The Balas branching test.*   The selection of the free variable to fix at 1 may significantly influence the algorithm efficiency. This is particularly true at the beginning, where a poor selection of a free variable could result in a needless enumeration of a large number of points which are not near a solution. A good rule designed to direct the search toward a solution has been given by Balas [1965].

For each free variable x_j we create the set M_j,

$$M_j = \{i : y_i - a_{ij} < 0\}$$

and calculate

$$v_j = \sum_{i \in M_j} (y_i - a_{ij})$$

or set $v_j = 0$ if M_j is empty. We determine which free variable, if set to 1, would reduce the total infeasibility most. By "total infeasibility" we mean the sum of the absolute values of the amount by which all constraints are violated. Thus we select the variable x_j such that it maximizes v_j.

If all the sets M_j are empty, a backtrack step is taken. An algorithm based on these criteria has been designed by Balas and is called the Balas zero–one additive algorithm.

1.5.2.   THE BALAS ZERO – ONE ADDITIVE ALGORITHM

To keep track of the enumeration progress two vectors are used: $\mathbf{u}^T = (u_1, u_2, \ldots, u_n)$ and $\mathbf{x}^T = (x_1, x_2, \ldots, x_n)$. If a partial solution consists of the variables $x_{j_1}, x_{j_2}, \ldots, x_{j_s}$ assigned in this order, the components of $\mathbf{x}$ corresponding to the assigned variables are either 0 or 1. The remaining variables in **x** corresponding to the free variables have values -1. The components of **u** are:

$$u_k = \begin{cases} j_k, & \text{if } x_{j_k} = 1 \text{ and its complement has not yet been considered} \\ -j_k, & \text{if } x_{j_k} = 1 \text{ or } 0 \text{ and its complement has been considered} \\ 0, & \text{if } k > s \end{cases}$$

Initial Step

Step 1. Check if **b** $\geqslant 0$. If yes, the optimal solution is $\hat{\mathbf{x}} = 0$. Otherwise, set $\mathbf{u}^T = (0, 0, \ldots, 0)$, $\mathbf{x}^T = (-1, -1, \ldots, -1)$, and $\hat{z} = \infty$ (a sufficiently large number).

Iterative Steps

Step 2. Calculate

$$y_i = b_i - \sum_{j \in J} a_{ij} x_{ij}$$

$$\bar{y} = \min y_i, \qquad i = 1, 2, \ldots, m$$

$$z = \sum_{j \in J} c_j x_j$$

where J is the set of indices of the assigned variables (partial solution). If $\bar{y} \geqslant 0$ and $z < \hat{z}$, then set $\hat{z} = z$ and $\hat{\mathbf{x}} = \mathbf{x}$. The vector $\hat{\mathbf{x}}$ is the new incumbent solution. Go to step 6 (backtracking). Otherwise, continue.

Step 3. Create a subset T of the free variables x_j, defined as

$$T = \{ j : z + c_j < \hat{z}, a_{ij} < 0 \text{ for } i \text{ such that } y_i < 0 \}$$

If $T = \varnothing$, then go to step 6. Otherwise, continue.

Step 4. *Infeasibility test.* Test if there is an index i such that $y_i < 0$ and $y_i - \sum_{j \in T} \min(0, a_{ij}) < 0$. If yes, go to step 6. Otherwise, continue.

Step 5. *Balas branching test.* For each free variable x_j create the set M_j,

$$M_j = \{ i : y_i - a_{ij} < 0 \}$$

If all sets M_j are empty, go to step 6. Otherwise, calculate for each free variable x_j,

$$v_j = \sum_{i \in M_j} (y_i - a_{ij})$$

where $v_j = 0$ if $M_j = \varnothing$. Add to the current partial solution the variable x_j that maximizes v_j. Return to step 2.

Step 6.   Backtracking

    a.   Modify the vector **u** by changing the sign of the rightmost positive component. All elements to the right are set to 0. That is the new partial solution. Return to step 2.

    b.   If there is no positive element in **u**, the implicit enumeration is complete. The optimal solution is the current vector $\hat{x}$, and the corresponding function value is $\hat{z}$. If $\hat{z} = \infty$, there is no feasible solution.

Example 1-5

To illustrate the algorithm, consider the following numerical case:

minimize
$$z = 5x_1 + 7x_2 + 10x_3 + 3x_4 + x_5$$

subject to
$$-x_1 + 3x_2 - 5x_3 - x_4 + 4x_5 \leqslant -2$$
$$2x_1 - 6x_2 + 3x_3 + 2x_4 - 2x_5 \leqslant 0$$
$$x_2 - 2x_3 + x_4 + x_5 \leqslant -1$$
$$x_1, x_2, \ldots, x_5 = 0 \text{ or } 1$$

Iteration 1

The subset T is $\{1, 3, 4\}$. Infeasibility test fails. The sets M_j and the values v_j for $j = 1, 2, \ldots, 5$ are

$$
\begin{aligned}
M_1 &= \{1, 2\}, & v_1 &= -4 \\
M_2 &= \{1, 3\}, & v_2 &= -7 \\
M_3 &= \{2\}, & v_3 &= -3 \\
M_4 &= \{1, 2, 3\}, & v_4 &= -5 \\
M_5 &= \{1, 3\}, & v_5 &= -8
\end{aligned}
$$

$$\max_{1 \leqslant j \leqslant 5} v_j = v_3 = -3$$

The variable $x_3 = 1$ is added to the partial solution. Other variables are free with their values set at 0.

Iteration 2

$T = \{2, 5\}$. Infeasibility test fails.

$$
\begin{aligned}
M_1 &= \{2\}, & v_1 &= -5 \\
M_2 &= \varnothing, & v_2 &= 0 \\
M_4 &= \{2\}, & v_4 &= -5 \\
M_5 &= \{1, 2\}, & v_5 &= -2
\end{aligned}
$$

The next variable added to the partial solution is $x_2 = 1$ since

$$\max v_j = 0, \qquad j = 1, 2, 4, 5$$

Iteration 3

The vector $\mathbf{x}^T = (0, 1, 1, 0, 0)$ is feasible and becomes an incumbent solution $\hat{\mathbf{x}}$ with $\hat{z} = 17$. Step 6 is executed (backtrack). We now test $x_3 = 1$, $x_2 = 0$, and the remaining variables are free.

Iteration 4

$T = \{5\}$. Infeasibility test is satisfied for $i = 2$ (the second constraint) and the backtrack step is taken. Now x_2 and x_3 are set at 0 and the remaining variables are free.

Iteration 5

$T = \{1, 4\}$. Infeasibility test is satisfied for $i = 3$ and the backtrack step is taken. The enumeration is complete since no backtracking is possible. The optimal solution is $\hat{x}_1 = 0$, $\hat{x}_2 = \hat{x}_3 = 1$, $\hat{x}_4 = \hat{x}_5 = 0$, and $\hat{z} = 17$. ∎

Computer Implementation of Balas Algorithm

The BALAS procedure is a Pascal version of the ALGOL 60 procedure also called *Balas* and published in Kucharczyk and Sysło [1975]. It solves the binary linear program:

minimize
$$z = \sum_{j=1}^{n} c_j x_j$$

subject to
$$\sum_{j=1}^{n} a_{ij} x_j \leqslant b_i, \qquad i = 1, 2, \ldots, m$$

$$x_j = 1 \text{ or } 0, \qquad j = 1, 2, \ldots, n$$

where $c_j \geqslant 0$. The detailed description of the method is given in the preceding section.

Global Constants

M   number of constraints, m
N   number of variables $x_1, x_2, \ldots, x_n$
N1   integer whose value is $n + 1$

Data Types

```
TYPE   ARRMN = ARRAY[1..M,1..N] OF INTEGER;
       ARRM  = ARRAY[1..M] OF INTEGER;
       ARRN  = ARRAY[1..N] OF INTEGER;
       ARRN1 = ARRAY[1..N1] OF INTEGER;
```

Procedure Parameters

```
PROCEDURE BALAS(
    M,N,INF    :INTEGER;
    VAR A      :ARRMN;
    VAR B      :ARRM;
    VAR C,X    :ARRN;
    VAR FVAL   :INTEGER;
    VAR EXIST  :BOOLEAN);
```

Input

M	number of constraints, m
N	number of variables $x_1, x_2, \ldots, x_n$
A[1..M,1..N]	array of the constraint matrix
B[1..M]	array of the right-hand-side vector
C[1..N]	array of the cost vector
INF	large positive integer

Output

X[1..N]	optimal solution vector
FVAL	optimal value of the objective function
EXIST	Boolean variable equal to TRUE if the optimal solution has been found, and equal to FALSE if there is no feasible solution; in the latter case, X is not meaningful

Example

Consider the following problem:

minimize

$$10x_1 + 14x_2 + 21x_3 + 42x_4$$

subject to

$$-8x_1 - 11x_2 - 9x_3 - 18x_4 \leqslant -12$$

$$-2x_1 - 2x_2 - 7x_3 - 14x_4 \leqslant -14$$

$$-9x_1 - 6x_2 - 3x_3 - 6x_4 \leqslant -10$$

$$x_1, x_2, x_3, x_4 = 0 \text{ or } 1$$

For this problem we have

$$N = 4, \quad M = 3, \quad N1 = 5,$$

$$A[1..3,1..4] = \begin{bmatrix} -8 & -11 & -9 & -18 \\ -2 & -2 & -7 & -14 \\ -9 & -6 & -3 & -6 \end{bmatrix}$$

$$B[1..3] = [-12, -14, -10]$$

$$C[1..4] = [10, 14, 21, 42]$$

The solution obtained is

$$EXIST = TRUE$$
$$X[1..N] = [1,0,0,1]$$
$$FVAL = 52$$

Pascal Procedure BALAS

```
PROCEDURE   BALAS(
    M,N,INF     :INTEGER;
    VAR   A     :ARRMN;
    VAR   B     :ARRM;
    VAR   C,X   :ARRN;
    VAR   FVAL  :INTEGER;
    VAR   EXIST :BOOLEAN);

LABEL   10,20;
VAR   ALFA,BETA,GAMMA,I,J,MNR,NR,P,R,R1,R2,S,T,Z:INTEGER;
    Y,W,ZR                                  :ARRM;
    II,JJ,XX                                :ARRN;
    KK                                      :ARRN1;
BEGIN
    FOR I:=1 TO M DO Y[I]:=B[I];
    Z:=1;
    FOR J:=1 TO N DO
    BEGIN
        XX[J]:=0;
        Z:=Z+C[J]
    END;
    FVAL:=Z+Z;
    S:=0; T:=0; Z:=0;
    KK[1]:=0;
    EXIST:=FALSE;
10:
    P:=0; MNR:=0;
    FOR I:=1 TO M DO
    BEGIN
        R:=Y[I];
        IF R < 0 THEN       (*  INFEASIBLE CONSTRAINT I  *)
```

```
BEGIN
    P:=P+1;
    GAMMA:=0;
    ALFA:=R;
    BETA:=-INF;
    FOR J:=1 TO N DO
        IF XX[J] <= 0 THEN
            IF C[J]+Z >= FVAL THEN
            BEGIN
                XX[J]:=2;
                KK[S+1]:=KK[S+1]+1;
                T:=T+1;
                JJ[T]:=J
            END
            ELSE
            BEGIN
                R1:=A[I,J];
                IF R1 < 0 THEN
                BEGIN
                    ALFA:=ALFA-R1;
                    GAMMA:=GAMMA+C[J];
                    IF BETA < R1 THEN BETA:=R1
                END
            END (* C[J]+Z < FVAL *);
        IF ALFA < 0 THEN GOTO 20;
        IF ALFA+BETA < 0 THEN
        BEGIN
            IF GAMMA+Z >= FVAL THEN GOTO 20;
            FOR J:=1 TO N DO
            BEGIN
                R1:=A[I,J];
                R2:=XX[J];
                IF R1 < 0 THEN
                BEGIN
                    IF R2 = 0 THEN
                    BEGIN
                        XX[J]:=-2;
                        FOR NR:=1 TO MNR DO
                        BEGIN
                            ZR[NR]:=ZR[NR]-A[W[NR],J];
                            IF ZR[NR] < 0 THEN  GOTO 20
                        END
                    END (* R2 = 0 *)
                END (* R1 < 0 *)
                ELSE
                    IF R2 < 0 THEN
                    BEGIN
                        ALFA:=ALFA-R1;
                        IF ALFA < 0 THEN GOTO 20;
                        GAMMA:=GAMMA+C[J];
                        IF GAMMA+Z >= FVAL THEN GOTO 20
                    END (* R1 >= 0, R2 < 0 *)
            END (* J *);
```

```
            MNR:=MNR+1;
            W[MNR]:=I;
            ZR[MNR]:=ALFA
      END (* ALFA + BETA < 0 *)
   END (* R < 0 *)
END (* I *);
IF P = 0 THEN
BEGIN              (*   UPDATING THE BEST SOLUTION  *)
   FVAL:=Z;
   EXIST:=TRUE;
   FOR J:=1 TO N DO
      IF XX[J] = 1 THEN X[J]:=1 ELSE X[J]:=0;
   GOTO 20
END;
IF MNR = 0 THEN
BEGIN
   P:=0;
   GAMMA:=-INF;
   FOR J:=1 TO N DO
      IF XX[J] = 0 THEN
      BEGIN
         BETA:=0;
         FOR I:=1 TO M DO
         BEGIN
            R:=Y[I];
            R1:=A[I,J];
            IF R < R1 THEN BETA:=BETA+R-R1
         END;
         R:=C[J];
         IF (BETA > GAMMA)OR(BETA = GAMMA)AND(R < ALFA) THEN
         BEGIN
            ALFA:=R;
            GAMMA:=BETA;
            P:=J
         END
      END (* XX[J] = 0 *);
   IF P = 0 THEN GOTO 20;
   S:=S+1;
   KK[S+1]:=0;
   T:=T+1;
   JJ[T]:=P;
   II[S]:=1; XX[P]:=1;
   Z:=Z+C[P];
   FOR I:=1 TO M DO
      Y[I]:=Y[I]-A[I,P]
END (* MNR = 0 *)
ELSE
BEGIN
   S:=S+1;
   II[S]:=0; KK[S+1]:=0;
   FOR J:=1 TO N DO
      IF XX[J] < 0 THEN
```

```
            BEGIN
              T:=T+1;
              JJ[T]:=J;
              II[S]:=II[S]-1;
              Z:=Z+C[J];
              XX[J]:=1;
              FOR I:=1 TO M DO
                Y[I]:=Y[I]-A[I,J]
            END;
      END;  (*  MNR <> 0  *)
      GOTO 10;
   20:                                (*  BACKTRACKING  *)
      FOR J:=1 TO N DO
        IF XX[J] < 0 THEN XX[J]:=0;
      IF S > 0 THEN
        REPEAT
          P:=T;
          T:=T-KK[S+1];
          FOR J:=T+1 TO P DO XX[JJ[J]]:=0;
          P:=ABS(II[S]);
          KK[S]:=KK[S]+P;
          FOR J:=T-P+1 TO T DO
          BEGIN
            P:=JJ[J];
            XX[P]:=2;
            Z:=Z-C[P];
            FOR I:=1 TO M DO
              Y[I]:=Y[I]+A[I,P]
          END;
          S:=S-1
        UNTIL (II[S+1] < 0) OR (S = 0);
      IF S > 0 THEN GOTO 10
    END;  (*END OF BALAS*)
```

Computational Performance of the Balas Method

A number of implicit enumeration algorithms have been designed to solve the linear zero–one programming problem. Recent comparative studies reported in the literature include:

1. Narula and Kindorf [1979], who compare algorithms by Balas, Hammer and Rudeanu [1968], Peterson [1967], Zionts [1972], Geoffrion [1969], and Zionts [1974].

2. Peterson [1967] compares several variants of the Balas additive algorithm.

3. Geoffrion [1969] compares his implicit enumeration algorithm with the Balas additive algorithm.

4. Land and Powell [1979] survey many commercial and academic codes for mixed integer programming and pure integer problems.

In their study, Narula and Kindorf tested unstructured problems with the number of variables n ranging from 30 to 100 and the density of matrix A equal 0.2,

0.4, 0.6, and 0.8. On the basis of their study they concluded that the Balas additive algorithm and the Zionts generalized additive algorithm (Zionts [1972]) are best among the six involved in their study. Furthermore, they recommended Balas's method for problems with density $d \leqslant 0.6$ and $n \geqslant 50$, and Ziont's method for problems with $d = 0.8$ and/or $n \leqslant 30$.

Our tests of the Balas method presented and the method of Hammer–Rudeanu published by Zieliński [1977] indicate that the Balas algorithm is faster in more cases. Table 1-5 summarizes a sample of our results. The time given is in milliseconds of the Amdahl 470 V/6 computer; m = number of constraints and n = number of variables.

Despite the good average performance of the Balas algorithm it is possible to construct pathological cases for which the algorithm execution time becomes exponential.

The following problem illustrates such a case:

minimize
$$x_{n+1}$$

subject to
$$2 \sum_{j=1}^{n} x_j + x_{n+1} = n$$

$$x_j = 0 \text{ or } 1 \quad \text{for all } j$$

where n is an odd positive integer. The problem has been solved using the BALAS procedure and the following execution times resulted:

n	T (msec of Amdahl 470 V/8)
7	8
9	29
11	115
13	452
15	1788

The function $T(n)$ is approximately $T(n) = 0.0625 \cdot 2^n$.

Table 1-5.   Performance of the Balas and the Hammer–Rudeanu Algorithms

m, n	BALAS	Hammer–Rudeanu	Problem Source
6, 12	58	11	Salkin [1975], problem 7.25
6, 12	17	10	Salkin [1975], problem 7.25 with $b_6 = -8$
3, 20	65	136	Salkin [1975], problem 7.26
10, 7	567	586	Balas [1965], problem 2
7, 10	4	6	Plane and McMillan [1971, p. 58]
3, 20	152	252	Balas [1967], problem 2
7, 10	8	10	Haldi [1964], problem IBM 1
15, 22	43,584	46,195	Haldi [1964], problem IBM 5

PROBLEMS

1-61.   (Glover and Sommer [1972])   Consider the following facility location problem [see (1-41)–(1-43)]:

$$m = n = 5$$
$$c_i = 0, \quad \text{for all } i$$
$$\mathbf{d}^T = (1, 1, 1, 1, 1)$$
$$\mathbf{p}^T = (2, 3, 2, 3, 2)$$
$$A = \begin{bmatrix} 93 & 70 & 48 & 68 & 81 \\ 45 & 89 & 97 & 85 & 96 \\ 92 & 93 & 58 & 37 & 99 \\ 55 & 103 & 55 & 57 & 38 \\ 74 & 60 & 78 & 54 & 52 \end{bmatrix}$$

Assume that (1-43) are equalities.

(a) Solve this problem without the integer restrictions.

(b) There are six combinations of facility locations (pairs of x_i) that can supply the customers. Fix the values of these pairs of x_i at 1 and the remaining x_i variables at 0 and obtain six linear programming problems. Find the optimal solution to the facilities location problem by solving all six linear programs.

1-62.   Consider the following zero–one problem:

minimize $3x_1 + 2x_2 + x_3 + x_4$

subject to $-x_1 + x_2 + 6x_3 + x_4 \leqslant 5$

$-x_1 - 2x_2 + 3x_3 - x_4 \leqslant -3$

$2x_1 + 2x_2 - x_3 - 8x_4 \leqslant -6$

$x_1, x_2, x_3, x_4 = 1 \text{ or } 0$

Solve this problem by performing a complete enumeration, and then by using the BALAS procedure.

1-63.   The eight queens problem.   In this problem we want to place four chess queens on a 4×4 chessboard so that no two queens attack each other. Formulate this problem as a zero–one integer programming problem and solve it using the BALAS procedure. (*Hint:* No two queens can be placed on a common row, column, or diagonal. The problem has 18 constraints in 16 variables.)

1-64.   The system design problem.   Using the BALAS procedure, solve the system design problem defined in Table 1-3. The system design problem is an integer programming problem, but assume that we know the following upper bounds for variables x_1 through x_{14}:

$$x_1 \leqslant 1, \quad x_2 \leqslant 4, \quad x_3 \leqslant 6, \quad x_4 \leqslant 6, \quad x_5 \leqslant 1, \quad x_6 \leqslant 2, \quad x_7 \leqslant 1,$$
$$x_8 \leqslant 1, \quad x_9 \leqslant 2, \quad x_{10} \leqslant 2, \quad x_{11} \leqslant 1, \quad x_{12} \leqslant 3, \quad x_{13} \leqslant 1, \quad x_{14} \leqslant 1$$

The optimal value of the objective function is 356. Make use of representation (1-40).

1-65. Consider the problem

minimize
$$x_{n+1}$$

subject to
$$2 \sum_{j=1}^{n} x_j + x_{n+1} = n$$

$$x_j = 0 \text{ or } 1 \quad \text{for all } j$$

where n is an odd positive integer. Draw the search tree generated by the BALAS procedure for $n = 3$ and 5.

1-66. Consider the problem

minimize
$$x_1$$

subject to
$$2 \sum_{j=1}^{n} x_j = n$$

$$x_j = 0 \text{ or } 1 \quad \text{for all } j$$

where n is an odd positive integer. Use the BALAS procedure and determine the nature of the time as a function of n which is necessary for the procedure to find that this problem has no feasible solution.

REFERENCES AND REMARKS

The references used in this section are as follows:

BALAS, E., An Additive Algorithm for Solving Linear Programs with Zero–One Variables, *Oper. Res.* 13(1965), 517–545

BALAS, E., Discrete Programming by the Filter Method, *Oper. Res.* 15(1967), 915–957

GEOFFRION, A. M., An Improved Implicit Enumeration Approach for Integer Programming, *Oper. Res.* 17(1969), 437–454

GLOVER, F., and D. SOMMER, Pitfalls of Rounding in Discrete Management Decision Problems, Management Science Report Series 72-2, University of Colorado, Boulder, 1972

HALDI, J., 25 Working Programming Test Problems, Working Paper No. 43, Stanford University, Stanford, Calif., 1964

HAMMER, P. L., and S. RUDEANU, *Boolean Methods in Operations Research and Related Areas*, Springer-Verlag, New York, 1968

KUCHARCZYK, J., and M. M. SYSŁO, *Optimization Algorithms in ALGOL 60*, Państwowe Wydawnictwo Naukowe, Warsaw, 1975 (in Polish)

LAND, A., and S. POWELL, Computer Codes for Problems of Integer Programming, *Ann. Discrete Math.* 5(1979), 221–269

NARULA, S. C., and J. R. KINDORF, Linear 0-1 Programming: A Comparison of Implicit Enumeration Algorithms, *Comput. Oper. Res.* 6(1979), 47–51

PETERSON, C. C., Computational Experience with Variants of the Balas Algorithm Applied to the Selection of R and D Projects, *Management Sci.* 13(1967), 736–750

PLANE, D. R., and C. MCMILLAN, JR., *Discrete Optimization, Integer Programming and Network Analysis for Management Decisions*, Prentice-Hall, Englewood Cliffs, N.J., 1971

SALKIN, H. M., *Integer Programming*, Addison-Wesley, Reading, Mass., 1975

ZIELIŃSKI, S., Solution of Linear Programming Problems in Zero–One Variables, *Zastos. Mat.* 16(1977), 122–131

ZIONTS, S., Generalized Implicit Enumeration Using Bounds on Variables for Solving Linear Programs with Zero–One Variables, *Naval Res. Logist. Quart.* 19(1972), 165–181

ZIONTS, S., *Linear and Integer Programming*, Prentice-Hall, Englewood Cliffs, N.J., 1974

Among computational studies of the specialized integer programming problems, we recommend

CHRISTOFIDES, N., and S. KORMAN, A Computational Survey of Methods for the Set Covering Problem, *Management Sci.* 21(1975), 591–599

FAYARD, D., and G. PLATEAU, Resolution of the 0–1 Knapsack Problem: Comparison of Methods, *Math. Programming* 8(1975), 272–307

For more information on these subjects the reader is referred to Chapter 2.

2

Packing and Covering

This chapter discusses special instances of the zero–one integer programming problems which, for their most popular and natural applications, are in general referred to as packing and covering problems. The first section presents the knapsack problem, and some covering problems are considered in the second section. Several real-world applications of these problems are presented in the text and as problems, and then discussed in the References and Remarks.

In spite of the simplicity of formulation, the problems of this chapter are computationally and theoretically very hard and no efficient (i.e., polynomial-time) algorithm has been found for their solution. On the other hand, these problems are highly structured and special-purpose methods based on such problem structure can be used.

Several common features can be identified in the presentation of the computational methods of this chapter. Initially, an attempt is made to reduce the problem size by determining the values of some variables. Then approximation algorithms are discussed which are polynomial time methods and produce solutions acceptable in practice. Finally, exact methods are presented, which in the worst case can behave very badly but perform reasonably on practical and random instances.

Several other problems can also be included in the family of packing and covering problems, for instance the cutting-stock and the loading problems formulated in Section 2.1. The bin-packing problem is yet another instance in the family. In this problem we are given a collection of unit bins (boxes) and a set of elements of size at most 1. The problem is to pack the elements into as few bins as possible. Perhaps the most interesting and practical applications of this problem are in scheduling and we postpone its discussion to Section 4.2.

2.1. THE KNAPSACK PROBLEM

2.1.1. PROBLEM FORMULATIONS AND APPLICATIONS

Many industrial problems can be formulated as the knapsack problem: for example, cargo loading, cutting stock, project selection, budget control, and so on. The most popular version of the problem contains only one linear constraint, but almost every integer linear programming problem and many other combinatorial problems can be reduced to it (see, e.g., Section 4.2). The knapsack problem arises also as a subproblem in several algorithms of pure and integer linear programming.

There are many different versions of the knapsack problem in the literature and as the standard one for our consideration we choose the *zero–one knapsack problem* (0–1 KP), which is to

maximize
$$\sum_{i=1}^{n} p_i x_i \tag{2-1}$$

subject to
$$\sum_{i=1}^{n} w_i x_i \leqslant V \tag{2-2}$$

$$x_i = 0 \text{ or } 1, \qquad i = 1, 2, \ldots, n \tag{2-3}$$

where p_i and w_i ($i = 1, 2, \ldots, n$) and V are integer numbers. In words, suppose that a knapsack has to be filled with different objects of profits p_i and of weights w_i without exceeding a prescribed total weight V. The problem is to find a feasible assignment of objects so that the total value of the objects in the knapsack is maximized.

The 0–1 KP is a special instance of the *bounded knapsack problem* (BKP), which is defined by (2-1), (2-2), and

$$0 \leqslant x_i \leqslant b_i, \quad x_i \text{ an integer}, \qquad i = 1, 2, \ldots, n$$

In the BKP, the knapsack may be filled in by at most b_i objects of type i. In the *general knapsack problem* (GKP), which is sometimes called *unbounded*, restriction (2-3) is relaxed by the following one:

$$x_i \geqslant 0, \quad x_i \text{ an integer}, \qquad i = 1, 2, \ldots, n$$

Without loss of generality (see Problem 2-1) we may assume that the parameters p_i, w_i ($i = 1, 2, \ldots, n$), and V in the problems above satisfy the conditions:

$$p_i \text{ and } w_i \text{ are positive integers}, \qquad i = 1, 2, \ldots, n \tag{2-4}$$
$$w_i \leqslant V, \qquad i = 1, 2, \ldots, n \tag{2-5}$$
$$\sum_{i=1}^{n} w_i > V \tag{2-6}$$

The problems 0–1 KP, BKP, and GKP are sometimes called *one-dimensional knapsack problems*, where *one-* refers to the number of linear constraints in the problems. The most popular special instances of one-dimensional knapsack problems are the value-independent KP (when $w_i = p_i$—see Section 2.1.6) and the change-making problem.

The *unbounded change-making problem* is to find the least number of coins of specified types (values) w_i which constitute exactly a given change V. We assume that each coin type is available in unlimited quantity. Formally, the problem is to

minimize
$$\sum_{i=1}^{n} x_i$$

subject to
$$\sum_{i=1}^{n} x_i w_i = V \tag{2-7}$$

$$x_i \geqslant 0, \ x_i \text{ an integer}, \qquad i = 1, 2, \ldots, n$$

Note that because of the equality constraint, a solution to this problem does not always exist, unless one of the available coins has value 1 (see also Problem 2-5).

The one-dimensional knapsack problems can be generalized in many ways. The most natural generalization is when the objects we have to carry may be put into m knapsacks, each of capacity V_j ($j = 1, 2, \ldots, m > 1$). Let x_{ij} be a 0–1 variable such that $x_{ij} = 1$ if the ith object is assigned to the jth knapsack. The 0–1 *many-knapsack problem* (0–1 MKP), also called the 0–1 *multiple* KP, is to

maximize $$\sum_{i=1}^{n} \sum_{j=1}^{m} p_i x_{ij}$$

subject to $$\sum_{i=1}^{n} w_i x_{ij} \leqslant V_j, \qquad j = 1, 2, \ldots, m$$

$$\sum_{j=1}^{m} x_{ij} \leqslant 1, \qquad i = 1, 2, \ldots, n$$

$$x_{ij} = 0 \text{ or } 1, \qquad i = 1, 2, \ldots, n; \quad j = 1, 2, \ldots, m$$

The first constraints mean that in a feasible assignment of objects no knapsack is overloaded and the second, that each object can be assigned to at most one knapsack. One can also easily formulate bounded and unbounded versions of this problem.

If before putting the objects into a knapsack we have to buy them, at cost c_i for the ith object, and we have a limited amount of money C, we may ask for such an assignment of objects to a knapsack which is not heavier than V and also costs no more than C. Therefore, the problem becomes

maximize $$\sum_{i=1}^{n} p_i x_i$$

subject to $$\sum_{i=1}^{n} w_i x_i \leqslant V$$

$$\sum_{i=1}^{n} c_i x_i \leqslant C$$

$$x_i = 0 \text{ or } 1, \qquad i = 1, 2, \ldots, n$$

In general, one may introduce several such constraints imposed on an assignment of objects to one knapsack. The problem that appears is called the 0–1 *multidimensional knapsack problem*. Evidently, one may also consider bounded and unbounded versions of this problem.

The knapsack problems are often referred to as the loading problems, but in fact, the *standard loading problem* is to allocate given objects with known volumes to boxes with constrained capacity, so as to minimize the number of boxes used. Let k_j

denote the capacity of the jth box and w_i be the volume of the ith object. The loading problem is defined as follows:

minimize
$$\sum_{j=1}^{m} y_j$$

subject to
$$\sum_{j=1}^{m} x_{ij} = 1, \qquad i = 1, 2, \ldots, n$$

$$\sum_{i=1}^{n} w_i x_{ij} \leqslant k_j y_j, \qquad j = 1, 2, \ldots, m$$

$$y_j, x_{ij} = 0 \text{ or } 1, \qquad i = 1, \ldots, n; \quad j = 1, \ldots, m$$

In a feasible loading, we have $x_{ij} = 1$ if the ith object is placed in the jth box, and $y_j = 1$ if the jth box is used.

The bounded and unbounded knapsack problem may also represent the problem of cutting one-dimensional objects (e.g., a length of paper, glass, and steel bar) into smaller pieces of given sizes and values so as to either maximize the total value of pieces cut off or to minimize the residual material.

This short review of possible modifications and generalizations of the 0–1 KP indicates that a variety of real-world problems can be modeled by problems belonging to a very rich family of knapsack-type problems.

All the problems above are easily seen to be special instances of the integer linear programming (ILP) problem. It may surprise the reader that there exists also the reverse relation between ILP problems and the 0–1 KP. Not only is the latter used as a subroutine in several algorithms for solving the former, but also most of the ILP instances can be transformed to an equivalent 0–1 knapsack instance.

Reduction of the ILP Problem to the Knapsack Problem

Every system of linear equations with integer coefficients can be transformed into a single linear equation which has the same set of nonnegative integer solutions as the system to which it corresponds. Therefore, the constraints of an ILP problem can first be transformed into a single constraint and then the problem can be solved as the single-constraint knapsack problem.

The basic step of the process of transformation aggregates two equations so that the nonnegative integer solution set does not alter. We will now describe such a method which assumes that all variables are bounded (for other methods, the reader is referred to Problem 2-6 and to the References and Remarks.)

Let us be given two integer equations in bounded variables

$$g_j(\mathbf{x}) = b_j - \sum_{i=1}^{n} a_{ji} x_i = 0, \qquad j = 1, 2$$

$$0 \leqslant x_i \leqslant u_i, \ x_i \text{ an integer}, \qquad i = 1, 2, \ldots, n \tag{2-8}$$

We will show how to determine the so-called *valid multipliers* λ_1 and λ_2 such that λ_1 and λ_2 are nonzero and the equation

$$\lambda_1 g_1(\mathbf{x}) + \lambda_2 g_2(\mathbf{x}) = 0 \tag{2-9}$$

implies that

$$g_1(\mathbf{x}) = g_2(\mathbf{x}) = 0$$

Since (2-9) is equivalent to $g_1(\mathbf{x}) + (\lambda_2/\lambda_1)g_2(\mathbf{x}) = 0$, we may ask for only one multiplier λ such that

$$g_1(\mathbf{x}) + \lambda g_2(\mathbf{x}) = 0 \tag{2-10}$$

implies that

$$g_1(\mathbf{x}) = g_2(\mathbf{x}) = 0$$

Using upper bounds on the variables, $g_j(\mathbf{x})$ ($j = 1, 2$) can be bounded as follows:

$$b_j - \sum_{i=1}^{n} a_{ji}^{+} u_i \leqslant g_j(\mathbf{x}) \leqslant b_j - \sum_{i=1}^{n} a_{ji}^{-} u_i \tag{2-11}$$

where $a_{ji}^{+} = \max\{a_{ji}, 0\}$ and $a_{ji}^{-} = \min\{a_{ji}, 0\}$. Now it is easy to prove the following theorem.

Theorem 2-1

The integer vector $\mathbf{x}$ satisfying $0 \leqslant \mathbf{x} \leqslant \mathbf{u}$ is a solution of (2-8) if and only if it is a solution of (2-10) with λ satisfying

$$\lambda > \max\left\{ b_1 - \sum_{i=1}^{n} a_{1i}^{-} u_i, \; -b_1 + \sum_{i=1}^{n} a_{1i}^{+} u_i \right\} \tag{2-12}$$

■

It is evident that (2-10) follows from (2-8). Therefore, we have to show that (2-10) with λ satisfying (2-12) implies (2-8). Suppose that $\mathbf{x}$ solves (2-10) and $g_2(\mathbf{x}) - K = 0$. We will show that if λ satisfies (2-12), then $K = 0$. Let λ be such a number. Multiplying the last equation by λ and subtracting from (2-10), we get $g_1(\mathbf{x}) + \lambda K = 0$. One can easily see that the bound imposed on λ guarantees that $|g_1(\mathbf{x})| < \lambda$ for every $\mathbf{x}$ that satisfies (2-8). Therefore, the last equation is satisfied only if $K = 0$. Hence $g_2(\mathbf{x}) = 0$ and $g_1(\mathbf{x}) = 0$ as well.

Example 2-1

Let us apply Theorem 2-1 to the following system of two equations:

$$4x_1 - x_2 + 2x_3 + x_4 + x_5 \quad\quad = \quad 0$$

$$x_1 \quad\quad - x_3 - x_4 \quad\quad - x_6 = -1$$

$$x_i = 0 \text{ or } 1, \quad\quad i = 1, 2, 3, 4, 5, 6$$

By the enumeration of all possible 0–1 vectors $(x_1, x_2, x_3, x_4, x_5, x_6)$, we find the only solutions of this system $(0, 1, 0, 1, 0, 0)$, $(0, 1, 0, 0, 1, 1)$, and $(0, 0, 0, 0, 0, 1)$. First, let us try two arbitrary multipliers $\lambda_1 = 1$ and $\lambda_2 = 1$. The aggregated equation is of the form

$$5x_1 - x_2 + x_3 \qquad + x_5 - x_6 = -1$$

and it has the three solutions above plus $(0, 1, 0, 0, 0, 0)$, $(0, 0, 0, 1, 0, 1)$, $(0, 1, 1, 0, 0, 1)$, $(0, 1, 1, 1, 0, 1)$, and $(0, 1, 0, 1, 1, 1)$.

However, applying Theorem 2-1, we find that λ must be greater than 8. For $\lambda = 9$, the valid aggregated equation has the form

$$13x_1 - x_2 - 7x_3 - 8x_4 + x_5 - 9x_6 = -9 \qquad \blacksquare$$

Once we know how to aggregate two equations into one, we can apply this process iteratively to an arbitrary system of equations. This approach, however, for solving ILP programs is of limited use due to a rapid growth in the values of the coefficients as successive constraints are aggregated into one.

Computational Methods

From the computational point of view, the knapsack problems are very hard. No polynomial-time algorithm has been found for any of them, and it is very unlikely that such an algorithm exists.

A variety of techniques have been proposed as solution methods for the knapsack problems and they can be classified as follows:

1. Reduction and approximation methods (greedy-type and Lagrangian)
2. Exact methods
   a. Network approach
   b. Dynamic programming
   c. Enumeration methods (backtracking, branch-and-bound, etc.)

In the next section we present several upper and lower bounds of the solution to the knapsack problem together with an algorithm for reducing the number of problem variables. Then we discuss approximation algorithms which mostly result from the greedy approach, and finally the exact methods of types b and c are presented.

2.1.2. LOWER AND UPPER BOUNDS

As we have already pointed out [see (2-4), (2-5), and (2-6)], we may assume without loss of generality that all parameters of the 0–1 KP are positive integers and satisfy $w_i \leqslant V$ $(i = 1, 2, \ldots, n)$ and $\sum_{i=1}^{n} w_i > V$. A transformation of an arbitrary knapsack instance to the equivalent one which satisfies these assumptions (see Problem 2-1) may be considered as the first reduction step of the problem since it often results in decreasing the number of variables. Similar reductions also exist for modifications and generalizations of the 0–1 KP (see Problem 2-2).

We shall assume also (except the subsection on dynamic programming) that the objects (variables) are ordered as

$$\frac{p_1}{w_1} \geqslant \frac{p_2}{w_2} \geqslant \ldots \geqslant \frac{p_n}{w_n} \tag{2-13}$$

The reason for such an assumption will soon become clear.

In what follows we will refer to the *knapsack problem*, which is the 0–1 knapsack problem (2-1)–(2-3) with the assumptions (2-4), (2-5), (2-6), and (2-13).

If the knapsack problem is considered as a linear programming problem, that is, when the constraint (2-3) is relaxed to $x_i \geqslant 0$ for $i = 1, 2, \ldots, n$, the following theorem can be proved easily.

Theorem 2-2

Vector $x_1 = V/w_1$, $x_i = 0$ $(i = 2, 3, \ldots, n)$ is an optimal solution to the knapsack problem with the constraint (2-3) relaxed to $x_i \geqslant 0$ $(i = 1, 2, \ldots, n)$. ∎

If all the variables are restricted to the interval $[0, 1]$, one can easily prove the following result due to Dantzig [1957].

Theorem 2-3 (Dantzig [1957])

The optimal solution to the knapsack problem with the constraint (2-3) relaxed to $0 \leqslant x_i \leqslant 1$ $(i = 1, 2, \ldots, n)$ is

$$x_i = 1, \qquad\qquad i = 1, 2, \ldots, r$$

$$x_i = 0, \qquad\qquad i = r + 2, \ldots, n$$

$$x_{r+1} = \left(V - \sum_{i=1}^{r} w_i \right) / w_{r+1}$$

where r is the largest index for which $\sum_{i=1}^{r} w_i \leqslant V$. ∎

The proofs of both theorems are left to the reader (Problems 2-7 and 2-8).

The last theorem gives rise to the following upper bound of the solution to the knapsack problem:

$$U_1 = \sum_{i=1}^{r} p_i + \left\lfloor \left(V - \sum_{i=1}^{r} w_i \right) p_{r+1} / w_{r+1} \right\rfloor \tag{2-14}$$

There have been several other upper bounds established and we shall make use of the following one proved by Martello and Toth [1977]. Let us define

$$U = \sum_{i=1}^{r} p_i + \left\lfloor \left(V - \sum_{i=1}^{r} w_i \right) p_{r+2}/w_{r+2} \right\rfloor$$

$$U' = \sum_{i=1}^{r} p_i + \left\lfloor p_{r+1} - \left(w_{r+1} - \left(V - \sum_{i=1}^{r} w_i \right) \right) p_r/w_r \right\rfloor$$

where r is the largest index for which $\sum_{i=1}^{r} w_i \leqslant V$. Then

$$U_2 = \max\{U, U'\} \tag{2-15}$$

is an upper bound to the solution of the knapsack problem and

$$U_2 \leqslant U_1 \tag{2-16}$$

To show that U_2 is an upper bound, note that an optimal solution to the problem can be obtained with either $x_{r+1} = 0$ or $x_{r+1} = 1$. In both cases, we again have the knapsack problem with $n - 1$ objects and weight limits V and $V - w_{r+1}$, respectively. Now applying Theorem 2-3 to the former subproblem, we obtain U. In the latter case, we have to remove from the knapsack some objects of total weight at least $w_{r+1} - (V - \sum_{i=1}^{r} w_i)$ and with the ratio of at least p_r/w_r.

Hence U' is the bound for the latter subproblem and U_2 is the bound for the original problem. The proof of inequality (2-16) is based mainly on assumption (2-13), and we leave it to the reader. Some other upper bounds are discussed in Problems 2-9 and 2-10.

A lower bound to the knapsack problem usually corresponds to a heuristic solution of the problem obtained by a method which is of a greedy type. Such a method fills the knapsack with objects which in a certain sense are most profitable. It can also be applied to a partially filled knapsack. Let J be a feasible assignment of objects already put into the knapsack, that is,

$$\sum_{i \in J} w_i \leqslant V$$

We may assume also that some objects are decided not to be assigned to the knapsack; let their set be denoted by K. Algorithm 2-1 is an implementation of the greedy method which finds LOWER_BOUND(J, K)—the value of the solution obtained by augmenting J with objects which do not belong to J and K. Therefore, the obtained solution with value LOWER_BOUND(J, K) is of the form $x_i = 1$ for $i \in J$ and $x_i = 0$ for $i \notin J$.

**Algorithm 2-1:   The Greedy Algorithm for Finding
a Heuristic Solution to the Knapsack Problem**

function LOWER _ BOUND(J, K);
       ⟨ ∗ LOWER _ BOUND(J, K) is the value of the greedy solution to the knapsack problem
       which first is reduced by assigning $x_i = 1$ for $i \in J$ and $x_i = 0$ for $i \in K$. Upon the
       exit, J is the set of variables which have value 1 in the obtained solution ∗⟩
begin
    $x \leftarrow V - \sum_{i \in J} w_i;$
    $y \leftarrow \sum_{i \in J} p_i;$
    for $i \leftarrow 1$ **to** n **do**
        if ($i \notin J \cup K$) **and** ($w_i \leqslant x$) **then**
        begin
            $J \leftarrow J \cup \{i\};$
            $x \leftarrow x - w_i;$
            $y \leftarrow y + p_i$
        end;
    LOWER _ BOUND $\leftarrow y$
end

Evidently, even for fixed subsets J and K, the value of LOWER_
BOUND(J, K) depends on the order in which objects are considered. However,
there exists no general rule which would guarantee that Algorithm 2-1 always
generates an optimal solution to the knapsack problem (see Problem 2-11). We may
only apply a heuristic argument based on Theorems 2-2 and 2-3, and order all
objects according to nonincreasing values of p_i/w_i, as we have already assumed by
(2-13). It is interesting that function LOWER_BOUND can be used to generate
approximate solutions with a required accuracy (see Section 2.1.4).

2.1.3.   REDUCTION ALGORITHM

Despite its simplicity, the knapsack problem, together with almost all of its special
instances and generalizations, is one of the most difficult problems of discrete
optimization. Formally, the knapsack problem is a member of the so-called NP-hard
class of problems. For this reason, before attempting to solve exactly its particular
instance, we should always try to simplify it by either imposing some additional
restrictions on the parameters or reducing the number of variables.

    As we shall see in the following section, some difficulties in solving the
knapsack problem are caused by the order of magnitude of its parameters. The
higher the value of the parameters, the more difficult it is to solve the problem.
Unfortunately, there is no method for reducing the magnitude of parameters. Hence

the knapsack instances resulting from the aggregation process of general integer linear problems are among the hardest instances of the knapsack problem.

If we are given an instance of the 0–1 KP (2-1), (2-2), and (2-3) with integer parameters, the first step is to transform it to an equivalent instance which satisfies the conditions (2-4), (2-5), and (2-6) (see Problem 2-1). This transformation may already result in some reduction of the number of variables, since if there is w_i such that $w_i > V > 0$, then in any feasible solution $x_i = 0$. We will now prove a theorem that can be used to find larger sets of variables which are either 0 or 1 in every optimal solution. Let $\bar{P}(x_i = \varepsilon_i)$ denote the optimal solution of the continuous knapsack problem with $x_i = \varepsilon_i$, where $\varepsilon_i = 0$ or $\varepsilon_i = 1$ ($i = 1, 2, \ldots, n$). Values $\bar{P}(x_i = \varepsilon_i)$ can be found by applying Theorem 2-3 to the problem that results from assigning $x_i \leftarrow \varepsilon_i$. Let P_0 denote the value of a feasible solution to the original problem. For every i, it is evident that if $\lfloor \bar{P}(x_i = \varepsilon_i) \rfloor < P_0$, then the profit $P(x_i = \varepsilon_i)$ of any feasible solution of the knapsack problem with $x_i = \varepsilon_i$ satisfies $P(x_i = \varepsilon_i) < P_0$. Therefore, the solution with value P_0 cannot be improved by any solution with $x_i = \varepsilon_i$. Hence we know that $x_i = 1 - \varepsilon_i$ in every optimal solution to the original problem. Note that the same arguments can be used when $\lfloor \bar{P}(x_i = \varepsilon_i) \rfloor$ is replaced by any upper bound $UB(x_i = \varepsilon_i)$ of the solution with $x_i = \varepsilon_i$. Due to inequality (2-16) this is even a stronger result since more variables may be reduced.

Thus we have proved the following theorem.

Theorem 2-4 (Ingargiola and Korsh [1973])

Let P_0 denote the value of a feasible solution to the knapsack problem. If $UB(x_i = \varepsilon_i) < P_0$, then $x_i = 1 - \varepsilon_i$ in every optimal solution of the problem.   ■

As a feasible solution we may take either the solution defined in Theorem 2-3 with $x_{r+1} = 0$ or the one obtained by LOWER_BOUND($\varnothing, \varnothing$).

A reduction method which is an implementation of Theorem 2-4 is presented in Algorithm 2-2. It finds two sets of objects (indices) I_0 and I_1 such that every optimal solution to the problem does not contain any object of I_0 and contains all objects of I_1. Additionally, *optsol = true* if the optimal solution has been found and *optsol = false*, otherwise.

The algorithm starts with finding index r, which is the smallest k such that $\sum_{i=1}^{k} w_i \geqslant V$. If $\sum_{i=1}^{r} w_i = V$, then evidently $I_1 = \{1, 2, \ldots, r\}$ and $I_0 = \{r + 1, \ldots, n\}$ form the optimal solution. Otherwise, Theorem 2-4 is applied. In the beginning LB becomes equal to LOWER_BOUND ($\varnothing, \varnothing$). It is clear that we do not need to compute $UB(x_i = 1)$ for $i < r$ and $UB(x_i = 0)$ for $i > r$ since they are equal to the upper bound of the solution to the original problem, and therefore they are greater than or equal to the final value of LB. Then we calculate all other upper bounds $UB(x_i = \varepsilon_i)$ and simultaneously an attempt is made to improve the lower bound LB. Final steps of Algorithm 2-2 are self-explanatory—they are straightforward implementation of Theorem 2-4.

It is easy to verify that Algorithm 2-2 has time complexity bounded by $O(n^2)$ since each variable needs a linear in n number of operations.

In the Pascal implementation of Algorithm 2-2 presented in the next section, the upper bounds are calculated by using bound U_2 [see (2-15)] and the lower bounds are found by the application of function LOWER_BOUND.

Algorithm 2-2: Reduction Procedure for the Knapsack Problem

procedure REDUCTION (I_0, I_1, *optsol*);
{∗ This procedure generates two subsets of objects such that in every optimal solution to the problem $x_i = 0$ for $i \in I_0$ and $x_i = 1$ for $i \in I_1$; *optsol* = *true* if $I_0 \cup I_1 = N = \{1, 2, \ldots, n\}$, that is, if $I_0 \cup I_1$ forms an optimal solution, and *optsol* = *false*, otherwise ∗}
begin

   $s \leftarrow$ largest k such that $\sum_{i=1}^{k} w_i < V$;

   $r \leftarrow s + 1$;

   if $\sum_{i=1}^{r} w_i = V$ **then**
   begin

      $I_1 \leftarrow \{1, 2, \ldots, r\}$;
      $I_0 \leftarrow \{r + 1, r + 2, \ldots, n\}$;
      optsol $\leftarrow$ *true*

   end
   else
   begin

      $LB \leftarrow$ LOWER_BOUND($\emptyset$, $\emptyset$);
      for $i \leftarrow 1$ **to** r **do**
      begin

         compute $UB(x_i = 0)$;
         {∗ Let J_i denote the set of variables which are equal to 1 in the solution generated by $UB(x_i = 0)$ ∗}
         $LB \leftarrow \max\langle LB, \text{LOWER\_BOUND}(J_i, \langle i \rangle)\rangle$

      end;
      for $i \leftarrow r$ **to** n **do**
      begin

         compute $UB(x_i = 1)$;
         {∗ Let J'_i denote the set of variables which are equal to 1 in the solution generated by $UB(x_i = 1)$ ∗}
         $LB \leftarrow \max\langle LB, \text{LOWER\_BOUND}(J'_i \cup \langle i \rangle, \emptyset)\rangle$

      end;
      $I_1 \leftarrow \langle i\!: UB(x_i = 0) < LB, i = 1, 2, \ldots, r \rangle$;
      $V \leftarrow V - \sum_{i \in I_1} w_i$;
      $I_0 \leftarrow \langle i\!: UB(x_i = 1) < LB, i = r, \ldots, n \rangle \cup \langle i\!: w_i > V, i = 1, 2, \ldots, n, i \notin I_1 \rangle$;
      if $I_0 \cup I_1 = N$ **then** *optsol* $\leftarrow$ *true*

```
          else
          begin
                optsol ←  ∑   wᵢ = V;
                        i∈N−I₀−I₁
                if optsol then I₁ ← N − I₀
          end
          end ⟨* else *⟩
     end ⟨* REDUCTION *⟩
```

Example 2-2

Let us apply Algorithm 2-2 to the following knapsack problem:

maximize $\quad\quad\quad 51x_1 + 20x_2 + 29x_3 + 6x_4 + 4x_5 + 6x_6$

subject to $\quad\quad\quad 25x_1 + 10x_2 + 15x_3 + 4x_4 + 3x_5 + 5x_6 \leqslant 44$

$$x_1 = 0 \text{ or } 1, \quad i = 1, 2, \ldots, 6$$

We have $s = 2$ since $25 + 10 < 44$ and $25 + 10 + 15 > 44$; therefore, $r = 3$. LOWER_BOUND$(\varnothing, \varnothing)$ produces a feasible solution $(1, 1, 0, 1, 1, 0)$ of value 81; hence $LB = 81$. The following table contains the values of the upper bounds $UB(x_i = \varepsilon_i)$ calculated by using (2-15) with $x_i = \varepsilon_i$, and LB is a current value of the lower bound.

i	ε_i	$UB(x_i = \varepsilon_i)$	LB
1	0	65	81
2	0	86	86
3	0	83	86
3	1	87	86
4	1	86	86
5	1	86	86
6	1	84	86

Now we may determine variables which have value 1 in every optimal solution. $I_1 = \{1, 3\}$, since $UB(x_1 = 0) = 65 < 86$ and $UB(x_3 = 0) = 83 < 86$. Therefore, the problem may be reduced by removing variables 1 and 3 and the reduced weight limit is 4 since the objects put into the knapsack (1 and 3) weigh 40. Next, we may reduce variables which do not belong to any optimal assignment of the knapsack. First, we remove those variables that satisfy $UB(x_i = 1) < LB$ $(i = 3, 4, 5, 6)$. Therefore, $x_6 = 0$. Finally, we remove objects that weight more than the current weight limit. Thus $x_2 = 0$. Hence $I_1 = \{1, 3\}$ and $I_0 = \{2, 6\}$, and the reduced problem is

maximize $\quad\quad\quad\quad\quad 6x_4 + 4x_5$

subject to $\quad\quad\quad\quad\quad 4x_4 + 3x_5 \leqslant 4$

$$x_4, x_5 = 0 \text{ or } 1 \quad\quad\quad ■$$

Computer Implementation of the Reduction Algorithm

The KNAPRED procedure when applied to the knapsack problem defined by (2-1) through (2-6) and (2-13) reduces its size by finding a set of variables which are either 0 or 1 in every optimal solution to the problem. The procedure is an implementation of Algorithm 2-2. The upper bounds are computed by using formula (2-15). Procedure UB0 computes $UB(x_i = 0)$ for $i = 1, 2, \ldots, r$ and UB1 computes $UB(x_i = 1)$ for $i = r, r + 1, \ldots, n$. The AUGMENT procedure augments a partial feasible solution obtained in the course of finding an upper bound to the best greedy solution. The KNAPRED procedure, to speed up the computation of upper bounds, utilizes some auxiliary quantities stored in arrays PR and WR which are partial sums of object profits and weights.

Global Constants

N   number of variables (objects) in the knapsack problem
N2   N + 2

Data Types

```
TYPE  ARRN   = ARRAY[1..N] OF INTEGER;
      ARR0N  = ARRAY[0..N] OF INTEGER;
      ARR0N2 = ARRAY[0..N2] OF INTEGER;
```

Procedure Parameters

```
PROCEDURE KNAPRED(
    N               :INTEGER;
    VAR P,W         :ARR0N2;
    VAR X           :ARRN;
    VAR V,PROFIT    :INTEGER;
    INF             :INTEGER;
    VAR OPTSOL      :BOOLEAN);
```

Input

N   number of variables of the problem
P[1..N] array of object profits
W[1..N] array of object weights
V   total weight limit of the knapsack
INF   maximal integer number available in the system that is used

Comment: Note that arrays P and W are declared as of type ARR0N2. The elements with indices 0, N + 1, and N + 2 are used in the procedure to unify the computation.

Output

X[1..N]   array that shows which variables may be reduced from the problem. Nonnegative elements keep their values in every optimal solution to the problem. Therefore, the variables such that X[I] = 0 or 1 may be removed.

V   total weight limit of the knapsack in the reduced problem, that is, V is equal to the input weight limit reduced by the weight of objects I such that X[I] = 1.

PROFIT   profit of the partial assignment, that is, the sum of P[I] over all I such that X[I] = 1.

OPTSOL   Boolean variable such that OPTSOL = TRUE if the reduced problem vanishes (i.e., the partial solution becomes optimal) and OPTSOL = FALSE, otherwise.

Example

Input to the procedure KNAPRED for the knapsack problem of Example 2-2 is of the form

$$N = 6$$
$$P[1..6] = [51,20,29,6,4,6]$$
$$W[1..6] = [25,10,15,4,3,5]$$
$$V = 44$$
$$INF = 1000$$

Arrays P and W are declared as of range 0..8 and their elements with indices 0, 7, and 8 are used only in the procedure.

The output obtained is

$$X[1..6] = [1,0,1,-1,-1,0]$$
$$V = 4$$
$$PROFIT = 80$$
$$OPTSOL = FALSE$$

Pascal Procedure KNAPRED

```
PROCEDURE KNAPRED(
    N             :INTEGER;
    VAR P,W       :ARRON2;
    VAR X         :ARRN;
    VAR V,PROFIT:INTEGER;
    INF           :INTEGER;
    VAR OPTSOL    :BOOLEAN);
```

```
VAR I,J,LB,PP,PPP,Q,QQ,R,S,VV,Y,Z,WW,WWW:INTEGER;
    B                                       :BOOLEAN;
    PR,WR                                   :ARRN;
    UB                                      :ARRON;

PROCEDURE AUGMENT(I,K:INTEGER);
   (*  THE PROCEDURE AUGMENTS A PARTIAL SOLUTION WHICH
       WEIGHS VV AND HAS PROFIT Q, STARTING WITH OBJECT K+1
       AND TRYING TO ADD AS MANY OBJECTS AS POSSIBLE
       ( EXCEPT THE I-TH OBJECT ) WHICH FOLLOW OBJECT K  *)
VAR KK,VVV:INTEGER;
BEGIN
   KK:=K+1;
   VVV:=VV;
   QQ:=Q;
   WHILE KK <= N DO
   BEGIN
      IF (VVV >= W[KK]) AND (KK <> I) THEN
      BEGIN
         VVV:=VVV-W[KK];
         QQ:=QQ+P[KK]
      END;
      KK:=KK+1
   END
END (*  AUGMENT  *);

PROCEDURE UBO(I:INTEGER);
   (* UBO(I) FINDS WEIGHT VV AND PROFIT Q OF THE GREEDY
      SOLUTION OBTAINED BY ASSUMING X[I]=0, WHERE I <= R,
      AND TAKING AS MANY AS POSSIBLE CONSECUTIVE OBJECTS.
      UPON EXIT, QQ IS THE PROFIT OF THE GREEDY SOLUTION
      AUGMENTED BY PROCEDURE AUGMENT.  *)
VAR WI:INTEGER;
    BB:BOOLEAN;
BEGIN
   WI:=W[I];
   S:=R-1;
   BB:=TRUE;
   WHILE (S <= N) AND BB DO
      IF WI >= WR[S] THEN S:=S+1
      ELSE BB:=FALSE;
   S:=S-1;
   VV:=WI-WR[S];
   Q:=PR[S]-P[I];
   AUGMENT(I,S)
END (*  UBO  *);

PROCEDURE UB1(I:INTEGER);
   (* UB1(I) FINDS WEIGHT VV AND PROFIT Q OF THE GREEDY
      SOLUTION OBTAINED BY ASSUMING X[I]=1,WHERE R <= I,
      AND TAKING AS MANY AS POSSIBLE CONSECUTIVE OBJECTS.
      UPON EXIT, QQ IS THE PROFIT OF THE GREEDY SOLUTION
      AUGMENTED BY PROCEDURE AUGMENT.  *)
```

```
VAR WI:INTEGER;
BEGIN
   WI:=W[I];
   S:=R;
   WHILE WR[S] < WI DO S:=S-1;
   VV:=WR[S]-WI;
   Q:=PR[S]+P[I];
   AUGMENT(I,R)
END (*  UB1  *);

BEGIN  (*  MAIN BODY  *)
   PP:=0;   WW:=0;   R:=1;
   WHILE WW+W[R] < V DO
   BEGIN
      WW:=WW+W[R];
      PP:=PP+P[R];
      R:=R+1
   END;
   WW:=V-WW;
   IF WW = W[R] THEN
   BEGIN (* THE GREEDY SOLUTION IS OPTIMAL  *)
      V:=0;   PROFIT:=0;
      FOR I:=1 TO R DO
      BEGIN
         X[I]:=1;
         PROFIT:=PROFIT+P[I]
      END;
      FOR I:=R+1 TO N DO X[I]:=0;
      OPTSOL:=TRUE
   END
   ELSE
   BEGIN     (*  REDUCTION ALGORITHM  *)
      W[O]:=1;   W[N+1]:=INF;   W[N+2]:=INF;
      P[O]:=P[1];   P[N+1]:=0;   P[N+2]:=0;
      VV:=WW;   Q:=PP;
      AUGMENT(0,R-1);
      LB:=QQ;
      WWW:=-WW;   PPP:=PP;             (* FINDING AUXILIARY *)
      WR[R-1]:=-WW;   PR[R-1]:=PP;     (* ARRAYS WR AND PR *)
      FOR I:=R TO N DO
      BEGIN
         WWW:=WWW+W[I];   WR[I]:=WWW;
         PPP:=PPP+P[I];   PR[I]:=PPP
      END;
      WWW:=WW+W[R-1];   PPP:=PP-P[R-1];
      FOR I:=R-2 DOWNTO 1 DO
      BEGIN
         WWW:=WWW+W[I];   WR[I]:=WWW;
         PPP:=PPP-P[I];   PR[I]:=PPP
      END;
      I:=1;   J:=-1;
      B:=TRUE;
      REPEAT    (*  FINDING UPPER AND LOWER BOUNDS  *)
```

```
      IF B THEN UBO(I)
      ELSE BEGIN UB1(I);    S:=S-1   END;
      Y:=TRUNC(VV*P[S+2]/W[S+2]);
      Z:=TRUNC(P[S+1]-(W[S+1]-VV)*P[S]/W[S]);
      IF Z > Y THEN Y:=Z;
      UB[I+J]:=Q+Y;
      IF QQ > LB THEN LB:=QQ;
      IF (I =R) AND B THEN
      BEGIN
         B:=FALSE;    J:=0;
         WR[R-1]:=WW+W[R-1];    WR[R]:=WW;
         PR[R-1]:=PP-P[R-1];    PR[R]:=PP
      END
      ELSE I:=I+1
UNTIL I > N;
PROFIT:=0;     (*  REDUCTION PROCESS  *)
J:=0;
FOR I:=1 TO N DO X[I]:=-1;
FOR I:=1 TO R DO
   IF UB[I-1] < LB THEN
   BEGIN
      J:=J+1;
      X[I]:=1;
      V:=V-W[I];
      PROFIT:=PROFIT+P[I]
   END;
I:=1;
WHILE I <= N DO
BEGIN
   IF (W[I] > V) AND (X[I] = -1) THEN
   BEGIN  X[I]:=0;  J:=J+1   END;
   I:=I+1
END;
I:=R;
WHILE (I <= N) AND (J < N) DO
BEGIN
   IF UB[I] < LB THEN
   BEGIN  X[I]:=0;  J:=J+1   END;
   I:=I+1
END;
OPTSOL:=J=N;
IF NOT OPTSOL THEN
BEGIN
   WW:=0;
   FOR I:=1 TO N DO
      IF X[I] = -1 THEN WW:=WW+W[I];
   OPTSOL:=WW=V;
   IF OPTSOL THEN
   BEGIN
      V:=0;
      FOR I:=1 TO N DO
         IF X[I] = -1 THEN
```

```
            BEGIN
                X[I]:=1;
                PROFIT:=PROFIT+P[I]
            END
          END
      END  (* J < N  *)
   END (*  WW <> W[R]  *)
END;   (*  KNAPRED  *)
```

2.1.4. APPROXIMATION ALGORITHMS

Since there is no known polynomial-time algorithm for the knapsack problem and none may well exist, restrictions on computing time and memory space suggest that we may be satisfied with an approximate solution to many instances of the problem.

In some cases we may also be content with a solution that is sufficiently close to the optimal one, for example, if problem parameters p_i, w_i ($i = 1, 2, \ldots, n$), and V themselves are only estimates of the real profits, weights, and the knapsack capacity. The purpose of this section is to discuss some heuristic algorithms for solving the knapsack problem and to show how they behave.

While discussing lower bounds, we presented the greedy algorithm (Algorithm 2-1). Problem 2-11 shows that for none of the three orderings of objects according to profits, weights, and their ratios, the greedy algorithm produces an optimal solution. Although it often yields quite good solutions, it cannot be determined exactly how far greedy solutions are from optimal ones. At the end of this section we will show how the greedy approach can be utilized to construct a series of increasingly accurate polynomial-time approximation algorithms.

Let A be an algorithm that generates a feasible solution to every instance $\mathcal{J}$ of the knapsack problem. Let $P_{\text{opt}}(\mathcal{J})$ and $P_A(\mathcal{J})$ denote, respectively, the profit of the optimal solution and the profit of the solution obtained by the algorithm A applied to $\mathcal{J}$. The first question that can be asked about algorithm A is whether for an arbitrary real number $\varepsilon > 0$ the algorithm A can generate solutions such that $P_{\text{opt}}(\mathcal{J}) - P_A(\mathcal{J}) \leqslant \varepsilon$ for every instance $\mathcal{J}$. An algorithm with this property would be called *absolute ε-approximate*. We shall now show that for the knapsack problem, getting an absolute ε-approximate algorithm working in polynomial-time is as difficult as getting an exact polynomial-time algorithm.

First, let us consider the following example:

Example 2-3

Let $\mathcal{J}_z$ denote the following instance of the problem $n = 2$, $(p_1, p_2) = (2, z)$, $(w_1, w_2) = (1, z)$, and $V = z$, where $z > 2$. Note that we have already $p_1/w_1 > p_2/w_2$. The optimal solution for $\mathcal{J}_z$ is $(x_1, x_2) = (0, 1)$ and its profit $P_{\text{opt}}(\mathcal{J}_z) = z$. However, the solution generated by the greedy algorithm is $(x_1, x_2) = (1, 0)$ and its profit $P_G(\mathcal{J}_z) = 2$. Hence $P_{\text{opt}}(\mathcal{J}_z) - P_G(\mathcal{J}_z) = z - 2$. Therefore, the greedy algorithm is not an absolute approximation algorithm, since there exists no constant ε such that $P_{\text{opt}}(\mathcal{J}_z) - P_G(\mathcal{J}_z) \leqslant \varepsilon$ for all $z > 2$. ∎

Assume now that there exists an absolute k-approximate algorithm A for the knapsack problem; that is, for every instance $\mathcal{J} = [n, p_i, w_i \ (i = 1, 2, \ldots, n), V]$, A finds a solution with value $P_A(\mathcal{J})$ such that $P_{\text{opt}}(\mathcal{J}) - P_A(\mathcal{J}) \leqslant k$. Consider now the instance $\mathcal{J}'$ in which profits are $(k + 1)p_i$ and the other parameters are the same as in $\mathcal{J}$. The instances $\mathcal{J}$ and $\mathcal{J}'$ have the same sets of feasible solutions, $P_{\text{opt}}(\mathcal{J}') = (k + 1)P_{\text{opt}}(\mathcal{J})$, and therefore $\mathcal{J}$ and $\mathcal{J}'$ have the same sets of optimal solutions. The algorithm A applied to instance $\mathcal{J}'$ generates the solution with value $P_A(\mathcal{J}')$ which satisfies

$$P_{\text{opt}}(\mathcal{J}') - P_A(\mathcal{J}') \leqslant k$$

Therefore,

$$(k + 1)\big(P_{\text{opt}}(\mathcal{J}) - P_A(\mathcal{J})\big) \leqslant k$$

and

$$P_{\text{opt}}(\mathcal{J}) - P_A(\mathcal{J}) \leqslant \frac{k}{k + 1}$$

and hence no absolute ε-approximate algorithm exists. Since $P_{\text{opt}}(\mathcal{J})$ and $P_A(\mathcal{J})$ are integers, we conclude that $P_{\text{opt}}(\mathcal{J}) = P_A(\mathcal{J})$. Thus we showed that an absolute k-approximate algorithm applied to the modified instance $\mathcal{J}'$ would generate an optimal solution to the original instance.

But there exist problems which do permit such algorithms; an example of the cutting-stock problem together with its absolute 1-approximate algorithm is presented in Problem 2-14.

Since the existence of an absolute ε-approximate algorithm for the knapsack problem working in polynomial time is very unlikely, we have to restrict our attention to a weaker concept of approximation. We will say that an algorithm A for solving the knapsack problem is ε-*approximate* if

$$\frac{P_{\text{opt}}(\mathcal{J}) - P_A(\mathcal{J})}{P_{\text{opt}}(\mathcal{J})} \leqslant \varepsilon \tag{2-17}$$

for all instances $\mathcal{J}$ of the problem. We are interested in such an algorithm working in polynomial time.

Example 2-4

For the instance $\mathcal{J}_z (z > 2)$ defined in Example 2-3, the greedy algorithm generates the solution $(x_1, x_2) = (1, 0)$, which satisfies

$$\frac{P_{\text{opt}}(\mathcal{J}_z) - P_G(\mathcal{J}_z)}{P_{\text{opt}}(\mathcal{J}_z)} = 1 - \frac{2}{z}$$

Therefore, the greedy algorithm is a 1-approximate algorithm for this family of instances.   ∎

We now present a sequence of heuristic algorithms for solving the knapsack problem which have the property that for every real number $\varepsilon > 0$ there exists an ε'-approximate algorithm in the sequence such that $\varepsilon' < \varepsilon$. The algorithms will work in time bounded by n^c, where c is a constant depending only on the desired accuracy ε; therefore, they are polynomial algorithms. The algorithms use function LOWER_BOUND(J, K) (see Algorithm 2-1) with $K = \varnothing$ and J running over some feasible subsets of objects that do not weigh more than the knapsack can carry. We remind the reader that LOWER_BOUND(J, K) is the value of the (greedy) solution obtained by filling in nonincreasing order of p_i/w_i that part of the knapsack which is left vacant after objects from J have been put into it, and objects from K are not considered. The kth algorithm in the sequence ($k \geqslant 1$), presented as Algorithm 2-3, generates the best greedy-type solution which can be obtained first by filling the knapsack with at most k objects (which do not weigh more than it can carry) and then augmenting it by the greedy strategy (implemented in function LOWER_BOUND).

Algorithm 2-3:   Approximation Algorithm for Solving the Knapsack Problem

```
function KNAPAPPROX(k);
    (* KNAPAPPROX(k) is the profit of the best feasible solution obtained first by filling the
       knapsack with at most k objects and then by augmenting them using the greedy
       strategy *)
begin
    Q ← 0;
    for all subsets I ⊂ {1, 2, ..., n} such that |I| ⩽ k and ∑_{i∈I} w_i ⩽ V do
        Q ← max{Q, LOWER _ BOUND(I, ∅)};
    KNAPAPPROX ← Q
end
```

Before discussing some properties of Algorithm 2-3, let us see how it works on a particular knapsack instance.

Example 2-5

Let us find KNAPAPPROX(k) for the following knapsack instance:

maximize $\qquad\qquad 200x_1 + 155x_2 + 115x_3 + 90x_4$

subject to $\qquad\qquad 50x_1 + \ \ 40x_2 + \ \ 30x_3 + 25x_4 \leqslant 95$

$$x_1 = 0 \text{ or } 1, \qquad i = 1, 2, 3, 4$$

The variables are already ordered properly since $200/50 \geqslant 155/40 \geqslant 115/30 \geqslant 90/25$. The greedy solution, that is, the solution generated by LOWER_

BOUND$(\varnothing, \varnothing)$ is $(1,1,0,0)$ and has profit 355. Let $k = 1$. Subsets of objects $I = \{1\}$ and $I = \{2\}$ give the same solution as $I = \varnothing$ since they are subsets of the solution produced by the greedy algorithm. For $I = \{3\}$ we obtain $(1,0,1,0)$ of the value 315 and for $I = \{4\}$—the solution is $(1,0,0,1)$ and has value 290. Therefore, KNAPAPPROX$(1) = 355$ and it corresponds to the solution $(1,1,0,0)$. Now let $k = 2$. Again, $I = \{1,2\}$ as a subset of the greedy solution gives $(1,1,0,0)$, the value 355, and for the other two object subsets each of which is feasible we obtain

I	$\{1,3\}$	$\{1,4\}$	$\{2,3\}$	$\{2,4\}$	$\{3,4\}$
LOWER_BOUND$(I, \varnothing)$	315	290	360	360	360
Solution	$(1,0,1,0)$	$(1,0,0,1)$	$(0,1,1,1)$	$(0,1,1,1)$	$(0,1,1,1)$

Therefore, KNAPAPPROX$(2) = 360$ and it is the value of the solution $(0,1,1,1)$. For $k = 3$, there is only one feasible subset $\{2,3,4\}$. The other subsets are infeasible. Finally, we have

k	0 (The greedy solution)	1	2	3	4
KNAPAPPROX(k)	355	355	360	360	360
Solution	$(1,1,0,0)$	$(1,1,0,0)$	$(0,1,1,1)$	$(0,1,1,1)$	$(0,1,1,1)$

The greedy solution and the solution for $k = 1$ are ε-approximate for every $\varepsilon \geqslant$ $(360 - 355)/360 = 0.014$. ∎

Sahni [1975] proved the following property of Algorithm 2-3.

Theorem 2-5 (Sahni [1975])

The solution generated by the function KNAPAPPROX(k) for $k > 0$ satisfies

$$\frac{P_{\text{opt}}(\mathcal{J}) - \text{KNAPAPPROX}(k)}{P_{\text{opt}}(\mathcal{J})} < \frac{1}{k+1}$$

for every knapsack instance $\mathcal{J}$. The computing time is $O(kn^{k+1})$ and the storage needed is $O(n)$, where n is the number of objects in $\mathcal{J}$. ∎

Note that this theorem says nothing about the greedy solutions which may be considered as corresponding to KNAPAPPROX(0). In fact, the greedy solutions are not ε-approximate, although they are often very "good" solutions. By the last theorem, an ε-approximate algorithm for solving the knapsack problem is KNAPAP-PROX(l), where l satisfies $1/(l+1) \leqslant \varepsilon$; therefore, $l \geqslant 1/\varepsilon - 1$. It was shown by Sahni that for a particular knapsack instance $\mathcal{J}$, we also have

$$\frac{P_{\mathrm{opt}}(\mathcal{J}) - \mathrm{KNAPAPPROX}(k)}{P_{\mathrm{opt}}(\mathcal{J})} < \min\left\{ \frac{1}{k+1}, \bar{p}_{k+1}/\mathrm{KNAPAPPROX}(k) \right\}$$

where $\bar{p}_{k+1}$ is the $(k+1)$st largest profit p_i in $\mathcal{J}$.

The computational experiments performed with ε-approximate algorithms by Sahni [1975] (see also Horowitz and Sahni [1978] and the computational results in Section 2.1.6) show that in most cases the function KNAPAPPROX with $k = 1$ and $k = 2$ may be expected to produce solutions that are very close to the optimal (within 1 to 5%). This motivates our decision to implement KNAPAPPROX(2).

Computer Implementation of the Function
KNAPAPPROX(2)

The KNAPAPPROX procedure finds a 0.33-approximate solution to the knapsack problem defined by (2-1) through (2-6) and (2-13). The implementation is almost self-explanatory (see the comments embedded in the procedure). First, the procedure finds the greedy solution. Then one- and two-object subsets are used as inputs for a greedy-type procedure LB to find if they can generate better solutions. Since the subsets of objects which are contained in the greedy solution lead to the same solution, we try to avoid as many such subsets as possible. The reader should have no difficulty in verifying that the time complexity of KNAPAPPROX is $O(N^3)$.

Global Constant

N   number of variables (objects) in the knapsack problem

Data Type

```
TYPE   ARRN = ARRAY[1..N] OF INTEGER;
```

Procedure Parameters

```
PROCEDURE KNAPAPPROX(

     N                :INTEGER;
     VAR P,W,X    :ARRN;
     VAR V,PROFIT :INTEGER;
     VAR EPS        :REAL);
```

Input

N	number of variables of the problem
P[1..N]	array of object profits
W[1..N]	array of object weights
V	total weight limit of the knapsack

Output

X[1..N]	array containing the solution vector
PROFIT	profit of the solution obtained
EPS	upper bound of the relative error of the solution, that is, EPS = min{0.33333,MAXP3/PROFIT}, where MAXP3 is the third largest profit

Example

For the knapsack problem of Example 2-5, the input is

$$N = 4$$
$$P[1..4] = [200,155,115,90]$$
$$W[1..4] = [50,40,30,25]$$
$$V = 95$$

and the output is

$$X[1..4] = [0,1,1,1]$$
$$PROFIT = 360$$
$$EPS = 0.33333$$

Pascal Procedure KNAPAPPROX

```
PROCEDURE KNAPAPPROX(
    N              :INTEGER;
    VAR P,W,X    :ARRN;
    VAR V,PROFIT:INTEGER;
    VAR EPS       :REAL);

VAR I,J,K,L,MAXP1,MAXP2,MAXP3,PP,Q,R,S,U,VV:INTEGER;

PROCEDURE LB(VAR G,H,Q,U:INTEGER);
  (* LB FINDS PROFIT Q AND RESIDUAL WEIGHT OF GREEDY TYPE
     SOLUTION WHICH IS ASSUMED TO CONTAIN OBJECTS G AND H *)
  VAR K:INTEGER;
```

```
BEGIN
   K:=0;
   REPEAT
      K:=K+1;
      IF (K <> G) AND (K <> H) AND (W[K] <= U) THEN
      BEGIN
         Q:=Q+P[K];
         U:=U-W[K]
      END
   UNTIL K=N
END (*  LOWER BOUND  *);

PROCEDURE MAX;
   (* MAX UPDATES MAXP1,MAXP2, AND MAXP3: LARGEST, SECOND
      AND THIRD LARGEST ELEMENTS OF THE PROFIT VECTOR *)
BEGIN
   IF P[I] > MAXP1 THEN
   BEGIN MAXP3:=MAXP2;  MAXP2:=MAXP1;  MAXP1:=P[I] END
   ELSE IF P[I] > MAXP2 THEN
        BEGIN MAXP3:=MAXP2;  MAXP2:=P[I] END
        ELSE IF P[I] > MAXP3 THEN MAXP3:=P[I]
END (*  MAX  *);

BEGIN  (*  MAIN BODY  *)
   I:=1;
   U:=V;
   PROFIT:=0;
   MAXP1:=0;  MAXP2:=0;  MAXP3:=0;
   WHILE W[I] <= U DO    (* FINDING GREEDY SOLUTION *)
   BEGIN
      U:=U-W[I];
      MAX;
      X[I]:=1;
      PROFIT:=PROFIT+P[I];
      I:=I+1
   END;
   I:=I-1;
   S:=I;
   REPEAT
      I:=I+1;
      IF W[I] <= U THEN
      BEGIN  U:=U-W[I];  X[I]:=1;  PROFIT:=PROFIT+P[I]  END
      ELSE X[I]:=0;
      MAX
   UNTIL I=N;
   Q:=PROFIT;
   (*  ONE ELEMENT SUBSETS OF OBJECTS  *)
   K:=0;  L:=0;          (* K AND L IDENTIFY THE OBJECTS WHICH *)
   FOR I:=S TO N DO      (* ARE ASSUMED TO BE IN A SOLUTION  *)
      IF X[I] <> 1 THEN
```

```
      BEGIN
         VV:=V-W[I];
         PP:=P[I];
         LB(I,I,PP,VV);
         IF PP > PROFIT THEN
         BEGIN PROFIT:=PP;  K:=I   END
      END;
   (* TWO ELEMENT SUBSETS OF OBJECTS *)
    R:=S;
    FOR I:=1 TO N-1 DO
    BEGIN
       IF I > S THEN R:=I;
       FOR J:=R+1 TO N DO
       BEGIN
          VV:=V-W[I]-W[J];
          IF VV >= 0 THEN
          BEGIN
             PP:=P[I]+P[J];
             LB(I,J,PP,VV);
             IF PP > PROFIT THEN
             BEGIN  PROFIT:=PP;  K:=I;  L:=J   END
          END
       END (*   J=R+1..N   *)
    END;   (*   I=1..N   *)
    IF PROFIT > Q THEN
    BEGIN
       IF K > 0 THEN
       BEGIN V:=V-W[K];  X[K]:=1   END;
       IF L > 0 THEN
       BEGIN V:=V-W[L];  X[L]:=1   END;
       FOR I:=1 TO N DO
          IF (I <> K) AND (I <> L) THEN
             IF W[I] <= V THEN
             BEGIN
                X[I]:=1;
                V:=V-W[I]
             END
             ELSE X[I]:=0
    END;
    EPS:=MAXP3/PROFIT;
    IF EPS > 0.33333 THEN EPS:=0.33333
END;   (*  KNAPAPPROX  *)
```

2.1.5. EXACT METHODS

As we mentioned in the introductory section, the exact methods for solving the knapsack problem are basically of the following type:

1. Network approach

2. Dynamic programming

3. Enumeration methods (backtracking, branch-and-bound, etc.)

A network approach is presented in Problems 2-17 and 2-18 and for further details we refer the reader to the literature discussed in the References and Remarks.

In the next two sections we first discuss a dynamic programming approach and then present an enumeration method together with its Pascal implementation.

Dynamic Programming Approach

In general, a dynamic programming algorithm finds the solution to a problem in a sequence of decisions. In the case of the knapsack problem, the solution may be viewed as a sequence of decisions related to the values of variables $x_1, x_2, \ldots, x_n$. First, we may decide whether $x_1 = 0$ or $x_1 = 1$, then consider x_2, and so on.

Let $\text{KNAP}(k, l, U)$ denote the following knapsack problem:

maximize
$$\sum_{i=k}^{l} p_i x_i$$

subject to
$$\sum_{i=k}^{l} w_i x_i \leqslant U$$

$$x_i = 0 \text{ or } 1, \qquad i = k, k + 1, \ldots, l - 1, l$$

Evidently, the 0–1 knapsack problem defined by (2-1)–(2-3) is $\text{KNAP}(1, n, V)$.

In this section the assumptions about the problem parameters (2-4) through (2-6) and (2-13) are not utilized by a dynamic programming approach, although in certain cases the problem may become easier (or at least smaller) after applying the transformation of Problem 2-1 and the reduction algorithm KNAPRED (see Algorithm 2-2) to it.

All dynamic programming algorithms for solving $\text{KNAP}(1, n, V)$ are based on the following property shared by its optimal solutions $x_1, x_2, \ldots, x_n$. If $x_1 = 0$, then $x_2, x_2, \ldots, x_n$ must be an optimal solution for the problem $\text{KNAP}(2, n, V)$. Otherwise, $x_1, x_2, \ldots, x_n$ would not be an optimal solution for $\text{KNAP}(1, n, V)$. Similarly, if $x_1 = 1$, then $x_2, x_3, \ldots, x_n$ must be an optimal solution for the problem $\text{KNAP}(2, n, V - w_1)$.

This property is a special case of the *principle of optimality* formulated by R. Bellman in 1957, which says that "an optimal sequence of decisions has the property that whatever the initial state and decisions are, the remaining decisions must constitute an optimal decision sequence with regard to the state resulting from the first decision."

Let us now express this property in a more mathematical way. Let $Q_j(U)$ denote the value of an optimal solution to $\text{KNAP}(j + 1, n, U)$. $Q_0(V)$ is the value of an optimal solution to $\text{KNAP}(1, n, V)$ and the principle of optimality may be written as follows:

$$Q_0(V) = \max\{Q_1(V), Q_1(V - w_1) + p_1\}$$

where the first term under max corresponds to $x_1 = 0$ and the second to $x_1 = 1$. In general, this property says that if $x_1, x_2, \ldots, x_n$ is an optimal solution to KNAP$(1, n, V)$, then for every j $(1 \leqslant j \leqslant n)$, partial solutions $x_1, x_2, \ldots, x_j$ and $x_{j+1}, \ldots, x_n$ must be optimal solutions to the problems KNAP$(1, j, \Sigma_{i=1}^{j} w_i x_i)$ and KNAP$(j + 1, n, V - \Sigma_{i=1}^{j} w_i x_i)$, respectively. Hence

$$Q_j(U) = \max\{Q_{j+1}(U), Q_{j+1}(U - w_{j+1}) + p_{j+1}\} \qquad (2\text{-}18)$$

where again the first term under max corresponds to $x_{j+1} = 0$ and the second to $x_{j+1} = 1$. Therefore, in general, we have (2-18) for $j = n - 1, n - 2, \ldots, 1, 0$.

The application of the principle of optimality results in a recurrence relations of type (2-18), and dynamic programming algorithms are design to solve such relations. In particular, to solve (2-18), we first assume $Q_n(U) = 0$ for all U, then calculate successively $Q_{n-1}(U)$ for all U from Q_n for all U, then $Q_{n-2}(U)$ for all U from $Q_{n-1}(U)$ for all U, etc. Finally, we evaluate $Q_0(V)$. Parameter U is an integer number from the range 0 and V, since all $Q_j(U)$ for $U \in (0, V)$ and $j = 1, 2, \ldots, n$ may be utilized in the calculations.

The way in which the optimality principle was formulated results in recurrence relations (2-18) and the solution method which may be called *forward*, since the decision on x_j is made in terms of the optimal decision sequence involving $x_{j+1}, \ldots, x_n$. In other words, we have to look forward to determine the value of a variable currently considered.

The principle of optimality may also be formulated in a *backward* fashion, where the sequence of decisions on the values of $x_1, x_2, \ldots, x_n$, leads to the following recurrence relations:

$$S_j(U) = \max\{S_{j-1}(U), S_{j-1}(U - w_j) + p_j\} \qquad (2\text{-}19)$$

for $j = 1, 2, \ldots, n$, where $S_j(U)$ is the value of the optimal solution to the problem KNAP$(1, j, U)$. The value of an optimal solution to our problem KNAP$(1, n, V)$ is $S_n(V)$, and may be obtained by using (2-19) for $j = 1, 2, \ldots, n$, beginning with $S_0(U) = 0$ for all $U \geqslant 0$ and $S_0(U) = -\infty$ for $U < 0$.

Example 2-6

Let us apply relations (2-19) to solve the following knapsack instance:

maximize $2x_1 + 3x_2 + 3x_3 + 4x_4$

subject to $x_1 + 2x_2 + 4x_3 + 5x_4 \leqslant 6$

                    $x_1, x_2, x_3, x_4 = 0$ or 1

Table 2-1 shows the values of function $S_j(U)$ for $j = 0, 1, 2, 3, 4$ and $U = 0, \ldots, 6$. The second number in each table entry shows how a particular $S_j(U)$ has been obtained; namely, it is 0 if $S_j(U) = S_{j-1}(U)$ and 1 if $S_j(U) = S_{j-1}(U - w_j) + p_j$. The optimal decisions on variables may be read from this table moving from

Table 2-1.   Values of $S_j(U)$ Computed for the Knapsack Instance of Example 2-6

U	j = 0		j = 1		j = 2		j = 3		j = 4	
0	0	—	0	0	0	0	0	0	0	0
1	0	—	2	1	2	0	2	0	2	0
2	0	—	2	1	3	1	3	0	3	0
3	0	—	2	1	5	1	5	0	5	0
4	0	—	2	1	5	1	5	0	5	0
5	0	—	2	1	5	1	5	0	5	0
6	0	—	2	1	5	1	6	1	6	0 or 1

bottom right to top left. Assume that we set $x_4 = 1$. To find x_3 we look at $U = V - w_4 = 1$ and $j = 3$, hence $x_3 = 0$. We also have $x_2 = 0$ for $U = 1$ and $j = 2$. Finally, for $U = 1$ and $j = 1$, we get $x_1 = 1$. Therefore, an optimal solution is $(1, 0, 0, 1)$. The reader may verify that starting with $x_4 = 0$ we would get $(0, 1, 1, 0)$ with the same value. Note that once such a table is computed it is possible to find optimal solutions to many knapsack instances which result from the original one either by considering the weight limit of the knapsack smaller than V or restricting the attention to the first k variables, where $1 \leqslant k \leqslant n$, or both.   ∎

Regarding the computational complexity of the dynamic programming algorithm which is an implementation of the relations (2-19) [or also (2-18) and several other of a similar type], we can observe that the computation time is proportional to the number of table entries and therefore is $O(nV)$. There are several improvements of this bound which take into account a special form of functions $S_j(U)$ and $Q_j(U)$. One can easily notice that those functions are stair-like; therefore, only their values at jump points have to be computed. Nevertheless, the computation time is still bounded by polynomial in n and V. It seems that we have arrived at a polynomial-time algorithm for the knapsack problem. However, this is not true, since nV is not a polynomial function in the size of the problem. The size of a problem, defined as the number of bits necessary to represent a given instance of the problem, for the knapsack problem, is bounded by $O(n \log_2 V)$. This follows from the fact that $w_i \leqslant V$ $(i = 1, 2, \ldots, n)$ and p_i may be assumed to be of the same order as w_i. Unfortunately, nV is not bounded by any polynomial in $n \log_2 V$, since when V tends to infinity, nV increases much faster than any power of $n \log_2 V$. Therefore, the computational complexity of the dynamic programming algorithms based on relations (2-18) and (2-19) is not polynomial in the size of the problem. It is, however, worth noting that for many practical knapsack instances, arising for instance in the area of scheduling (see Section 4.2), the knapsack limit V is relatively small and the dynamic programming algorithms turn out to be quite efficient.

The dynamic programming approach can also be applied to the bounded knapsack program (see Problem 2-19) and can easily be generalized to handle a nonlinear objective function and a nonlinear constraint.

Implicit Enumeration Algorithms

In this section we present an enumeration method for solving the knapsack problem. Such a method is also referred to as backtracking or branch-and-bound. There have been several enumeration methods designed for the knapsack problem. We have chosen a method that illustrates basic steps of almost all enumeration methods, and at the same time is one of the fastest methods. The method has been first proposed by Horowitz and Sahni [1974] and then improved by Martello and Toth [1977].

An implicit enumeration algorithm finds a solution to a problem instance by a systematic search of the solution space of the instance. The solution space of an n-object knapsack problem consists of the 2^n vectors of 0's and 1's which correspond to the 2^n distinct ways of assigning zero and one values to the problem variables.

An enumeration algorithm uses a tree structure of the solution space. There are many possible tree structures of the family of the 2^n zero–one vectors. We will utilize a tree in which the variables correspond to successive tree levels and every tree node has at most two sons. We assume also that a left son always corresponds to one and a right son to zero values of a variable. Thus the sons of the tree root correspond to x_1, where $x_1 = 1$ is in the left node and $x_1 = 0$ in the right node. In general, every tree node corresponds to an assignment of the values to those variables which lie on the path from the tree root to this node. Notice that such a tree structure of the solution space is independent of the problem instance being solved; therefore, it can be called a *static tree* organization (see Problem 2-23, where a dynamic tree organization is defined).

Having decided how the solution space is organized in the form of a tree, the problem may be solved by a systematic generation of the tree nodes. However, not all nodes correspond to feasible solutions of the problem and only some feasible nodes correspond to the optimal solution. Since the size of the tree (the number of nodes) is exponential, a method for generating the solution should avoid, as much as possible, generating tree nodes which are either infeasible or do not lead to a better solution than that found so far. However, even a tree containing only feasible nodes, that may lead to a better complete solution, can be of unreasonable size. We chose therefore a method for generating tree nodes which is very space-economical and yet does not ignore any tree node that may lead to an optimal solution. The method first generates the left son and then the right son, and as soon as a new node is generated, the search moves to this node. In terms of the problem variables, the method first assigns one to a variables and then moves to the next variable. When the search in the left son (subtree) is completed, it moves to the right one. Such a method of searching the solution space is called *depth-first search*.

To reduce generation of some nodes which cannot improve the best current solution, a *bounding function* is used. Let *PROF* denote the profit of the best solution determined and its elements by $x_1, x_1, \ldots, x_n$. In the beginning, we assume $PROF = -1$ and $x_i = 0$ for $i = 1, \ldots, n$. A bounding function can be derived from an upper bound on the profit of the best feasible solution which can be generated by augmenting the current solution. The value of the bounding function can be obtained by applying Theorem 2-3 to the problem in which the integer constraint on variables y_i not yet considered are relaxed to $0 \leqslant y_i \leqslant 1$. If this upper bound is not greater than $PROF$, then the further search from this tree node down may be aborted and the search may either move to the right son of the current node, if it was in the left one or go back to the first variable with value 1. Such a method of visiting the nodes of the solution tree can be fully described by means of a current solution vector $y_1, y_2, \ldots, y_n$. The vector shows the variables that have been considered and their values.

Notice, that if a tree node has a feasible left son, both have the same value of the bounding function. Hence the bounding function should be used only after a series of successful moves to feasible left sons.

The algorithm terminates when the search comes back to the tree root seeking a variable with value 1, and $y_1 = 0$.

The method as described above can also be formulated in terms of two main actions: forward and backward moves. A *forward move* inserts into the knapsack the largest set of new consecutive objects subject to the weight constraint. The variable corresponding to the first object which does not fit is set to zero and it is tested if further forward moves could lead to a solution improving the current optimal one. If so, a new forward move is performed; otherwise, a backward move follows. If during the forward move, the last object has been considered, it is tested whether the current solution improves the best current solution. If so, $PROF$ and $(x_1, x_2, \ldots, x_n)$ are updated.

A *backward move* removes from the current solution the object with the largest index (the corresponding problem variable becomes 0) and the search returns to a forward move.

We now describe in detail the bounding function and its implementation. Let P and W denote, respectively, the profit and the weight of the current solution, and assume that k is the index of the last object considered. Therefore, P and W correspond to the partial feasible solution stored as $y_1, y_2, \ldots, y_k$. Now, to find whether the partial solution can be augmented to a solution improving the best current solution, the constraint $y_i = 0$ or 1 imposed on further variables $y_{k+1}, \ldots, y_n$ is relaxed to $0 \leqslant y_i \leqslant 1$ $(i = k + 1, \ldots, n)$. Therefore, the following problem is solved:

maximize
$$P + \sum_{i=k+1}^{n} p_i y_i$$

subject to
$$\sum_{i=k+1}^{n} w_i y_i \leqslant V - W \qquad (2\text{-}20)$$

$$0 \leqslant y_i \leqslant 1, \qquad i = k + 1, \ldots, n$$

This is a linear programming problem whose solution is given in Theorem 2-3 and which can be solved by the greedy algorithm.

Algorithm 2-4 presents the BOUND function, which calculates the value of the solution to problem (2-20). Additionally, Q and U are, respectively, the profit and the weight of the integer part of the solution and l is the index of the first object that does not fit into the knapsack (if no object remains, $l = n + 1$).

Figure 2-1 illustrates the solution obtained by call BOUND(P, W, k, Q, U, l). The complete description of the method is given in Algorithm 2-5.

Algorithm 2-4: Function BOUND for Solving Problem (2-20)

```
function BOUND(P, W, k, Q, U, l);
     {* This function solves problem (2-20). Q and U are the profit and the weight of the
        integer part of the solution and l is the index of the first object that does not fit into
        the knapsack *}
begin
     Q ← P; U ← W;
     l ← k + 1; B ← true;
     while (l ⩽ n) and B do
     begin
          B ← U + w_l ⩽ V;
          if B then
          begin
               Q ← Q + p_l; U ← U + w_l;
               y_l ← 1; l ← l + 1
          end
     end {* while *};
     if B then BOUND ← Q
     else BOUND ← Q + ⌊(V − U)p_l/w_l⌋
end {* BOUND *}
```

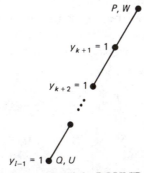

Figure 2-1 *Action of the BOUND function.*

Algorithm 2-5:   Enumeration Algorithm of Horowitz and Sahni [1974]

```
begin
    k ← 0; PROF ← −1;
    P ← 0; W ← 0;
    repeat ⟨∗ until the search is completed, i.e., k = 0 ∗⟩
        repeat ⟨∗ until a significant forward move can be made ∗⟩
            B ← BOUND(P, W, k, Q, U, I) ≤ PROF;
            if B then
            begin ⟨∗ backward move to the last object in the knapsack ∗⟩
                find the largest j such that yⱼ = 1 and j ≤ k;
                k ← j;
                if k > 0 then
                begin ⟨∗ removal of the kth object from the knapsack ∗⟩
                    yₖ ← 0;
                    P ← P − pₖ; W ← W − wₖ
                end
            end ⟨∗ B ∗⟩
        until (not B) or (k = 0);
        if (k > 0) or (PROF = −1) then
        begin ⟨∗ forward move ∗⟩
            P ← Q; W ← U;
            k ← I;
            if k > n then
            begin ⟨∗ a new better solution has been found ∗⟩
                PROF ← P;
                k ← n;
                xᵢ ← yᵢ (i = 1, 2, ..., n)
            end
            else yₖ ← 0
        end ⟨∗ forward move ∗⟩
    until k = 0
end
```

Example 2-7

Let us illustrate Algorithms 2-4 and 2-5 with the following knapsack instance.

maximize $\qquad\qquad 2x_1 + 3x_2 + 6x_3 + 3x_4$

subject to $\qquad\qquad x_1 + 2x_2 + 5x_3 + 4x_4 \leqslant 5$

$$x_1, x_2, x_3, x_4 = 0 \text{ or } 1$$

In the beginning, $P = W = 0$, $k = 0$ and $PROF = −1$. The first call of function BOUND returns a partial solution $y_1 = y_2 = 1$, of profit $Q = 5$, weight $U = 3$, and

$l = 3$. The value of the function is $\lfloor Q + (V - U)p_3/w_3 \rfloor = \lfloor 5 + 2\frac{6}{5} \rfloor = 7$. Since $PROF = -1$, the forward move follows, that is $P \leftarrow Q$, $W \leftarrow U$, $k \leftarrow l$ and $y_3 \leftarrow 0$, and the algorithm again calls BOUND. However, the assignment of 1 to y_4 would be infeasible. Thus y_4 becomes 0. In the next call of BOUND, l becomes $5 > n = 4$, and since the function value is still not greater than -1, the current solution is the best solution found so far. Therefore, **x** becomes $\mathbf{y}^T = (1, 1, 0, 0)$, $PROF$ becomes $P = 5$, and $k = 4$.

The next step of the algorithm moves the search to the last object assigned to the knapsack. This object is removed; therefore, $y_2 \leftarrow 0$, $P \leftarrow 5 - 3 = 2$, and $W \leftarrow 3 - 2 = 1$. The next call of BOUND attempting to put into the knapsack object 3 fails since $U + w_3 = 6 \not\leq 5$. However, the call of BOUND which follows produces the solution $\mathbf{y}^T = (1, 0, 0, 1)$ of profit 5, which does not improve the best current solution. Then the search moves to y_1, which becomes 0 and again BOUND is called. After assigning $y_2 \leftarrow 1$, it returns value 6. Since this is not smaller than or equal to $PROF = 5$, the forward move is restored, and the next call to BOUND returns $5 \leqslant PROF$. Now the backward move follows in which $y_2 \leftarrow 0$. The next forward move assigns $y_3 \leftarrow 1$ and returns 6, which leads to $y_4 \leftarrow 0$ and to the solution $(0, 0, 1, 0)$ with profit 6, which improves the best current solution. Finally, $y_3 \leftarrow 0$, and BOUND assigns $y_4 \leftarrow 1$ and returns 3. Since there are no other objects in the knapsack to be removed, the algorithm terminates. Figure 2-2 shows the search tree constructed in the course of the algorithm ∎

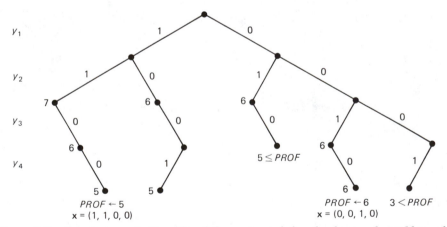

Figure 2-2 *Search tree of Algorithm 2-5 constructed for the knapsack problem of Example 2-7. The numbers attached to tree branches are the values of the corresponding variables and the numbers in the tree nodes are the values of the function* BOUND.

Now we will describe some improvements of the foregoing algorithm due to Martello and Toth [1977] which result in one of the most efficient enumeration algorithms for solving the knapsack problem. In the beginning, the algorithm constructs the greedy solution and saves it. Then a depth-first search enumeration is performed.

A forward move is split into two phases: building and saving. A *building phase* determines the largest set of new consecutive objects which can be added to the current contents of the knapsack. A *saving phase* updates the current solution by introducing into the knapsack the objects assigned by the most recent building phase. If at least one of the objects following those just introduced can still be added to the current contents of the knapsack, a new building phase is performed. Otherwise, if it is worthwhile, the best current solution is updated and a backward move follows. A saving phase is executed only if an upper bound does not eliminate the possibility that the current contents of the knapsack can be augmented to an assignment which is better than the best current solution. Otherwise, a backward move is executed.

Building and saving phases use some auxiliary quantities that speed up the computations and decisions. Let $M_j = \min\{w_k: j < k \leqslant n\}$. Thus given the residual capacity V^* of the knapsack, if $M_j \leqslant V^*$, then at least one more object can be put in. Some parts of the current solution which can be utilized again in a building phase are saved in vectors $\bar{p}$, $\bar{w}$, and $\bar{z}$. They are used in the following way. Suppose that a current solution has been obtained by putting into the knapsack objects from the ith through the lth. Then, when trying to put in the objects from an jth ($i < j \leqslant l$), if the current solution before the ith object has not been altered, it is possible to augment the new current solution by the objects from the jth through the lth.

A backward move first finds the last object in the current solution. If such an object does not exist, the algorithm terminates since the solution space has been exhausted. Otherwise, the last object is removed from the knapsack and a forward move follows.

A standard forward move is performed only if the residual knapsack capacity of the current solution preceding the backward move allows the introduction of at least one of the objects following the kth one. Otherwise, a special forward move is performed (step 7) which is based on the following argument. The current solution could be improved only if the kth object (the one just removed) is replaced by one or more objects which weigh no more than the residual knapsack capacity and their inclusion may lead to an improved solution.

The detail description of the Martello–Toth algorithm is given in the form of the set of steps with **goto** statements. Then we demonstrate how the algorithm can be implemented without using any **goto** statements. Finally, a Pascal implementation is presented.

Initialization

Step 1.   Find the greedy solution, that is, the largest index l such that $\sum_{j=1}^{l} w_j \leqslant V$. If the last relation is satisfied with the equality sign, the optimal solution is given by $x_j = 1$ ($j = 1, \dots, l$), $x_j = 0$ ($j = l + 1, \dots, n$). Stop. Otherwise, let $P^* \leftarrow \sum_{j=1}^{l} p_j$ and $W^* = \sum_{j=1}^{l} w_j$, and compute $M_j = \min\{w_k: j < k \leqslant$

$n\}, j = 1, \ldots, n - 1$, where $M_n = V + 1$. Set $UB \leftarrow U_2$ [see (2-15)], $Q \leftarrow 0$, $PROF \leftarrow 0$, $y_j \leftarrow 0$ ($j = 1, \ldots, n$), $i \leftarrow 0$, $l^* \leftarrow n$, and $V^* \leftarrow V$. UB is the upper bound of the solution. Vector $(x_1, \ldots, x_n)$ is the best current solution of profit $PROF$. The current solution is stored in $(y_1, \ldots, y_n)$ and its profit is Q. Go to step 4 to save the current solution.

Forward Moves (Building and Saving Phases)

The building phase constitutes steps 2 and 3.

Step 2. *Trying to insert the ith object into the current solution.* If $w_i \leqslant V^*$ (V^* is the residual capacity), go to step 3. Otherwise, if $PROF \geqslant Q + \lfloor V^* p_{i+1}/w_{i+1} \rfloor$, go to step 6 since the current solution cannot be improved, else set $i \leftarrow i + 1$ and repeat step 2.

Step 3. *Building a new current solution.* Find the largest index $l \leqslant n$ for which $W^* = \overline{w}_i + \sum_{j=\overline{z}_i}^{l} w_j \leqslant V^*$. If $\overline{w}_i + w_{\overline{z}_i} > V^*$, then set $l \leftarrow \overline{z}_i - 1$. Let $P^* = \overline{p}_i + \sum_{j=\overline{z}_i}^{l} p_j$. Two cases may occur:

(a) $W^* < V^*$ and $l < n$: If $PROF \geqslant Q + P^* + \lfloor (V^* - W^*) p_{l+1}/w_{l+1} \rfloor$ then go to step 6, since the current solution cannot be improved. Otherwise, save the current solution, that is, go to step 4.

(b) $W^* = V^*$ or $l = n$: If $PROF \geqslant Q + P^*$, then go to step 6 for the same reason as above. Otherwise, since a better solution has been found, set $PROF \leftarrow Q + P^*$ and update the optimal solution $x_j \leftarrow y_j$ ($j = 1, \ldots, i - 1$), $x_j \leftarrow 1$ ($j = i, \ldots, l$), $x_j \leftarrow 0$ ($j = l + 1, \ldots, n$). If $PROF = UB$, then stop since no better solution exists; if not, go to step 6.

Comment: Notice that the current solution is not updated in the latter case since for a present value of i, P^* cannot become larger. This avoids the useless backward moves related to the variables from the ith through the lth.

The saving phase constitutes steps 4 and 5.

Step 4.   *Saving the current solution.* Set $Q \leftarrow Q + P^*$, $V^* \leftarrow V^* - W^*$

$$y_j \leftarrow 1 \text{ for } j = i, i + 1, \ldots, l$$

Compute $\overline{p}_i \leftarrow P^*$, $\overline{w}_i \leftarrow W^*$, $\overline{z}_i \leftarrow l + 1$;

$$\overline{p}_j \leftarrow \overline{p}_{j-1} - p_{j-1}, \overline{w}_j \leftarrow \overline{w}_{j-1} - w_{j-1}, \overline{z}_j \leftarrow l + 1 \text{ for } j = i + 1, \ldots, l;$$
$$\overline{p}_j \leftarrow 0, \overline{w}_j \leftarrow 0, \overline{z}_j \leftarrow j \text{ for } j = l + 1, \ldots, l^*;$$

Set $l^* \leftarrow l$.
Three possibilities exist:

(a) $l = n - 1$: that is, all the objects have been considered with the last one, which cannot be added to the current knapsack. Set $i \leftarrow n$ and go to step 5.

(b) $l = n - 2$: that is, $n - 1$ objects have been considered. Therefore, we check the last object and proceed to step 5. If $V^* \geq w_n$, then set $Q \leftarrow Q + p_n$, $V^* \leftarrow V^* - w_n$, $y_n \leftarrow 1$. Set $i \leftarrow n - 1$ and go to step 5.

(c) $l < n - 1$: Set $i \leftarrow l + 2$. If $V^* < M_{i-1}$, then go to step 5; otherwise, go to step 2.

Step 5. *Updating the best current solution.* If $PROF < Q$, then set $PROF \leftarrow Q$ and $x_j \leftarrow y_j$ for $j = 1, 2, \ldots, n$. If $PROF = UB$, then stop. Otherwise (i.e., if $PROF \geq Q$ or $PROF \neq UB$), if $y_n = 1$, then set $Q \leftarrow Q - p_n$, $V^* \leftarrow V^* + w_n$, $y_n \leftarrow 0$. Go to step 6.

Backward Moves

Step 6. *Backtrack.* Find the largest $k < i$ for which $y_k = 1$. If no such k exists (i.e., either $i = 1$ or $y_1 = \ldots = y_{i-1} = 0$), then stop. Otherwise, set $R \leftarrow V^*$ and remove the kth object from the knapsack (i.e., set $Q \leftarrow Q - p_k$, $V^* \leftarrow V^* + w_k$, $y_k \leftarrow 0$). If $R \geq M_k$, then set $i \leftarrow k + 1$ and go to step 2. Otherwise, set $i \leftarrow k$, $m \leftarrow k + 1$ and go to step 7.

Step 7. *Trying to substitute the mth element for the kth one.* If $m > n$ or $PROF \geq Q + \lfloor V^* p_m / w_m \rfloor$, then go to step 6. Otherwise set $D \leftarrow w_m - w_k$ and $T \leftarrow R - D$. Three possibilities exist:

(a) $D = 0$: Set $m \leftarrow m + 1$ and repeat step 7.

(b) $D > 0$: If $T < 0$ or $PROF \geq Q + p_m$ then set $m \leftarrow m + 1$ and repeat step 7.

   Otherwise, a better solution has been found; therefore,

$$PROF \leftarrow Q + p_m$$

$$x_j \leftarrow y_j \, (j = 1, 2, \ldots, k), \, x_j \leftarrow 0 \, (j = k + 1, \ldots, n)$$

$$x_m \leftarrow 1$$

If $PROF = UB$, then stop.
Otherwise, set $R \leftarrow T$, $k \leftarrow m$, $m \leftarrow m + 1$ and repeat step 7.

(c) $D < 0$: If $T < M_m$, then set $m \leftarrow m + 1$ and repeat step 7. Otherwise, if $PROF \geq Q + p_m + \lfloor T p_m / w_m \rfloor$, then go to step 6. Otherwise, set

$$Q \leftarrow Q + p_m, \, V^* \leftarrow V^* - w_m, \, y_m \leftarrow 1$$

$$i \leftarrow m + 1$$

$$\bar{p}_m \leftarrow p_m, \, \bar{w}_m \leftarrow w_m, \, \bar{z}_m \leftarrow m + 1$$

$$\bar{p}_j \leftarrow \bar{w}_j \leftarrow 0, \, \bar{z}_j \leftarrow j \text{ for } j = m + 1, \ldots, l^*$$

$$l^* \leftarrow m, \text{ go to step 2.}$$

Example 2-8

Let us apply the Martello–Toth algorithm to the knapsack problem of Example 2-7.

Step 1.   The greedy solution is formed by the first two objects; hence $l = 2$, $P^* = 5$, and $W^* = 3$, and it is not an optimal solution, since $W^* \neq V$. Auxiliary variables are $\mathbf{M}^T = (2, 4, 4, 6)$, $UB = 6$, $Q = PROF = 0$, $\mathbf{y}^T = (0, 0, 0, 0)$, $i \leftarrow 1$, $l^* \leftarrow 4$, and $V^* \leftarrow 5$.

Step 4.   *Saving the current solution*

$$Q \leftarrow 5, V^* \leftarrow 2, \mathbf{y}^T \leftarrow (1, 1, 0, 0)$$
$$\overline{\mathbf{p}}^T \leftarrow (3, 2, 0, 0), \overline{\mathbf{w}}^T \leftarrow (5, 3, 0, 0), \overline{\mathbf{z}}^T \leftarrow (3, 3, 3, 4), l^* \leftarrow 2.$$

(b) $l = n - 2$: $V^* \not\geq w_4$, $i \leftarrow 3$.

Step 5.   *Saving the best current solution*

$$PROF < Q; \text{ hence } PROF \leftarrow 5, \mathbf{x}^T \leftarrow (1, 1, 0, 0), PROF \neq UB, y_4 \neq 1$$

Step 6.   *Backtracking*

$$k = 2. R \leftarrow 2, V^* \leftarrow 4, Q \leftarrow 2, \mathbf{y}^T \leftarrow (1, 0, 0, 0)$$
$$R \not\geq M_2: i \leftarrow 2, m \leftarrow 3$$

Step 7.   *Substitution*

$$m \not> n \text{ and } PROF \not\geq Q + \lfloor V^* p_m / w_m \rfloor$$
$$D \leftarrow 3, T \leftarrow -1$$

(b) $T < 0$: $m \leftarrow 4$.

Step 7.

$$PROF = Q + \lfloor V^* p_m / w_m \rfloor$$

Step 6.   *Backtracking*

$$k = 1. R \leftarrow 4, V^* \leftarrow 5, Q \leftarrow 0, \mathbf{y}^T \leftarrow (0, 0, 0, 0)$$
$$R \geqslant M_1: i \leftarrow 2$$

Step 2.   *Forward move*

$$w_2 \leqslant V^*$$

Step 3.   *Building a new current solution*

$$l = 2, P^* \leftarrow 3, W^* \leftarrow 2$$
$$W^* < V^* \text{ and } l < n: PROF \not\geq Q + P^* + \lfloor (V^* - W^*) p_{l+1} / w_{l+1} \rfloor$$

Step 4. *Saving the current solution*

$$Q \leftarrow 3, V^* \leftarrow 3, \mathbf{y}^T \leftarrow (0,1,0,0)$$
$$\bar{\mathbf{p}}^T \leftarrow (\ ,3,\ ,\ ), \bar{\mathbf{w}}^T \leftarrow (\ ,2,\ ,\ ), \bar{\mathbf{z}}^T \leftarrow (\ ,3,\ ,\ ), l^* \leftarrow 2$$

(b) $l = n - 2$: $V^* \not\geq w_4$, $i \leftarrow 3$.

Step 5. *Saving the best current solution*

$$PROF \not< Q, y_4 \neq 1$$

Step 6. *Backtracking*

$$k = 2. R \leftarrow 3, V^* \leftarrow 5, Q \leftarrow 0, \mathbf{y}^T \leftarrow (0,0,0,0)$$
$$R \not\geq M_2 : i \leftarrow 2, m \leftarrow 3$$

Step 7. *Substitution*

$$m \not> n \text{ and } PROF \not\geq Q + \lfloor V^* p_m / w_m \rfloor$$
$$D \leftarrow 3, T \leftarrow 0$$

(b) $T \not< 0$ and $PROF \not\geq Q + p_m$. A better solution has been found.

$$PROF \leftarrow 6, \mathbf{x}^T \leftarrow (0,0,1,0)$$
$$PROF = UB; \text{ STOP} \qquad\blacksquare$$

The search tree of the algorithm for the instance of Example 2-8 is shown in Fig. 2-3.

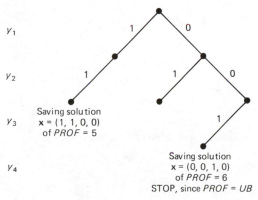

Figure 2-3 *Search tree of Algorithm 2-6 constructed for the knapsack problem of Example 2-8.*

Unlike the foregoing description of the algorithm, its Pascal implementation which follows does not use **goto** statements. First, a greedy solution is computed and some auxiliary variables are evaluated. Then, the main part of the implementation is embedded in the **repeat** statement, which is performed until a Boolean variable STOP becomes TRUE. The execution of a particular step of the algorithm depends on the value of the related Boolean variable. Algorithm 2-6 shows the structural frame of the implementation and for the details the reader is referred to the implementation itself.

Algorithm 2-6:   Structure of the Pascal Implementation of the Martello – Toth Algorithm

```
begin
        calculate the greedy solution, auxiliary variables Mj (j = 1,..., n), UB, and initialize the
            current solution and the optimal current solution;
        STOP ← FALSE;
    repeat
            while STEP2 do
            begin
                .
              · {* steps 2 and 3 — building a new solution *}
                .
            end;
            if STEP4 do
            begin
                .
              · {* step 4 — saving the current solution *}
                .
              · {* step 5 — saving the best current solution *}
            end;
            if STEP56 do
            begin
                .
              · {* step 6 — backward move *}
                .
                    while STEP7 do
                    begin
                        .
                      · {* step 7 — the process of substitution *}
                        .
                    end
            end
    until STOP
end
```

Computer Implementation
of the Martello–Toth Algorithm

The KNAPBACKTRACK procedure solves the knapsack problem defined by (2-1) through (2-6) and (2-13), and is a Pascal implementation of the algorithm due to Martello and Toth described in the preceding section. The procedure is a modified version of the FORTRAN program presented in Martello and Toth [1979b].

Global Constants

N   number of variables (objects) in the knapsack problem
N1   N + 1

Data Types

```
TYPE   ARRN  = ARRAY[1..N] OF INTEGER;
       ARRN1 = ARRAY[1..N1] OF INTEGER;
```

Procedure Parameters

```
PROCEDURE KNAPBACKTRACK(
    N                    :INTEGER;
    VAR P,W              :ARRN1;
    VAR X                :ARRN;
    V                    :INTEGER;
    VAR PROFIT,COUNT  :INTEGER);
```

Input

N   number of variables of the problem
P[1..N]   array of object profits
W[1..N]   array of object weights
V   total weight limit of the knapsack

Output

X[1..N]   array containing the solution vector
PROFIT   profit of the optimal solution stored in X
COUNT   number of forward moves (iterations of step 2) performed by the procedure

Example

For the knapsack instance of Example 2-7 the input has the form

$$N = 4$$
$$P[1..4] = [2, 3, 6, 3]$$
$$W[1..4] = [1, 2, 5, 6]$$
$$V = 5$$

The output obtained is

$$X[1..4] = [0, 0, 1, 0], \quad \text{PROFIT} = 6, \quad \text{COUNT} = 2$$

Pascal Procedure KNAPBACKTRACK

```
PROCEDURE KNAPBACKTRACK(
   N                :INTEGER;
   VAR P,W          :ARRN1;
   VAR X            :ARRN;
   V                :INTEGER;
   VAR PROFIT,COUNT:INTEGER);



VAR D,I,J,K,L,LL,LIM,M,PP,Q,R,T,WW  :INTEGER;
    B,STEP2,STEP4,STEP56,STEP7,STOP:BOOLEAN;
    MIN,PD,WD,Y,ZD                  :ARRN;

PROCEDURE WORKVAR;
VAR W1,P1:INTEGER;
BEGIN   (*  SAVING CURRENT SOLUTION  *)
   FOR J:=I TO L DO  Y[J]:=1;
   P1:=PP;    PD[I]:=P1;
   W1:=V-WW;  WD[I]:=W1;
   ZD[I]:=L+1;
   FOR J:=I+1 TO L DO
   BEGIN
      P1:=P1-P[J-1];  PD[J]:=P1;
      W1:=W1-W[J-1];  WD[J]:=W1;
      ZD[J]:=L+1
   END;
   FOR J:=L+1 TO LL DO
   BEGIN
      PD[J]:=0;  WD[J]:=0;  ZD[J]:=J
   END;
   V:=WW;   Q:=Q+PP;
   LL:=L
END (*  WORKVAR  *);
```

```
BEGIN   (*  MAIN BODY  *)
   COUNT:=1;
   PP:=0;  WW:=V;  L:=0;
   WHILE WW >= W[L+1] DO
   BEGIN
      L:=L+1;
      PP:=PP+P[L];
      WW:=WW-W[L]
   END;
   STOP:=WW=0;
   IF STOP THEN
   BEGIN  (*  THE GREEDY SOLUTION IS OPTIMAL  *)
      PROFIT:=PP;
      FOR J:=1 TO L DO X[J]:=1;
      FOR J:=L+1 TO N DO X[J]:=0
   END
   ELSE
   BEGIN     (*  INITIALIZATION  *)
      R:=V+1;  MIN[N]:=R;
      FOR J:=N DOWNTO 2 DO
      BEGIN
         IF W[J] < R THEN R:=W[J];
         MIN[J-1]:=R
      END;
      P[N+1]:=0;  W[N+1]:=V+1;
      LIM:=PP+TRUNC(WW*P[L+2]/W[L+2]);
      R:=PP+TRUNC(P[L+1]-(W[L+1]-WW)*P[L]/W[L]);
      IF R > LIM THEN LIM:=R;
      FOR J:=1 TO N DO Y[J]:=0;
      PROFIT:=0;  Q:=0;
      I:=1;  LL:=N;
      STEP2:=FALSE;  STEP4:=TRUE;
      REPEAT
         WHILE STEP2 DO
         BEGIN    (* STEPS 2 AND 3 - BUILDING A NEW SOLUTION  *)
            COUNT:=COUNT+1;
            IF W[I] <= V THEN
            BEGIN  (*  STEP 3  *)
               STEP2:=FALSE;
               PP:=PD[I];  WW:=V-WD[I];
               L:=ZD[I]; B:=TRUE;
               WHILE (L <= N) AND B DO
               BEGIN
                  B:=W[L] <= WW;
                  IF B THEN
                  BEGIN
                     PP:=PP+P[L];  WW:=WW-W[L];
                     L:=L+1
                  END
               END;
               L:=L-1;
               IF (WW > 0) AND (L < N) THEN
```

```
        BEGIN
            STEP56:=PROFIT >= Q+PP+TRUNC(WW*P[L+1]/W[L+1]);
            STEP4:= NOT STEP56
        END
        ELSE
        BEGIN
            STEP56:=TRUE;   STEP4:=FALSE;
            IF PROFIT < Q+PP THEN
            BEGIN
                PROFIT:=Q+PP;
                FOR J:=1 TO I-1 DO X[J]:=Y[J];
                FOR J:=I TO L DO    X[J]:=1;
                FOR J:=L+1 TO N DO X[J]:=0;
                STOP:=PROFIT=LIM;
                STEP56:= NOT STOP
            END
        END (* WW = 0 OR L = N  *)
    END (*  STEP 3  *)
    ELSE
    BEGIN
        STEP56:=PROFIT >= Q+TRUNC(V*P[I+1]/W[I+1]);
        STEP4:=FALSE;
        STEP2:= NOT STEP56;
        IF STEP2 THEN I:=I+1
    END
END (*  WHILE STEP2  *);
IF STEP4 THEN
BEGIN   (*  STEP 4 - SAVING CURRENT SOLUTION  *)
    WORKVAR;
    IF L < N-2 THEN
    BEGIN
        I:=L+2;
        STEP56:=V < MIN[I-1];
        STEP2:= NOT STEP56;   STEP7:=FALSE;
    END
    ELSE
    BEGIN
        STEP56:=TRUE;
        IF L=N-2 THEN
        BEGIN
            IF V >= W[N] THEN
            BEGIN
                Q:=Q+P[N];   V:=V-W[N];
                Y[N]:=1
            END;
            I:=N-1
        END
        ELSE I:=N
    END   (*  L >= N-2  *);
    IF STEP56 THEN
    BEGIN  (* STEP 5 - SAVING CURRENT OPTIMAL SOLUTION *)
        IF PROFIT < Q THEN
```

```
     BEGIN  (* BETTER SOLUTION HAS BEEN FOUND  *)
        PROFIT:=Q;
        FOR J:=1 TO N DO X[J]:= Y[J];
        STOP:=PROFIT=LIM;
        STEP56:= NOT STOP
     END;
     IF STEP56 AND (Y[N] = 1) THEN
     BEGIN
        Q:=Q-P[N];  V:=V+W[N];
        Y[N]:=0
     END
  END  (*  STEP56  *)
END  (*  STEP4  *);
IF STEP56 THEN
BEGIN  (* STEPS 6 AND 7 - BACKTRACKING  *)
   STOP:=TRUE;  K:=I;
   WHILE STOP AND (K > 1) DO
   BEGIN  K:=K-1;  STOP:=Y[K]=0  END;
   STEP7:= NOT STOP;
   IF STEP7 THEN
   BEGIN
      R:=V;  Y[K]:=0;
      Q:=Q-P[K];  V:=V+W[K];
      STEP2:=R >= MIN[K];
      STEP7:= NOT STEP2;
      IF STEP2 THEN I:=K+1
      ELSE BEGIN  I:=K;  M:=K+1  END;
   END;
   WHILE STEP7 DO
   BEGIN  (* STEP 7 - PROCESS OF SUBSTITUTION  *)
      STEP56:=(M > N) OR (PROFIT >= Q+TRUNC(V*P[M]/W[M]));
      STEP7:= NOT STEP56;  STEP2:=STEP7;  STEP4:=STEP7;
      IF STEP7 THEN
      BEGIN
         D:=W[M]-W[K]; T:=R-D;
         IF D = 0 THEN M:=M+1
         ELSE
            IF D > 0 THEN
            BEGIN
               IF (T < 0) OR (PROFIT >= Q+P[M]) THEN
                  M:=M+1
               ELSE
               BEGIN
                  PROFIT:=Q+P[M];
                  FOR J:=1 TO K DO X[J]:=Y[J];
                  FOR J:=K+1 TO N DO X[J]:=0;
                  X[M]:=1;
                  STOP:=PROFIT=LIM;
                  STEP7:= NOT STOP;
                  IF STEP7 THEN
                  BEGIN
                     R:=T;  K:=M;
                     M:=M+1
```

```
                         END
                    END
               END  (*  D > 0  *)
               ELSE
               BEGIN  (*  D < 0  *)
                    IF  T < MIN[M] THEN M:=M+1
                    ELSE
                    BEGIN
                         STEP7:=FALSE;
                         STEP56:=Q+P[M]+TRUNC(T*P[M]/W[M])
                                   <= PROFIT;
                         STEP2:= NOT STEP56;  STEP4:=STEP2;
                         IF STEP2 THEN
                         BEGIN
                              Q:=Q+P[M];  V:=V-W[M];
                              Y[M]:=1;  I:=M+1;
                              PD[M]:=P[M];  WD[M]:=W[M];
                              ZD[M]:=M+1;
                              FOR J:=M+1 TO LL DO
                              BEGIN
                                   PD[J]:=0;  WD[J]:=0;
                                   ZD[J]:=J
                              END;
                              LL:=M
                         END  (*  STEP2  *)
                    END  (*  T >= MIN[M]  *)
               END  (*  D < 0  *)
          END  (*  IF STEP7  *)
        END  (*  WHILE STEP7  *)
      END  (*  IF STEP56  *)
    UNTIL STOP
  END  (*  NOT STOP  *)
END;  (*  KNAPBACKTRACK  *)
```

2.1.6. COMPUTATIONAL RESULTS

In this section we present the results of some computational experiments carried out with the algorithms described in previous sections.

Since the computation times are strongly dependent on the correlation between profits p_i and weights w_i, five uniformly distributed randomly generated data sets, with different degrees of correlation, have been considered:

Uncorrelated:	UCR1:	$p_i, w_i \in [0, 100]$	$i = 1, 2, \ldots, n$
	UCR2:	$p_i, w_i \in [50, 100]$	$i = 1, 2, \ldots, n$
Correlated:	WCR1:	$p_i = 1.2 w_i$	$i = 1, 2, \ldots, n$
	WCR2:	$p_i = w_i$	$i = 1, 2, \ldots, n$
	SCR:	$p_i = w_i + 10$	$i = 1, 2, \ldots, n$

Table 2-2. Computational Results from the Application of the
HEAPSORT and KNAPRED Procedures[a]

	Data	n						
		50	100	150	200	300	400	500
Sorting time (HEAPSORT)	All	2	6	10	13	21	30	39
Reduction time (KNAPRED)	UCR1 + V1	5	14	26	44	90	154	232
	UCR2 + V1	6	15	30	49	102	175	270
	UCR1 + V2	3	9	16	25	46	74	110
	WCR2 + V1	5	17	32	56	113	193	300
	WCR1 + V1	6	20	40	70	130	230	385
	WCR2 + V2	5	9	17	28	56	90	145

[a]Entries are average times (in milliseconds on Amdahl 470 V/8) for 100 and 10 problems, respectively.

Two different values of V were used:

$$\text{V1:} \quad V = 0.5 \sum_{i=1}^{n} w_i \qquad \text{V2:} \quad V = 0.8 \sum_{i=1}^{n} w_i$$

The algorithms require the objects to be sorted according to nonincreasing values of p_i/w_i. Table 2-2 shows the average sorting times of the HEAPSORT procedure,* which depend mainly on the number of objects. This table gives also computational times of the procedure KNAPRED. One can notice higher times for the shorter interval (UCR2) and 50% reduction for the greater total knapsack capacity (V2). The percentage of remaining objects was highly dependent on all factors of the data. There was no reduction at all in the case of the value-independent instances (WCR2). For weakly correlated data, the percentage of remaining objects was in the range 30 to 100%, and in the range 5 to 10% and 5 to 30% for uncorrelated instances with the total capacity V1 and V2, respectively.

Table 2-3 shows the results from the application of the approximation algorithm. All the solutions obtained by the KNAPAPPROX procedure were within 2% of the optimal solution value.

Tables 2-4 compare the enumeration algorithms of Horowitz–Sahni (HS) and of Martello–Toth (MT). The algorithms have been run both directly and with previously applied the reduction procedure (R).

The use of the reduction algorithm is always beneficial to the algorithm HS, whereas the algorithm MT definitely does not benefit from its application. The

---

*We used a modified version of the *heapsort* procedure published in N. Wirth, *Algorithms + Data Structures = Programs*, Prentice-Hall, Englewood Cliffs, N.J., 1976, p. 75.

Table 2-3.   Computational Results from the Application of the KNAPAPPROX Procedure[a]

n	10	30	50	100	150	200
Average Time[b]	1	40	180	1400	4600	11,000
Performance	16/40	10/30	4/30	2/10	1/10	1/10

[a] The third row shows the ratio of the number of instances that have not been solved optimally to the number of all instances run.
[b] In milliseconds on Amdahl 470 V/8.

Table 2-4.   Computational Results from the Application of the KNAPRED (Algorithm R) and KNAPBACKTRACK (Algorithm MT) Procedures and Algorithm 2-5 (Algorithm HS) to Random Instances of the Knapsack Problem[a]

Data:	UCR1 + V1				UCR2 + V2			
Algorithm: n	HS	MT	R + HS	R + MT	HS	MT	R + HS	R + MT
50	6	2	7	6	10	3	5	5
100	20	6	16	16	31	5	11	9
250	103	12	70	68	175	13	65	65
500	392	27	250	240	627	18	120	111

Data:	SCR + V1			
Algorithm: n	HS	MT	R + HS	R + MT
10	2	1	3	2
30	74	22	74	23
50	749	117	748	117
100	5400	409	5380	411

Data:	WCR1 + V1				WCR2			
					V1		V2	
Algorithm: n	HS	MT	R + HS	R + MT	HS	MT	HS	MT
50	3	1	9	7	7	2	15	2
100	7	1	25	21	13	1	26	2
250	32	4	130	110	60	4	139	5
500	115	9	460	400	225	7	430	10

[a] Entries are average times (in milliseconds on Amdahl 470 V/8) for 10 problems.

magnitude of the total knapsack capacity can influence the computational time as much as 100% in the case of the value-independent instances (WCR2); however, the range of parameters (UCR1 and UCR2) influenced only slightly the performance (the computational results are not included here).

In all the cases tested, the algorithm of Martello and Toth turned out to be superior to that of Horowitz and Sahni.

PROBLEMS

2-1.   **(a)** Show that every instance of the 0–1 knapsack problem

$$\max\left\{\sum_{i=1}^{n} p_i x_i : \sum_{i=1}^{n} w_i x_i \leqslant V, \ x_i = 0 \text{ or } 1, \ i = 1, 2, \ldots, n\right\}$$

where p_i, w_i $(i = 1, 2, \ldots, n)$ and V are arbitrary integer numbers, can be transformed into an equivalent instance of the following 0–1 knapsack problem:

$$\max\left\{\sum_{i=1}^{m} q_i y_i : \sum_{i=1}^{m} v_i y_i \leqslant W, \ y_i = 0 \text{ or } 1, \ i = 1, 2, \ldots, m\right\}$$

where $m \leqslant n$

$$\sum_{i=1}^{m} v_i > W > 0$$

$$v_i \leqslant W, q_i > 0, v_i > 0, i = 1, 2, \ldots, m$$

(b) Write a Pascal subroutine that performs the transformation above.
(c) Apply the transformation above to the following instance:

maximize $-x_1 + 2x_2 - 3x_3 + x_4 + 5x_5$

subject to $2x_1 - 2x_2 - x_3 + 2x_4 - 3x_5 \leqslant 10$

$$x_i = 0 \text{ or } 1, \qquad i = 1, 2, \ldots, 5$$

2-2.   Similarly to Problem 2-1(a), show that without loss of generality, the following assumptions can be made about the parameters of the 0–1 many-knapsack problem:

$$p_i > 0, \quad w_i > 0, \qquad i = 1, 2, \ldots, n$$
$$V_j > 0, \qquad\qquad\qquad j = 1, 2, \ldots, m$$

$$\min_i \{w_i\} \leqslant \min_j \{V_j\}$$

$$\max_i \{w_i\} \leqslant \max_j \{V_j\}$$

$$\sum_{i=1}^{n} w_i > \max_j \{V_j\}$$

2-3. Show how to transform the bounded and the unbounded single-knapsack problems into equivalent 0–1 single-knapsack problems. (*Hint:* Replace each integer variable by a finite set of binary variables.)

2-4. Show that the 0–1 many-knapsack problem is a special instance of the 0–1 multidimensional knapsack problem.

2-5. The change-making problem (2-7). Assume that $w_1 > w_2 > \ldots > w_n$.

   (a) Show that if $w_n > 1$, there exists an instance of the problem (i.e., a finite set of coin types and V) for which there is no solution to (2-7).

   (b) Show that if $w_n = 1$, there is always a solution.

   (c) When $w_n = 1$, a greedy algorithm assigns change by using the coin types in the order $w_1, w_2, \ldots, w_n$. When the ith coin type is considered, as many coins of this type as possible are chosen. Write a Pascal routine based on this strategy and show that this algorithm does not necessarily produce optimal solutions.

2-6. (Kendall and Zionts [1977])

   (a) Prove the following fact: The constraints

$$\sum_{i=1}^{n} a_{ji} x_i - b_j = 0 \quad \left[g_j(\mathbf{x}) = 0 \right] \qquad j = 1, 2$$

where $0 \leqslant x_i \leqslant u_i$, x_i an integer, $i = 1, 2, \ldots, n$, are equivalent to

$$\lambda_1 g_1(\mathbf{x}) + \lambda_2 g_2(\mathbf{x}) = 0$$

if the multipliers λ_1 and λ_2 satisfy the following conditions:

1. For any integer solution values $0 \leqslant x_i \leqslant u_i$, $i = 1, 2, \ldots, n$, either λ_1 does not divide $g_2(\mathbf{x})$ or λ_2 does not divide $g_1(\mathbf{x})$.
2. λ_1 and λ_2 are relatively prime integers.

(b) Check whether this theorem yields smaller coefficients for the equations of Example 2-1.

2-7. Prove Theorem 2-2.

2-8. Prove Theorem 2-3 (see Dantzig [1957] for the original proof and Horowitz and Sahni [1978, p. 160] for a combinatorial proof).

2-9. (Hudson: see Martello and Toth [1979a]) Let s be the largest index for which

$$\sum_{i=1}^{s} w_i \leqslant V - w_{r+1}$$

where r is determined in Theorem 2-3. Define

$$U'' = p_{r+1} + \sum_{i=1}^{s} p_i + \left\lfloor \left(V - w_{r+1} - \sum_{i=1}^{s} w_i \right) p_{s+1}/w_{s+1} \right\rfloor$$

and

$$U_3 = \max\{U_1, U''\}$$

where U_1 is defined by (2-14). Prove that U_3 is an upper bound for the solution of the knapsack problem and

$$U_3 \leqslant U_2$$

2-10. (Müller–Merbach [1978]) Let p_i^* be the dual variables of the optimal continuous solution of Theorem 2-3, that is,

$$p_i^* = p_i - w_i p_{r+1}/w_{r+1} \qquad \text{for} \quad i = 1, 2, \ldots, r$$
$$p_i^* = -p_i + w_i p_{r+1}/w_{r+1} \qquad \text{for} \quad i = r + 2, \ldots, n$$

Let

$$U_4 = \sum_{i=1}^{r} p_i + \left\lfloor \left(V - \sum_{i=1}^{r} w_i \right) p_{r+1}/w_{r+1} \right.$$
$$\left. - \min\left\{ \left(V - \sum_{i=1}^{r} w_i \right) p_{r+1}/w_{r+1}, \min_i \{ p_i^* : i \neq r + 1 \} \right\} \right\rfloor$$

(a) Prove that U_4 is an upper bound for the solution of the knapsack problem.

(b) Show that $U_4 \leqslant U_1$, but U_4 can be less, equal, or greater than both U_2 and U_3.

2-11. Construct an instance of the knapsack problem for which Algorithm 2-1, with $J = \varnothing$ and $K = \varnothing$, does not produce an optimal solution for the following orderings of items:

(a) According to nonincreasing profits p_i.

(b) According to nonincreasing p_i/w_i.

(c) According to nondecreasing weights w_i.

2-12. (Fayard and Plateau, and Hansen: see Lauriere [1978])   Let $\bar{P}$ denote the value of the optimal solution to the continuous knapsack problem (see Theorem 2-3), and P_0 be the value of the best known feasible solution. The values

$$q_i = p_i - w_i \frac{p_{r+1}}{w_{r+1}} \qquad (i = 1, 2, \ldots, n, i \neq r + 1)$$

are the *marginal* or (*shadow*) costs calculated for the pivot element x_{r+1}.

(a) Prove that if for $i \neq r + 1$,

$$|q_i| \geq \bar{P} - P_0$$

then x_i has the same value in the optimal integer solution as in the optimal continuous solution.

(b) Show that the test described in Theorem 2-4 eliminates at least as many variables as the test above; that is, for $i \in \{1, 2, \ldots, n\}$,

$$|q_i| \geq \bar{P} - P_0 \quad \text{implies that} \quad \lfloor \bar{P}(x_i = \varepsilon_i) \rfloor < P_0$$

2-13. Apply Algorithm 2-2 to the knapsack instance of Example 2-2 with $V = 42$ and $V = 43$.

2-14. (Brucker [1979])   Given are two bars each of length L from which we have to cut off as many as possible smaller bars of length $l_1, l_2, \ldots, l_n$. Prove that the following heuristic algorithm generates absolute 1-approximate solutions.

```
begin
        order bars as to satisfy l_i ≤ l_{i+1} (i = 1, 2, ..., n - 1);
        i ← 1;
        for j ← 1, 2 do
        begin
            s ← 0;
            while (s + l_i ≤ L) and (i ≤ n) do
            begin
                cut length l_i from bar j;
                s ← s + l_i;
                i ← i + 1
            end
        end
    end
```

(*Hint:* Let p and q denote the number of bars cut by the algorithm from two bars each of length L and from one bar of length $2L$. Show first that $p \leqslant k \leqslant q$, where k is the optimal solution, and then prove that $q - 1 \leqslant p$.)

2-15.   (Brucker [1979])   The following algorithm B finds a heuristic solution for the unbounded single-constrained knapsack problem.

```
begin
    order the objects according to nonincreasing p_i/w_i;
    s ← 0;
    for i ← 1 to n do
    begin
        x_i ← ⌊(V − s)/w_i⌋;
        s ← s + w_i x_i
    end
end
```

(a) Prove that it is a $\frac{1}{2}$-approximate algorithm. [*Hint:* Show first that for every instance $\mathcal{I}$, $P_{\text{opt}}(\mathcal{I}) \leqslant p_1(V/w_1)$ and $p_1 \lfloor V/w_1 \rfloor \leqslant P_B(\mathcal{I})$.]

(b) Design a sequence of unbounded knapsack instances for which (2-17) in the case of algorithm B approaches $\frac{1}{2}$ for large n.

2-16.   Show that the algorithm of Problem 2-15 is not an ε-approximate algorithm for the 0–1 knapsack problem.

2-17.   Shapiro [1968a] was the first who formulated the knapsack problem as a shortest path problem. Let us consider the unbounded knapsack problem with an equality constraint. We assume that all weights are integer and satisfy $w_i \neq w_j$ for $i \neq j$. The directed graph corresponding to such a problem has $V + 1$ nodes $v_0, v_1, v_2, \ldots, v_m$ where $m = V$ and (v_k, v_l), where $0 \leqslant k < l \leqslant V$, is an edge if there exists i $(1 \leqslant i \leqslant n)$ such that $l - k = w_i$. An edge (v_k, v_l) which satisfies $l - k = w_i$ has the weight $-p_i$.

(a) Show that the network constructed above is acyclic and the optimal solution to the knapsack problem is given by a shortest path from 0 to V_m.

(b) Why can this method not be applied to the bounded and binary knapsack problems?

(c) Show how to modify the network for an inequality constraint.

(d) Construct the network for the following instance of the unbounded knapsack problem with an inequality constraint:

$$n = 3, p_1 = 3, p_2 = 2, p_3 = 4$$

$$w_1 = 3, w_2 = 4, w_3 = 2, V = 7$$

2-18.   Show how using the dynamic programming approach to solve the knapsack problem a network can be assigned to the bounded knapsack problem.

2-19.   Dynamic programming algorithm for solving the bounded knapsack problem

    (a) Define the backward recurrence relations [similar to (2-18)] for the bounded knapsack problem (assume that all problem parameters are nonnegative). Consider also the bounds on variables resulting from the linear constraint.

    (b) Apply the relations derived above to the following knapsack instance:

$$\text{maximize} \qquad 3x_1 + x_2 + 2x_3 + 6x_4$$
$$\text{subject to} \qquad 4x_1 + 3x_2 + 2x_3 + 3x_4 \leqslant 16$$
$$0 \leqslant x_1 \leqslant 5, \quad 0 \leqslant x_2 \leqslant 1, \quad 0 \leqslant x_3 \leqslant 5, \quad 0 \leqslant x_4 \leqslant 3$$

    (c) Write a Pascal subroutine for solving the bounded knapsack problem which implements the dynamic recurrence relations derived in part (a).

2-20.   **(a)** Show how to modify the recurrence relations (2-18), (2-19), and that derived in Problem 2-18 for the knapsack problems with an equality constraint.

    (b) Apply the modified relations to the instance of Problem 2-19(b).

2-21.   Implement the recurrence relations (2-19) in the form of a Pascal procedure.

2-22.   Solve the knapsack instance given in Example 2-2 by using both implicit enumeration algorithms: Algorithm 2-5, and its improved version.

2-23.   (Greenberg and Hegerich [1970])   The implicit enumeration methods presented in the text work on static search trees; that is, the ith variable corresponds to the ith level of a search tree and the order of variables is given a priori and does not depend on the problem parameters.

    A method utilizing a *dynamic search tree* works as follows. First the knapsack problem with the relaxed constraints $0 \leqslant x_i \leqslant 1$ (for all $i = 1, 2, \ldots, n$) is solved. If the solution is integer ($x_i = 0$ or 1), it is also an optimal solution to the original problem. Otherwise, let x_i be the variable that satisfies $0 < x_i < 1$ in the solution to the relaxed problem. Now, the family of all solutions to the original problem is partitioned into two subfamilies (subtrees), the left with $x_i = 0$ and the right with $x_i = 1$. Thus the x_i corresponds to the first level of the search tree; the left subtree corresponds to $x_i = 0$ and the right to $x_i = 1$. In general, at each node of the tree, the greedy algorithm is applied to solve the relaxed problem with the restrictions corresponding to the assignments of variables made along the path from the root to this node. If the solution is all-integer, an optimal solution for this node has been found. If not, there exists one x_j such that $0 < x_j < 1$ and again the left son of the node corresponds to $x_j = 0$ and the right to $x_j = 1$.

    A candidate node for branching is chosen in a depth-first manner, similarly as in the algorithms described in the text.

(a) Work out a complete description of the algorithm. In particular, justify the partitioning scheme and the assignment of $x_i = 0$ to the left branches.

(b) Apply the algorithm to the knapsack instances of Examples 2-2 and 2-6 (notice that in the latter case the variables are not properly ordered).

(c) Implement the algorithm in Pascal and compare its performance with that of the KNAPBACKTRACK procedure.

REFERENCES AND REMARKS

In almost every book on integer programming a separate chapter is devoted to the knapsack problem. We recommend especially

GARFINKEL, R. S., and G. L. NEMHAUSER, *Integer Programming*, Wiley, New York, 1972, Chapter 6

SALKIN, H. M., *Integer Programming*, Addison-Wesley, Reading, Mass., 1975, Chapter 10

In the book

HOROWITZ, E., and S. SAHNI, *Fundamentals of Computer Algorithms*, Computer Science Press, Potomac, Md., 1978

the 0–1 knapsack problem is used to illustrate different algorithmic techniques, such as the greedy method, dynamic programming, backtracking, branch-and-bound, and approximate algorithms. We also refer the reader to the current review paper

MARTELLO, S., and P. TOTH, The 0–1 Knapsack Problem, in N. Christofides, A. Mingozzi, P. Toth, and C. Sandi (eds.), *Combinatorial Optimization*, Wiley, New York, 1979a, pp. 237–279

Although the knapsack problem is one of the simplest instances of integer linear programming problems, it may represent many practical situations such as cargo loading, cutting stock, capital budgeting, and several others. The loading problem is discussed in

EILON, S., and N. CHRISTOFIDES, The Loading Problem, *Management Sci.* 17(1971), 259–268

See also the following paper dealing with the loading problems in which liquid materials are loaded into tanks:

CHRISTOFIDES, N., A. MINGOZZI, and P. TOTH, Loading Problems, in N. Christofides, A. Mingozzi, P. Toth, and C. Sandi (eds.), *Combinatorial Optimization*, Wiley, New York, 1979, pp. 339–369

The *cutting-stock problem* is another way of looking at the loading problem. The problem is to find the most profitable pattern of cutting some material (e.g., fabric, glass, steel) into pieces of different shapes, dimensions, and values. The main reference is a series of papers by P. C. Gilmore and R. E. Gomory who developed the theory of knapsack functions based on the dynamic programming approach. See the following paper and its references:

GILMORE, P. C., and R. E. GOMORY, The Theory and Computation of Knapsack Functions, *Oper. Res.* 14(1966), 1045–1079

The simplest version of the loading and cutting-stock problem, known as the bin-packing problem, is discussed in Problems 4-86 and 4-87.

The *capital budgeting problem* in its simplest form is to find among various competing investment possibilities a subset that maximizes the total payoff and does not exceed the available funds. The standard text for this problem is

WEINGARTNER, H., *Mathematical Programming and the Analysis of Capital Budgeting Problems*, Prentice-Hall, Englewood Cliffs, N.J., 1963

See also

WEINGARTNER, H., Capital Budgeting and Interrelated Projects: Survey and Synthesis, *Management Sci.* 12(1968), 485–516

A recursive algorithm for the change-making problem (with $w_n = 1$) is given in

CHANG, S. K., and A. GILL, Algorithmic Solution of the Change-Making Problem, *J. ACM* 17(1970), 113–122

and a dynamic programming method is proposed in

WRIGHT, J. W., The Change-Making Problem, *J. ACM* 22(1975), 125–128

Some computational results with different methods for solving the change-making problem are reported in Section A4 of Martello and Toth [1979a].

G. B. Mathews in 1897 showed how to aggregate two equations with positive integer coefficients into a single equation which has the same set of nonnegative integer solutions (see Salkin [1975, p. 393] for the formulation and proof of Mathews's theorem). Then, based on this result, Elmaghraby and Wing (as reported in Salkin [1975]) gave the rules for transforming the system of inequality constraints in bounded integer variables into a single constraint equation.

The basic idea behind all the aggregation methods is to find an equivalent single-constrained problem that is much easier to solve, and several other aggregation techniques have been proposed that tend to produce constraints with smaller coefficients. One of the best transformation procedures appeared in

GLOVER, F., New Results for Reducing Linear Programs to Knapsack Problems, Management Science Report Series 72-7, University of Colorado, Boulder, Apr., 1972

See also Salkin [1975, pp. 398–401] for the description of Glover's algorithm.

Problem 2-6 is taken from the following paper, which integrates some of the previous results and discusses their implementations:

KENDALL, K. E., and S. ZIONTS, Solving Integer Programming Problems by Aggregating Constraints, *Oper. Res.* 25(1977), 346–351

The relations between linear and discrete programming problems have been discussed for the first time in

DANTZIG, G., Discrete-Variable Extremum Problems, *Oper. Res.* 5(1957), 266–277

Several other upper bounds for the 0–1 knapsack problem are discussed in Martello and Toth [1979a]; see also

MÜLLER - MERBACH, H., An Improved Upper Bound for the Zero–One Knapsack Problem: A Note on the Paper by Martello and Toth, *European J. Oper. Res.* 2(1978), 212–213

The greedy method for the knapsack problem implemented in Algorithm 2-1 is described in Chapter 4 of Horowitz and Sahni [1978].

The first reduction method for the knapsack problem appeared in

INGARGIOLA, G. P., and J. F. KORSH, A Reduction Algorithm for Zero–One Single Knapsack Problems, *Management Sci.* 20(1973), 460–463

and then has been improved by Toth and by Dembo and Hammer (see Martello and Toth [1979a]). For comparison of the efficiency of different reduction methods, see

FAYARD, D., and G. PLATEAU, Reduction Algorithms for Single and Multiple Constraints 0–1 Linear Programming Problems, presented at the conference "Methods of Mathematical Programming," Zakopane, Poland, 1977

LAURIERE, M., An Algorithm for the 0/1 Knapsack Problem, *Math. Programming* 14(1978), 1–10

Approximation algorithms for the knapsack problem and other NP-hard and NP-complete problems are discussed in Chapter 12 of Horowitz and Sahni [1978]. An excellent bibliography on these algorithms is

GAREY, M. R., and D. S. JOHNSON, Approximation Algorithms for Combinatorial Problems: An Annotated Bibliography, in J. F. Traub (ed.), *Algorithms and Complexity: New Directions and Recent Results*, Academic Press, New York, 1976, pp. 41–52

Algorithm 2-3 appeared in

SAHNI, S., Approximate Algorithms for the 0/1 Knapsack Problem, *J. ACM* 22(1975), 115–124

Another ε-approximate algorithm for the knapsack problem whose computing time is a polynomial both in n and in $1/\varepsilon$ has been presented in (such an algorithm is called a *fully polynomial time approximation scheme*)

IBARRA, O. H., and C. E. KIM, Fast Approximation Algorithms for the Knapsack and Sum of Subset Problems, *J. ACM* 22(1975), 463–468

and improved in

LAWLER, E. L., Fast Approximation Algorithms for Knapsack Problems, *Math. Oper. Res.* 4(1979), 339–356

Approximation algorithms for the knapsack problem and its variations are also presented in

BRUCKER, P., NP-Complete Operations Research Problems and Approximation Algorithms, *Z. Oper. Res.* 23(1979), 73–94

The first formulation of the unbounded knapsack problem as a shortest-path problem (discussed in Problem 2-17) appeared in

SHAPIRO, J. F., and H. M. WAGNER, Finite Renewal Algorithm for the Knapsack and Turnpike Models, *Oper. Res.* 15(1967), 319–341

See also

FRIEZE, A. M., Shortest Path Algorithms for Knapsack Type Problems, *Math. Programming* 11(1976), 150–157

SHAPIRO, J. F., Dynamic Programming Algorithms for the Integer Programming Problems I: The Integer Programming Problem Viewed as a Knapsack Type Problem, *Oper. Res.* 16(1968a), 103–121

SHAPIRO, J. F., Shortest Route Methods for Finite State Space Deterministic Dynamic Programming Problems, *SIAM J. Appl. Math.* 16(1968b), 1232–1250

The network formulation of the bounded knapsack problem is the subject of Problem 2-18, and it can be derived from the dynamic programming recursion.

A general reference for dynamic programming techniques and their applications is

BELLMAN, R., and S. DREYFUS, *Applied Dynamic Programming*, Princeton University Press, Princeton, N.J., 1962

We also suggest Chapter 5 of Horowitz and Sahni [1978], where an improved algorithm based on relations (2-19) is presented, and Martello and Toth [1979a]. One of the most efficient dynamic programming algorithms for the knapsack problem has appeared recently in

TOTH, P., Dynamic Programming Algorithms for the Zero–One Knapsack Problem, *Computing* 25(1980), 29–45

The first enumeration method for the knapsack problem was a breadth-first branch-and-bound algorithm proposed in

KOLESAR, P. J., A Branch and Bound Algorithm for the Knapsack Problem, *Management Sci.* 13(1967), 723–735

The memory and time requirements of Kolesar's algorithm have been significantly reduced by a depth-first branch-and-bound algorithm which is the subject of Problem 2-23 and appeared in

GREENBERG, H., and R. L. HEGERICH, A Branch Search Algorithm for the Knapsack Problem, *Management Sci.* 16(1970), 327–332

Algorithm 2-5 is due to Horowitz and Sahni (see Horowitz and Sahni [1978]) originally published in

HOROWITZ, E., and S. SAHNI, Computing Partitions with Applications to the Knapsack Problem, *J. ACM* 21(1974), 277–292

The same algorithm has been independently obtained by several other authors (see comments in Martello and Toth [1979a]). The revised version of Algorithm 2-5 has been presented in

MARTELLO, S., and P. TOTH, An Upper Bound for the Zero–One Knapsack Problem and a Branch and Bound Algorithm, *European J. Oper. Res.* 1(1977), 169–175

(see also Martello and Toth [1979a]). The KNAPBACKTRACK procedure is a Pascal version (without **goto** statements) of the FORTRAN subroutine published in

MARTELLO, S., and P. TOTH, Algorithm 37. Algorithm for the Solution of the 0–1 Single Knapsack Problem, *Computing* 21(1979b), 81–86

A very efficient algorithm for large knapsack problems which does not require the objects to be sorted has been proposed in

BALAS, E., and E. ZEMEL, An Algorithm for Large Zero–One Knapsack Problems, *Oper. Res.* 28(1980), 1132–1154

Recently, a new sophisticated FORTRAN code for solving the knapsack problem appeared in

FAYARD, D., and G. PLATEAU, Algorithm 47. An Algorithm for the Solution of the 0–1 Knapsack Problem, *Computing* 28(1982), 269–287.

The algorithm implemented there combines several ideas elaborated by its authors in early works. The computational results reported in this paper show that the code is superior to that presented in Martello and Toth [1979b].

Computational experiments with algorithms for knapsack problems are reported in almost every paper presenting a new algorithm. We recommend particularly the paper by Martello and Toth [1979a].

2.2. COVERING PROBLEMS

2.2.1. PROBLEMS FORMULATIONS AND APPLICATIONS

Formulations

The problems discussed in this section are most naturally formulated in terms of sets.

Let $M = \{1, 2, \ldots, m\}$ be a *base* set and

$$\mathscr{P} = \{P_1, P_2, \ldots, P_n\} \quad \text{or simply} \quad \mathscr{P} = \{P_j\}_{j \in N}$$

where $N = \{1, 2, \ldots, n\}$, be a collection of subsets of M; that is, every P_j satisfies $P_j \subseteq M$. Let $F_i = \{j \colon i \in P_j, \ j = 1, 2, \ldots, n\}$ for all $i = 1, 2, \ldots, m$; therefore, F_i consists of those sets P_j which contain element i. Let h_j denote $|P_j|$, the cardinality of P_j.

A subfamily $\mathscr{Q} \subseteq \mathscr{P}$ is a *set packing* if all members of $\mathscr{Q}$ are mutually disjoint. If a set packing covers all elements of the base set M, then it is called a *set partitioning*. A *set covering* is a subfamily that covers all elements of M (members of a set covering are not necessarily disjoint).

Let **c** be a nonnegative (cost or profit) weight function (vector) defined on N; that is, every subset P_j may cost or produce a profit.

A *set-packing problem* (SP) is to find a set packing of maximum weight (profit), the *set-partitioning problem* (SPP) is to find a set partitioning of minimum weight (cost), and the *set-covering problem* (SC) is to find a set covering of minimum weight (cost).

One may define these problems in a uniform way as follows: Find a maximum-profit (or minimum-cost) collection $\mathcal{Q}$ of members of $\mathcal{P}$ such that every element of M is contained in at most exactly or at least one of the members of $\mathcal{Q}$. Occasionally, we will refer to these problems as *covering problems*. Equivalently, covering problems can be formulated as special integer programming problems.

Let $A = [a_{ij}]$ be an $m \times n$ element-subset incidence matrix of $\mathcal{P}$. That is, rows of A correspond to elements of M, columns of A correspond to members of $\mathcal{P}$, and $a_{ij} = 1$ if $i \in P_j$ and $a_{ij} = 0$ otherwise. Let R_i ($i = 1, 2, \ldots, m$) denote the ith row of A and let C_j ($j = 1, 2, \ldots, n$) denote the jth column of A. Therefore, R_i and C_j may be considered as the characteristic vectors of F_i and P_j, respectively. This allows us to refer to sets as *columns* and to base elements as *rows*, and we will use these terms interchangeably.

Now the covering problems can be formulated as follows:

maximize
$$c^T x$$

subject to
$$A x \leqslant e \tag{SP}$$
$$x_j = 0 \text{ or } 1, \quad j = 1, 2, \ldots, n$$

minimize
$$c^T x$$

subject to
$$A x = e \tag{SPP}$$
$$x_j = 0 \text{ or } 1, \quad j = 1, 2, \ldots, n$$

minimize
$$c^T x$$

subject to
$$A x \geqslant e \tag{SC}$$
$$x_j = 0 \text{ or } 1 \quad j = 1, 2, \ldots, n$$

where e is the m-dimensional vector of 1's.

The set-partitioning problem can be viewed either as the problem of finding a set packing which covers all elements of the base set or the problem of covering the base set by mutually disjoint subsets. Notice that unlike the other two problems, the set-partitioning problem may not have a feasible solution. Similarly to the knapsack problem, we may assume that the weight vector c is nonnegative (see Problems 2-24 and 2-25). There are also possible some relaxations of the constraints which follow from the linear programming theory. However, we do not discuss them here, since the computational methods presented in this section are purely combinatorial. The reader interested in this topic is referred to Problems 2-26, 2-27, and 2-28 and to the literature discussed in the References and Remarks.

These three covering problems are closely related to each other in a sense that one can be brought to the form of another. Such transformations are illustrated in Problems 2-29, 2-30, and 2-31.

The most popular special instances of the covering problems are those defined on special discrete structures such as graphs and networks. The interested reader is referred to Problems 2-32 through 2-41 and also to the graph-theoretic literature. The problem of packing edges in a network known as the matching problem is discussed extensively in Section 3.7.

There exists also a reverse relation between the set-packing problem and the edge-packing problem on networks which allows one to convert the former problem into the latter one (see Problem 2-38).

After discussing some applications of the covering problems, the remainder of this section is focused mainly on the set-partitioning problem. First, we present general reduction rules for the SPP and the SC problems, and their implementation for the former problem. Unlike the knapsack problem and the other two problems of this section, the set-partitioning problem cannot have a polynomial-time approximation algorithm, since the problem of finding if it has a feasible solution is NP-complete. However, we present two approximation algorithms for the SC problem (see Problems 2-48 and 2-49). Finally, an implicit enumeration method for the SPP problem is presented, which can easily be modified for the SC problem.

Applications

Numerous real-world situations can be modeled as covering problems. They include such applications as airline crew scheduling, truck delivery, political redistricting, facility location, circuit design, and even symbolic logic.

The *airline crew scheduling problem* involves generating rotations for the flying crews. In the beginning, an airline works out a flight schedule and then cabin and flight personnel have to be assigned to each flight. These assignments have to be consistent with several restrictions and regulations, and the airlines intend to minimize the cost of service. Flight legs can usually be combined into round trips called *rotations*, starting at a crew base and lasting not longer than a fixed period of time.

The schedule of flights may be used to generate all feasible rotations, whose number, even for medium-size airlines, goes in thousands. Each rotation can be assigned the cost of the crew operating, and the problem is to find a partition of all flight legs into rotations which are of minimum total cost to the airlines. In terms of covering problems, the base set consists of flight legs and the collection of subsets $\mathscr{P}$ is formed by all feasible rotations. Therefore, the constraint matrix is the incidence matrix of legs versus rotations and the problem is equivalent to SPP with such a matrix.

In several situations, which can be observed on board, a crew flies deadhead as passengers, due to loosened restrictions which can be modeled by the set-covering problem. However, in such an approach several difficulties arise. First, the cost of a rotation involving a deadhead differs from that without deadheading, and second, deadheading is usually permitted both on an arbitrary craft and on other airlines.

The importance and the degree of difficulty in solving the airline crew scheduling problem can be appreciated if we realize that even medium-size airlines have to schedule hundreds of legs, which can be combined into thousands of rotations.

A ground version of the airline crew scheduling problem is called the *delivery problem* and concerns the optimal (minimum-cost) transportation of goods by a carrier from warehouse to a number of customers. All customer orders are first combined into feasible round trips of the carrier, and then the corresponding set (customer)-partitioning problem is constructed and solved to find an optimal way of meeting all orders. Again, if transits are allowed, the problem can be modeled as that of finding a covering of all orders by a set of round trips.

Some *facility location problems* can be also formulated as covering problems. In the simplest form, a graph is given, and a subset of nodes must be found such that every node is either in the subset or is adjacent to one of the subset nodes. In general, for a distance function defined on the edge set, we may ask for a subset of nodes which are at distance at most d from the other nodes. The covering problems on networks are the subject of Problems 2-23 through 2-39.

Another group of applications are the problems of designing logic or switching circuits and fault testing. In *designing a switching circuit*, we are given the set of k inputs of 0–1 symbols and one binary output which depends only on a function of the 2^k possible input vectors. Given is also the switching function or equivalently a truth table which specifies the output for each of the 2^k inputs. The problem is to construct a cheapest circuit which realizes the function and is built only of the "and" and "or" gates. This problem can be reformulated as that which asks for a simplest disjunctive normal form equivalent to a given formula of Boolean algebra.

The *problem of fault testing* is to find a cheapest set of tests which will be able to discover a network (or circuit) failure. A test in this case corresponds to a path in the network, and can discover only faults which are at the nodes of the path. Therefore, the problem is to find an optimal family of paths between specified nodes which cover all nodes of the network.

2.2.2.   REDUCTION ALGORITHMS

It is a common technique in dealing with hard discrete problems to reduce their size by applying a simple, polynomial-time reduction algorithm. Such a method has been described and implemented for the knapsack problem (see Section 2.1.3). We shall now present a collection of reduction rules for the SC and SPP problems and then implement them for SPP.

The reductions are described by referring to the constraint matrix A of the problem, although the implementation also makes use of the list representations of the lines of the matrix. The matrix of a current problem (possibly after some reductions) is defined by two sets M and N which are, respectively, the set of uncovered elements (rows) and the index set of unused sets (columns). Rows and columns of a current matrix contain those elements which belong to columns and rows in N and M, respectively. Set D, empty in the beginning, contains those columns which appear in every optimal solution; therefore, after the reductions, columns in $\{1, 2, \ldots, n\} - (N \cup D)$ can be removed from the problem.

For the sake of clarity, some reduction rules for the SC and SPP are described separately; even when they differ only slightly.

Rule 1R (SC and SPP)   Zero Rows

If R_i is a zero vector for any i, no feasible solution exists.

> *Comment:*   Notice that if the initial matrix of the set-covering problem has no such row, the problem has a solution. Therefore, in the case of SC, rule 1R has to be applied only once, in the beginning. However, it is not difficult to see that for some instances of the SPP problem, zero rows may appear after a sequence of other reductions.

Rule 1C (SC and SPP)   Zero Columns

Delete all zero columns. Therefore, if $C_j = 0$, then $N \leftarrow N - \{j\}$.

> *Comment:*   This reduction has no influence on the existence of the solution. It is always useful to check at the beginning if there is no empty subset in the input. Then, for the SPP problem, a zero column never appears again during the reduction process, since if a column has an 1 and the corresponding row is deleted, then also the column is removed from the problem. However, in the SC problem, some further reductions of rows may introduce zero columns.

Rule 2 (SC)   Essential Columns

If $R_i = \mathbf{e}_t^T$, where $\mathbf{e}_t$ is the tth identity vector, then set P_t is in every cover of M. Therefore, $D \leftarrow D \cup \{t\}$ and $N \leftarrow N - \{t\}$. Then row R_i as well as all rows R_l such that $a_{lt} = 1$ can be deleted from the problem. Hence $M \leftarrow M - \{i\} - \{l: a_{lt} = 1, l \in M\}$.

Rule 2 (SPP)   Essential Columns

Additionally to the reduction described in rule 2(SC), every column C_k which has 1 in the deleted row R_l can be also deleted in order not to overcover row R_l. Therefore, we also set $N \leftarrow N - \{k: a_{lk} = 1 \text{ and } a_{lt} = 1, k \in N, l \in M\}$, where t is determined in rule 2(SC).

> *Comment:*   This reduction may introduce zero rows, and reduce some other rows to those containing single 1's.

Rule 3R (SC)   Dominating Rows

If $R_l \geqslant R_k$, then R_l may be deleted since every column which covers row k covers also row l. Therefore, $M \leftarrow M - \{l\}$.

Rule 3R (SPP)   Dominating Rows

Additionally to rule 3R(SC), any column C_h such that $a_{lh} = 1$ and $a_{kh} = 0$ can be deleted since it cannot belong to any feasible partition; hence

$$N \leftarrow N - \{h: a_{lh} = 1 \text{ and } a_{kh} = 0, h \in N\}$$

where k and l are determined in rule 3R(SC).

Rule 3C (SC)   Dominating Columns

If $C_l \leqslant C_k$ and $c_l \geqslant c_k$, then column C_l can be deleted since column C_k may cover all elements covered by C_l at not greater cost. Therefore,

$$N \leftarrow N - \{l: C_l \leqslant C_k \text{ and } c_l \geqslant c_k, l, k \in N\}$$

Rule 3C (SPP)   Dominating Columns

Similarly to rule 3C(SC), we have

$$N \leftarrow N - \{l: C_l = C_k \text{ and } c_l \geqslant c_k, l, k \in N\}$$

> *Comment:*   Notice that once all identical columns are checked for the dominance, no other reduction can introduce new dominating columns.

Rule 4 (SC)   Dominating Columns—General Case

If there exists a column C_l and a subset of columns $K \subset N$ such that $l \notin K$, $C_l \leqslant \sum_{j \in K} C_j$, and $c_l \geqslant \sum_{j \in K} c_j$, then C_l may be deleted, since columns in K may cover all elements covered by C_l at not greater cost; hence $N \leftarrow N - \{l\}$. See Problem 2-43 for some refinements of this rule. When K is restricted to one element subsets, we have rule 3(SC).

Rule 4(SPP)   Dominating Columns—General Case

Similarly to rule 4(SC), we can delete column C_l if there exists a subset $K \subset N$ such that $l \notin K$, $C_l = \sum_{j \in K} C_j$ and $c_l \geqslant \sum_{j \in K} c_j$. Rule 3(SPP) is a special case of this rule when $|K| = 1$.

> *Comment:*   In general, application of rule 4 may be very time consuming, since its verification is equivalent to another instance of the set-partitioning problem. To demonstrate this, all rows with 0's in C_l can be deleted as well as all columns with 1's in the deleted rows. Let C_j' denote the reduced columns, and M' and N' be the sets of rows and columns still in the problem. Now the problem is to find if M' can be partitioned into some columns in N' at cost not greater than c_l. This is a decision version of the set-partitioning problem which is NP-complete; therefore, it is unlikely that it can be solved in polynomial time. The reduced problem can be further simplified

by considering weights. Assume that there are no two identical columns among $\{C'_j : j \in N'\}$, and let c_{j_1} and c_{j_2} be the smallest and the second smallest weights in N'. It is obvious that if $c_l < c_{j_1} + c_{j_2}$, then the reduced problem has no solution. Otherwise, we may further reduce the problem by deleting those columns C'_k which satisfy $c_l < c_{j_1} + c_k$.

Rule 5(SPP)   Infeasible Columns

If a column C_t is assigned to a solution, we have to remove from the problem those columns C_j which have 1 at the same rows as C_t has. If it leads to a zero row, then evidently C_t must be excluded from every feasible solution (see Problem 2-44 for an equivalent formulation of this rule).

Comment: This rule has no counterpart for the set-covering problem.

Example 2-9

Let us apply the rules above to reduce the instance of the SP problem (assume that all costs are equal to 1) shown in Fig. 2-4. The instance has neither an empty set (a zero column) nor a zero row. First we reduce identical columns (rule 3C); therefore, $N \leftarrow \{2, 3, 4, 5, 6\}$. Then we apply the other rules in the order 2, 3R, and 5.

	C_1	C_2	C_3	C_4	C_5	C_6
R_1	1	0	1	0	0	1
R_2	0	1	0	1	0	0
R_3	1	0	1	1	1	1
R_4	0	1	0	0	0	0
R_5	1	0	0	0	1	1

Figure 2-4   *An instance of the SPP problem.*

Reduction by rule 2 transforms $D \leftarrow \{2\}$, $M \leftarrow M - \{2, 4\} = \{1, 3, 5\}$, and $N \leftarrow N - \{2, 4\}$. Now the reduced instance has the following form:

	C_3	C_5	C_6
R_1	1	0	1
R_3	1	1	1
R_5	0	1	1

In rule 3R, first we have $R_3 \geqslant R_1$; therefore, $M \leftarrow M - \{3\} = \{1, 5\}$ and $N \leftarrow N - \{5\} = \{3, 6\}$, since $a_{15} = 0$ and $a_{35} = 1$. Then $R_1 \geqslant R_5$. Hence $M \leftarrow M - \{1\} = \{5\}$ and $N \leftarrow N - \{3\} = \{6\}$. Now the instance has only one row, 5 and only one column, 6, and $a_{56} = 1$.

Rule 5 cannot reduce the instance any further. We repeat the rules in the same order, and now rule 2 first assigns $D \leftarrow D \cup \{6\}$, and then transforms the instance to an empty one ($M = N = \varnothing$). Therefore, $D = \{2, 6\}$ is the optimal set partitioning. ∎

Our Pascal implementation of the reduction rules for the SPP problem contains all the rules except rule 4, which, as explained, calls for a solution of another hard instance of the same problem.

As can be seen from Example 2-9 (see also Problem 2-4), some reductions may transform an instance to another one, to which some other reduction rules (sometimes the same) can be applied again. We leave to the reader to provide suitable SPP instances which will illustrate and confirm the following properties of the rules (see Problem 2-46). The properties of the reduction rules for the SC problem are also left as an exercise (Problem 2-47).

1. Rule 1C(SPP): Zero columns (i.e., empty sets) once removed from the problem never appear again.

2. Rule 3C(SPP): Similarly, dominating columns should be eliminated only from the original instance, since no other reduction can introduce such columns.

3. Rule 1R(SPP): Problem reductions resulting from the application of rules 2, 3R, and 5 may introduce zero rows, so rule 1R has to be applied after each of them to check if the problem is still feasible.

4. Rules 2, 3R, and 5(SPP): The application of one of them may transform an instance to another one which will allow some further reductions. Each of the rules may introduce a zero row.

5. Finally, each of the rules 1C, 2, 3R, and 5 or a sequence of them may remove all rows from the problem. In this case, the set D contains the optimal solution of the problem, that is, the minimum-cost set-partitioning of the base set.

Algorithm 2-7 combines the rules 1C, 2, 3R, 3C, and 5 according to their properties described above. Two Boolean variables are used to indicate the state of the finally reduced problem: *nonfeasible* becomes *true* if a zero row appears in the reduction process; otherwise, it is *false*; and *optimal* is *true* if the problem vanishes, and *false*, otherwise.

Algorithm 2-7:   Reduction Algorithm for the Set-Partitioning Problem

```
begin
    nonfeasible ← false;
    optimal ← false;
    if there exists a zero row then
        nonfeasible ← true;
```

```
        else
        begin
            delete zero columns (rule 1C);
            delete dominating columns (rule 3C);
            repeat
                apply rule 2;
                if not (problem vanished or there is a zero row) then
                begin
                    apply rule 3R;
                    if not (problem vanished or there is a zero row) then
                        apply rule 5
                end
            until problem vanishes or there is a zero row or none of the
                rules 2, 3R, and 5 has been successfully applied in the
                last run of this statement;
            if problem vanished or there is a zero row then
                if all elements are covered then
                    optimal ← true
                else nonfeasible ← true
        end ⟨∗ the original instance has no zero row ∗⟩
    end
```

Computer Implementation of the Reduction Algorithm

The SETPARTRED procedure is an implementation of the reduction algorithm for the set-partitioning problem given in Algorithm 2-7. The computer representation of an instance of the problem is illustrated in Fig. 2-5. Assume that initially the base set has M elements and the family of its subsets has N members. Note that in the word description of the algorithms we use M and N to denote the subsets of elements

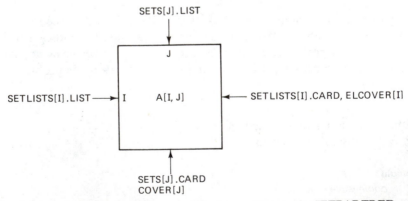

Figure 2-5 *Representation of the set-partitioning problem in the* SETPARTRED *procedure.*

$M \subset \{1, 2, \ldots, m\}$ and $N \subset \{1, 2, \ldots, n\}$, and in the implementations, M and N denote the maximal ranges of the elements in the corresponding subsets. Array A[1..M,1..N] represents the element-subset incidence matrix of the instance. Additionally, the sets $\{P_j\}$ and the index sets $\{F_i\}$ are represented as linked lists SETS[1..N].LIST and SETLISTS[1..M].LIST, respectively. SETS[1..N].CARD and SETLISTS[1..M].CARD are the cardinalities of the corresponding lists.

The advantage of using the double representation of the problem is at least threefold: Matrix elements can be accessed randomly (i.e., in a constant time), matrix lines (rows and columns) can be compared easily element by element, and the sets $\{P_j\}$ and the index sets $\{F_i\}$ can be scanned in the time proportional to their cardinalities.

The dimensions of the current problem, that is, the number of rows and the number of columns still in the problem, are represented by MR and NR, respectively. Initially, MR = M and NR = N, and then if one of these numbers becomes zero, the problem vanishes.

The current problem state is described by using two arrays ELCOVER[1..M] and COVER[1..N] (see Fig. 2-5), which assume the following values:

$$\text{ELCOVER[I]} = \begin{cases} 0, & \text{if element(row) I has not been covered yet} \\ 1, & \text{otherwise} \end{cases} \quad (2\text{-}21)$$

$$\text{COVER[J]} = \begin{cases} -1, & \text{if set(column) J is still in the problem} \\ 0, & \text{if set J has been removed from the problem} \\ 1, & \text{if set J has been removed from the problem as} \\ & \text{a member of every optimal set partitioning} \end{cases}$$

$$(2\text{-}22)$$

Therefore, initially all elements of ELCOVER are 0 and all elements of COVER are -1. During the reduction process, MR is the number of 0's in ELCOVER and NR is the number of -1's in COVER. After the reduction is successfully completed (i.e., there is no zero row), the quantities MR, NR, ELCOVER, and COVER define the reduced problem which has MR rows I where ELCOVER[I] = 0, and NR columns J where COVER[J] = -1. If MR = 0, an optimal solution has been obtained.

Sets J such that COVER[J] = 1 constitute a partial set partitioning which is contained in every feasible solution, and sets J for which COVER[J] = 0 can be removed since they do not belong to any feasible solution of the problem at hand. For some rows I we may have ELCOVER[I] = 1 even if no column has been assigned to a partial solution. Such rows may result from the application of the row dominance test 3R. The SETPARTRED procedure is a straightforward implementation of Algorithm 2-7. Rules 1R, 1C, 2, 3R, 3C, and 5 are implemented in procedures EMPTYLIST, EMPTYSET, SPRED2(BR2), SPRED3R(BR3R), SPRED3C, and SPRED5(BR5), respectively, where BR2, BR3R, and BR5 are

Boolean variables which assume value TRUE if the corresponding rule has reduced the problem, and FALSE otherwise.

Global Constants

M number of elements in the base set
N number of subsets in the family $\mathcal{P}$

Data Types

```
TYPE  ARRM       = ARRAY[1..M] OF INTEGER;
      ARRN       = ARRAY[1..N] OF INTEGER;
      ARRMN      = ARRAY[1..M,1..N] OF 0..1;
      POINTEL    = @ELEMENTLIST;
      ELEMENTLIST = RECORD
                        ELEM: INTEGER;
                        NEXT: POINTEL
                    END;
      SETDATA    = RECORD
                        CARD : INTEGER;
                        LIST  : POINTEL
                    END;
      INDEXSET   = ARRAY[1..M] OF SETDATA;
      SETFAMILY  = ARRAY[1..N] OF SETDATA;
```

Comment: In our Pascal codes, character @ stands for the pointer arrow ↑.

Procedure Parameters

```
PROCEDURE SETPARTRED(
    M,N                        :INTEGER;
    VAR MR,NR                  :INTEGER;
    VAR A                      :ARRMN;
    VAR SETS                   :SETFAMILY;
    VAR SETLISTS               :INDEXSET;
    VAR C                      :ARRN;
    VAR NONFEASIBLE,OPTIMAL    :BOOLEAN;
    VAR ELCOVER                :ARRM;
    VAR COVER                  :ARRN;
    VAR COST                   :INTEGER);
```

Input

M	number of elements in the base set
N	number of subsets in the family $\mathcal{P}$
A[1..M,1..N]	array containing the 0–1 constraint matrix of the instance
SETS[1..N]	array of N records representing the family of sets $\mathcal{P}$. Each record consists of two fields; one is designated to hold the cardinality of a set and the other is a linked list of the set elements
SETLISTS[1..M]	array of M records representing index sets $\{F_I\}$. Each record I consists of two fields; one holds the cardinality of F_I and the other is a linked list of indices of sets which contain element I
C[1..N]	array of the costs

Output

MR	number of elements of the base set in the reduced problem; MR is equal to the number of zero elements in ELCOVER
NR	number of members of the family $\mathcal{P}$ in the reduced problem; NR is equal to the number of -1's in COVER
NONFEASIBLE	Boolean variable which has value TRUE if during the process of reduction it has been discovered that the problem has no feasible solution (zero row appeared), and FALSE otherwise
OPTIMAL	Boolean variable which has value TRUE if an optimal solution has been found by the reduction process, and FALSE otherwise; in the former case, all elements of ELCOVER are 1's and COVER has no -1's
ELCOVER[1..M] COVER[1..N]	two arrays that define the reduced problem [see (2-21) and (2-22) for the description of their values]
COST	cost of the partial solution consisting of those sets J for which COVER[J] = 1

Example

The input and output of procedure SETPARTRED are illustrated with the instance whose array A[1..M,1..N] is shown in Fig. 2-4. We have

$$M = 5, \quad N = 6, \quad C[1..6] = [1,1,1,1,1,1]$$

SETS[J]

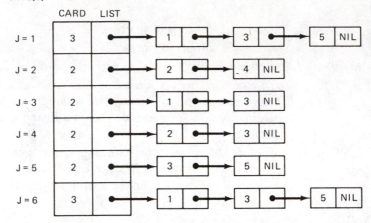

SETLISTS[I]

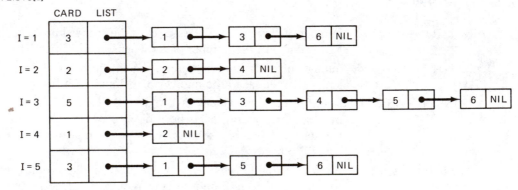

The solution is

$$NONFEASIBLE = FALSE, \quad OPTIMAL = TRUE$$
$$ELCOVER[1..5] = [1,1,1,1,1]$$
$$COVER[1..6] = [0,1,0,0,0,1]$$
$$COST = 5$$

Pascal Procedure SETPARTRED

```
PROCEDURE SETPARTRED(
   M,N                    :INTEGER;
   VAR MR,NR              :INTEGER;
   VAR A                  :ARRMN;
   VAR SETS               :SETFAMILY;
   VAR SETLISTS           :INDEXSET;
```

```
    VAR C                       :ARRN;
    VAR NONFEASIBLE,OPTIMAL:BOOLEAN;
    VAR ELCOVER                 :ARRM;
    VAR COVER                   :ARRN;
    VAR COST                    :INTEGER);


VAR I,J                     :INTEGER;
    BR2,BR3R,BR5,CONTINUE:BOOLEAN;
    NSL                     :ARRM;
    NS                      :ARRN;

FUNCTION MN:BOOLEAN;
BEGIN  (*  MN IS TRUE IF THE PROBLEM HAS NOT VANISHED  *)
   MN:=(MR > 0) AND (NR > 0)
END;  (* MN *)

PROCEDURE EMPTYSET;
VAR I:INTEGER;
BEGIN  (* EMPTYSET REMOVES EMPTY SETS  *)
    FOR I:=1 TO N DO
        IF (COVER[I] = -1) AND (NS[I] = 0) THEN
        BEGIN
            NR:=NR-1;
            COVER[I]:=0
        END
END;  (* EMPTY SET *)

FUNCTION EMPTYLIST:BOOLEAN;
   (*  EMPTYLIST IS TRUE IF IN THE CURRENT PROBLEM SOME
       ELEMENT CANNOT BE COVERED BY ANY SET  *)
VAR I:INTEGER;
    B:BOOLEAN;
BEGIN
   B:=FALSE;
   I:=1;
   WHILE NOT B AND (I <= M) DO
   BEGIN
       B:=(NSL[I] = 0) AND (ELCOVER[I] = 0);
       I:=I+1
   END;
   EMPTYLIST:=B
END;  (* EMPTY LIST *)


PROCEDURE DELETESET(VAR J,NR:INTEGER;VAR NSL:ARRM;VAR COVER:ARRN);
VAR L:INTEGER;
    P:POINTEL;
BEGIN  (*  THE PROCEDURE DELETES SET J FROM THE PROBLEM  *)
   NR:=NR-1;
   COVER[J]:=0;
   P:=SETS[J].LIST;
   WHILE P <> NIL DO
```

```
    BEGIN
       L:=P@.ELEM;
       NSL[L]:=NSL[L]-1;
       P:=P@.NEXT
    END
END;   (* DELETE SET J *)



PROCEDURE SPRED2(VAR BR2:BOOLEAN);
VAR I,K,L,T:INTEGER;
    P,Q      :POINTEL;
BEGIN   (* REDUCTION 2 - ROWS WITH SINGLE 1 *)
   BR2:=FALSE;
   FOR I:=1 TO M DO
      IF (ELCOVER[I] = 0) AND (NSL[I] = 1) THEN
      BEGIN
         BR2:=TRUE;
         MR:=MR-1;
         ELCOVER[I]:=1;
         P:=SETLISTS[I].LIST;
         REPEAT
            T:=P@.ELEM;
            P:=P@.NEXT
         UNTIL COVER[T] = -1;
         NR:=NR-1;
         COVER[T]:=1;
         COST:=COST+C[T];
         P:=SETS[T].LIST;
         WHILE P <> NIL DO
         BEGIN
            L:=P@.ELEM;
            IF ELCOVER[L] = 0 THEN
            BEGIN
               MR:=MR-1;
               ELCOVER[L]:=1;
               Q:=SETLISTS[L].LIST;
               WHILE Q <> NIL DO
               BEGIN
                  K:=Q@.ELEM;
                  IF COVER[K] = -1 THEN DELETESET(K,NR,NSL,COVER);
                  Q:=Q@.NEXT
               END
            END;
            P:=P@.NEXT
         END   (* SET T *)
      END   (* ROW I WITH SINGLE 1  *)
END;   (*  SPRED2  *)



PROCEDURE SPRED3R(VAR BR3R:BOOLEAN);
VAR H,I,J,K,L:INTEGER;
    B         :BOOLEAN;
```

```
BEGIN  (*  REDUCTION 3R - DOMINATING ROWS  *)
   BR3R:=FALSE;
   FOR I:=1 TO M-1 DO
      IF ELCOVER[I] = O THEN
      BEGIN
         J:=I;
         REPEAT
            J:=J+1;
            L:=0;
            IF ELCOVER[J] = O THEN
            BEGIN
               IF NSL[I] <= NSL[J] THEN
               BEGIN K:=I;  L:=J END
               ELSE BEGIN K:=J; L:=I END;
               H:=1;
               B:=TRUE;
               WHILE (H <= N) AND B DO
               BEGIN
                  IF COVER[H] = -1 THEN
                      B:=A[L,H] >= A[K,H];
                  H:=H+1
               END;
               IF B THEN
               BEGIN      (*  ROW L CAN BE DELETED  *)
                  BR3R:=TRUE;
                  MR:=MR-1;
                  ELCOVER[L]:=1;
                  FOR H:=1 TO N DO
                     IF COVER[H] = -1 THEN
                        IF A[L,H] = 1 THEN
                        BEGIN
                           NS[H]:=NS[H]-1;
                           IF A[K,H] = O THEN
                              DELETESET(H,NR,NSL,COVER)
                        END
               END
            END  (*  ROW J  *)
         UNTIL (J = M) OR ( B AND (L = I))
      END  (*  ROW I  *)
END;  (*  SPRED3R  *)

PROCEDURE SPRED3C;
VAR H,I,J,K,L:INTEGER;
    B        :BOOLEAN;
BEGIN  (*  REDUCTION 3C - COST DOMINATING COLUMNS  *)
   FOR J:=1 TO N-1 DO
      IF COVER[J] = -1 THEN
      BEGIN
         I:=J;
         REPEAT
            I:=I+1;
            L:=0;
            IF COVER[I] = -1 THEN
```

```
                BEGIN
                    IF C[J] <= C[I] THEN
                    BEGIN K:=J; L:=I END
                    ELSE BEGIN K:=I; L:=J END;
                    H:=1;
                    B:=TRUE;
                    WHILE (H <= M) AND B DO
                    BEGIN
                        IF ELCOVER[H] = 0 THEN
                            B:=A[H,K] = A[H,L];
                        H:=H+1
                    END;
                    IF B THEN    (*  SET L CAN BE DELETED  *)
                        DELETESET(L,NR,NSL,COVER)
                END  (* SET I *)
            UNTIL (I = N) OR (B AND (L = J))
        END  (* COLUMN J *)
END;  (* SPRED3C  *)

PROCEDURE SPRED5(VAR BR5:BOOLEAN);
  (*  REDUCTION 5 - INFEASIBLE COLUMNS  *)
VAR J:INTEGER;
FUNCTION REMOVE(T,NRR:INTEGER;NL:ARRM;COV:ARRN):BOOLEAN;
  (*  REMOVE IS TRUE IF SET T CAN BE REMOVED SINCE ITS
      PRESENCE IN THE SOLUTION WOULD CAUSE INFEASIBILITY  *)
VAR I,J:INTEGER;
    B :BOOLEAN;
    P,Q:POINTEL;
BEGIN
    COV[T]:=1;
    P:=SETS[T].LIST;
    WHILE (P <> NIL) AND (NRR > 0) DO
    BEGIN
        I:=P@.ELEM;
        Q:=SETLISTS[I].LIST;
        WHILE (Q <> NIL) AND (NRR > 0) DO
        BEGIN
            J:=Q@.ELEM;
            IF COV[J] = -1 THEN
                DELETESET(J,NRR,NL,COV);
            Q:=Q@.NEXT
        END;
        P:=P@.NEXT
    END;
    B:=TRUE;
    I:=1;
    WHILE (I <= M) AND B DO
    BEGIN
        B:=(ELCOVER[I] = 1) OR (NL[I] > 0);
        I:=I+1
    END;
    REMOVE:=NOT B
END;  (*  REMOVE  *)
```

```
BEGIN
    BR5:=FALSE;
    FOR J:=1 TO N DO
        IF COVER[J] = -1 THEN
            IF REMOVE(J,NR,NSL,COVER) THEN
            BEGIN
                BR5:=TRUE;
                DELETESET(J,NR,NSL,COVER)
            END
END;  (*  SPRED5  *)

BEGIN  (*  MAIN BODY  *)
    COST:=0;
    FOR I:=1 TO M DO
    BEGIN
        ELCOVER[I]:=0;
        NSL[I]:=SETLISTS[I].CARD
    END;
    FOR J:=1 TO N DO
    BEGIN
        COVER[J]:=-1;
        NS[J]:=SETS[J].CARD
    END;
    MR:=M;   NR:=N;
    NONFEASIBLE:=FALSE;
    OPTIMAL:=FALSE;
    IF EMPTYLIST THEN
        NONFEASIBLE:=TRUE
    ELSE
    BEGIN
        EMPTYSET;
        SPRED3C;
        REPEAT
            BR2:=FALSE;     BR3R:=FALSE;
            BR5:=FALSE;     CONTINUE:=TRUE;
            SPRED2(BR2);
            IF BR2 THEN
                CONTINUE:=MN AND NOT EMPTYLIST;
            IF CONTINUE THEN
            BEGIN
                SPRED3R(BR3R);
                IF BR3R THEN
                    CONTINUE:=MN AND NOT EMPTYLIST;
                IF CONTINUE THEN
                BEGIN
                    SPRED5(BR5);
                    IF BR5 THEN
                        CONTINUE:=MN AND NOT EMPTYLIST
                END
            END
        UNTIL NOT ((BR2 OR BR3R OR BR5) AND CONTINUE);
        IF NOT CONTINUE THEN
            IF MR = 0 THEN
```

```
        BEGIN
           OPTIMAL:=TRUE;
           FOR J:=1 TO N DO
               IF COVER[J] = -1 THEN COVER[J]:=0
        END
        ELSE NONFEASIBLE:=TRUE
    END  (*  NOT EMPTYLIST  *)
END;  (*  SETPARTRED  *)
```

2.2.3. IMPLICIT ENUMERATION METHOD FOR THE SET-PARTITIONING PROBLEM

Almost every implicit enumeration algorithm for solving the set-partitioning problem greatly benefits from the special structure of the problem which usually results in very tight equality constraints.

First, the reduction method as described in the preceding section can be applied in order to reduce the number of rows and columns and test if the instance at hand has no zero rows.

The enumeration of all possible solutions can be carried out very economically by first rearranging columns. Let B_i denote the subset of columns which have the first nonzero element in row i. Clearly, if there is no zero row, then block B_1 is nonempty, but other blocks may be empty. Also, it is clear that every feasible solution of the problem contains at most one column from each block. Therefore, a search tree of an instance (which is not necessarily a binary tree—why?) can have at most n' levels, where n' is the number of nonempty blocks B_i. We have $n' \leqslant m$ and $n' \leqslant n$. For the sake of simplicity, let us assume that the columns are already properly grouped in blocks $B_1, B_2, \ldots, B_m$.

Let Y denote a current partial solution; that is, no element of M is covered by more that one column of Y. Let $M(Y)$ denote the set of elements covered by Y. The algorithm consists of two main moves: forward and backward. A *forward move* first augments a current solution Y with a column which has already been tested for feasibility and dominance, and then moves the search to a next candidate (column) for inclusion, if such a column exists. A *backward move* follows when a partial solution cannot be augmented to a feasible one. It removes from Y the last added column and searches for a next column in the same block as the column just removed or in the blocks next to the end whose columns are in Y.

Before a forward move is called, some feasibility and dominance tests are carried out to avoid decisions which may lead to either an infeasible solution or to a solution worse than the best solution currently known. For a given partial solution Y, a next column which can be added belongs to one of the blocks B_k, where $k \geqslant i$ and i is the smallest element uncovered by Y. First, feasibility test checks if all elements of M uncovered by Y can be covered by some columns in $\{B_k : k \geqslant i\}$.

Example 2-10

As an illustration, consider an instance of the SPP problem with the following parameters: $m = 9$, $n = 11$, $\mathbf{c}^T = (3, 3, 1, 2, 2, 2, 3, 3, 2, 1, 1)$, and

$$A = \begin{bmatrix}
1 & 1 & 1 & 0 & 0 & 0 & 0 & 0 & 0 & 0 & 0 \\
0 & 0 & 0 & 1 & 1 & 0 & 0 & 0 & 0 & 0 & 0 \\
0 & 0 & 0 & 1 & 0 & 1 & 1 & 1 & 0 & 0 & 0 \\
0 & 1 & 0 & 0 & 0 & 0 & 0 & 0 & 0 & 0 & 0 \\
1 & 0 & 0 & 0 & 0 & 0 & 0 & 1 & 1 & 0 & 0 \\
0 & 0 & 0 & 0 & 1 & 0 & 0 & 0 & 0 & 0 & 0 \\
0 & 0 & 0 & 0 & 0 & 1 & 1 & 1 & 1 & 0 & 0 \\
0 & 0 & 0 & 0 & 0 & 0 & 0 & 0 & 0 & 1 & 0 \\
1 & 1 & 0 & 0 & 0 & 0 & 1 & 0 & 0 & 0 & 1
\end{bmatrix}$$

If $Y = \{1\}$, then element 4 cannot be covered by any column in blocks B_k ($k \geqslant 2$). Similarly, if $Y = \{2, 4\}$, then element 6 cannot be covered by any column in B_k ($k \geqslant 5$).   ∎

To speed up this test we introduce an $m \times m$ matrix *blockcover* whose ith column contains 1's in the entries which can be covered by some columns in blocks $B_k(k \geqslant i)$. A further speed-up can be gained by introducing a Boolean vector *bcbool* and a variable *bctotal* such that $bcbool_i = true$ if all elements $l \geqslant i$ can be covered by some columns in blocks $B_k(k \geqslant i)$, and $bctotal = true$ if all entries of *bcbool* are *true*. Hence if *bctotal* is *true* or $bcbool_i$ is *true*, we do not need to scan the ith column of *blockcover*. Notice that if $bcbool_1 = false$, the constraint matrix of the instance has a zero row.

Example 2-10 (cont.)

Matrix *blockcover*, vector *bcbool*, and variable *bctotal* have the following values:

$$blockcover = \begin{bmatrix}
1 \\
1 & 1 & & & & & & & 0 \\
1 & 1 & 1 \\
1 & 0 & 0 & 0 \\
1 & 1 & 1 & 1 & 1 \\
1 & 1 & 0 & 0 & 0 & 0 \\
1 & 1 & 1 & 1 & 1 & 0 & 0 \\
1 & 1 & 1 & 1 & 1 & 1 & 1 & 1 \\
1 & 1 & 1 & 1 & 1 & 1 & 1 & 1 & 1
\end{bmatrix}$$

$$bcbool = (T \quad F \quad F \quad F \quad F \quad F \quad F \quad T \quad T), \qquad bactotal = F$$

where T stands for *true* and F for *false*.   ∎

When a particular column C_j is considered, another feasibility test checks if C_j has no 1's covered already by a partial solution Y.

The two feasibility tests above are based entirely on the structure of equality constraints. For their detailed realization the reader is referred to functions FEASIBLE and FIT, respectively, in the Pascal code.

The computations start with no initial feasible solution since an instance of the SPP problem may not have any. However, once a feasible solution has been found, its cost can be used to eliminate some partial solutions which are not likely to improve the best current solution.

Let $c(Y)$ denote the cost of a partial solution Y and c^* be the cost of the best current solution. It is clear that every column C_j satisfying $c(Y) + c_j \geqslant c^*$ can be dismissed from being considered as a candidate for augmenting Y.

Further dominance considerations can be based on the following relaxation. Suppose that a next column to be added to a partial solution Y is one of the columns $C_r, C_{r+1}, \ldots, C_n$. Let $m_r(Y)$ denote the number of elements uncovered by Y, that is, $m_r(Y) = m - |M(Y)|$. It is clear that every feasible augmentation of Y corresponds to a feasible solution of the following problem:

minimize
$$z_r = \sum_{j=r}^{n} c_j y_j$$

subject to
$$\sum_{j=r}^{n} h_j y_j = m_r(Y), \qquad \text{(SPP}_r)$$

$$y_j = 0 \text{ or } 1, \qquad j = r, r + 1, \ldots, n$$

Notice that the linear constraint of the SPP_r problem is the sum of the equality constraints of SPP restricted to columns from r through n. We may reduce this problem further by removing those columns C_j ($r \leqslant j \leqslant n$) for which either $h_j > m_r(Y)$ or C_j overcovers some elements covered by Y, for they are not candidates for any feasible augmentation of Y.

The SPP_r problem belongs to the large family of knapsack problems discussed in the preceding section. The difference between the 0–1 knapsack problem presented there and SPP_r is in that the former problem maximizes the profit subject to an inequality constraint. This is, however, a minor difference and we can still apply to SPP_r almost the same solution methods as presented for the knapsack problem of Section 2.1.

In the context of the SPP problem, we want to know if the optimal solution z_r^* to SPP_r satisfies the inequality

$$c(Y) + z_r^* \geqslant c^* \qquad (2\text{-}23)$$

If so, the further forward search with Y as a partial solution can be abandoned and we can backtrack, because Y cannot be augmented to a feasible solution better than that of weight c^*. Since solving the knapsack problem exactly is very hard, we shall apply a less powerful test which instead of the optimal value z_r^* assumes in (2-23) a lower bound $\bar{z}_r$ to z_r^*.

Similarly to the maximization knapsack problem and its upper bounds, one can introduce various lower bounds to the solution of the minimization problem. First, assume that the columns C_j ($j = r, r+1, \ldots, n$) are ordered according to nondecreasing ratios c_j/h_j. Then relaxing the constraints imposed on the variables y_j, $0 \leqslant y_j \leqslant 1$, we may get lower bounds for SPP, similar to the upper bounds for the 0–1 knapsack problem given in Theorems 2-2 and 2-3. However, there is a significant drawback in using the latter bound, since all the columns are already grouped into blocks according to their smallest elements. Therefore, we decide to use only the bound following from Theorem 2-2, which needs only the smallest ratio $c_l/h_l = \min\{c_j/h_j : r \leqslant j \leqslant n\}$, and we ask the reader to implement the bound based on Theorem 2-3 and compare its effects on random problem instances (Problem 2-55).

Let $ch_r = \min\{c_j/h_j : r \leqslant j \leqslant n\}$ for $r = 1, 2, \ldots, n$ and $mc_i = \min\{c_j : j \in B_i\}$ for $i = 1, 2, \ldots, m$.

Now we summarize the dominance consideration. If i is the first (smallest) element uncovered by a partial solution Y, then we may immediately backtrack if either

$$c(Y) + mc_i \geqslant c^* \tag{2-24}$$

or

$$c(Y) + m(Y) \cdot ch_k \geqslant c^* \tag{2-25}$$

where k is the index of the first column in the block B_i. If the search can continue in the block B_i, we skip feasible columns C_j which satisfy either

$$c(Y) + c_j \geqslant c^* $$

or

$$c(Y) + c_j + \big(m(Y) - h_j\big) \cdot ch_{j+1} \geqslant c^* \tag{2-26}$$

for they cannot augment Y and lead to a solution better than that of weight c^*.

There are several heuristic arguments which may be applied to order the columns within blocks; for instance, according to nondecreasing costs c_j, nonincreasing set cardinalities h_j, or nondecreasing ratios c_j/h_j. Our implementation uses the last one, but can easily be modified to use any other. We refer the reader to computational results (Sec. 2.2.4) to see how two of these heuristics may influence the computational time (see also Problem 2-54).

In Algorithm 2-8 we present two procedures which implement the forward and backward moves. In both cases, the moves are more elaborate than described above.

They return either a massage that the search has been completed (*move = false*) or a column which is a candidate for a next forward move. A complete description of the implicit enumeration method is given in Algorithm 2-9.

There have been many improvements proposed to this basic algorithm; some of them are described in Problems 2-53 through 2-56 and in the References and Remarks.

Algorithm 2-8: Procedures BACKWARDMOVE and FORWARDMOVE of the Implicit Enumeration Algorithm for the SPP Problem

procedure BACKWARDMOVE($Y, i, j, move$);
  {∗ The procedure moves the search back to the first block B_j, which is present in a partial solution Y and has an unexplored column C_j. A Boolean variable *move* becomes *true* if such a column exists, and *false* otherwise ∗}

procedure FORWARDMOVE($Y, j_1, i, j_2, move$);
  {∗ The procedure adds column C_{j_1} to the current solution Y. If Y becomes a feasible solution of the problem, the best current solution is updated and BACK-WARDMOVE($Y, i, j_2, move$) follows. Otherwise, the feasibility tests and two domination tests [(2-24) and (2-25)] are applied to find a feasible column C_{j_2} in block B_j. A Boolean variable *move* becomes *true* if a feasible column exists, and *false* otherwise. ∗}
begin
    $Y \leftarrow Y \cup \{j_1\}$;   {∗ updating the current solution Y ∗}
    $c(Y) \leftarrow c(Y) + c_{j_1}$;
    $m(Y) \leftarrow m(Y) - h_{j_1}$;
    if $m(Y) = 0$ **then**
    begin   {∗ a feasible solution of the problem has been found ∗}
        nonfeasible ← *false*;
        update the current best solution;
        BACKWARDMOVE($Y, i, j_2, move$)
    end
    else
    begin
        find the smallest uncovered element i;
        if currently uncovered elements cannot be covered
            by columns in blocks $B_l(l \geqslant i)$ **then**
            BACKWARDMOVE($Y, i, j_2, move$)
        else
            if (2-24) **or** (2-25) holds **then**
                BACKWARDMOVE($Y, i, j_2, move$)
            else *move* ← *true*
    end
end {∗ FORWARDMOVE ∗}

Algorithm 2-9:   Implicit Enumeration Algorithm for the Set-Partitioning Problem

```
      begin
            calculate auxiliary quantities used in the feasibility and domination tests;
            c* ← inf; {* inf is a very large number *}
            nonfeasible ← true;
            if there is no zero row then
            begin
                  Y ← ∅;  c(Y) ← 0;  m(Y) ← m;
                  FORWARDMOVE(Y, 1, i, j, move);
                  while move do
                  begin
                        repeat
                              repeat
                                    find a column C_j' (j' ≥ j) in block B_i which does not overcover
                                          solution Y and c(Y) + c_j' < c*;
                                    if such a column does not exist then
                                          BACKWARDMOVE(Y, i, j, move)
                              until (not move) or (C_j' exists);
                              if C_j' exists and satisfies (2-26) then j ← j' + 1
                        until (not move) or ((C_j' exists) and (C_j' does not satisfy (2-26)));
                        if (C_j' exists) and (C_j' does not satisfy (2-26)) then
                              FORWARDMOVE(Y, j', i, j_2, move)
                  end {* while move = true *}
            end {* the original problem has no zero row *}
      end
```

Example 2-10 (cont.)

Let us apply Algorithm 2-9 to solve our illustrative problem. Since $bcbool_1$ is *true*, the instance has no zero rows. After initialization, procedure FORWARDMOVE is called. It assigns $Y \leftarrow \{1\}$ and then finds that such a partial solution cannot be augmented to cover element 4; therefore, BACKWARDMOVE is called. The next attempt is $Y \leftarrow \{2\}$ and since the elements uncovered by Y can be covered by some further columns, a forward move continues. Next we get $Y \leftarrow Y \cup \{4\} = \{2, 4\}$, and the first feasibility test finds that element 6 cannot be covered by any column from the next blocks. We again backtrack and try $Y = \{2, 5\}$. Now, the first uncovered element is 3 and column C_6 is a feasible candidate; hence $Y = \{2, 5, 6\}$. Then the first uncovered element is 5 but the only column in B_5 is infeasible. Therefore, backward move follows: C_6 is removed from the solution and another column in B_3 is tried. Column C_7 is infeasible and column C_8 gives rise to an improved feasible solution. The only uncovered element is 8 and it can be covered by column C_{10}. Therefore, we reach a complete solution $\{2, 5, 6, 10\}$ which costs 9. After updating the current best solution, BACKWARDMOVE goes to the first block. However, assignment $Y \leftarrow \{3\}$ is infeasible since no column in blocks B_k ($k \geq 2$) can cover element 4. Then the

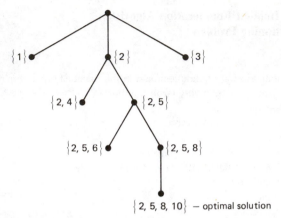

Figure 2-6 *Search tree for the instance of Example 2-10. The sets in the nodes are partial solutions encountered by the algorithm.*

algorithm terminates. The search tree of the algorithm for our example is shown in Fig. 2-6.

The reader may easily check that the SETPARTRED reduction procedure finds an optimal solution for our instance.   ■

Computer Implementation of the Enumeration Algorithm

The SETPARTBACKTRACK procedure finds the optimal solution to the set-partitioning problem, if any exists. The computer representation of the problem is the same as in the SETPARTRED procedure, except that SETPARTBACKTRACK does not use the constraint matrix A explicitly and it assumes that elements in SETS[J].LIST and in SETLISTS[I].LIST are ordered for every J and I. Therefore, SETS[J].LIST points to the smallest-value element of set J.

The procedure is a straightforward implementation of Algorithms 2-8 and 2-9. Two additional functions and one procedure are declared and used. Procedure SAVESOLUTION saves the current best solution and is called when all base elements are covered. Functions FEASIBLE and FIT are the implementations of feasibility tests. The function FEASIBLE(I) assumes the value TRUE if currently uncovered elements can be covered by some columns in blocks I, I + 1,..., M, and FALSE otherwise; whereas function FIT(J) becomes TRUE when column J does not overcover the elements covered by a current partial solution, and FALSE otherwise.

In the beginning, the procedure calculates the quantities which are then used by the feasibility and dominance tests. Arrays BLOCKCOVER[1..M,1..M], BCBOOL[1..M], and variable BCTOTAL have been already described (see Example 2-10); they are used in function FEASIBLE.

The columns (sets) are first grouped into blocks and then ordered within blocks according to their parameters. There are M blocks, BLOCKSIZE[I] is the number of columns in block I, and BLOCK[I] is the index of the first column in block I, provided that BLOCK[I] $\neq$ BLOCK[I + 1]. If these two numbers are equal, block I is empty. We assume that BLOCK[M + 1] = N + 1. Arrays ORDER[1..N] and INVORDER[1..N] determine the order of columns. Initially, columns are grouped into blocks and then they are rearranged by taking into account the set costs and cardinalities. In our code we use the procedure HEAP-SORT(K, J, RATIO, ORDER), which orders columns within blocks according to their ratios c_j/h_j. Using the heapsort algorithm,* this procedure orders K consecutive columns according to their keys stored in array RATIO. The columns to be sorted are identified by their indices stored in array ORDER starting with the one of index J. The ordered columns are again placed in array ORDER. Array IN-VORDER defines the inverse order; that is, ORDER[INVORDER[I]] = I. MINCBLOCK[I] is the minimum cost of columns in block I, and MINRATIO[J] is equal to ch_J.

Further comments on the implementation are embedded in the procedure.

Global Constants

M   number of elements in the base set
N   number of subsets in the family $\mathcal{P}$
M1   M + 1
N1   N + 1

Data Types

```
TYPE  ARRM        = ARRAY[1..M] OF INTEGER;
      ARRM1       = ARRAY[1..M1] OF INTEGER;
      ARRMB       = ARRAY[1..M] OF BOOLEAN;
      ARRN        = ARRAY[1..N] OF INTEGER;
      ARRN1       = ARRAY[1..N1] OF INTEGER;
      ARRNR       = ARRAY[1..N] OF REAL;
      ARRMM       = ARRAY[1..M,1..M] OF INTEGER;
      ARRN1R      = ARRAY[1..N1] OF REAL;
      POINTEL     = @ELEMENTLIST;
      ELEMENTLIST = RECORD
                        ELEM : INTEGER;
                        NEXT : POINTEL
                    END;
```

---

*We used a modified version of the *heapsort* procedure published in N. Wirth, *Algorithms + Data Structures = Programs*, Prentice-Hall, Englewood Cliffs, N.J., 1976.

```
SETDATA        = RECORD
                     CARD : INTEGER;
                     LIST  : POINTEL
                END;
INDEXSET       = ARRAY[1..M] OF SETDATA;
SETFAMILY      = ARRAY[1..N] OF SETDATA;
```

Comment: In our Pascal codes, character @ stands for the pointer arrow ↑.

Procedure Parameters

```
PROCEDURE SETPARTBACKTRACK(
    M,N,INF            : INTEGER;
    VAR SETS           : SETFAMILY;
    VAR SETLISTS       : INDEXSET;
    VAR C              : ARRN;
    VAR NONFEASIBLE : BOOLEAN;
    VAR COST,COUNT  : INTEGER;
    VAR COVER          : ARRN);
```

Input

M, N, SETS, SETLISTS, and C have the same meaning as in the procedure SETPARTRED (see its input); additionally

INF   maximal integer number available in the system that is used

Comment: Notice that the constraint matrix A is not used as an input. The procedure assumes that the elements (ELEM) of lists SETS[1..N].LIST and SETLISTS[1..M].LIST are in increasing order.

Output

NONFEASIBLE   Boolean variable which is TRUE if the problem has no feasible solution, and FALSE otherwise

COST   cost of optimal set partitioning, if any exists

COUNT   number of forward moves performed by the procedure

COVER[1..N]   array containing the optimal solution; COVER[I] = 1, if set I belongs to the optimal cover, and 0 otherwise

Example

The input and the output for the SETPARTBACKTRACK procedure are illustrated with the instance of Example 2-10. We have

$$M = 9, \quad N = 11, \quad INF = 1000$$
$$C[1..11] = [3,3,1,2,2,2,3,3,2,1,1]$$

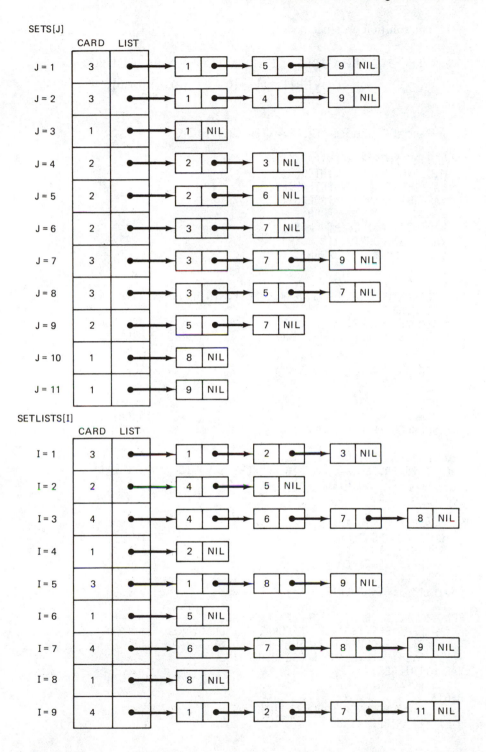

The solution obtained is

$$\text{NONFEASIBLE} = \text{FALSE}, \quad \text{COST} = 9, \quad \text{COUNT} = 4$$

$$\text{COVER}[1..11] = [0,1,0,0,1,0,0,1,0,1,0]$$

Pascal Procedure SETPARTBACKTRACK

```
PROCEDURE SETPARTBACKTRACK(
   M,N,INF          :INTEGER;
   VAR SETS         :SETFAMILY;
   VAR SETLISTS     :INDEXSET;
   VAR C            :ARRN;
   VAR NONFEASIBLE  :BOOLEAN;
   VAR COST,COUNT   :INTEGER;
   VAR COVER        :ARRN);


VAR CURRENTCOST,G,H,I,J,JK,K,L,MCOV,MIN,N1,OPTK:INTEGER;
   CC,MINR                                     :REAL;
   B,MOVE,TOTALBC                              :BOOLEAN;
   BLOCKSIZE,ELCOV,MINCBLOCK                   :ARRM;
   BLOCK                                       :ARRM1;
   BCBOOL                                      :ARRMB;
   COV,COV1,OPTCOV                             :ARRN;
   ORDER,INVORD                                :ARRN1;
   RATIO                                       :ARRNR;
   MINRATIO                                    :ARRN1R;
   BLOCKCOVER                                  :ARRMM;

PROCEDURE SAVESOLUTION;
   (* THIS PROCEDURE SAVES A CURRENT SOLUTION AS THE BEST
       CURRENT SOLUTION  *)
VAR H:INTEGER;
BEGIN
   NONFEASIBLE :=FALSE;
   COST:=CURRENTCOST;
   FOR H:=1 TO K DO
       OPTCOV[H]:=COV[H];
   OPTK:=K
END; (*  SAVE SOLUTION  *)

FUNCTION FEASIBLE(I:INTEGER):BOOLEAN;
   (*  FEASIBLE IS TRUE IF CURRENTLY UNCOVERED ELEMENTS
       CAN BE COVERED BY SOME UNUSED SETS IN BLOCKS
       I,I+1,...,M AND FALSE OTHERWISE  *)
VAR G:INTEGER;
    B:BOOLEAN;
BEGIN
   IF TOTALBC OR BCBOOL[I] THEN
       FEASIBLE:=TRUE
```

```
    ELSE
    BEGIN
       G:=I;
       REPEAT
          B:=(ELCOV[G] = 0) AND (BLOCKCOVER[G,I] = 0);
          G:=G+1
       UNTIL  B OR (G > M);
       FEASIBLE:=NOT B
    END
END;  (*  FEASIBLE  *)


FUNCTION FIT(J:INTEGER):BOOLEAN;
  (*  FIT IS TRUE IF SET J DOES NOT OVERCOVER ELEMENTS COVERED
      BY A CURRENT PARTIAL SOLUTION, AND FALSE OTHERWISE *)
VAR B:BOOLEAN;
    P:POINTEL;
BEGIN
   IF SETS[J].CARD > MCOV THEN
      FIT:=FALSE
   ELSE
   BEGIN
      B:=TRUE;
      P:=SETS[J].LIST;
      WHILE (P <> NIL) AND B DO
      BEGIN
         B:=ELCOV[P@.ELEM] = 0;
         P:=P@.NEXT
      END;
      FIT:=B
   END
END;  (*  FIT  *)


PROCEDURE BACKWARDMOVE(VAR MOVE:BOOLEAN);
  (*  PROCEDURE MOVES THE SEARCH TO THE FIRST FROM THE END
      BLOCK WHICH STILL CONTAINS SOME UNUSED SETS.
      VARIABLE MOVE BECOMES TRUE IF SUCH A BLOCK EXISTS
      AND FALSE OTHERWISE  *)
VAR P:POINTEL;
BEGIN
   IF K = 0 THEN
      MOVE:=FALSE
   ELSE
   BEGIN
   REPEAT
      J:=COV[K];
      I:=COV1[K];
      K:=K-1;
      P:=SETS[J].LIST;
      WHILE P<> NIL DO
```

```
         BEGIN
            ELCOV[P@.ELEM]:=0;
            P:=P@.NEXT
         END;
         CURRENTCOST:=CURRENTCOST-C[J];
         MCOV:=MCOV+SETS[J].CARD;
         MOVE:=INVORD[J] < BLOCK[I+1]-1
      UNTIL (K = 0) OR MOVE;
      IF MOVE THEN
      BEGIN
         JK:=INVORD[J]+1;
         J:=ORDER[JK]
      END
    END (*  K <> 0  *)
END;  (*  BACKWARD MOVE  *)

PROCEDURE FORWARDMOVE;
  (*  PROCEDURE ADDS SET J TO A CURRENT PARTIAL COVER.
      IF ALL ELEMENTS ARE COVERED, PROCEDURE SAVESOLUTION
      UPDATES THE BEST CURRENT SOLUTION AND BACKWARDMOVE
      IS CALLED. OTHERWISE, SEARCH FOR A FEASIBLE SET IS
      CONTINUED.  *)
VAR P:POINTEL;
BEGIN
   COUNT:=COUNT+1;
   K:=K+1;
   COV[K]:=J;
   COV1[K]:=I;
   P:=SETS[J].LIST;
   WHILE P <> NIL DO
   BEGIN
      ELCOV[P@.ELEM]:=1;
      P:=P@.NEXT
   END;
   CURRENTCOST:=CURRENTCOST+C[J];
   MCOV:=MCOV-SETS[J].CARD;
   IF MCOV = 0 THEN        (*  ALL ELEMENTS ARE COVERED - A NEW *)
   BEGIN                   (*  BETTER SOLUTION HAS BEEN FOUND  *)
      SAVESOLUTION;
      BACKWARDMOVE(MOVE)
   END
   ELSE
   BEGIN
      WHILE ELCOV[I] = 1 DO
         I:=I+1;
      IF NOT FEASIBLE(I) THEN
         BACKWARDMOVE(MOVE)
      ELSE
```

```
        BEGIN
            JK:=BLOCK[I];
            J:=ORDER[JK];
            IF (CURRENTCOST+MINCBLOCK[I] >= COST) OR
                (TRUNC(CURRENTCOST+MCOV*MINRATIO[JK]) >= COST) THEN
                    BACKWARDMOVE(MOVE)
            ELSE MOVE:=TRUE
        END;  (* FEASIBLE *)
    END
END;  (*  FORWARD MOVE  *)


BEGIN  (*  MAIN BODY  *)
  (*  INITIALIZATION  *)
   N1:=N+1;                      (* ORDERING SETS  *)
   FOR I:= 1 TO M DO
       BLOCKSIZE[I]:=0;
   FOR J:=1 TO N DO
   BEGIN
       RATIO[J]:=C[J]/SETS[J].CARD;
       L:=SETS[J].LIST@.ELEM;
       BLOCKSIZE[L]:=BLOCKSIZE[L]+1
   END;
   L:=1;
   FOR I:=1 TO M DO
   BEGIN
       BLOCK[I]:=L;
       L:=L+BLOCKSIZE[I]
   END;
   BLOCK[M+1]:=L;
   FOR I:=1 TO M DO ELCOV[I]:=BLOCK[I];
   FOR J:=1 TO N DO
   BEGIN
       L:=SETS[J].LIST@.ELEM;
       K:=ELCOV[L];
       ORDER[K]:=J;
       ELCOV[L]:=K+1
   END;
   ORDER[N1]:=N1;
   INVORD[N1]:=N1;
   J:=1;  I:=1;
   WHILE J < N DO
   BEGIN
       K:=BLOCKSIZE[I];
       IF K > 1 THEN
           HEAPSORT(K,J,RATIO,ORDER);
       I:=I+1;
       J:=BLOCK[I]
   END;
   FOR J:=1 TO N DO
       INVORD[ORDER[J]]:=J;
   FOR I:=1 TO M DO                (* AUXILIARY VARIABLES FOR  *)
```

```
BEGIN                            (* DOMINANCE TESTS  *)
   MIN:=INF;
   FOR J:=BLOCK[I] TO BLOCK[I+1]-1 DO
   BEGIN
      K:=C[ORDER[J]];
      IF K < MIN THEN MIN:=K
   END;
   MINCBLOCK[I]:=MIN
END;
MINR:=INF;
MINRATIO[N1]:=MINR;
FOR J:=N DOWNTO 1 DO
BEGIN
   CC:=RATIO[ORDER[J]];
   IF CC < MINR THEN MINR:=CC;
   MINRATIO[J]:=MINR
END;
TOTALBC:=BLOCK[M] <= N;        (* AUXILIARY VARIABLES FOR  *)
BCBOOL[M]:=TOTALBC;            (* FEASIBILITY TESTS  *)
IF TOTALBC THEN BLOCKCOVER[M,M]:=1
ELSE BLOCKCOVER[M,M]:=0;
L:=M;
B:=TOTALBC;
FOR I:=M-1 DOWNTO 1 DO
   IF BLOCKSIZE[I] = 0 THEN
   BEGIN
      B:=FALSE;
      BCBOOL[I]:=FALSE;
      TOTALBC:=FALSE;
      BLOCKCOVER[I,I]:=0;
      FOR H:=I+1 TO M DO
         BLOCKCOVER[H,I]:=BLOCKCOVER[H,L];
      L:=I
   END  (*  EMPTY BLOCK  *)
   ELSE
      IF B THEN
      BEGIN
         BCBOOL[I]:=TRUE;
         BLOCKCOVER[I,L]:=1
      END
      ELSE
      BEGIN
         BLOCKCOVER[I,I]:=1;
         K:=1;
         FOR H:=I+1 TO M DO
         BEGIN
            G:=BLOCKCOVER[H,L];
            BLOCKCOVER[H,I]:=G;
            K:=K+G
         END;
         N1:=M-I+1;
         J:=BLOCK[I];
         WHILE (J < BLOCK[I+1]) AND (K < N1) DO
```

```
        BEGIN
            H:=ORDER[J];
            P:=SETS[H].LIST@.NEXT;
            WHILE (P <> NIL) AND (K < N1) DO
            BEGIN
                G:=P@.ELEM;
                IF BLOCKCOVER[G,I] = O THEN
                BEGIN
                    BLOCKCOVER[G,I]:=1;
                    K:=K+1
                END;
                P:=P@.NEXT
            END;
            J:=J+1
        END;  (*  BLOCK I  *)
        B:=K=N1;
        BCBOOL[I]:=B;
        L:=I
    END;  (*  NON EMPTY BLOCK AND NOT TOTALBC  *)

(*  END OF INITIALIZATION  *)

COUNT:=O;
NONFEASIBLE:=TRUE;
FOR L:=1 TO M DO ELCOV[L]:=0;
IF FEASIBLE(1) THEN
BEGIN
    I:=1;            (*  I - BLOCK NUMBER  *)
    K:=O;            (*  K - CARDINALITY OF PARTIAL SOLUTION *)
    J:=ORDER[1];     (*  J - CURRENT SET NUMBER (ORIGINAL) *)
    JK:=1;           (*  JK - INDEX OF SET J IN ORDER  *)
    COST:=INF;
    CURRENTCOST:=0;
    MCOV:=M;         (*  MCOV - NUMBER OF UNCOVERED ELEMENTS *)
    FORWARDMOVE;
    WHILE MOVE DO  (*  MOVE IS TRUE IF SEARCH IS NOT EXHAUS- *)
    BEGIN          (*  TED, AND FALSE OTHERWISE  *)
        REPEAT
            REPEAT
                B:=FALSE;
                WHILE (JK < BLOCK[I+1]) AND (NOT B) DO
                BEGIN
                    B:=(CURRENTCOST+C[J] < COST) AND FIT(J);
                    IF NOT B THEN
                    BEGIN
                        JK:=JK+1;
                        J:=ORDER[JK]
                    END
                END;
                IF JK = BLOCK[I+1] THEN
                    BACKWARDMOVE(MOVE)
            UNTIL B OR NOT MOVE;
            IF B THEN
```

```
                B:=CURRENTCOST+C[J]+(MCOV-SETS[J].CARD)
                    *MINRATIO[JK+1] < COST;
            IF MOVE AND NOT B THEN
                BEGIN  JK:=JK+1;  J:=ORDER[JK]  END
          UNTIL B OR NOT MOVE;
          IF B THEN FORWARDMOVE
       END  (*  WHILE MOVE  *);
       IF NOT NONFEASIBLE THEN
       BEGIN
            FOR J:=1 TO N DO COVER[J]:=0;
            FOR J:=1 TO OPTK DO
                COVER[OPTCOV[J]]:=1
       END
   END  (*  NO NULL ROW  *)
END;  (*  SETPARTBACKTRACK  *)
```

2.2.4. COMPUTATIONAL RESULTS

The SETPARTRED and SETPARTBACKTRACK procedures have been tested on a number of standard (published) and random instances of the SPP problem. In particular, we performed a computational experiment with two instances of the problem using the procedures of this section and two other procedures: GOMORY and BALAS of Chapter 1. GOMORY is designed to solve the general integer linear programming (ILP) problem and BALAS, the general 0–1 ILP problem. The two test problems were the 5×31 instance from Pierce [1968] and the 15×32 instance from Lemke et al. [1971].

Table 2-5 shows the running times of the procedures used. As one could expect, the special-purpose algorithm (SETPARTBACKTRACK) was superior to

Table 2-5. Computational Results from the Application of the Three Procedures to the Two Literature Instances of the Set-Partitioning Problem

Procedure	5×31		15×32	
	Time (msec)[a]	COUNT	Time (msec)[a]	COUNT
GOMORY	22	6	148	30
BALAS	36	—	2438	—
SETPARTBACKTRACK				
Columns ordered according to c_j/h_j	3	4	23	178
Columns ordered according to c_j	4	18	26	208

[a] Time on Amdahl 470 V/8.

the general algorithms. However, it is surprising that the GOMORY procedure performed better than the BALAS procedure.

The SETPARTRED procedure left unchanged the 5×31 problem (due to a very dense constraint matrix), whereas in the 15×32 problem, the number of sets has been reduced from 32 to 18 and the number of the elements to be covered from 15 to 13. However, the total computational time of both the SETPARTRED and SETPARTBACKTRACK procedures was 55 msec, twice that of the latter procedure applied to the nonreduced instance.

In conclusion, we strongly support the idea of designing and using special-purpose methods, such as the one implemented in the SETPARTBACKTRACK procedure, which usually surpass the general-purpose methods. The SETPARTRED procedure has been designed to work in time bounded by a polynomial function in the problem size. However, it may take longer to reduce the problem than to solve it. Nevertheless, the application of the reduction method is strongly recommended, especially for dropping from the problem formulation those sets and elements which are either inconsistent with the other sets (since they cannot belong to any feasible solution) or redundant. As a result of the problem reduction we may find that the original problem is infeasible.

For further computational results obtained with the methods for covering problems, the reader is referred to the literature discussed in the References and Remarks.

PROBLEMS

2-24. Show that without loss of generality we may assume that all weights in the set-covering problem are positive (i.e., $c_j > 0$ for $j = 1, 2, \ldots, n$).

2-25. (a) Devise a transformation of the set-partitioning problem with arbitrary weights to an equivalent set-partitioning problem with positive weights. (*Hint:* Multiply the ith constraint by a constant b_i and subtract from the objective function for $i = 1, 2, \ldots, m$.)

    (b) Show how the transformation above can be used to find a lower bound for the solution of the original problem.

2-26. Prove that the constraints $x_j = 0$ or 1 ($j = 1, 2, \ldots, n$) in the SPP and SC problems are equivalent to $x_j \geq 0$ and integer ($j = 1, 2, \ldots, n$).

2-27. Let SC_{LP} denote the set-covering problem with the constraints on variables relaxed to $x_j \geq 0$ ($j = 1, 2, \ldots, n$). Show that the set-covering problem has a feasible solution if and only if SC_{LP} has a feasible solution. Provide a counterexample that a similar fact for the SPP problem does not hold in general.

2-28. Prove that the round-up solution of the SC_{LP} problem (i.e., when nonzero variables are replaced by 1's) is a feasible solution of the SC problem.

2-29.   Show that the set-partitioning problem can be brought to the form of the set-covering problem. [*Hint:* Prove and apply the following relation:

$$\min\{\mathbf{c}^T\mathbf{x} : A\mathbf{x} = \mathbf{e}, \; x_j = 0 \text{ or } 1 \; (j = 1, 2, \ldots, n)\}$$

$$= -Mm + \min\{\bar{\mathbf{c}}^T\mathbf{x} : A\mathbf{x} \geqslant \mathbf{e}, \; x_j = 0 \text{ or } 1 \; (j = 1, 2, \ldots, n)\}$$

where M is a sufficiently large number (e.g., $M > \sum_{j=1}^{n} c_j$) and $\bar{\mathbf{c}}^T = \mathbf{c}^T + Me^T A$.]

2-30.   **(a)** Show that the set-packing problem is a special instance of the set-partitioning problem.

   (b) Show also the converse relation; that is, the set-partitioning problem can be transformed to the form of the set-packing problem. [*Hint:* Prove and apply the following relation:

$$\min\{\mathbf{c}^T\mathbf{x} : A\mathbf{x} = \mathbf{e}, \; x_j = 0 \text{ or } 1 \; (j = 1, 2, \ldots, n)\}$$

$$= Mm + \max\{\tilde{\mathbf{c}}^T\mathbf{x} : A\mathbf{x} \leqslant \mathbf{e}, \; x_j = 0 \text{ or } 1 \; (j = 1, 2, \ldots, n)\}$$

where M is a sufficiently large number and $\tilde{\mathbf{c}}^T = Me^T A - \mathbf{c}^T$.]

2-31.   Illustrate the transformations described in Problems 2-29 and 2-30 on the following matrix A and cost vector $\mathbf{c}$:

$$A[1..4, 1..6] = \begin{bmatrix} 1 & 1 & 0 & 0 & 0 & 0 \\ 1 & 0 & 1 & 0 & 1 & 0 \\ 0 & 0 & 0 & 1 & 1 & 0 \\ 0 & 1 & 0 & 1 & 0 & 1 \end{bmatrix}$$

$$c[1..6] = \begin{bmatrix} 5 & 3 & 7 & 1 & 2 & 0 \end{bmatrix}$$

2-32.   A *node packing* (or an *independent node set*) in a graph G is a subset of nodes which are mutually nonadjacent in G. The *node-packing problem* is to find a maximum-weight node packing in G_c. Show that this problem is a special case of the set-packing problem.

2-33.   A *matching* (or an *edge packing*) in G is a subset of mutually nonadjacent edges. A matching E' is *perfect* if every node of G is incident with an edge in E'. Demonstrate that not every graph, even with an even number of nodes, has a perfect matching. The (*perfect*) *matching problem* is to find a maximum-weight (perfect) matching in G_d. Show that the matching problem is a special instance of the set-packing problem, and that the perfect matching problem is a special case of the set-partitioning problem.

2-34.   A subset of nodes $W \subseteq V$ is a *node cover* of G if every edge of G has at least one end node in W. The *node-covering problem* is to find a node cover of minimum weight with respect to function c. When $c(v) = 1$ for every vertex

$v \in V$, the problem is referred to as the *cardinality node-covering problem*.

(a) Demonstrate that the node-covering problem is a special case of the set-covering problem.

(b) An approximation solution for the cardinality node-covering problem can be constructed by finding a maximal matching $M \subset E$ of G, that is, a maximal (in the set-theoretic sense) set of mutually nonadjacent edges. Let W be the set of end nodes of edges in M. Show that:

  (1) W is a node cover of G,

  (2) $|W| \leqslant 2|W_{\text{opt}}|$ (i.e., the cover W is at most twice as large as the optimal solution).

See Problem 2-51 for some approximation algorithms for the weighted node-covering problem.

2-35. The following greedy algorithm also generates a node-covering *cover* of a graph G:

```
begin
    cover ← ∅;
    while E ≠ ∅ do
    begin
        find a node v of maximum degree in the current graph;
        cover ← cover ∪ {v};
        V ← V − {v};
        E ← E − {(v, u) : u ∈ V}
    end
end
```

Does this algorithm always generate a minimum-cardinality node cover?

2-36. **(a)** Formulate *the edge-covering problem* which calls for a subset of edges covering all nodes of a graph. Show that this is also a special instance of the set-covering problem.

(b) Prove that a minimum cardinality edge cover of a graph can be obtained from a maximum cardinality matching by adding a single edge to each node which has no edge incident to it in the matching.

(c) Work out a similar transformation of a minimum cardinality edge cover to a maximum cardinality matching.

2-37. (Gallai, 1959) Let α_0, β_0, α_1, and β_1 be the cardinalities of a maximum node packing, minimum node cover, maximum matching, and minimum edge cover in a graph G, respectively. Prove that if G is a connected graph with n nodes, then

$$\alpha_0 + \beta_0 = \alpha_1 + \beta_1 = n$$

2-38. Let $\mathcal{P} = \{P_j\}_N$ be a family of subsets of $M = \{1, 2, \ldots, m\}$. The intersection graph $L(\mathcal{P})$ of $\mathcal{P}$ over M has the node set corresponding to $\mathcal{P}$ and two nodes are adjacent if the corresponding subsets overlapped. Let c be a nonnegative weight function defined on $\mathscr{P}$. Show that the set-packing problem for $\mathscr{P}$ can be transformed to the node packing problem on $L(\mathscr{P})$.

Is this transformation also valid for the set-partitioning problem? Discuss the necessary modifications.

2-39. Set up the set-covering problems which when solved give the solution to the node- and edge-covering problems for the graph shown in Fig. 2-7.

2-40. Set up the set-covering problem whose solution would give a minimum-cardinality set of edges that disconnect all paths from node 7 to node 2 in Fig. 2-7.

Comment: Problems 2-32 through 2-40 deal with the covering problems defined on graphs and networks. We assume that $G = (V, E)$ is an undirected graph and that c and d are two nonnegative weight functions defined on V and E, respectively. Let G_c and G_d denote a graph (network) with only one function, c or d, respectively. For other graph-theoretic terms the reader is referred to Section 3.1.

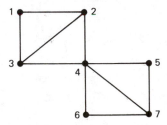

Figure 2-7   *Graph for Problems* 2-39 *and* 2-40.

2-41. Formulate the graph-coloring problem (see Section 4.1 for definitions) as a covering problem. Construct an instance of the latter problem corresponding to the coloring of the graph in Fig. 4-1. Why is this approach to coloring of little practical use?

2-42. Let $x_1, x_2, \ldots, x_n$ be a set of Boolean variables and S_i be the disjunction of those variables which correspond to 1's in the ith row of an $m \times n$ matrix A. Show that $\mathbf{x}^T = (x_1, x_2, \ldots, x_n)$ is a feasible solution to the set-covering problem with matrix A if and only if $\mathbf{x}$ satisfies the following equation in Boolean variables:

$$\bigcap_{i=1}^{m} S_i = 1$$

How can this approach be used to:

(a) Generate all minimal (in the set-theoretic sense) coverings?
(b) Solve the minimum-cardinality set-covering problem?
(c) Solve the minimum-weight set-covering problem?

2-43. **(a)** Transform rule 4C(SC) to an equivalent decision set-covering problem.
(b) Then implement it as a Pascal subroutine which will use one of the approximation algorithms for the set-covering problem discussed in Problems 2-48 and 2-49.

2-44. The equivalent form of the reduction rule 5(SPP). Prove that if the inner product of columns C_t and C_j is nonzero for every $j \in F_i$ for some $i \in M$, column C_t can be removed from the problem.

2-45. Reduce the instance of Example 2-4 by applying the reduction rules in the order 2, 3R, 5, and 3C.

2-46. Confirm properties 1 through 5 of the reduction rules for the SPP problem by providing suitable instances of the problem.

2-47. **(a)** Verify which of properties 1 through 5 of the reduction rules for the SPP problem also hold for the SC problem, and if necessary, modify them.

(b) Based on part (a), design an algorithm for reducing the SC problem and implement it as a Pascal procedure.

2-48. Chvátal [1979] proposed an approximation algorithm for the SC problem which constructs a cover sequentially according to nondecreasing ratios h_j/c_j (the ratios count the number of elements covered by P_j per unit cost).

Algorithm 2-10: Chvátal's Algorithm

```
begin
    cover ← ∅;
    repeat
        find a set Pₖ maximizing hⱼ/cⱼ;
        cover ← cover ∪ {k};
        for all nonempty Pⱼ do Pⱼ ← Pⱼ − Pₖ
    until all Pⱼ are empty
end
```

Note that when all costs are equal, the sets P_j are considered in the nondecreasing order of their cardinalities.

(a) Apply Algorithm 2-10 to the following instance of the SC problem:

$$P_j = \{j\} \ (j = 1, 2, \ldots, m), \ P_{m+1} = \{1, 2, \ldots, m\},$$

$$c_j = 1/(m - i + 1) \ (j = 1, 2, \ldots, m + 1).$$

Break ties by taking a set with a smaller index. Compare the solution obtained with the optimal one.

(b) Implement this algorithm as a Pascal procedure. What is the complexity of your procedure in terms of the problem parameters?

2-49. Another approximation algorithm for the set-covering problem has been proposed by Bar-Yehuda and Even [1981]:

Algorithm 2-11: Algorithm due to Bar-Yehuda and Even

```
begin
    Q ← M;
    cover ← ∅;
    for all j ∈ N do c'_j ← c_j;
    repeat
        let i ∈ Q;
        r ← min{c'_l : l ∈ F_i};
        {* let k be such that c'_k = r *}
        for all l ∈ F_i do c'_l ← c'_l − r;
        cover ← cover ∪ {k};
        Q ← Q − P_k
    until Q = ∅
end
```

(a) Apply Algorithm 2-11 to the instance defined in Problem 2-48(a).

(b) Implement this algorithm to run in time proportional to $\sum_{j=1}^{n} h_j$.

2-50. Compare the efficiency of your implementations of Algorithms 2-10 and 2-11 on random instances of the SC problem. Test also your procedures on the examples corresponding to some random instances of the node-covering problem (see Problem 2-34).

2-51. Simplify Algorithms 2-10 and 2-11 for the node-covering problem. Then implement them in Pascal and profile on a set of random networks.

2-52. An optimization version of the unweighted set-partitioning problem can be formulated as follows:

$$\text{find a set covering } N' \subset N \text{ which minimizes } \sum_{j \in N'} h_j$$

In other words, an optimal solution is a set covering with the least possible overlapping. Implement the following approximation algorithm for the problem above:

```
begin
    cover ← ∅;
    uncov ← M;
    repeat
        find a nonempty set P_k minimizing |P_j − uncov|/|P_j ∩ uncov|;
        cover ← cover ∪ {k};
        uncov ← uncov − P_k;
        for all nonempty P_j do P_j ← P_j − P_k
    until uncov = ∅
end
```

Find a family of instances for which this algorithm fails to produce the optimal solution.

2-53. Modify the SETPARTBACKTRACK procedure by including a rearrangement of rows such that $i \leqslant k$ implies that

$$\sum_{j=1}^{n} a_{ij} \leqslant \sum_{j=1}^{n} a_{kj}$$

Test the effects of this change on your set of inputs.

2-54. Compare the performance of the SETPARTBACKTRACK with orderings of columns within blocks according to:

(a) Nondecreasing costs c_j.

(b) Nonincreasing set-cardinalities h_j.

(c) Nondecreasing ratios c_j/h_j.

2-55. Modify the SETPARTBACKTRACK procedure by incorporating the lower bound for SPP, resulting from Theorem 2-3 in the place of the bound following from Theorem 2-2 actually used. Run both versions of the procedure on random data to compare their efficiency.

2-56. (Pierce and Lasky [1973]) An alternative heuristic way of ordering columns in the implicit enumeration method is as follows. Assume that the columns are ordered according to nondecreasing ratios c_j/h_j. Then assign the first column to the first group, and each other to the first group for which it has at least one nonzero element with every other column in the group. If there is no such a group, create a new one. Justify this method by applying the dominance considerations based on the knapsack relaxation of the problem.

2-57. Modify Algorithms 2-8 and 2-9 to work for the set-covering problem and implement them in Pascal.

2-58. All algorithms described in Section 2.2 can be modified by utilizing binary (or logic) operations on entire 0–1 vectors (e.g., rows and columns of matrix A). Implement the reduction algorithm and the implicit enumeration method making use of the binary (bit) representation of all 0–1 vectors and matrices and performing the bit- and bit-vector-wise operations whenever possible.

REFERENCES AND REMARKS

The main references for the covering problems are the standard texts on integer programming

GARFINKEL, R. S., and G. L. NEMHAUSER, *Integer Programming*, Wiley, New York, 1972, Chapter 8

SALKIN, H. M., *Integer Programming*, Addison-Wesley, Reading, Mass., 1975, Chapter 11

and review papers

BALAS, E., and M. W. PADBERG, Set Partitioning—A Survey, *SIAM Rev.* 18(1976), 710–760

BALINSKI, M., Integer Programming: Methods, Uses, Computation, *Management Sci.* 12(1965), 253–313

PADBERG, M. W., Covering, Packing, and Knapsack Problems, *Ann. Discrete Math.* 4(1979), 265–287

Relations between different covering problems are discussed in Salkin [1975], Balas and Padberg [1976], and Padberg [1979]. Balas and Padberg [1976] also review covering problems on graphs and networks. However, we refer the reader to the graph-theoretic literature for an exhaustive treatment of this special case (see also Chapter 3).

The computational complexity of various covering problems on sets and networks is discussed in

LENSTRA, J. K., and A. H. G. RINNOOY KAN, Complexity of Packing, Covering, and Partitioning Problems, in A. Schrijver (ed.), *Packing and Covering in Combinatorics*, Mathematical Centre Tracts 106, Mathematisch Centrum, Amsterdam, 1979, pp. 275–291

The applications of covering problems are extensively presented in Balinski [1965], Salkin [1975], and Balas and Padberg [1976]. In the last paper, references are grouped according to the applications they discuss. Some successful applications of the set-partitioning problem to solving crew scheduling problems and interactive routing problems have been recently reported in the literature; see, respectively,

MARSTEN, R. E., and F. SHEPARDSON, Exact Solution of Crew Scheduling Problems Using the Set Partitioning Model: Recent Successful Applications, *Networks* 11(1981), 165–177

CULLEN, F. H., J. J. JARVIS, and H. D. RATLIFF, Set Partitioning Based Heuristics for Interactive Routing, *Networks* 11(1981), 125–143

Linear programming and the integer liner programming approach to covering problems are presented in Salkin [1975] and Balas and Padberg [1976].

The reduction rules for the covering problems are discussed in almost every text on these problems. In particular, for the set-partitioning problem, the rules have been proposed in

GARFINKEL, R. S., and G. L. NEMHAUSER, The Set-Partitioning Problem: Set Covering with Equality Constraints, *Oper. Res.* 17(1969), 845–856 and then improved in

GUHA, D. K., The Set Covering Problem with Equality Constraints, *Oper. Res.* 21(1973), 348–351

The reduction rules for the set-covering problem are discussed in Balinski [1965] and Salkin [1975].

The implicit enumeration algorithm as it is described in the text has appeared independently in at least two papers:

PIERCE, J. F., Application of Combinatorial Programming to a Class of All–Zero–One Integer Programming Problems, *Management Sci.* 15(1968), 191–209

and in Garfinkel and Nemhauser [1969] (see also Garfinkel and Nemhauser [1972]). The former paper has been improved in a subsequent work:

PIERCE, J. F., and J. S. LASKY, Improved Combinatorial Programming Algorithms for a Class of All–Zero–One Integer Programming Problems, *Management Sci.* 19(1973), 528–543

Further improvements have been proposed and their efficiency compared in

CHRISTOFIDES, N., and S. KORMAN, A Computational Survey of Methods for the Set Covering Problem, *Management Sci.* 21(1975), 591–599

A large variety of enumeration methods utilize relations between the covering problems and their linear programming relaxations; see, for instance,

LEMKE, C. E., H. M. SALKIN, and K. SPIELBERG, Set Covering by Single-Branch Enumeration with Linear-Programming Subproblems, *Oper. Res.* 19(1971), 998–1022

SALKIN, H. M., and R. D. KONCOL, Set Covering by an All Integer Algorithm: Computational Experience, *J. ACM* 20(1973), 189–193

and also Salkin [1975].

Gallai's theorem which is the subject of Problem 2-37 and other relations between the numbers assigned to packings, covers, and matchings in graphs are discussed in

HARARY, F., *Graph Theory*, Addison-Wesley, Reading, Mass., 1969, Chapter 10

An approach to covering problems utilizing Boolean operations as presented in Problem 2-42 is exchaustively discussed in the monograph

HAMMER, P. L., and S. RUDEANU, *Boolean Methods in Operations Research and Related Areas*, Springer-Verlag, Berlin, 1968

Such an approach is impractical even for medium-size problems, but it becomes more efficient if all Boolean operations can be performed on entire 0–1 vectors. See also

LAWLER, E. L., Covering Problems: Duality Relations and a New Method of Solution, *SIAM J. Appl. Math.* 14(1966), 1115–1132

The approximation algorithms for the set-covering problem presented in Problems 2-48 and 2-49 appeared in

BAR-YEHUDA, R., and S. EVEN, A Linear-Time Approximation Algorithm for the Weighted Vertex Cover Problem, *J. Algorithms* 2(1981), 198–203

CHVÁTAL, V., A Greedy Heuristic for the Set-Covering Problem, *Math. Oper. Res.* 4(1979), 233–235

Chvátal proved that the cost of the solution *cover* generated by Algorithm 2-10 can exceed the cost of an optimal cover by a factor of at most $\sum_{j=1}^{d} 1/j$, where d is the size of the largest set P_j. Bar-Yehuda and Even showed that the cost of the cover generated by their algorithm (Algorithm 2-11) is at most f times the cost of an optimal solution, where f is the largest number of sets that contain a certain element (i.e., $f = \max\{|F_i| : i \in M\}$). Although this bound is usually inferior to the bound of Chvátal's algorithm, it is very attractive for the node-covering problem (defined in Problem 2-34) for which $f = 2$ [compare with Problem 2-34(b)].

Chvátal's algorithm is a generalization of the algorithm proposed by Johnson and Lovász for the equal-weight set-covering problem; see

JOHNSON, D.S., Approximation Algorithms for Combinatorial Problems, *J. Comput. System Sci.* 9(1974), 256–278

LOVÁSZ, L., On the Ratio of Optimal Integral and Fractional Covers, *Discrete Math.* 13(1975), 383–390

Problem 2-52 is from Johnson's paper.

Yet another heuristic method for the set-covering problem is presented in

HOCHBAUM, D. S., Approximation Algorithm for the Set Covering and Vertex Cover Problems, *SIAM J. Comput.* 11(1982), 555–556

The approach proposed in this paper makes use of the polynomial-time algorithm for solving linear programs and produces solutions which are bounded similarly as those generated by the algorithm due to Bar-Yehuda and Even.

3

Optimization on Networks

A large number of transportation, distribution, and communication systems are designed and studied through network models. Some examples are highway networks, urban transportation networks, telephone networks, gas pipeline systems, electrical power distribution systems, and computer networks. In such real-life problems we find that the resulting graphs are so large that their analysis without the aid of the computer is practically impossible. Our ability to solve practical problems by network techniques is only as good as our ability to handle large networks on the computer. Thus efficient algorithms for solving network problems are of great practical importance. In this chapter we present a number of efficient algorithms for solving some of the more frequently encountered network optimization problems, but first a few definitions.

A *graph* $G = (V, E)$ consists of a finite set of *vertices* $V = \{v_1, v_2, \ldots, v_n\}$ and a finite set E of *edges* $E = \{e_1, e_2, \ldots, e_m\}$ (see Fig. 3-1). To each edge e there corresponds a pair of distinct vertices (u, v) which e is said to be *incident* on. While drawing a graph we represent each vertex by a dot and each edge by a line segment joining its two *end vertices*. A graph is said to be a *directed graph* (or *digraph* for short) (see Fig. 3-2) if the vertex pair (u, v) associated with each edge e (also called *arc*) is an ordered pair. Edge e is then said to be *directed from* vertex u to vertex v, and the direction is shown by an arrowhead on the edge. A graph is *undirected* if the end vertices of all edges are unordered (i.e., edges have no direction). A *network* is a directed or undirected graph in which a real number is assigned to each edge. This number is often referred to as the *weight* of that edge. In a practical network this number (weight) may represent the driving distance, the construction cost, the transit time, the reliability, the transition probability, the carrying capacity, or any other such attribute of the edge.

Two edges are said to be in *parallel* if they have the same pair of end vertices (and additionally, if they have the same direction in case of a directed graph). Throughout this chapter we will assume that the networks under consideration have no parallel edges. (This assumption gives us some simplicity without any cost in generality.) Thus we can refer to each edge by its end vertices.

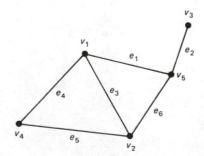

Figure 3-1   *Undirected graph with 5 vertices and 6 edges.*

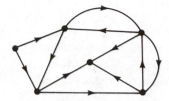

Figure 3-2   *Digraph with 6 vertices and 11 edges.*

Throughout this chapter we use the letters n and m to denote the number of vertices and number of edges, respectively, in a network. In this chapter a vertex will often be referred to as a *node* (a term more popular in applied fields).

3.1. COMPUTER REPRESENTATION OF A NETWORK

Weight Matrix

The simplest and perhaps the most popular computer representation of a network is the *weight matrix*. The weight matrix of an *n*-node network is an $n \times n$ matrix $W = [w_{ij}]$ in which the (i, j)th entry w_{ij} is the weight of (i, j), the edge from node i to node j in the network G. If there is no edge (i, j) in G, the corresponding element w_{ij} is usually set to ∞ (in practice, some very large number). The diagonal entries are usually set to zero (or to some other value depending on the application and algorithm). It is easy to see that the weight matrix of an undirected network is always symmetric. A network and its weight matrix are shown in Fig. 3-3. Boxed numbers next to the edges are their weights.

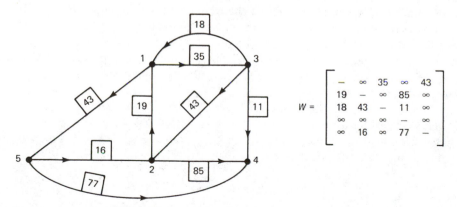

Figure 3-3   *A network and its weight matrix.*

List of Edges

It is clear that a weight matrix representation will require n^2 words of computer memory to store a network. (For an undirected network one can manage with about half as many words.) If the network is sparse [i.e., the number of edges m is much less than $n(n - 1)$, the number of all possible edges], then we would be wasting a good deal of memory space in simply storing ∞'s. In such a case it is often more efficient to simply list the m edges as triplets of end nodes and the weights. This representation can also be implemented by three arrays: $B = (b_1, b_2, \ldots, b_m)$, $D = (d_1, d_2, \ldots, d_m)$, and $Z = (z_1, z_2, \ldots, z_m)$. An edge is going from node b_i to node d_i and has a weight of z_i. For example, the network of Fig. 3-3 would be represented as

$$B = (1, 1, 2, 2, 3, 3, 3, 5, 5)$$
$$D = (3, 5, 1, 4, 1, 2, 4, 2, 4)$$
$$Z = (35, 43, 19, 85, 18, 43, 11, 16, 77)$$

Note that we have listed edges in ascending order of their initial node. But they could have been listed in some other prescribed order, say, in ascending order of their weights. This method of representation requires $3m$ words of computer memory in contrast with n^2 words of memory for the weight matrix.

Linked Adjacency List

In many instances the processing of the network becomes more efficient if all edges emanating from a node are grouped together (but without wasting the memory space in storing entire rows of the weight matrix with many ∞ entries). With such grouping one needs to represent each edge by its terminal node only (and its weight, of course) since its initial node is implicitly defined. This representation is conveniently implemented by means of n linked lists, one for each node i. All edges emanating from the ith node are linked together. Thus each edge requires three fields: one for the destination node of the edge, one for the weight of the edge, and one for the pointer to the next edge in the list (emanating from the same vertex i). Each linked list has a header which contains the pointer pointing to the first edge emanating from the corresponding node. These headers are kept sequentially in an array, because in a typical processing we start at a random node k (and access all its immediate successors).

Figure 3-4 shows a linked adjacency list representation of the network of Fig. 3-3. This representation requires $3m + n$ words of storage. The storage required is slightly more than the storage for a list of edges, but the linking of edges gives us a flexibility, the ease with which one can create and delete edges.

Forward Star

This is a variation of linked-adjacency-list representation. If the network processing does not require any adding or deleting of edges, one could do away with all the pointer fields for each of the edges in the previous representation and simply store

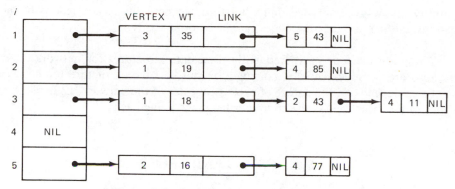

Figure 3-4 *Linked adjacency list of Fig. 3-3.*

the edges sequentially. (The set of edges emanating from each node v is called the *forward star* of v. All edges in the forward star of node k are stored immediately after the edges in the forward star of node $k - 1$.) For example, the forward star representation of the network in Fig. 3-3 would be as shown in Fig. 3-5 using three arrays. Observe that all edges emanating from node i are: $(i, \text{END\_VERTEX}[\text{POINTER}[i]])$, $(i, \text{END\_VERTEX}[\text{POINTER}[i] + 1])$, ..., $(i, \text{END\_VERTEX}[\text{POINTER}[i + 1] - 1])$. Note that the case $i = n$ is taken care of by adding a dummy $(n + 1)$th cell to array POINTER, which contains the value $(m + 1)$.

Analogous to the forward star, a network can also be represented by the *backward star*, where all edges entering *into* node k are grouped together.

For a directed network of n nodes and m edges, the forward-star representation requires $(n + 1) + 2m$ words of computer memory. We have thus gained m words of storage (over the linked-list representation) at the expense of some

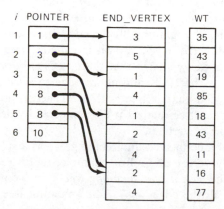

Figure 3-5 *Forward-star representation of Fig. 3-3.*

flexibility. (Now since all edges must be stored sequentially in contiguous memory locations, adding and deleting edges cannot be done easily.)

These are the four most frequently used data structures for networks. Other variations are possible. There is no single overall best data structure for networks. The choice depends on the nature of processing, size of the network, its density, and on the computer as well as the computer language being used.

3.2.   PATHS AND TREES

In a digraph if there is an edge (x, y)—directed from vertex x to vertex y—then y is called an *immediate successor* of x, and x is called an *immediate predecessor* of y. If the edge (x, y) is undirected, then the two vertices x and y are said to be *adjacent* to each other. A *simple path* or *path* for short from vertex v_1 to v_k in a graph G is a sequence of successive edges $(v_1, v_2), (v_2, v_3), \ldots, (v_{k-2}, v_{k-1}), (v_{k-1}, v_k)$, sometimes written as $(v_1, v_2, \ldots, v_k)$, in which all vertices $v_1, v_2, \ldots, v_k$ are distinct. In a digraph this path is said to be *directed from* v_1 to v_k; in an undirected graph this path is said to be *between* v_1 and v_k. The *weight of a path* p in a network is the sum of the weights of all edges in p. For example, in Fig. 3-3, the sequence $(5, 2)$, $(2, 1)$, $(1, 3)$ is a path from vertex 5 to vertex 3, which has three edges and its weight is 70 ($= 16 + 19 + 35$). A single edge (a, b) is clearly a path from a to b. A *cycle* or *circuit* is defined just like a path except that the first vertex v_1 and the last vertex v_k are the same. Thus in Fig. 3-3, the sequence of edges $(2, 1)$, $(1, 5)$, $(5, 2)$ is a cycle. A graph that contains no cycle is called *acyclic*.

In general, in a network there will be many paths from one given vertex s to another specified vertex t. Among all the paths from s to t the one with smallest weight is called the *shortest path* from s to t. Clearly, there may be several shortest paths from s to t.

A *subgraph* of a graph $G = (V, E)$ is a graph whose vertices and edges are in G. An undirected graph G is said to be *connected* if there is at least one path between every pair of vertices v_i and v_j in G. A directed graph G is connected if the undirected graph obtained by ignoring all the edge directions in G is connected.

A connected, undirected acyclic graph is called a *tree*, and a set of trees is called a *forest*. It can be easily shown that there is exactly one path between every pair of vertices in a tree. Therefore, adding an edge to a tree results in exactly one cycle. It can also be seen that a tree with n vertices has exactly $n - 1$ edges. Removing any edge from a tree renders it disconnected by destroying the only path between at least one pair of vertices.

A *spanning tree* of a connected undirected graph is a subgraph of G which is a tree and contains every vertex of G. For example, in Fig. 3-6, a spanning tree is shown in bold lines. The *weight of a spanning tree* T in a connected network is the sum of the weights of all the edges in T. The spanning tree in Fig. 3-6 (shown in

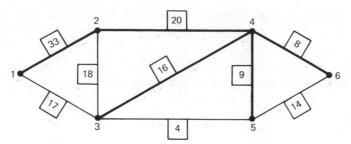

Figure 3-6 *A graph and a spanning tree.*

bold) has a weight of 86 (= 33 + 20 + 16 + 8 + 9). In general, a given connected network has many spanning trees, and we are often interested in finding one with minimum weight (among all spanning trees of G). Such a tree is called a *minimum spanning tree* (MST).

3.3. SHORTEST PATH PROBLEMS

Shortest-path problems are the most fundamental and also the most commonly encountered problems in the study of transportation and communication networks. There are many types of shortest-path problems. For example, we may be interested in determining the shortest path (i.e., the most economical path or fastest path, or minimum-fuel-consumption path) from one specified node in the network to another specified node; or we may need to find shortest paths from a specified node to all other nodes; or perhaps shortest paths between all pairs of nodes in the network. Sometimes, one wishes to find a shortest path from one given node to another given node that passes through certain specified intermediate nodes. In some applications, one requires not only the shortest but also the second and third shortest path. There are occasions when the actual shortest path is not required, but only the shortest distance is required. Thus the shortest-path problems constitute a large class of problems. This is particularly true if we generalize this class to include related problems, such as the longest-path problems, the most-reliable-path problems, the largest-capacity-path problems, and various routing problems. Therefore, it is not surprising that the number of papers, books, reports, dissertations, and surveys dealing with the subject of shortest paths runs into several hundreds.

Obviously, in this section we cannot deal with all types of shortest-path problems and algorithms. We will confine ourselves to two very basic and most important shortest-path problems: (1) how to determine shortest distance (and a shortest path) from a specified node s to another specified node t, and (2) how to determine shortest distances (and paths) from every node to every other node in the network. As we will see later, several other problems can be solved using these two basic algorithms.

We will assume in this section that the given network is directed and has a weight associated with every edge. (In practice, these weights may represent distances, travel time, costs, fuel consumption, or any other such attribute.) The algorithms can be easily adapted for undirected networks, by simply replacing every undirected edge (i, j) with a pair of directed edges (i, j) and (j, i)—each with weight equal to the weight of the original undirected edge. Similarly, if the given graph has no weights associated with its edges, all edges are assigned unit weights.

3.3.1. SINGLE-SOURCE PATHS, NONNEGATIVE WEIGHTS

Let us first consider an algorithm due to Dijkstra [1959] for finding a shortest path (and its weight) from a specified node s (*source* or *origin*) to another specified node t (*target* or *sink*) in a network G in which all edge weights are nonnegative. The basic idea behind Dijkstra's algorithm is to fan out from s and proceed toward t (following the directed edges), labeling the nodes with their distances from s, measured so far. The label of a node u is made *permanent* once we know that it represents the shortest possible distance from s (to u). All nodes not permanently labeled have *temporary* labels.

We start by giving a permanent label 0 to source node s, because zero is the distance of s from itself. All other nodes get labeled ∞, temporarily, because they have not been reached yet. Then we label each immediate successor v of source s, with temporary labels equal to the weight of the edge (s, v). Clearly, the node, say x, with smallest temporary label (among all immediate successors) is the node closest to s. Since all edges have nonnegative weights, there can be no shorter path from s to x. Therefore, we make the label of x permanent. Next, we find all immediate successors of node x, and shorten their temporary labels if the path from s to any of them is shorter by going through x (than it was without going through x). Now, from among all temporarily labeled nodes we pick the one with smallest label, say node y, and make its label permanent. This node y is the second closest node from s. Thus, at each iteration, we reduce the values of temporary labels whenever possible (by selecting a shorter path through the most recent permanently labeled node), then select the node with smallest temporary label and make it permanent. We continue in this fashion until the target node t gets permanently labeled. In order to distinguish the permanently labeled nodes from the temporarily labeled ones, we will keep a Boolean vector *final* of order n. When the ith node becomes permanently labeled, the ith element of this vector changes from *false* to *true*. Another array, *dist*, of order n will be used to store labels of nodes. A variable *recent* will be used to keep track of most recent node to be permanently labeled.

Assuming that the network is given in the form of a weight matrix $W = [w_{ij}]$, with ∞ weights for nonexistent edges, and nodes s and t are specified, this algorithm (which is called Dijkstra's shortest-path or the label-setting algorithm) may be described more precisely as follows:

Algorithm 3-1:   Dijkstra's Shortest-Path Algorithm

INITIALIZATION:
    for all $v \in V$ **do**
    begin
        $dist(v) \leftarrow \infty$;
        $final(v) \leftarrow false$;
        $pred(v) \leftarrow -1$
    end;
    $dist(s) \leftarrow 0$;
    $final(s) \leftarrow true$;
    $recent \leftarrow s$;
    ⟨∗ node s is permanently labeled with 0. All other nodes are temporarily labeled with
        ∞. Node s is the most recent node to be permanently labeled ∗⟩

ITERATION:
    while $final(t) = false$ **do**
    begin
        for every immediate successor v of $recent$ if **not** $final(v)$ **do**
        begin ⟨∗ update temporary labels ∗⟩
            $newlabel \leftarrow dist(recent) + w_{recent, v}$;
            if $newlabel < dist(v)$ **then**
            begin
                $dist(v) \leftarrow newlabel$;
                $pred(v) \leftarrow recent$
            end ⟨∗ relabel v if there is a shorter path via node $recent$ and make $recent$ the
                predecessor of v on the shortest path from s ∗⟩
        end;
        let y be the node with smallest temporary label, which is $\neq \infty$;
        $final(y) \leftarrow true$;
        $recent \leftarrow y$
        ⟨∗ y, the next closest node to s gets permanently labeled ∗⟩
    end

Note that the array *pred* keeps track of the immediate predecessor of a node in the shortest path from s to t. Initially, its entries are set to -1. At the end of the execution the shortest path length is given by the value $dist(t)$. The actual path can be obtained by tracing backward from t to s the predecessors in the *pred* array. That is the sequence of nodes

$$s, pred\,(\,pred(\ldots\,)),\ldots, pred(\,pred(t)), pred(t), t$$

is a shortest path from s to t. In case there is no path from s to t in the given network, $dist(t)$ will remain ∞. This condition will occur if and only if at some point

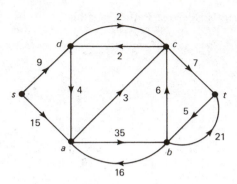

Figure 3-7 *Directed network.*

all temporarily labeled nodes have ∞ weights. Upon detecting this condition, we must exit from the outer loop in the ITERATION step and stop.

A complete Pascal implementation of Algorithm 3-1 will be given shortly, but first let us see how it works by means of a small network shown in Fig. 3-7. Initially, node *s* is assigned a permanent label of 0, and the other five nodes get temporary labels of ∞. Next, the immediate successors of *s*, namely, *a* and *d*, get their temporary labels reduced from ∞ to 15 and 9, respectively. Now, since *d* is the node with smallest temporary label, the label of *d* becomes permanent. Next, the immediate successors of *d*, which are *a* and *c*, get their temporary label reduced to 13

	s	*a*	*b*	*c*	*d*	*t*
Initialization	⓪	∞	∞	∞	∞	∞
Iteration 1	⓪	15	∞	∞	9	∞
	⓪	15	∞	∞	⑨	∞
Iteration 2	⓪	13	∞	11	⑨	∞
	⓪	13	∞	⑪	⑨	∞
Iteration 3	⓪	13	∞	⑪	⑨	18
	⓪	⑬	∞	⑪	⑨	18
Iteration 4	⓪	⑬	48	⑪	⑨	18
	⓪	⑬	48	⑪	⑨	⑱

Figure 3-8 *Shortest path from s to t in Fig. 3-7.*

and 11, respectively. And so on. The labels of all nodes, as they undergo the changes during the execution of this algorithm, are shown in Fig. 3-8. Note that the permanent labels are circled and the temporary ones are not.

Time and Storage Complexities of Dijkstra's Shortest-Path Algorithm

The time required in the initialization step is primarily in setting up the three arrays *dist*, *final*, and *pred*, each of length n (where n is the number of nodes in the network). It is thus proportional to n.

The outer loop in the iteration may, in the worst case, be executed $(n - 1)$ times. This will be the case if the target node t is farthest from s (and thus gets a permanent label last). For each execution of the loop, one has to examine one row of W, the weight matrix (the row corresponding to node *recent*) and update arrays *dist* and *pred*, thus requiring time proportional to n. In addition, the number of comparisons required in finding the smallest temporary label in the ith iteration is $(n - i - 1)$. Thus the total execution time for Dijkstra's algorithm is $O(n^2)$ in the worst case, when the network is given in the form of the weight matrix W.

Notice that for a given n the computation time is independent of the number of edges the network may have. This is because it is tacitly assumed that the network is complete—each missing edge is simply given a very large weight, ∞. It is clear that if the network is complete or if it is given in the form of a weight matrix W, an $O(n^2)$ algorithm is the best one can hope to achieve (see Problem 3-18).

If the network is sparse [i.e., the number of edges is much smaller than $n(n - 1)$], it is possible to reduce the actual computation time by selecting a different data structure (e.g., adjacency lists) and by considering for updating only those nodes that are immediate successors of *recent*. An important issue in implementation of Dijkstra's algorithm is how to keep the list of temporarily labeled nodes. The simplest method is to keep them in an unsorted list, and find the one with smallest label in each iteration. This is how we have implemented in the DIJKSTRA procedure. In a complete graph (or if weight matrix is used to represent the network, as we have in the DIJKSTRA procedure) the number of comparisons required in selecting the minimum of all $(n - i)$ temporary labels in the ith iteration will still be $(n - i - 1)$. Since

$$\sum_{i=1}^{n-2} (n - i - 1) = O(n^2)$$

the computation time for this portion of the algorithm will remain proportional to $O(n^2)$. There are several techniques of reducing the computation time for finding the minimum temporary label. Some of these are discussed in Problems 3-9 and 3-10.

The memory requirement is primarily that of storing the given weight matrix W and the three arrays *dist*, *final*, and *pred*. These require $n^2 + 3n$ words of storage space.

Remarks

1. Dijkstra's shortest-path algorithm, as presented here, finds a single shortest path from s to t. Sometimes, there are several shortest paths from s to t—all of the same lengths, of course. If all shortest paths are required, the foregoing procedure would have to be modified. This modification for producing *all* shortest paths from s to t is discussed in Problem 3-2.

2. In this algorithm, had we continued the labeling until every node got a permanent label (rather than stopping at the permanent labeling of the destination node t), we would have gotten an algorithm for shortest paths from starting node s to all other nodes. Since in each execution of the loop one node gets permanently labeled, we have to change simply the **while** loop to

    for $x \leftarrow 1$ **to** n **do**

    to obtain shortest-path lengths from s to all nodes (reachable from s). Even this modified algorithm will require $O(n^2)$ time.

    If we are interested in finding a shortest path from every node to every other node, in a network with positive-weight edges, we can do so by invoking this modified Dijkstra's algorithm n times—once from each node. This will, of course, require $O(n^3)$ time.

3. If we take a shortest path from the starting node s to each of the other nodes (which are reachable from s), the union of these paths will be a directed tree T rooted at node s. Every path in T from s is the (unique) shortest path in the network. Such a tree is called a *shortest-path tree*—not to be confused with the minimum spanning tree (as defined in Section 3.2). In transportation engineering, such a tree is also referred to as the *skim tree*.

    In order to construct a skim tree we continue labeling until all nodes get permanently labeled (as discussed in Remark 2), and each time a node x gets permanently labeled, the edge ($pred(x)$, x) is included in the skim tree.

    For example, the skim tree of Fig. 3-7 (from source node s) is given in Fig. 3-9. The computational cost of obtaining a skim tree is also $O(n^2)$.

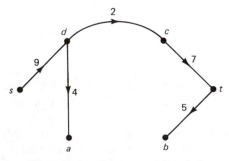

Figure 3-9   *Skim tree of Fig.* 3-7.

4.   If the given digraph *G* is not weighted, every edge in *G* has a weight of 1, and matrix *W* is the same as the *adjacency matrix* of *G*. Then the problem is somewhat simpler (Problems 3-5, 3-6, and 3-8).

Computer Implementation of Dijkstra's Shortest-Path Algorithm

The DIJKSTRA procedure computes the shortest distance from a given node S to another specified node T in an N-node network, given as a weight matrix W. It also computes the PRED array from which an actual shortest path from S to T can be traced, by moving back from T along the predecessors, as explained earlier in this section. If there is no path from the source S to the sink T, Boolean variable PATH is set to FALSE. The nonexistent edges have INF weights in matrix W. This procedure follows closely the method outlined in Algorithm 3-1.

Global Constant

N   number of nodes in the given network

Data Types

```
TYPE   ARRNN     = ARRAY[1..N,1..N] OF INTEGER;
       ARRN      = ARRAY[1..N] OF INTEGER;
       BOOLARRN  = ARRAY[1..N] OF BOOLEAN;
```

Procedure Parameters

```
PROCEDURE DIJKSTRA(
    N,INF,S,T      :INTEGER;
    VAR W          :ARRNN;
    VAR PATH       :BOOLEAN;
    VAR FINAL      :BOOLARRN;
    VAR DIST,PRED  :ARRN);
```

Input

N	number of nodes in the given network
INF	integer much larger than largest edge weight
S	source node
T	sink node
W[1..N,1..N]	weight matrix of the given network

Output

PATH	Boolean variable such that if PATH = FALSE, then there is no path from the source S to the sink T

FINAL[1..N]   Boolean array. If a node V has been permanently labeled, FINAL[V] = TRUE. Otherwise, FINAL[V] = FALSE

DIST[1..N]   array that contains shortest distance from S to nodes that have been permanently labeled. (Note the exception: If PATH = FALSE, then DIST[T] = INF, although FINAL[T] = TRUE.)

PRED[1..N]   array with which an actual shortest path from S can be traced

Example

For the network shown in Fig. 3-7, input data and the initial states of the arrays DIST and PRED are as follows (note that the diagonal elements of W are immaterial and can be set to any value and ∞'s are equal to at least INF):

$$N = 6, \quad INF = 200, \quad S = 1, \quad T = 6,$$

$$W[1..6,1..6] = \begin{array}{c} \\ s \\ a \\ b \\ c \\ d \\ t \end{array} \begin{array}{cccccc} s & a & b & c & d & t \\ \left[\begin{array}{cccccc} - & 15 & \infty & \infty & 9 & \infty \\ \infty & - & 35 & 3 & \infty & \infty \\ \infty & 16 & - & 6 & \infty & 21 \\ \infty & \infty & \infty & - & 2 & 7 \\ \infty & 4 & \infty & 2 & - & \infty \\ \infty & \infty & 5 & \infty & \infty & - \end{array} \right] \end{array}$$

$$DIST[1..6] = [0, \infty, \infty, \infty, \infty, \infty]$$

$$PRED[1..6] = [-1, -1, -1, -1, -1, -1]$$

The solution obtained is

$$PATH = TRUE$$

$$FINAL[1..6] = [TRUE, TRUE, FALSE, TRUE, TRUE, TRUE]$$

$$DIST[1..6] = [0, 13, 48, 11, 9, 18]$$

$$PRED[1..6] = [-1, 5, 2, 5, 1, 4]$$

Pascal Procedure DIJKSTRA

```
PROCEDURE DIJKSTRA(
    N,INF,S,T    : INTEGER;
    VAR W        : ARRNN;
    VAR PATH     : BOOLEAN;
    VAR FINAL    : BOOLARRN;
    VAR DIST,PRED: ARRN);


VAR  U,V,Y,RECENT,NEWLABEL,TEMP: INTEGER;
BEGIN
    FOR V:= 1 TO N DO
```

```
BEGIN
    DIST[V] := INF;              (*  INF = WEIGHT OF A  *)
    FINAL[V]:= FALSE;            (*  NONEXISTENT EDGE   *)
    PRED[V] := -1
END;
DIST[S]:= 0; FINAL[S]:= TRUE;
PATH:= TRUE; RECENT:= S;
(* INITIALIZATION OVER *)

WHILE NOT FINAL[T] DO
BEGIN
    FOR V:= 1 TO N DO                    (* FIND NEW LABEL *)
        IF (W[RECENT,V] < INF) AND (NOT FINAL[V]) THEN
        BEGIN
            NEWLABEL:= DIST[RECENT] + W[RECENT,V];
            IF NEWLABEL < DIST[V] THEN
            BEGIN
                DIST[V]:= NEWLABEL;
                PRED[V]:= RECENT
            END
        END;
    TEMP:= INF;
    FOR U:= 1 TO N DO      (* FIND SMALLEST LABELED VERTEX *)
        IF (NOT FINAL[U]) AND (DIST[U] < TEMP) THEN
        BEGIN
            Y:= U;
            TEMP:= DIST[U]
        END;
    IF TEMP < INF THEN                (* THERE IS A PATH *)
    BEGIN
        FINAL[Y]:= TRUE;
        RECENT   := Y
    END
    ELSE                   (* THERE IS NO PATH FROM S TO T *)
    BEGIN
        PATH:= FALSE;
        FINAL[T]:= TRUE
    END
END (* WHILE *)
END (* DIJKSTRA *);
```

3.3.2.   SINGLE-SOURCE PATHS, ARBITRARY WEIGHTS

In Dijkstra's shortest-path algorithm (Algorithm 3-1), it was assumed that all edge weights w_{ij} were nonnegative numbers. If some of the edge weights are negative, Algorithm 3-1 will not work. (Negative weights in a network may represent costs and positive ones profit.) The reason for the failure is that once the label of a node is made permanent, it cannot be changed in future iterations (Problem 3-4).

.In order to handle a network that has both positive and negative weights, we must ensure that no label is considered permanent until the program halts. Such an algorithm was indeed proposed by Bellman [1958] and Moore [1957] and efficiently implemented by several subsequent researchers. The Bellman–Moore method is often called a *label-correcting method*, in contrast to Dijkstra's *label-setting method*. The Bellman–Moore algorithm may be described as below. (Note the similarity and differences between Algorithms 3-1 and 3-2.)

Like Dijkstra's algorithm, the label of the starting node s is set to zero and that of every other node is set to ∞, a very large number. That is, the initialization consists of

> *dist*(s) $\leftarrow$ 0;
> **for** all $v \neq s$ **do** *dist*(v) $\leftarrow \infty$

In the iterative step, *dist*(v) is always updated to the currently known distance from s to v, and the predecessor *pred*(v) of v is also updated to be the predecessor node of v on the currently known shortest path from s to v. More compactly, the iteration may be expressed as follows:

> **while** there exists an edge (u, v) such that
>   *dist*(u) + $w_{u,v}$ < *dist*(v) **do**
> **begin**
>   *dist*(v) $\leftarrow$ *dist*(u) + $w_{u,v}$;
>   *pred*(v) $\leftarrow u$
> **end**

Several implementations of this basic iterative step have been studied, experimented with, and reported in the literature. Depending on the nature of the given network, some are more efficient than others. One very efficient implementation, which is often faster than even Dijkstra's algorithm for large sparse networks, works as follows.

We maintain a queue of "nodes to be examined". Initially, this queue, Q, contains only the starting node s. The node u from the front of the queue is "examined" (as follows) and deleted. Examining u consists of considering all edges (u, v) going out of u. If the length of the path to node v (from s) is reduced by going through u, that is,

> **if**   *dist*(u) + $w_{u,v}$ < *dist*(v)
> **then** ⟨∗ *dist*(v) is reset to the smaller value ∗⟩
>   *dist*(v) $\leftarrow$ *dist*(u) + $w_{u,v}$;
>   *pred*(v) $\leftarrow u$

Moreover, this node v is added to the queue (if it is not already in the queue) as a node to be examined later. Note that v enters the queue only if *dist*(v) is decremented as above and if v is currently not in the queue.

Observe that in Dijkstra's method a node was "examined" only once, and it was the most recent node to be permanently labeled, that is, the temporary node closest to the origin, s. In Bellman-Moore's method a node may enter (and leave) the queue several times—each time a shorter path is discovered. This source of inefficiency can be reduced by the following heuristic suggested by d'Esopo and refined by Pape [1980]. Place node v at the tail of the queue if v has never been reached before (i.e., v has never entered the queue before) and place v at the head of the queue if it has already entered, been examined, and removed from the queue. The reason for placing v at the head of the queue (if v was once entered in the queue, examined, and removed) is to reexamine it immediately and shorten labels of all nodes that were reached via v. This reduces the number of times a node may reenter the queue (and therefore be reexamined).

The label-correcting algorithm suggested by Moore [1957] and Bellman [1958] and improved by d'Esopo and Pape [1980] may be described more precisely and compactly as shown below:

Algorithm 3-2:   Moore–Bellman–d'Esopo's Shortest-Path Algorithm

```
INITIALIZATION:
    for all v ∈ V do
    begin
        dist(v) ← ∞;
        pred(v) ← −1
    end;
    dist(s) ← 0;
    initialize Q to contain s only;
    head ← s;


ITERATION:
    while Q is not empty do
    begin
        delete the head node u from Q;
        for each edge (u, v) that starts at u do
        begin
            newlabel ← dist(u) + w_{u, v};
            if newlabel < dist(v) then
            begin
                dist(v) ← newlabel;
                pred(v) ← u;
                if v was never in Q then
                    insert v at the tail of Q
                else if v was in Q but is not currently in Q then
                    insert v at the head of Q
            end
        end
    end
```

As pointed out earlier, Algorithm 3-2 works correctly even if the network contains some negative-weight edges, (unlike Algorithm 3-1). Since each insertion of a node v into the queue Q is preceded by a decrement of $dist(v)$, the algorithm is guaranteed to terminate, provided that the given network has no cycle in which the sum of all edge weights is negative. (The problem has no solution if a negative-weight cycle exists, because then one could continue minimizing the distance by going around and around this cycle forever. In fact, an algorithm for detecting negative-weight cycles will be discussed in the next section.) It can easily be shown that in this algorithm $dist(v)$'s do indeed converge to the shortest distance from s.

You may have observed that Algorithm 3-2 gives shortest distance (and path) from s to not only one specified node t but to all nodes in the network. Thus the output is always the skim tree and distances along the skim tree from s.

Figure 3-10 illustrates how Algorithm 3-2 works. Array $dist$ and the queue Q are shown for various iterations as the algorithm progresses. Array $pred$ has not been shown in order to keep the figure from getting cluttered. Notice that only node a has been examined twice. When it entered the queue the second time, it was placed at the head of the queue. The reader is encouraged to use Algorithm 3-1 on this example and notice the difference between the two algorithms. It would also be instructive to observe how much additional computation would be done if we simplified Algorithm 3-2 to add nodes only at the tail of the queue, regardless of their past entries (and exits) into the queue.

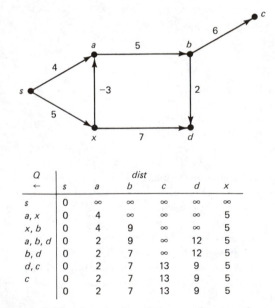

Q				$dist$			
$\leftarrow$	s	a	b	c	d	x
s	0	∞	∞	∞	∞	∞
a, x	0	4	∞	∞	∞	5
x, b	0	4	9	∞	∞	5
a, b, d	0	2	9	∞	12	5
b, d	0	2	7	∞	12	5
d, c	0	2	7	13	9	5
c	0	2	7	13	9	5
	0	2	7	13	9	5

Figure 3-10 *A network and shortest distances from s.*

Performance Analysis

Both Algorithms 3-1 and 3-2 are fast and efficient. If the input network has edges with both positive and negative weights, there is no choice. We must use Algorithm

3-2. If the network has only positive-weight edges, we have a choice: If we wish to find the shortest distance and path from s to only t (and not from s to all other nodes), then probably Dijkstra's algorithm will be faster. Performance will depend on when node t gets permanently labeled. It may be the second closest node to s or it may be the farthest node from s. Thus an exact analysis is difficult for a general network.

An analytic expression for execution time of Algorithm 3-2 is even more difficult to derive, because the number of times various nodes reenter the queue is not easily predictable for an arbitrary network. If each node entered the queue exactly once, the time complexity will be proportional to the number of edges in the network, and in that case Algorithm 3-2 would outperform Algorithm 3-1. Pathological networks can be constructed for which the worst-case computational time for Algorithm 3-2 becomes exponential in the number of nodes n (see Problems 3-15 and 3-16). In practice, however, such networks do not occur. For Algorithm 3-1, on the other hand, no matter what the input network is (as long as all the edges have positive weights), the computation time never exceeds $O(n^2)$.

Extensive experiments of Denardo and Fox [1979], Dial et al. [1979], and Van Vliet [1978] show that if the input network has low number of edges to number of nodes ratio, Algorithm 3-2 is faster than almost every other known algorithm for finding shortest paths from a given node to all other nodes.

In brief, if the given network is small and dense, Algorithm 3-1 (Dijkstra's) is faster, particularly if only a path s to t is required. If the network is large and sparse, and if we need shortest paths from s to all other nodes (or the skim tree), then Algorithm 3-2 (Moore–Bellman–d'Esopo's) is better. The crossover points will depend on the computer being used and the characteristics of the input network.

Computer Implementation of PDM Shortest-Path Algorithm

The PDM (for Pape, d'Esopo, and Moore) procedure computes the shortest distances from a given node S to all other nodes in an N-node network. It also computes the PRED array with which an actual shortest path from S to every other node can be traced (if one exists). As discussed earlier, the union of these paths forms a skim tree rooted at S. Since this algorithm works best for sparse networks, the input network in this procedure is assumed to be in a forward-star form.

Global Constants

N   number of nodes in the given network

N1   N + 1

M   number of edges in the given network

Data Types

```
TYPE   ARRN  = ARRAY[1..N] OF INTEGER;
       ARRN1 = ARRAY[1..N1] OF INTEGER;
       ARRM  = ARRAY[1..M] OF INTEGER;
```

Procedure Parameters

```
PROCEDURE PDM(
    N,M,S,INF          :INTEGER;
    VAR POINTER        :ARRN1;
    VAR ENDV,WT        :ARRM;
    VAR DIST,PRED      :ARRN);
```

Input

N	number of nodes in the given network
M	number of edges in the given network
S	source node
INF	integer much larger than largest edge weight
POINTER[1..N1]	array of pointers in forward-star form
ENDV[1..M]	array of edges in forward-star form
WT[1..M]	array of weights of edges in forward-star form

Output

DIST[1..N]   array which contains shortest distances from S

PRED[1..N]   array with which actual shortest paths from S to the other nodes
can be traced

Example

For the network shown in Fig. 3-7, the data representation in forward-star
form and the initial states of the arrays DIST and PRED are as follows (nodes are
ordered in s, a, b, c, d, t):

$$N = 6, \quad M = 12, \quad S = 1, \quad INF = 200$$
$$POINTER[1..7] = [1, 3, 5, 8, 10, 12, 13]$$
$$ENDV[1..12] = [2, 5, 3, 4, 2, 4, 6, 5, 6, 2, 4, 3]$$
$$WT[1..12] = [15, 9, 35, 3, 16, 6, 21, 2, 7, 4, 2, 5]$$
$$DIST[1..6] = [0, \infty, \infty, \infty, \infty, \infty]$$
$$PRED[1..6] = [-1, -1, -1, -1, -1, -1]$$

The solution obtained is

$$DIST[1..6] = [0, 13, 23, 11, 9, 18]$$
$$PRED[1..6] = [-1, 5, 6, 5, 1, 4]$$

Pascal Procedure PDM

```
PROCEDURE PDM(
    N,M,S,INF           : INTEGER;
    VAR POINTER         : ARRN1;
    VAR ENDV,WT         : ARRM;
    VAR DIST,PRED       : ARRN);
  VAR U,V,HEAD,NEXT,FIRST,LAST,J,NEWLABEL,TEMP: INTEGER;
      QUEUE: ARRN;
BEGIN
    FOR V:= 1 TO N DO
    BEGIN
        DIST[V]:= INF;                        (* INF = WEIGHT OF A *)
        PRED[V]:= -1;                         (* NONEXISTENT EDGE  *)
        QUEUE[V]:= -1
    END;
    DIST[S]:= 0; HEAD:= S; U:= S;
    QUEUE[HEAD]:= INF;              (* V IS THE LAST ITEM IN QUEUE *)
(* INITIALIZATION OVER *)          (* IF QUEUE(V) = INF.          *)
    WHILE U <> INF DO               (* QUEUE IS EMPTY IF U = INF *)
    BEGIN
        NEXT := QUEUE[U];
        QUEUE[U]:= 0;
        FIRST:= POINTER[U];
        LAST := POINTER[U+1] - 1;
        FOR J:= FIRST TO LAST DO      (* FOR EACH ARC (U,V) THAT *)
        BEGIN                         (* STARTS AT U DO          *)
            V:= ENDV[J];
            NEWLABEL:= DIST[U] + WT[J];
            IF DIST[V] > NEWLABEL THEN            (* CHANGE LABEL *)
            BEGIN
                PRED[V]:= U;
                DIST[V]:= NEWLABEL;
                TEMP:= NEXT;
                IF QUEUE[V] < 0 THEN        (* V WAS NEVER IN QUEUE *)
                BEGIN
                    QUEUE[HEAD]:= V;
                    HEAD:= V;
                    QUEUE[HEAD]:= INF;
                    IF TEMP = INF THEN TEMP:= V;
                    NEXT:= TEMP
                END
                ELSE IF QUEUE[V] = 0 THEN   (* V WAS IN QUEUE, BUT *)
                BEGIN                        (* NOT IN QUEUE NOW    *)
                    QUEUE[V]:= TEMP;
                    NEXT:= V
                END
                ELSE    NEXT:= TEMP          (* V IS IN QUEUE *)
            END (* DIST[V] > NEWLABEL *)
        END (* J *);
        U:= NEXT
    END (* WHILE *)
END (* PDM *);
```

3.3.3. SHORTEST PATHS BETWEEN ALL PAIRS OF NODES

So far in this section we have considered only the single-source shortest-path problems—Algorithm 3-1 was devised to find a shortest path from one specified node, s, to another specified node, t, and Algorithm 3-2 was for finding a shortest path from one specified node, s, to every other node in the network. Let us now consider the problem of finding a shortest path between every pair of nodes in the network. Clearly, in an n-node directed network there are $n(n-1)$ such paths—one for each ordered pair of distinct nodes—and $n(n-1)/2$ such paths in an undirected network. One could, of course, solve this problem by repeated application of Algorithm 3-1 or 3-2, once for each node in the network taken as the source node s. However, Algorithm 3-1 does not work for networks with negative-weight edges (Problem 3-4), whereas Algorithm 3-2 may take an exponential amount of time in the worst-case input (Problems 3-15 and 3-16). We will instead consider a different algorithm for finding shortest paths between all pairs of nodes, which is due to Floyd [1962]. It requires computation time proportional to n^3, and allows some of the edges to have negative weights, as long as no cycles of negative weight exist.

The algorithm works by inserting one or more nodes into paths, whenever it is advantageous to do so. Starting with the $n \times n$ weight matrix $W = [w_{ij}]$ of direct distances between the nodes of the given network G, we construct a sequence of n matrices $W^{(1)}, W^{(2)}, \ldots, W^{(n)}$. Matrix $W^{(l)}$, $1 \leqslant l \leqslant n$, may be thought of as the matrix whose (i, j)th entry $w_{ij}^{(l)}$ gives the length of the shortest path among all paths from i to j with nodes $1, 2, \ldots, l$ allowed as intermediate nodes. Matrix $W^{(l)} = [w_{ij}^{(l)}]$ is constructed as follows:

$$w_{ij}^{(0)} = w_{ij}$$
$$w_{ij}^{(l)} = \min\left\{w_{ij}^{(l-1)}, w_{il}^{(l-1)} + w_{lj}^{(l-1)}\right\} \qquad \text{for } l = 1, 2, \ldots, n \qquad (3\text{-}1)$$

In other words, in iteration 1, node 1 is inserted in the path from node i to node j if $w_{ij} > w_{i1} + w_{1j}$. In iteration 2, node 2 can be inserted, and so on.

Suppose, for example, that the shortest path from node 7 to 3 is 7-4-1-9-5-3. The following replacements occur:

Iteration 1: $w_{49}^{(0)}$ is replaced by $(w_{41}^{(0)} + w_{19}^{(0)})$.

Iteration 4: $w_{79}^{(3)}$ is replaced by $(w_{74}^{(3)} + w_{49}^{(3)})$.

Iteration 5: $w_{93}^{(4)}$ is replaced by $(w_{95}^{(4)} + w_{53}^{(4)})$.

Iteration 9: $w_{73}^{(8)}$ is replaced by $(w_{79}^{(8)} + w_{93}^{(8)})$.

Once the shortest distance is obtained in $w_{73}^{(9)}$, the value of this entry will not be altered in subsequent operations.

We assume as usual that the weight of a nonexistent edge is ∞, that $x + \infty = \infty$, and that $\min\{x, \infty\} = x$ for all x. It can be easily seen that all distance matrices

$W^{(l)}$ calculated from (3-1) can be overwritten on W itself. Hence, Floyd's algorithm may be stated now as follows:

Algorithm 3-3:   Shortest Distances between All Pairs of Nodes

```
for l ← 1 to n do
    for i ← 1 to n do
        if wᵢₗ ≠ ∞ then
            for j ← 1 to n do wᵢⱼ ← min{wᵢⱼ, wᵢₗ + wₗⱼ}
```

If the network has no negative-weight cycle, the diagonal entries $w_{ii}^{(n)}$ represent the length of shortest cycles passing through node i. The off-diagonal entries $w_{ij}^{(n)}$ are the shortest distances. Notice that negative weight of an individual edge has no effect on this algorithm as long as there is no cycle with a net negative weight. If the network has negative-weight cycles, then some diagonal entries in $W^{(n)}$ are negative and the off-diagonal entries do not represent distances. Why?

Note that Algorithm 3-3 does not actually list the paths, it only produces their lengths. Obtaining paths is slightly more involved than in Algorithms 3-1 and 3-2, where a predecessor vector *pred* was sufficient. Here the paths can be constructed from a *path matrix* $P = [p_{ij}]$ (also called *optimal policy matrix*), in which p_{ij} is the second to the last node along the shortest path from i to j—the last node being j. The path matrix P is easily calculated in Algorithm 3-3. Initially, we set

$$p_{ij} \leftarrow i, \qquad \text{if} \quad w_{ij} \neq \infty$$
$$p_{ij} \leftarrow 0, \quad \cdot \quad \text{if} \quad w_{ij} = \infty$$

In the lth iteration if node l is inserted between i and j; that is, if $w_{il} + w_{lj} < w_{ij}$, then we set $p_{ij} \leftarrow p_{lj}$. At the termination of the execution, the shortest path $(i, v_1, v_2, \ldots, v_q, j)$ from i to j can be obtained from matrix P as follows:

$$v_q = p_{ij}$$
$$v_{q-1} = p_{i, v_q}$$
$$v_{q-2} = p_{i, v_{q-1}}$$
$$\vdots$$
$$i = p_{i, v_1}$$

Performance Analysis

The storage requirement is n^2, no more than for storing the weight matrix itself. All the intermediate matrices as well as the final distance matrix are overwritten on W itself. Another n^2 storage space would be required if it generated the path matrix P also.

The computation time for Algorithm 3-3 is clearly $O(n^3)$ regardless of the density of the given network.

Computer Implementation of Floyd's Shortest-Path Algorithm

The FLOYD procedure produces the shortest-distance matrix W and the path matrix P in a given network with N nodes. If there is a negative cycle, the Boolean variable NEGACYCLE is set to TRUE.

Global Constant

N   number of nodes in the given network

Data Type

```
TYPE   ARRNN = ARRAY[1..N,1..N] OF INTEGER;
```

Procedure Parameters

```
PROCEDURE FLOYD(
     N,INF              :INTEGER;
     VAR NEGACYCLE :BOOLEAN;
     VAR W,P            :ARRNN);
```

Input

N	number of nodes in the given network
INF	integer much larger than largest edge weight
W[1..N,1..N]	weight matrix of the given network

Output

NEGACYCLE	Boolean variable such that if the network contains a negative cycle, then NEGACYCLE = TRUE, and NEGACYCLE = FALSE, otherwise
W[1..N,1..N]	shortest-distance matrix from every node to every node
P[1..N,1..N]	path matrix with which actual shortest path from every node to every node can be traced

Example

For the network shown in Fig. 3-7, we have N = 6, INF = 300, and the array W and the initial state of the array P are as follows (nodes are ordered in *s, a, b, c, d, t*):

$$W[1..6,1..6] = \begin{bmatrix} \infty & 15 & \infty & \infty & 9 & \infty \\ \infty & \infty & 35 & 3 & \infty & \infty \\ \infty & 16 & \infty & 6 & \infty & 21 \\ \infty & \infty & \infty & \infty & 2 & 7 \\ \infty & 4 & \infty & 2 & \infty & \infty \\ \infty & \infty & 5 & \infty & \infty & \infty \end{bmatrix},$$

$$P[1..6,1..6] = \begin{bmatrix} 0 & 5 & 6 & 5 & 1 & 4 \\ 0 & 5 & 6 & 2 & 4 & 4 \\ 0 & 5 & 6 & 3 & 4 & 4 \\ 0 & 5 & 6 & 5 & 4 & 4 \\ 0 & 5 & 6 & 5 & 4 & 4 \\ 0 & 5 & 6 & 3 & 4 & 4 \end{bmatrix}$$

The solution obtained is

$$\text{NEGACYCLE} = \text{FALSE}$$

$$W[1..6,1..6] = \begin{bmatrix} \infty & 13 & 23 & 11 & 9 & 18 \\ \infty & 9 & 15 & 3 & 5 & 10 \\ \infty & 12 & 18 & 6 & 8 & 13 \\ \infty & 6 & 12 & 4 & 2 & 7 \\ \infty & 4 & 14 & 2 & 4 & 9 \\ \infty & 17 & 5 & 11 & 13 & 18 \end{bmatrix}$$

$$P[1..6,1..6] = \begin{bmatrix} 0 & 1 & 0 & 0 & 1 & 0 \\ 0 & 0 & 2 & 2 & 0 & 0 \\ 0 & 3 & 0 & 3 & 0 & 3 \\ 0 & 0 & 0 & 0 & 4 & 4 \\ 0 & 5 & 0 & 5 & 0 & 0 \\ 0 & 0 & 6 & 0 & 0 & 0 \end{bmatrix}$$

Pascal Procedure FLOYD

```
PROCEDURE FLOYD(
   N,INF       : INTEGER;
   VAR NEGACYCLE: BOOLEAN;
   VAR W,P      : ARRNN);



VAR  I,J,L,NEWLABEL: INTEGER;
BEGIN
   FOR I:= 1 TO N DO
      FOR J:= 1 TO N DO
         IF W[I,J] <> INF THEN P[I,J]:= I
         ELSE P[I,J]:= 0;
```

```
      NEGACYCLE:= FALSE;
(* INITIALIZATION OVER *)

   L:= 1;
   WHILE (L <= N) AND (NOT NEGACYCLE) DO
   BEGIN
      FOR I:= 1 TO N DO
      BEGIN
         IF W[I,L] <> INF THEN
            FOR J:= 1 TO N DO
               IF W[L,J] <> INF THEN
               BEGIN
                  NEWLABEL:= W[I,L] + W[L,J];
                  IF W[I,J] > NEWLABEL THEN
                  BEGIN
                     W[I,J]:= NEWLABEL;
                     P[I,J]:= P[L,J]
                  END
               END (* J *);
         NEGACYCLE:= NEGACYCLE OR (W[I,I] < 0)
      END (* I *);
      L:= L + 1
   END (* WHILE *)
END (* FLOYD *);
```

3.3.4. COMPARATIVE PERFORMANCES OF SHORTEST-PATH ALGORITHMS

Of the three shortest-path algorithms discussed in this section, the choice depends on the problem at hand. Dijkstra's algorithm should be used for a dense network without any negative edge. It is extremely efficient, and even its n-fold version can be used for finding paths between all node pairs—as long as the network is dense and has no negative edges. But the PDM procedure will, on the average, outperform DIJKSTRA if the given network is very sparse, such as a grid network. Floyd's algorithm is only for finding paths between all pairs of nodes when the network is

Table 3-1 Computing Times for All-Pair Shortest-Path Algorithms on Complete Networks

Number of Nodes	Run Time (sec)[a]		
	DIJKSTRA	PDM	FLOYD
40	0.527	0.708	0.646
60	1.767	2.489	2.156
80	4.208	6.038	5.078
100	8.052	11.601	9.862

[a] On Amdahl 470 V/8.

dense. Even then, unless there is a negative edge, the n-fold Dijkstra on the average will outperform Floyd. Table 3-1 gives the run times of the three procedures for the all-pairs problem on complete networks (i.e., there is an edge from every node to every other node). The number of nodes vary from 40 to 100. Note that the DIJKSTRA and PDM procedures have been called n times—once for each node.

PROBLEMS

3-1. Prove that the value of the permanent label of a node in Algorithm 3-1 is the distance of that node from s, the starting node. (*Hint:* Use the induction on the number of nodes with permanent labels.)

3-2. Modify Algorithm 3-1 so that it finds all shortest paths from s to t in time $O(n^2)$. (*Hint:* Instead of the *pred* values, maintain a network P in which any path in P from s to any other node in P is a shortest path. Consider *recent* to be a set of all nodes whose labels became permanent on the previous pass.)

3-3. Extend the idea in Problem 3-2 so that all shortest paths from s to all other nodes are found in time $O(n^2)$.

3-4. Give an example of a digraph in which some of the edges have negative weights but where there is no cycle with a total negative weight, such that direct application of Algorithm 3-1 produces an incorrect result.

3-5. Rewrite Dijkstra's shortest-path procedure for finding the shortest paths from s to all other nodes in a unweighted graph (i.e., every edge has weight 1), as efficiently as you can.

3-6. Rewrite the DIJKSTRA procedure assuming that the input network is represented as adjacency lists and every edge has weight 1. Make sure that the computation time of your procedure is $O(m)$. (*Hint:* Maintain a queue of temporarily labeled vertices.)

3-7. *Breadth-first search* (BFS) explores the edges of a connected graph $G = (V, E)$ and produces a BFS tree as follows: Start at an arbitrarily chosen node s and traverse all edges incident on s; let these edges be $(s, a_1), (s, a_2), \ldots, (s, a_k)$. Then explore the edges incident on $a_1, a_2, \ldots, a_k$. Let these edges be $(a_1, a_{1,1}), (a_1, a_{1,2}), \ldots, (a_1, a_{1,t_1}), (a_2, a_{2,1}), (a_2, a_{2,2}), \ldots, (a_2, a_{2,t_2}), \ldots, (a_k, a_{k,t_k})$. The edges incident on nodes $a_{1,1}, a_{1,2}, \ldots, a_{k,t_k}$ are then explored, and so on. This process continues until all the edges have been explored. (If G is not connected, the process can be repeated to produce a BFS forest.) Devise an algorithm to explore an undirected graph according to this scheme and produce a BFS tree. Observe the similarity to Problem 3-6.

3-8. Use the BFS method on a connected unweighted (i.e., every edge has weight 1) graph to determine all nodes at a given distance d from a specified vertex s.

3-9. Rewrite the DIJKSTRA procedure such that in the process of finding the smallest temporarily labeled node in each iteration you consider only those nodes which have temporary labels (rather than all n nodes as done in the text). (*Hint:* Maintain an array H of node numbers such that in the kth iteration, the first $n - k$ entries in H are temporarily labeled. The remaining k entries in H are permanent, and these need not be considered.) Run experiments to determine reduction in computation time.

3-10. Further refine the procedure obtained in Problem 3-9 so that the list of all temporarily labeled nodes are kept continuously sorted. This reduces the amount of work done in finding the smallest temporarily labeled node in each iteration. (One of the methods to accomplish this is to maintain a heap.)

3-11. In some applications (e.g., in critical path analysis, see Section 4.2.1) we need the longest paths (rather than the shortest) from s to all other nodes in a directed network G. Will the maximization procedures analogous to the minimization procedures in Algorithms 3-1, 3-2, and 3-3 yield longest paths? Why?

3-12. Will your answer to Problem 3-11 be different if the given network G were acyclic (i.e., without cycles)? Give reasons.

3-13. In the beginning of Section 3.3 it was mentioned that for the purpose of solving shortest-path problems an undirected network can be converted into a directed network by replacing each undirected edge with a pair of oppositely directed edges. This conversion is trivial if all edges in the network have positive weights. Modify Algorithm 3-2 to find shortest paths in an undirected network G in which some of the edges have negative weights (but G has no undirected cycle of negative weight). What is the time complexity of the modified algorithm?

3-14. Like Floyd's algorithm, an algorithm due to G. B. Dantzig [1967] finds all shortest distances (paths) in a network. Find out the details of Dantzig's algorithm, and compare it with Floyd's in number of operations it requires.

3-15. When the following network is given as an input to Algorithm 3-2, how many times does node 1 get inserted into (and deleted from) Q?

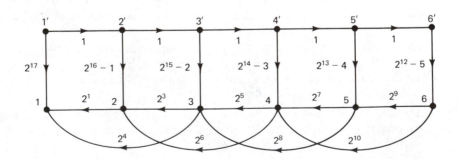

3-16. Generalize the network given in Problem 3-15 to one with $n = 2k$ nodes and $2n - 4$ edges. Derive a closed-form expression for the number of times node 1 gets inserted into Q, and show that for such a network Algorithm 3-2 requires an exponential (in n) number of operations.

3-17. *Transitive closure* $R = [r_{ij}]$ (also called the *reachability matrix*) of a digraph is defined as follows:

$$r_{ij} = \begin{cases} 1, & \text{if there is a path from node } i \text{ to node } j \text{ with} \\ & \text{one or more edges} \\ 0, & \text{otherwise} \end{cases}$$

Write a procedure for finding the reachability matrix starting from the adjacency matrix of a given digraph using:

(a) Dijkstra's method.

(b) Floyd–Warshall method.

Compare their efficiencies by running the two programs for digraphs of various sizes and densities.

3-18. Assume that a digraph is given in the form of its adjacency matrix A. Show that in the worst-case one has to check at least:

(a) $n(n - 1)/2$ elements of A to determine if G has a cycle.

(b) $(n^2 - 1)/4$ elements of A to determine if there is a path from a given node i to another given node j.

(*Hint:* See Holt and Reingold [1972].)

REFERENCES AND REMARKS

In the introduction and Section 3.2 we gave a brief introduction to graphs, networks, paths, and trees. There are many good books on graph theory where additional material on these topics can be obtained. We suggest the following, among others:

DEO, N., *Graph Theory with Applications to Engineering and Computer Science*, Prentice-Hall, Englewood Cliffs, N.J., 1974

HARARY, F., *Graph Theory*, Addison-Wesley, Reading, Mass., 1969

MINIEKA, E., *Optimization Algorithms for Networks and Graphs*, Dekker, New York, 1978

SWAMY, M. N. S., and K. THULSIRAMAN, *Graphs, Networks, and Algorithms*, Wiley, New York, 1981

Different data structures for computer representation of networks were discussed in Section 3.1. More on this subject can be found in

GOTLIEB, C. C., and L. R. GOTLIEB, *Data Type and Structures*, Prentice-Hall, Englewood Cliffs, N.J., 1978, Chapter 7

HOROWITZ, E., and S. SAHNI, *Fundamentals of Computer Algorithms*, Computer Science Press, Potomac, Md., 1978, Chapter 2

HOROWITZ, E., and S. SAHNI, *Fundamentals of Data Structures*, Computer Science Press, Potomac, Md., 1976, Chapter 6

REINGOLD, E. M., J. NIEVERGELT, and N. DEO, *Combinatorial Algorithms: Theory and Practice*, Prentice-Hall, Englewood Cliffs, N.J., 1977, Chapters 2 and 8

In Section 3.3, three algorithms for three different types of shortest-path problems were given. In fact, depending on the application at hand, one may ask a myriad of shortest-path questions about the network under consideration. A lucid survey in

DREYFUS, S. E., An Appraisal of Some Shortest-Path Algorithms, *Oper. Res.* 17(1969), 395–412

has considered the following five types of shortest-path problems: (1) determining the shortest path between two specified nodes of a network; (2) determining the shortest paths between all pairs of nodes of a network; (3) determining the second, third, and so on, shortest path; (4) determining the fastest path through a network with travel times depending on the departure time; and (5) finding the shortest path between specified end points that passes through specified intermediate nodes.

A finer and a more systematic classification is given in

DEO, N., and C. Y. PANG, Shortest Path Algorithms: Taxonomy and Annotation, Technical Report CS-80-057, Computer Science Dept., Washington State University, Pullman, Mar. 1980

In this report all known shortest-path algorithms have been classified in three different ways: according to (1) the problem type (i.e., questions being asked about the given network), (2) the input-type (i.e., salient features of the given network), and (3) the type of underlying techniques used to solve the problem. This report also contains 222 references on shortest-path problems (culled out of a larger body of literature), of which 79 are annotated in detail. Another good bibliography on shortest-path problems is

PIERCE, A. R., Bibliography on Algorithms for Shortest Path, Shortest Spanning Tree and Related Circuit Routing Problems (1956–1974), *Networks* 5(1975), 129–149

Algorithm 3-1 originally appeared in the following paper:

DIJKSTRA, E. W., A Note on Two Problems in Connexion with Graphs, *Numer. Math.* 1(1959), 269–271

This paper was preceded by the following works on the single-course shortest-path problem:

BELLMAN, R. E., On a Routing Problem, *Quart. Appl. Math.* 16(1958), 87–90

FORD, L. R., JR., Network Flow Theory, Report P-923, Rand Corporation, Santa Monica, Calif., Aug. 1956

MOORE, E. F., The Shortest Path through a Maze, *Proc. Int. Symp. on the Theory of Switching*, Part II, Apr. 2–5, 1957. *The Annals of the Computation Laboratory of Harvard University 30*, Harvard University Press, Cambridge, Mass., 1959, pp. 285–292

The Ford–Bellman–Moore algorithm as modified by d'Esopo and refined by Pape is presented as Algorithm 3-2. A FORTRAN version of this is given in

PAPE, U., Algorithm 562: Shortest Path Lengths, *ACM Trans. Math. Software* 5(1980), 450–455

Some extensive studies that have compared the performances of the two algorithms are reported in

DENARDO, E. V., and B. L. FOX, Shortest-Route Methods: 1. Reaching, Pruning, and Buckets, *Oper. Res.* 27(1979), 161–186

DIAL, R. B., F. GLOVER, D. KARNEY, and D. KLINGMAN, A Computational Analysis of Alternative Algorithms and Labeling Techniques for Finding Shortest Path Trees, *Networks* 9(1979), 215–248

VAN VLIET, D., Improved Shortest Path Algorithm for Transportation Networks, *Transportation Res.* 12(1978), 7–20

Van Vliet's paper contains a FORTRAN listing of Dijkstra's algorithm (Algorithm 3-1).

Floyd's algorithm (Algorithm 3-3) was published as an ALGOL procedure in

FLOYD, R. W., Algorithm 97: Shortest Path, *Comm. ACM* 5(1962), 345

which in turn was based on Warshall's algorithm for transitive closure presented in

WARSHALL, S., A Theorem on Boolean Matrices, *J. ACM* 9(1962), 11–12

Another $O(n^3)$ algorithm for finding shortest paths between all pairs of nodes is due to G. B. Dantzig, and can be found in

DANTZIG, G. B., All Shortest Routes in a Graph, in P. Rosenstiehl (ed.), *Theory of Graphs*, Gordon and Breach, New York, 1967, pp. 91–93

The order of the number of operations required in finding the transitive closure (Problem 3-17) of a digraph is the same for both *n*-fold Dijkstra's method and Floyd–Warshall method—both being $O(n^3)$. The actual computation times for digraphs of different sizes and densities of these two algorithms (as well as some others) have been compared in

SYSŁO, M. M., and J. DZIKIEWICZ, Computational Experiences with Some Transitive Closure Algorithms, *Computing* 15(1975), 33–39

Although the worst-case complexities of Floyd–Warshall, Dantzig, and *n*-fold Dijkstra algorithms turn out to be $O(n^3)$, it has been possible to devise algorithms for which the average-case complexity is lower. For example, in

SPIRA, P. M., A New Algorithm for Finding All Shortest Paths in a Graph of Positive Arcs in Average Time $O(n^2 \log^2 n)$, *SIAM J. Comput.* 2(1973), 28–32

an average-case complexity of $O(n^2 \log^2 n)$ has been achieved by first sorting all edge weights and then applying Dijkstra's algorithm *n* times—once from each node. Similarly, it has been shown in

FREDMAN, M. L., New Bounds on the Complexity of the Shortest Path Problem, *SIAM J. Comput.* 5(1976), 83–89

that if all edge weights are independently and identically distributed the shortest paths between all pairs of nodes can be found in $O(n^2 \log^2 n)$ time on the average. A more recent and further improvement is reported in

BLONIARZ, P., A Shortest Path Algorithm with Expected Time $O(n^2 \log n \log^* n)$, *Proc. 12th Ann. ACM Symp. on Theory of Computing*, Los Angeles, 1980, pp. 378–384

For a good look at the mathematical structure of various shortest-path problems, one should see

MAHR, B., A Bird's Eye View to Path Problems, in H. Noltemeier (ed.), *Theoretical Concepts in Computer Science*, Springer-Verlag, Berlin, 1981, pp. 335–353

It was mentioned in passing in Section 3.3 that answering almost any "nontrivial" question on a given graph requires us to look at almost every edge (or all n^2

entries in the adjacency matrix) at least once. Problem 3-18 is to prove this intuitive statement for two simple problems. The proofs are given in

HOLT, R. C., and E. M. REINGOLD, On the Time Required to Detect Cycles and Connectivity in Graphs, *Math. Systems Theory* 6(1972), 103–106

There are some "exceptions" to the statement made in the preceding paragraph. For a discussion of properties that such exceptions (i.e., problems that do not require our looking at every edge or the $O(n^2)$ entries in the adjacency matrix, at least once) must have, see

RIVEST, R. L., and J. VUILLEMIN, On Recognizing Graph Properties from Adjacency Matrices, *Theoret. Comput. Sci.* 3(1976/77), 371–384

3.4.   MINIMUM SPANNING TREE PROBLEM

The problem of finding a minimum spanning tree (MST) in a large, connected, undirected network arises in many practical situations. Connecting a set of terminals, using the smallest total length of telephone wires is one such example. (A network may have more than one MST.)

3.4.1.   KRUSKAL'S ALGORITHM

There are two classical algorithms for finding an MST. One that may be called the smallest-edge-first, due to J. B. Kruskal [1956], works as follows: First we sort all the edges in the given network by weight, in nondecreasing order. Then one by one the edges are examined in order, smallest to the largest. If an edge e_i, upon examination, is found to form a cycle (when added to edges already selected) it is discarded. Otherwise, e_i is selected to be included in the minimum spanning tree T. The construction stops when the required $n - 1$ edges have been selected or when all m edges have been examined. If the given network is disconnected, we would get a *minimum spanning forest* (instead of tree). More formally, Kruskal's method may be stated as follows

```
    T ← ∅;
    while |T| < (n − 1) and E ≠ ∅ do
    begin
        e ← smallest edge in E;
        E ← E − {e};
        if (T ∪ {e}) has no cycle then T ← T ∪ {e}
    end;
    if |T| < n − 1 then write 'network disconnected'
```

Although the algorithm outlined above is simple enough (in fact, surprisingly simple), we do need to work out some implementation details and select an appropriate data structure, for achieving an efficient execution on the computer.

There are two crucial implementation details that we must consider in this algorithm. If we initially sort all m edges in the given network, we may be doing a lot of unnecessary work. For example, in a 100-node (undirected) network, there could be as many as 4950 edges. To sort all 4950 edges, while we may examine only the first couple of hundreds, is wasteful. All we really need is to be able to determine the next smallest edge in the network at each iteration. Therefore, in practice, the edges are only partially sorted and kept as a heap with smallest edge at the root of the heap.* The initial construction of the heap would require $O(m)$ computational steps, and the next smallest edge from a heap can be obtained in $O(\log m)$ steps. With this improvement, the sorting cost is $O(m + p \log m)$, where p is the number of edges examined before MST is constructed. Typically, p is much smaller than m.

The second crucial detail is how to maintain the edges selected (to be included in the MST) so far, such that the next edge to be examined can be efficiently tested for a cycle formation.

We observe that as edges are examined and included in T, a forest of disconnected trees (i.e., subtrees of the final spanning tree) is produced. The edge e being examined will form a cycle if and only if both its end nodes belong to the same subtree in T. Thus to ensure that the edge currently being examined does not form a cycle, it is sufficient to check if it connects two different subtrees in T.

An efficient way to accomplish this is to group the n nodes of the given network into disjoint subsets defined by the subtrees (formed by the edges included in T, so far). Consider, for example, the six-node network in Fig. 3-6. Initially, the subsets will be just $\{1\}, \{2\}, \{3\}, \{4\}, \{5\}, \{6\}$—the singleton subsets. After the smallest edge $(3, 5)$ is selected, the subsets become $\{1\}, \{2\}, \{3, 5\}, \{4\}, \{6\}$. Next, after edge $(4, 6)$ is selected, the subsets become $\{1\}, \{2\}, \{3, 5\}, \{4, 6\}$. In the third iteration edge $(4, 5)$ is selected, and the two subsets are merged, yielding $\{1\}, \{2\}, \{3, 4, 5, 6\}$. The next smallest edge is $(4, 6)$, but because both of its end nodes are in the same connected component (represented by subset $\{3, 4, 5, 6\}$), it forms a cycle and therefore edge $(4, 6)$ is discarded. And so on.

Thus if we maintain the partially constructed MST by means of subsets of nodes, we can add a new edge by forming the UNION of two relevant subsets, and we can check for cycle formation by FINDing if the two end nodes of the edge, being examined, are in the same subset.

The subsets can themselves be kept as rooted trees. The root is an element of the subset and is used as a name to identify that subset. The FIND subprocedure is called twice—once for each end node of edge e—to determine the sets to which the two end nodes belong. If they are different, the UNION subprocedure will merge the two subsets. (If they are the same subset, edge e will be discarded.)

---

*A *heap* is a binary tree in which the weight of every node is smaller than or equal to the weights of its sons.

These subsets, kept as rooted trees, are implemented by keeping an array of *father* pointers for each of the *n* elements. *Father* of a root, of course, is null. [It could be any number as long as it is not the valid label of any node (i.e., cannot be an integer 1 through *n*). In fact, it is useful to assign *father(root)* = −number of nodes in the tree.] While taking the UNION of two subsets, we merge the smaller subset into the larger one by pointing the *father* pointer in the root of the smaller subset to the root of the larger subset.

Some of these details are shown in the following elaboration of Kruskal's procedure.

Algorithm 3-4: Kruskal's Minimum Spanning Tree Algorithm

```
INITIALIZATION:
    set father array to −1; {* n nodes form singleton sets *}
    form initial heap of m edges;
    ecount ← 0; {* number of edges examined so far *}
    tcount ← 0; {* number of edges in T so far *}
    T ← 0;

ITERATION:
    while tcount < (n − 1) and ecount < m do
    begin
        e ← edge (u, v) from top of heap;
        ecount ← ecount + 1;
        remove e from heap;
        restore heap;
        r1 ← FIND(u);
        r2 ← FIND(v);
        if r1 ≠ r2 then
        begin
            T ← T ∪ {e};
            tcount ← tcount + 1
            UNION(r1, r2)
        end
    end;
    if tcount < (n − 1) then write 'network disconnected'
```

Note that $r1$ and $r2$ are the roots identifying the sets to which u and v belong. In the complete Pascal procedure (to be given shortly) the FATHER fields of the roots are utilized to store the size of the subsets (instead of storing the null pointer). The size is then utilized in the UNION subprocedure. Initially, the FATHER fields are set to −1 to indicate the singleton sets.

Data Structure

Since we need to examine only the edges (one by one in order of their increasing weights) it would be wasteful to store the input network in the form of its weight matrix, if the given network is sparse (which usually is the case for large networks, in practice). Therefore, we will use the list of edges. That is, the *m* edges of the network will be represented by three arrays of size *m*—two for the two end nodes and one for the weights. (Alternatively, we could have used a single array of *m* records in Pascal, as Pascal supports the record structure. It would have saved some statements in the program, but would have made the program less transparent to a non-Pascal programmer.) These three arrays in the following program are ENDV1, ENDV2, and WEIGHT.

Similarly, the output tree (the MST) will be represented by two arrays TEDGE1 and TEDGE2 representing the end nodes of the $(n - 1)$ edges selected in the MST. A scalar TWEIGHT will give the total weight of the tree obtained.

A complete Pascal procedure, called KRUSKAL, which implements the minimum spanning tree algorithm, outlined in Algorithm 3-4, follows. It uses three subprocedures, namely, HEAP, FIND, and UNION, as discussed. Determining the computational complexity of this procedure is left as an exercise (Problem 3-25).

Computer Implementation of Kruskal's Minimum Spanning Tree Algorithm

The KRUSKAL procedure finds a minimum spanning tree (forest) in a given undirected network with N nodes and M edges. If the network is disconnected, a Boolean variable CONNECT is set to FALSE; otherwise, it is TRUE. The minimum spanning tree (forest) is stored in arrays TEDGE1 and TEDGE2. TWEIGHT is the total weight of the minimum spanning tree (forest).

Global Constants

N   number of nodes in the given network
N1   N-1 (number of edges in a minimum spanning tree)
M   number of edges in the given network

Data Types

```
TYPE   ARRN  = ARRAY[1..N] OF INTEGER;
       ARRN1 = ARRAY[1..N1] OF INTEGER;
       ARRM  = ARRAY[1..M] OF INTEGER;
```

Procedure Parameters

```
PROCEDURE KRUSKAL(
     N,M                         :INTEGER;
     VAR ENDV1,ENDV2,WEIGHT :ARRM;
     VAR CONNECT                 :BOOLEAN;
     VAR TEDGE1,TEDGE2           :ARRN1;
     VAR TWEIGHT                 :INTEGER);
```

Input

N	number of nodes in the given network
M	number of edges in the given network
ENDV1[1..M]	one end node of the edges
ENDV2[1..M]	the other end nodes of the edges
WEIGHT[1..M]	weights of the edges

Output

CONNECT	Boolean variable such that if the network is disconnected, CONNECT = FALSE, otherwise, CONNECT = TRUE
TEDGE1[1..N1]	one end node of the edges in the MST
TEDGE2[1..N1]	the other end nodes of the edges in the MST
TWEIGHT	total weight of the MST

Example

For the network shown in Fig. 3-11, we have $N = 7$, $M = 11$, and the network is represented using three arrays ENDV1, ENDV2, and WEIGHT as follows:

$$ENDV1[1..11] = [1, 1, 2, 2, 2, 3, 3, 4, 4, 5, 6]$$
$$ENDV2[1..11] = [2, 3, 3, 4, 7, 4, 5, 5, 6, 6, 7]$$
$$WEIGHT[1..11] = [43, 27, 28, 20, 24, 26, 14, 19, 18, 24, 22]$$

The solution obtained is

$$CONNECT = TRUE$$
$$TEDGE1[1..6] = [3, 4, 4, 2, 6, 1]$$
$$TEDGE2[1..6] = [5, 6, 5, 4, 7, 3]$$
$$TWEIGHT = 120$$

Pascal Procedure KRUSKAL

```
PROCEDURE KRUSKAL(
   N,M                     : INTEGER;
   VAR ENDV1,ENDV2,WEIGHT: ARRM;
   VAR CONNECT             : BOOLEAN;
   VAR TEDGE1,TEDGE2       : ARRN1;
   VAR TWEIGHT             : INTEGER);


VAR I,LAST,U,V,R1,R2,ECOUNT,TCOUNT: INTEGER;
    FATHER                         : ARRN;
```

```
PROCEDURE HEAP(
   FIRST,LAST : INTEGER);

VAR J,K,TEMP1,TEMP2,TEMP3: INTEGER;
BEGIN
   J:= FIRST;
   WHILE J <= TRUNC(LAST/2) DO
   BEGIN
      IF (2*J < LAST) AND (WEIGHT[2*J+1] < WEIGHT[2*J]) THEN
         K:= 2 * J + 1
      ELSE
         K:= 2 * J;
      IF WEIGHT[K] < WEIGHT[J] THEN
      BEGIN
         TEMP1    :=ENDV1[J]; TEMP2    :=ENDV2[J]; TEMP3    :=WEIGHT[J];
         ENDV1[J]:=ENDV1[K]; ENDV2[J]:=ENDV2[K]; WEIGHT[J]:=WEIGHT[K];
         ENDV1[K]:=TEMP1;    ENDV2[K]:=TEMP2;    WEIGHT[K]:=TEMP3;
         J:= K
      END
      ELSE J:= LAST
   END (* WHILE *)
END (* HEAP *);


FUNCTION FIND(
   I : INTEGER): INTEGER;

VAR PTR: INTEGER;
BEGIN
   PTR:= I;
   WHILE FATHER[PTR] > 0 DO
      PTR:= FATHER[PTR];
   FIND:= PTR
END (* FIND *);


PROCEDURE UNION(
   I,J : INTEGER);

VAR X: INTEGER;
BEGIN
   X:= FATHER[I] + FATHER[J];
   IF FATHER[I] > FATHER[J] THEN
   BEGIN
      FATHER[I]:= J;
      FATHER[J]:= X
   END
   ELSE
   BEGIN
      FATHER[J]:= I;
      FATHER[I]:= X
   END
END (* UNION *);
```

```
BEGIN
    FOR I:= 1 TO N DO      (* INITIALIZE ARRAY FATHER *)
        FATHER[I]:= -1;
    FOR I:= TRUNC(M/2) DOWNTO 1 DO  (* INITIAL HEAP *)
        HEAP(I,M);
    LAST    := M;
    ECOUNT := 0; TCOUNT := 0;
    TWEIGHT:= 0; CONNECT:= TRUE;
(* INITIALIZATION OVER *)

    WHILE ((TCOUNT < N-1) AND (ECOUNT < M)) DO
    BEGIN
        (* REMOVE EDGE (U,V) FROM TOP OF HEAP AND EXAMINE *)
        ECOUNT:= ECOUNT + 1;
        U:= ENDV1[1];
        V:= ENDV2[1];
        (* CHECK IF U AND V ARE IN THE SAME TREE *)
        R1:= FIND(U);
        R2:= FIND(V);
        IF R1 <> R2 THEN      (* INCLUDE (U,V) IN MST *)
        BEGIN
            TCOUNT:= TCOUNT + 1;
            UNION(R1,R2);
            TEDGE1[TCOUNT]:= U;
            TEDGE2[TCOUNT]:= V;
            TWEIGHT:= TWEIGHT + WEIGHT[1]
        END;
        (* REDUCE HEAP SIZE BY 1 AND RESTORE HEAP *)
        ENDV1[1]:= ENDV1[LAST];
        ENDV2[1]:= ENDV2[LAST];
        WEIGHT[1]:= WEIGHT[LAST];
        LAST:= LAST - 1;
        HEAP(1,LAST)
    END (* WHILE *);
    IF TCOUNT <> N-1 THEN CONNECT:= FALSE
END (* KRUSKAL *);
```

3.4.2. PRIM'S ALGORITHM

A second algorithm, discovered by R. C. Prim in 1957 and independently by Dijkstra [1959], is called the *nearest-neighbor method*. In this method one starts with an arbitrary node s and joins it to its nearest neighbor, say y. That is, of all edges incident on node s, edge (s, y), with the smallest weight, is made part of the MST. Next, of all the edges incident on s or y we choose one with minimum weight that leads to some third node, and make this edge part of the MST. We continue this process of "reaching out" from the partially constructed tree (so far) and bringing in the "nearest neighbor" until all nodes reachable from s have been incorporated into the tree.

As an example, let us use this method to find the minimum spanning tree of the network given in Fig. 3-11. Suppose that we start at node 1. The nearest

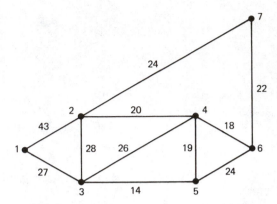

Figure 3-11 *Sample network for* MST *algorithms.*

neighbor of node 1 is node 3. Therefore, edge $(1, 3)$ becomes part of the MST. Next, of all the edges incident on nodes 1 and 3 (and not included in the MST so far) we select the smallest, which is edge $(3, 5)$ with weight 14. Now the partially constructed tree consists of two edges $(1, 3)$ and $(3, 5)$. Among all edges incident at nodes 1, 3, and 5, edge $(5, 4)$ is the smallest, and is therefore included in the MST. The situation at this point is shown in Fig. 3-12. Clearly, $(4, 6)$, with weight 18, is the next edge to be included. Finally, edges $(4, 2)$ and $(6, 7)$ will complete the desired MST.

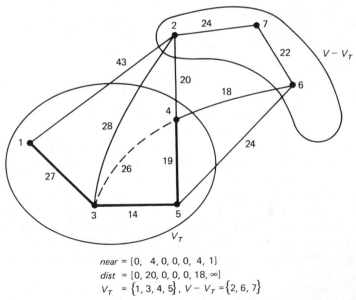

$$near = [0, \quad 4, 0, 0, 0, \quad 4, \ 1]$$
$$dist \ = [0, 20, 0, 0, 0, 18, \infty]$$
$$V_T \ = \{1, 3, 4, 5\}, \ V - V_T = \{2, 6, 7\}$$

Figure 3-12 *Partially constructed* MST *for the network of Fig. 3-11.*

The main computational task in this algorithm is that of finding the next edge to be included into the MST at each iteration. For each efficient execution of this task we will maintain an array $near(u)$ for each node u not yet in the tree (i.e., $u \in V - V_T$). $Near(u)$ is that node in V_T which is closest to u. (Note that V is the set of all nodes in the network and V_T is the subset of V included in MST thus far.) Initially, we set $near(s) \leftarrow 0$ to indicate that s is in the tree, and for every other node v, $near(v) \leftarrow s$.

For convenience, we will maintain in another array $dist(u)$ the actual distance (i.e., edge weight) to that node in V_T which is closest to u. In order to determine which node is to be added to the set V_T next, we compare all nonzero values in $dist$ array and pick the smallest. Thus $n - i$ comparisons are sufficient to identify the ith node to be added. Initially, since s is the only node in V_T, $dist(u)$ is set to w_{su}. As the algorithm proceeds, these two arrays are updated in each iteration [see Fig. 3-12 for illustration].

A more complete description of the nearest-neighbor algorithm is given in Algorithm 3-5. It is assumed that the input is given in the form of an $n \times n$ weight matrix W (in which nonexistent edges have ∞ weights). Set $V = \{1, 2, \ldots, n\}$ is the set of nodes of the network. V_T and E_T are the sets of nodes and edges of the partially formed (minimum spanning) tree. Node set V_T is identified by zero entries in array $near$.

A complete Pascal procedure, called PRIM, which implements the minimum spanning tree algorithm as outlined in Algorithm 3-5 with $s = 1$, follows. Computational complexity of this algorithm has been left as an exercise (Problem 3-26).

Algorithm 3-5: Prim's Minimum Spanning Tree Algorithm

INITIALIZATION:
    choose starting node s arbitrarily;
    $near(s) \leftarrow 0$;
    for every node i other than s **do**
    begin
        $near(i) \leftarrow s$;
        $dist(i) \leftarrow w_{s,i}$
    end;
    $V_T \leftarrow \{s\}$; {* set of nodes in MST so far *}
    $E_T \leftarrow \emptyset$; {* set of edges in MST so far *}

ITERATION:
    while $|V_T| < n$ **do**
    begin
        $u \leftarrow$ vertex in $(V - V_T)$ with smallest value of $dist(u)$;
        if $dist(u) \geqslant \infty$ **then** write 'graph disconnected' and exit;
        $E_T \leftarrow E_T \cup \{(u, near(u))\}$;
        $V_T \leftarrow V_T \cup \{u\}$;
        for $x \in (V - V_T)$ **do**

> **if** $w_{ux} < dist(x)$ **then**
> **begin**
> $dist(x) \leftarrow w_{ux};$
> $near(x) \leftarrow u$
> **end**

end

Computer Implementation of Prim's Minimum Spanning Tree Algorithm

The PRIM procedure finds a minimum spanning tree in a given undirected connected network with N nodes. The minimum spanning tree is stored in the arrays TEDGE1 and TEDGE2. TWEIGHT is the total weight of the minimum spanning tree. If the network is disconnected, a Boolean variable CONNECT is set to FALSE.

Global Constants

N   number of nodes in the given network
N1   N − 1

Data Types

```
TYPE   ARRNN = ARRAY[1..N,1..N] OF INTEGER;
       ARRN  = ARRAY[1..N] OF INTEGER;
       ARRN1 = ARRAY[1..N1] OF INTEGER;
```

Procedure Parameters

```
PROCEDURE PRIM(
    N,INF                :INTEGER;
    VAR W                :ARRNN;
    VAR CONNECT          :BOOLEAN;
    VAR TEDGE1,TEDGE2 :ARRN1;
    VAR TWEIGHT          :INTEGER);
```

Input

N   number of nodes in the given network
INF   integer much larger than the largest edge weight
W[1..N,1..N]   weight matrix of the given network output

Output

CONNECT   Boolean variable such that if the network is disconnected, CONNECT = FALSE and otherwise, CONNECT = TRUE

TEDGE1[1..N1]   one end node of the edges in the MST
TEDGE2[1..N1]   the other end nodes of the edges in the MST
TWEIGHT   total weight of the MST

Example

For the network shown in Fig. 3-11, we have $N = 7$, $INF = 1000$, and the network is represented in the form of an $N \times N$ array W as follows:

$$W[1..7,1..7] = \begin{bmatrix} \infty & 43 & 27 & \infty & \infty & \infty & \infty \\ 43 & \infty & 28 & 20 & \infty & \infty & 24 \\ 27 & 28 & \infty & 26 & 14 & \infty & \infty \\ \infty & 20 & 26 & \infty & 19 & 18 & \infty \\ \infty & \infty & 14 & 19 & \infty & 24 & \infty \\ \infty & \infty & \infty & 18 & 24 & \infty & 22 \\ \infty & 24 & \infty & \infty & \infty & 22 & \infty \end{bmatrix}$$

The solution obtained is

$$CONNECT = TRUE$$
$$TEDGE1[1..6] = [1, 3, 5, 4, 4, 6]$$
$$TEDGE2[1..6] = [3, 5, 4, 6, 2, 7]$$
$$TWEIGHT = 120$$

Pascal Procedure PRIM

```
PROCEDURE PRIM(
   N,INF              : INTEGER;
   VAR W              : ARRNN;
   VAR CONNECT        : BOOLEAN;
   VAR TEDGE1,TEDGE2: ARRN1;
   VAR TWEIGHT        : INTEGER);


VAR U,K,MIN,TCOUNT: INTEGER;
    NEAR,DIST      : ARRN;
BEGIN
   NEAR[1]:= 0;     (* VERTEX 1 IS IN T *)
   FOR I:= 2 TO N DO
   BEGIN
      NEAR[I]:= 1;
      DIST[I]:= W[1,I]
   END;
   TCOUNT:= 0;
   TWEIGHT:= 0;
   CONNECT:= TRUE;
(* INITIALIZATION OVER *)

   WHILE (TCOUNT < N-1) AND (CONNECT = TRUE) DO
```

```
    BEGIN    (* LET U BE THE VERTEX SUCH THAT NEAR[U] <> 0 *)
             (* AND W[U,NEAR[U]] IS MINIMUM.                *)
      MIN:= INF;
      FOR K:= 2 TO N DO
          IF NEAR[K] <> 0 THEN
             IF DIST[K] < MIN THEN
             BEGIN
                U:= K;
                MIN:= DIST[K]
             END;
      IF DIST[U] >= INF THEN CONNECT:= FALSE
      ELSE
      BEGIN
         TCOUNT:= TCOUNT + 1;
         TEDGE1[TCOUNT]:= NEAR[U];
         TEDGE2[TCOUNT]:= U;
         TWEIGHT:= TWEIGHT + DIST[U];
         NEAR[U]:= 0;
         FOR K:= 2 TO N DO          (* UPDATE NEAR *)
             IF NEAR[K] <> 0 THEN
                IF W[K,NEAR[K]] > W[K,U] THEN
                BEGIN
                   DIST[K]:= W[K,U];
                   NEAR[K]:= U
                END
      END (* ELSE *)
   END (* WHILE *)
END (* PRIM *):
```

3.4.3. COMPARATIVE PERFORMANCE OF MST ALGORITHMS

If all the edge weights are distinct in the given network G, there is a unique minimum spanning tree in G, and both the smallest-edge-first and the nearest-neighbor algorithms will produce this tree. However, the order in which the $n - 1$ edges are selected will, in general, be different. If the network has edges with identical weights, there may be more than one minimum spanning tree in it—all of the same weight. In that case, the two algorithms may produce different spanning trees (both MSTs).

Many variations of these two algorithms are possible and have been suggested and studied in the literature. These variations are based primarily on data structures and implementational details.

It is not possible to state categorically that either one of the algorithms is better than the other under all circumstances. Depending on the size of the given network, its density, the distribution of its edge weights, and its structure, either algorithm may outperform the other. As a rule of thumb, for smaller graphs (say, up to 100 nodes), Prim's algorithm is superior, particularly if the graph is dense. For large, sparse graphs Kruskal's algorithm is better. The following timing figures for the two procedures run on an Amdahl 470 V/8 computer will give some idea of the execution times on some random networks.

Number of Edges	Execution Times (msec)	
	KRUSKAL	PRIM
Dense networks with 80 nodes		
790	64	52
1580	95	51
3160	186	50
Sparse networks with 100 nodes		
200	44	78
300	66	80

As expected, the execution time of the PRIM procedure depends only on the number of nodes in the network and not on its density. On the other hand, the time for the KRUSKAL procedure increases as the number of edges is increased in a network with the same number of nodes. The KRUSKAL procedure required 1.3 seconds on a grid (very sparse) network of 5000 nodes and 10,000 edges on the same computer.

There is yet a third MST algorithm, which is attributed to M. Sollin. It works as follows: First the smallest edge incident on each node is found; these edges form part of the minimum spanning tree (see Problem 3-27). There are at least $\lceil n/2 \rceil$ such edges. (Why?) The connected components formed by these edges are collapsed into "supernodes". (There are no more than $\lfloor n/2 \rfloor$ such nodes at this point.) The process is repeated on "supernodes" and then on the resulting "supersupernodes," and so on, until only a single node remains. This will require at most $\lfloor \log_2 n \rfloor$ steps, because at each step the number of nodes is reduced at least by a factor of 2. The implementational details are left as an exercise in Problems 3-28 and 3-29.

PROBLEMS

3-19. In an undirected network G, let T be a minimum spanning tree.

  (a) Let e be an edge of G that does not belong to T, and let p be the unique path in T between the end vertices of edge e. Show that weight $w(e)$ of e cannot be less than the weight of any edge in p.

  (b) If the weights of all edges in G are distinct, show that no edge belonging to T can be the largest edge of any cycle (not necessarily fundamental) in G.

3-20. Prove that the smallest-edge-first method, as used in Algorithm 3-4, does indeed produce a minimum spanning tree in a connected network:

  (a) Prove that it produces a spanning tree.

  (b) Prove that the tree produced is minimum.

3-21.   Solve Problem 3-20 for Algorithm 3-5.

3-22.   Modify Algorithm 3-4 appropriately so that it finds a maximum (rather than minimum) spanning tree in a given undirected, weighted graph. Repeat the exercise for Algorithm 3-5.

3-23.   Show that in the KRUSKAL procedure the time required for each call of the subprocedure HEAP(I, M) is $O(\log_2(M/I))$. Then show that the time required in forming the initial heap of edges is $O(M)$.

3-24.   Show that for a worst-case input, the FIND function will be called 2M times, and the UNION procedure will be called (N − 1) times. Also show that each call of FIND(I) will require no more than $O(\log_2(I))$ operations.

3-25.   From the analysis in Problems 3-23 and 3-24, obtain an expression for the (worst-case input) time complexity of the KRUSKAL procedure.

3-26.   Perform an analysis of the time complexity of Algorithm 3-5 (i.e., of the PRIM procedure).

3-27.   Let e_v, $v \in V$, be the shortest edge incident with node v.
  (a) Show that for every v there exists a MST containing e_v.
  (b) Does $E' = \{e_v : v \in V\}$ contain a cycle?
  (c) Is E' always a spanning tree of G?
  (d) Is there a MST containing E'?

3-28.   Using set-theoretic FIND and UNION operations as given in the KRUSKAL procedure in this section, implement Sollin's MST algorithm requiring $O(m \log n)$ computation time.

3-29.   Improve the algorithm in Problem 3-28 so that it requires $O(m \log \log n)$ time.

3-30.   Define a *1-tree* of an undirected graph to be a spanning tree with one extra edge added (this edge causes a cycle). Design, analyze, and prove to be correct an algorithm to find a minimum 1-tree of a weighted undirected graph.

3-31.   In Section 3.3 the concept of shortest-path tree or skim tree was introduced (see Fig. 3-9), which is very different from a minimum spanning tree. Draw an undirected network in which *no* shortest-path tree is a minimum spanning tree.

3-32.   Write a procedure similar to Kruskal's to generate a (any one) spanning tree in (an undirected, unweighted, connected) graph G. What is the worst-case complexity of your algorithm?

3-33.   Maximal connected subgraphs of a graph are called its *connected components*. Devise an $O(m)$ algorithm to produce all connected components of a given undirected graph. [*Hint:* Use breadth-first search (see Problem 3-7) using a forward-star or linked adjacency list as input and a queue. Observe how similar it is to Prim's algorithm. In each iteration, instead of just the nearest neighbor, *all new* neighbors of the current node u are being brought into the tree.]

3-34. Write an $O(n^2)$ procedure for finding the transitive closure matrix of an undirected graph. (*Hint:* First obtain connected components.)

3-35. Let T be a spanning tree in a connected undirected graph G. Let e be an edge in G that does not belong to T. The unique cycle produced by adding e to T is called a *fundamental cycle* of G with respect to T. Show that there are $\mu = m - n + 1$ such fundamental cycles. Write an algorithm to produce a set of fundamental cycles.

3-36. Consider the algorithm that attempts to find a minimum spanning tree T^* by deleting largest edges from the given connected network G while keeping it connected. That is,

```
let w(e₁) ≥ w(e₂) ≥ ... ≥ w(eₘ);
T* ← G;
for e₁, e₂, ..., eₘ do
    if T* − ⟨eᵢ⟩ is connected then
        T* ← T* − ⟨eᵢ⟩
```

Prove that T^* is indeed a minimum spanning tree of G.

3-37. In a complete undirected network (one in which there is an edge between every pair of nodes) of n labeled nodes, show that there are n^{n-2} distinct labeled spanning trees. Draw all 16 spanning trees for the case $n = 4$.

REFERENCES AND REMARKS

The smallest-edge-first method of finding a minimum spanning tree given in Algorithm 3-4 is due to

KRUSKAL, J. B., On the Shortest Spanning Subtree of a Graph and the Traveling Salesman Problem, *Proc. Amer. Math. Soc.* 7(1956), 48–50

The nearest-neighbor minimum spanning tree algorithm (Algorithm 3-5) was first published in

PRIM, R. C., Shortest Connection Networks and Some Generalizations, *Bell System Tech. J.* 36(1957), 1389–1401

and independently in

DIJKSTRA, E. W., A note on Two Problems in Connexion with Graphs, *Numer. Math.* 1(1959), 269–271

A thorough study of the relative efficiencies of these two algorithms and their different implementations is given in

KERSCHENBAUM, A., and R. VAN SLYKE, Computing Minimum Spanning Trees Efficiently, *Proc. 25th Ann. Conf. of the ACM*, 1972, pp. 518–527

The third method, mentioned only briefly at the end of Section 3.4.3 (see also Problems 3-27 through 3-29) is attributed to M. Sollin by Berge and Ghouila-Houri. It can be found in

BERGE, C., and A. GHOUILA - HOURI, *Programming, Games and Transportation Networks*, Wiley, New York, 1965

as well as in

GOODMAN, S. E., and S. T. HEDETNIEMI, *Introduction to the Design and Analysis of Algorithms*, McGraw-Hill, New York, 1977, pp. 265–269

Asymptotically, more efficient MST algorithms are given in

CHERITON, D., and R. E. TARJAN, Finding Minimum Spanning Trees, *SIAM J. Comput.* 5(1976), 724–742

YAO, A. C. C., An $O(|E|\log\log|V|)$ Algorithm for Finding Minimum Spanning Trees, *Inform. Process. Lett.* 4(1975), 21–23

But our computational experience indicates that except for very large networks one is better off using the simpler algorithms of Prim and Kruskal with efficient implementations.

The three basic minimum spanning tree algorithms discussed in this section are special cases of a general method of designing algorithms called the *greedy method*, which makes the best local choice without considering the global optimality. For the MST problem, a local optimum turns out to be the global optimum because the spanning trees have matroidal properties. The general discussion of the greedy method can be found in

HOROWITZ, E., and S. SAHNI, *Fundamentals of Computer Algorithms*, Computer Science Press, Potomac, Md., 1978, Chapter 4

The greedy method and its relationship to matroids are described in

LAWLER, E. L., *Combinatorial Optimization. Networks and Matroids*, Holt, Rinehart and Winston, New York, 1976, Chapter 7

SWAMY, M. N. S., and K. THULSIRAMAN, *Graphs, Networks, and Algorithms*, Wiley, New York, 1981, Chapter 10

3.5.   MAXIMUM FLOW PROBLEM

Sending objects from one place (a node in a network) to another place (another node) in an optimal fashion is of great practical importance. For example, the movement of water, oil, or gas through a network of pipelines, the flow of electrical power through a grid, the transmission of information through communication channels, the shipment of finished goods from factory to a distributor, the movement of mail from posting points to their destinations, and the movement of people from their homes to places of employment can all be regarded as flows through networks.

Although there are hundreds of different flow situations, there are a few standard problems that are encountered in most of them. We will consider two of the most common problems: (1) how to maximize the amount of flow (quantity moving per unit of time) from a given node s (called the *source*) to another given node t (called the *sink*), in a network in which each edge e_i has a specified capacity $c(e_i)$—the maximum flow that can pass through edge e_i; and (2) how to determine the least costly way to ship a specified quantity from s to t in a network in which each edge e_i has a specified capacity as well as specified transportation cost per unit flow through e_i.

In this section we consider the maximum-flow problem, and the minimum cost flow problem is discussed in Section 3.6.

The maximum-flow problem (called "max-flow" for short) may be stated formally as follows: Let $G = (V, E)$ be a directed network (i.e., weighted digraph) in which the weight associated with every edge (i, j) represents its capacity c_{ij}. A *flow pattern* (or simply *flow*) is an assignment of a nonnegative number f_{ij} to every edge (i, j) such that the following conditions are satisfied:

1.   For every edge (i, j) in G,

$$0 \leqslant f_{ij} \leqslant c_{ij} \qquad (3\text{-}1)$$

2.   There is a specified node s in G, called the *source*, for which

$$\sum_i f_{si} - \sum_i f_{is} = \mathcal{F} \qquad (3\text{-}2)$$

where the summation $\sum_i f_{ij}$ is over all incoming edges to node j and $\sum_j f_{ij}$ is over all outgoing edges from node i. Quantity $\mathcal{F}$ is called the value of the flow or *flow value*.

3.   There is another specified node t in G, called the *sink* or *destination*, for which

$$\sum_i f_{ti} - \sum_i f_{it} = -\mathcal{F} \qquad (3\text{-}3)$$

4.   For each of the remaining nodes j, called *intermediate* nodes,

$$\sum_i f_{ji} - \sum_i f_{ij} = 0 \tag{3-4}$$

Condition (3-1) states that the flow through any edge does not exceed its capacity. Condition (3-4) states that the aggregate flow that enters each node, other than s or t, will be equal to the aggregate flow leaving that node. The other two conditions state that the net flow out of s as well as the net flow into the sink equals $\mathcal{F}$. Thus it is quite appropriate to call quantity $\mathcal{F}$ the *value of the flow* through the network. Condition (3-3) can, in fact, be derived from (3-2) and (3-4) and is therefore not an independent condition. The max-flow problem is: Given G, s, t, and c_{ij}'s, find a flow pattern [i.e., values of f_{ij}'s for all edges (i, j) in G] such that the flow value $\mathcal{F}$ is maximum while all four conditions are satisfied. Such a flow is called a *max-flow*. [We have tacitly assumed that the nodes have no capacity restrictions, that is, the flow through the network is limited only by edge capacities. In practice, this is no real restriction: A node that has a limited handling capacity of, say, x can be split into two with a hypothetical edge between them having the edge capacity x. Similarly, we have assumed that the given network is directed. An undirected network can be converted easily into a directed one by replacing every undirected edge with two oppositely directed edges, each with the capacity of the original edge (see Problems 3-40 and 3-42.]

The max-flow problem was first posed and solved by Ford and Fulkerson in a 1956 paper. It was shown later that the original Ford–Fulkerson algorithm can take infinitely many steps to converge to the optimal solution in worst cases. This weakness was overcome by a modification suggested by Edmonds and Karp in 1969, and the modified algorithm requires $O(nm^2)$ computational steps for an n-node, m-edge network. In 1970, Dinic proposed another algorithm, fundamentally different from Ford and Fulkerson's approach. His algorithm requires $O(n^2m)$ steps. In the past 10 years a number of more efficient algorithms have been discovered, based on Dinic's basic approach. On the whole, Dinic's algorithm as improved by Malhotra et al. [1978] is the simplest and probably the most efficient, particularly when the network is dense. We will therefore only discuss the Dinic–Malhotra–Kumar–Maheshwari (DMKM) algorithm in detail. Others are discussed in the References and Remarks at the end of this section.

Max-Flow Algorithm

The main idea behind the DMKM max-flow algorithm (in fact, in a broad sense, behind all max-flow algorithms) is to start pushing a certain amount of material from source s progressively through to sink t. Then look at the "useful" or "unsaturated" edges in the network (edges that can be used for additional flow from s to t) and push some more. Continue pushing until no more material can be pushed through.

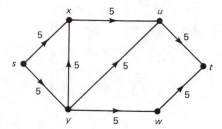

Figure 3-13   *Flow network.*

This pushing must, however, be done systematically and carefully; otherwise, an unwary pusher may fall into a trap. Consider, for example, Fig. 3-13. The capacity of each edge is 5, as shown in the figure. Suppose that we push 5 units of flow from s to y, then from y to u, and finally from u to t. Now, with this flow pattern, there seems to be no path from s to t through which flow can be added. One may, therefore (wrongly), conclude that the maximum flow through this network is 5. But it is clear that a flow value of 10 units can be sustained from s to t in this network—5 units through path (s, x, u, t) and 5 through path (s, y, w, t).

Thus, unless we are careful, we could end up with a wrong result or spend an inordinately long time in computing. This makes a max-flow algorithm somewhat involved. We will, therefore, present it in two installments. First, we will develop a subprocedure for a restricted type of directed network, called a layered network, and then use it to solve the general max-flow problem.

Saturating Flow in a Layered Network

A *layered network* $G_L = (V_L, E_L)$ is an acyclic network, in which all the nodes V_L are partitioned into "layers," $V_1, V_2, \ldots, V_l$. The first layer, V_1, consists of only the source s. Layer V_2 consists of all nodes that are immediate successors of s (i.e., at a distance 1 from s). V_3 consists of all nodes that are immediate successors of nodes in V_2. And so on. The ith layer V_i consists of all nodes at a distance $(i - 1)$ from s. Finally, $V_l = \{t\}$. Thus every node in a layered network lies on a path from s to t, and all paths from s to t are of the same length $l - 1$. Every edge in a layered network goes from some node in the ith layer to some node in the $(i + 1)$th layer, and there is no edge between two nodes within the same layer. A simple example of a layered network is shown in Fig. 3-14.

An edge (x, y) in a flow network (not necessarily layered) is said to be *saturated* if $f_{xy} = c_{xy}$. A path p from s to t is said to be saturated if at least one of the edges in p is saturated. A flow pattern through a network is said to be a *saturating flow* if every path from s to t is saturated. Clearly, every max-flow must be a saturating flow; for otherwise we could send an additional quantity from s to t through an unsaturated path. However, every saturating flow need not be a max-flow. For example, in Fig. 3-14 flow pattern $f_{sa} = f_{ac} = f_{ct} = 2$, and 0 for other

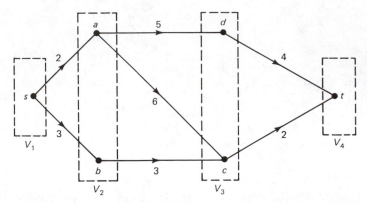

Figure 3-14   *Layered network G (numbers are edge capacities).*

edges, is a saturating flow, but not a max-flow. The maximum total flow value through the network is 4 units.

We will now discuss an efficient procedure for obtaining a saturating flow in a given layered network G_L. For each node v in the given layered network, we determine the maximum additional flow that can be forced through v. Let this number be called the *potential poten(v)* of node v. The potential is determined by first finding the total additional flow that can go into v [called *inpotential, inpot(v)*], and the total additional flow that can possibly flow out of v [called *outpotential outpot(v)*] and then taking the smaller of the two as the potential *poten(v)*. Clearly, node v can handle no more than *poten(v)* amount of flow. This potential is calculated for all nodes except s and t. For the source s, the potential is *poten(s)* = *outpot(s)* and for the sink t, *poten(t)* = *inpot(t)*. Next, we find node r which has the smallest potential among all nodes of G_L, and call r the *reference node*.

For the layered network of Fig. 3-14, the potential of various nodes are *poten(s)* = 5, *poten(a)* = 2, *poten(b)* = 3, *poten(c)* = 2, *poten(d)* = 4, *poten(t)* = 6. The nodes with minimum potential are a and c, either of which could serve as the reference node.

It is not difficult to see that we can readily establish a flow of *poten(r)* units through the network from s to t via r as follows: Imagine that we have *poten(r)* units available at r. We distribute *poten(r)* unit to the outgoing edges from r—taking these edges one by one and saturating them until all units are finished—thus pushing *poten(r)* units to the next higher layer. The flow now reaching the next layer can be distributed among their outgoing edges and pushed to the next layer. This process of pushing *poten(r)* units to the next higher layer is continued until we reach sink t. We can never get stuck with too much supply at our hands, because *poten(r)* is the minimum of all potentials. Similarly, we can "pull" *poten(r)* units to the reference node r from immediate predecessors of r, which in turn will pull this amount from their predecessors, and so on, until this amount is pulled from source s.

(Observe how absence of cycles in a layered network is essential for the foregoing argument to hold.)

Now, we delete all saturated edges from G_L, since they will not affect flow in later iterations. Similarly, a node that has had either all its incoming or outgoing edges saturated (and therefore deleted) is also deleted from the network. Deletion of a node causes the deletion of all its incoming and outgoing edges. The effect of these deletions will be reductions of inpotentials and outpotentials of various nodes in the network. We update these. This completes one iteration. We proceed to the next iteration with the reduced network and updated node potentials. Since in any iteration all incoming or outgoing edges at the reference node will become saturated, at least one node gets deleted in each iteration. Thus there can be at most n iterations for an n-node network. The work done in each iteration in distributing the flow, recomputation of flow potentials, and finding the reference node will not exceed the number of nodes in the network during that iteration. Therefore, the total work done in finding a saturating flow in an n-node layered network is bounded by $O(n^2)$. (The effort in deletion of edges will be proportional to the number of edges deleted in each iteration. Since no edge is deleted twice, total deletion effort is proportional to m.)

Let us use the procedure we have just outlined to obtain a saturating flow through the network in Fig. 3-15. Initial values of node potentials are calculated as shown in Fig. 3-15. Since h has the smallest potential, it should be the reference node for the first iteration. We push through a flow of 4 units through (h, t). We pull a flow of 3 units through (d, h), (a, d), and (s, a). Similarly, another flow of 1 unit is pulled through (e, h), (a, e), and (s, a). This flow now saturates node h and edge (a, d); therefore, we delete these. This flow also causes the following reduction: the potential of s is reduced by 4, the in and out potential of a is reduced by 4; the in and out potential of d is reduced by 3; the in and out potential of e is reduced by 1; and the potential of t is reduced by 8. The remainder of the nodes remain unaffected.

For the second iteration, node c has the smallest potential with value $poten(c)$ $= 7$. In the second iteration we add a flow of 7 units through edges (s, c), (c, d), (d, f), and (f, t). Node c gets deleted. Similarly, in the third iteration d becomes the reference node with $poten(d) = poten(f) = 2$, and an additional flow of 2 units is pushed through. Nodes f and g are also deleted together with d. In the fourth iteration, nodes e and k both have the minimum potential $poten(e) = poten(k) = 10$. We arbitrarily pick e as the reference node. At the end of the fourth iteration there is no path left from s to t in the reduced graph and we are through. The node potentials through various iterations are shown in Table 3-2. Figure 3-16 shows the saturating flow pattern so obtained. The value of the saturating flow through the network is 23 units.

The preceding method of obtaining a saturating flow pattern $F_L = [fl_{ij}]$ in a given layered network $G_L = (V_L, E_L)$ from source s to sink t is due to Malhotra et al. [1978]. Let us express it more precisely as the SATURATE procedure, which in

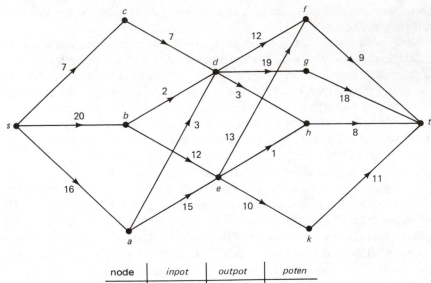

node	inpot	outpot	poten
s	--	43	43
a	16	18	16
b	20	14	14
c	7	7	7
d	12	34	12
e	27	24	24
f	25	9	9
g	19	18	18
h	4	8	4
k	10	11	10
t	46	--	46

Figure 3-15   *A flow network and its initial node potentials.*

turn uses three other procedures: (1) REFNODE(G_L, l, r, *rlayer*), for finding the reference node r and its layer number, *rlayer*; (2) PUSH(i), for pushing a specified amount of flow forward from the ith layer to the $(i + 1)$st; and (3) PULL(j), for pulling a specified amount of flow from the $(j - 1)$st layer to the jth:

Algorithm 3-6(a):   Finding Saturating Flow Pattern F_L from s to t in a Layered Network G_L

```
procedure SATURATE (G_L, l, s, t, F_L);
    (* the procedure produces F_L, the saturating flow pattern in the given layered network
        G_L, with l layers, from s to t *)
INITIALIZATION:
    for all v ∈ V_L do
        compute inpot(v) and outpot(v);
```

for all edges $(i, j) \in E_L$ **do**
    $fl_{ij} \leftarrow 0$; {∗ initialize flow pattern to 0 ∗}
  $flag \leftarrow true$; {∗ G_L is unsaturated ∗}

ITERATION:
  while $flag = true$ **do** {∗ **while** G_L is unsaturated ∗}
  begin
    REFNODE$(G_L, l, r, rlayer)$; {∗ find reference node and its layer number ∗}
    if $poten(r) \neq 0$ **then**

  begin
    $inflow(r) \leftarrow outflow(r) \leftarrow poten(r)$;
    for all $v \in V_L - \langle r \rangle$ **do**
      $inflow(v) \leftarrow outflow(v) \leftarrow 0$;
    for $k \leftarrow rlayer$ **to** $(l - 1)$ **do** PUSH(k);
    for $j \leftarrow rlayer$ **downto** 2 **do** PULL(j)
  end;
  if $(poten(r) \neq 0)$ **or** $((r \neq s)$ **and** $(r \neq t))$ **then**
  begin
    $V_L \leftarrow V_L - \langle r \rangle$;
    adjust inpotentials of immediate successors of r and
        outpotentials of immediate predecessors of r
  end
  else $flag \leftarrow false$ {∗ G_L is saturated ∗}
end

Table 3-2. Node Potentials through Various Iterations

Node	Initial Potential	Potential after Iteration 1	Potential after Iteration 2	Potential after Iteration 3
s	43	39	32	30
a	16	12	12	12
b	14	14	14	12
c	7	⑦	—	—
d	12	9	②	—
e	24	23	23	10
f	9	9	2	—
g	18	18	18	—
h	④	—	—	—
k	10	10	10	10
t	46	38	31	11

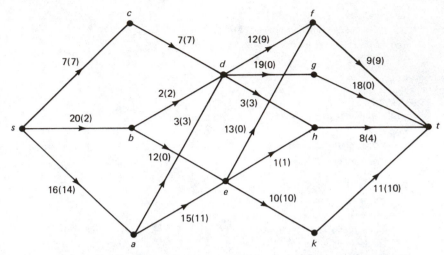

Figure 3-16   *Network with a saturating flow (edge flows in parentheses).*

For a given layered network $G_L = (V_L, E_L)$ with l layers in which *inpot* and *outpot* for every node has been computed, the REFNODE subprocedure finds the reference node r and the layer number *rlayer* to which r belongs. This procedure may be expressed as follows:

Algorithm 3-6(b):   Finding Reference Node r and Its Layer Number *rlayer* in a Layered Network G_L

```
procedure REFNODE (GL, l, r, rlayer);
INITIALIZATION:
     poten(s) ← outpot(s);
     poten(t) ← inpot(t);
     if poten(s) < poten(t) then
          r ← s; rlayer ← 1
     else
          r ← t; rlayer ← l;

ITERATION:
     for v ∈ VL − {s, t} do
     begin
          poten(v) ← min⟨inpot(v), outpot(v)⟩;
          if poten(v) < poten(r) then
          begin
               r ← v;
               rlayer ← layer to which v belongs
          end
     end
```

In a given layered network $G_L = (V_L, E_L)$ with l layers such that $V_L = V_1 \cup V_2 \cup \ldots \cup V_l$, and also given the amount of *outflow* from each node in layer V_i, the PUSH(i) procedure advances the outflows from ith layer to $(i + 1)$st. It updates the node potentials in the two layers.

Algorithm 3-6(c):   Pushing Outflows
from the ith Layer to the $(i + 1)$st

```
procedure PUSH(i);
    for u ∈ Vᵢ such that outflow(u) ≠ 0 do
    begin
        for v ∈ succ(u) do  {* for all immediate successors of u *}
        begin
            avacap ← cᵤᵥ − fᵤᵥ;  {* available capacity of edge (u, v) *}
            if avacap > 0 then
            begin  {* send min⟨avacap, outflow(u)⟩ through (u, v) *}
                if avacap > outflow(u) then
                    avacap ← outflow(u);
                fᵤᵥ ← fᵤᵥ − avacap;
                outflow(u) ← outflow(u) − avacap;
                outflow(v) ← outflow(v) + avacap;
                outpot(u) ← outpot(u) − avacap;
                inpot(v) ← inpot(v) − avacap
            end
        end
    end
end
```

Almost identical to procedure PUSH, we have procedure PULL, which updates the flow pattern into the edges going from layer $(j - 1)$ to layer j (in order to pull flow from V_{j-1} to V_j). It also updates the potentials in both layers. It may be described as follows:

Algorithm 3-6(d):   Pulling Inflows
from the $(j - 1)$st Layer to the jth

```
procedure PULL(j);
    for u ∈ Vⱼ such that inflow(u) ≠ 0 do
    begin
        for v ∈ pred(u) do  {* for all immediate predecessors of u *}
        begin
            avacap ← cᵥᵤ − fᵥᵤ;  {* available capacity of edge (v, u) *}
            if avacap > 0 then
```

```
      begin  ⟨* send min⟨avacap, inflow(u)⟩ through (v, u) *⟩
         if avacap > inflow(u) then
              avacap ← inflow(u);
         f_vu ← f_vu + avacap;
         inflow(u) ← inflow(u) − avacap;
         inflow(v) ← inflow(v) + avacap;
         outpot(v) ← outpot(v) − avacap;
         inpot(u) ← inpot(u) − avacap
      end
   end
end
```

Extracting a Layered Network

Thus far we have developed a procedure for finding a saturating flow (not neces-sarily a maximum flow) through a layered network. Let us now return to the original problem of finding a max-flow pattern through an arbitrary network. We will solve this problem in stages. First, we extract a layered network G_L from the given flow network G with zero flow through it. Then we saturate this layered network G_L to obtain a saturating flow pattern $F = [f_{ij}]$. Next, we consider the original network G with the flow pattern F just obtained, and extract another layered network G'_L. We saturate G'_L and obtain a flow pattern $F' = [f'_{ij}]$. Now the cumulative flow pattern through G is an algebraic sum, $F + F'$. Once again we extract a layered network G_L'' from G with flow $F + F'$. This process is continued until it is no longer possible to obtain a layered network (i.e., there is no "usable" path left from s to t). When this happens the cumulative flow pattern obtained is the maximum flow through the network.

   To construct a layered network from a given network with an existing pattern we must first understand the concept of a "usable" edge. A *usable edge* (x, y) is said to exist from node x to y if (1) either the network has an edge (x, y) whose capacity c_{xy} is greater than the currently assigned flow f_{xy}, or (2) if there is an edge (y, x) in the network with some nonzero flow f_{yx} through it. In both cases, the effective flow from x to y can be augmented: in case (1) by adding more forward flow, and in case (2) by canceling some reverse flow.

   The layering algorithm extracts a layered network from s to t by labeling s with 1 (i.e., putting s in first layer V_1). Then every node connected directly to s by a usable edge is labeled 2 (i.e., put in set V_2). All nodes that are immediate successors (through usable edges) of nodes with label 2 are labeled 3 (i.e., put in layer 3, V_3). This process is continued until sink t gets a label. All nodes that remain unlabeled are deleted together with edges incident on them. Moreover, since node t is the only node allowed in V_l, all nodes that may have gotten labeled l before t got labeled must also be deleted.

   It is clear that this labeling procedure is like Dijkstra's shortest-path algorithm described in Section 3.3.1. It labels each node by its edge distance from s (i.e., number of edges in the shortest path from s) plus 1.

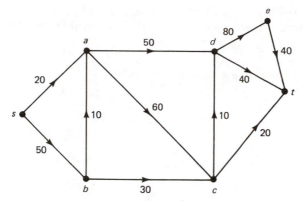

Figure 3-17 *A network G and its edge capacities.*

Figure 3-17 shows a network with no flow through its edges. The edge capacities are shown next to the edges. Figure 3-18 shows the layered network extracted in accordance with the method outlined. Observe how all nodes and edges that do not lie on a shortest path from s to t are deleted. In practice we can accomplish this by first labeling all nodes with a negative number (say, -1). At the end of the labeling all those nodes that remained negative are then deleted, by traveling backward from t to s.

We now give a more precise description of the algorithm for extracting a layered network. The input is a network $G = (V, E, C, F)$ with specified edge capacities c_{ij}'s and existing flows f_{ij}'s. The output is the layered network $G_L = (V_L, E_L)$ such that the node set $V_L = V_1 \cup V_2 \cup \ldots \cup V_l$; and for every node x in V_L the lists of all its immediate predecessors $pred(x)$ and all its immediate successors $succ(x)$ in G_L are produced. If there is no path (of usable edges) from s to t, the

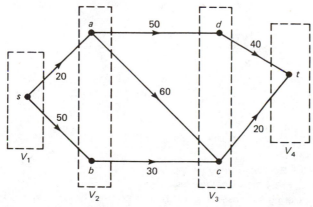

Figure 3-18 *Layered network G_L extracted from Fig. 3-17.*

layering is not possible and the Boolean variable *stpath* is set *false*. In that case the existing flow pattern *F* has achieved the maximum possible flow through *G*.

Algorithm 3-6(e)   Extracting a Layered Network

```
procedure LAYER(G_L, l, stpath)
        {* the procedure extracts layered network G_L = (V_L, E_L) with l layers from a network G
           with flow pattern F. If no s − t path exists, stpath is set false *}
INITIALIZATION:
        stpath ← true; {* path from s to t exists *}
        for all v ∈ V do
        begin
             label(v) ← −1;
             succ(v) ← ∅; {* successors and predecessors of nodes in layered network *}
             pred(v) ← ∅
        end;
        V_1 ← {s};
        label(s) ← 1;
        E_L ← ∅; {* E_L is edge set of layered network *}
        i ← 1; {* current layer number is i *}

ITERATION: {* forward traversal and labeling *}
        while (V_i ≠ ∅) and (label(t) = −1) do
        begin
             V_{i+1} ← ∅;
             for all x ∈ V_i do
             begin
                  for every edge (x, y) out of x if f_xy < c_xy
                        and for every edge (y, x) entering x if
                        f_yx > 0 and (label(y) = −1 or label(y) = i + 1) do
                  begin
                       V_{i+1} ← V_{i+1} ∪ {y};
                       label(y) ← i + 1;
                       succ(x) ← succ(x) ∪ {y};
                       pred(y) ← pred(y) ∪ {x};
                       E_L ← E_L ∪ {(x, y)}
                  end
             end;
             i ← i + 1
        end;
        l ← i;

ITERATION: {* Backward traversal and pruning *}
        if label(t) = −1 then stpath ← false {* exit *}
             {* layering not possible since there is no useful path from s to t *}
        else {* remove stray edges and nodes *}
```

```
begin
    j ← i;
    while j ≠ 1 do
    begin
        for all w ∈ Vⱼ do
            if (succ(w) = ∅) and (w ≠ t) then
            begin
                for all x ∈ pred(w) do
                begin
                    E_L ← E_L − {(x, w)};
                    succ(x) ← succ(x) − {w}
                end;
                Vⱼ ← Vⱼ − {w};
                pred(w) ← ∅
            end;
        j ← j − 1
    end
end
```

Complete Max-Flow Algorithm

Now the entire max-flow algorithm may be described as follows:

Algorithm 3-6(f): Max-Flow Algorithm

```
read the given network G = (V, E, C);
    {* C = [cᵢⱼ] the edge capacities *}
for all (u, v) ∈ E do f_uv ← 0;
call the LAYER procedure;
while stpath = true do
begin
    call the SATURATE procedure; {* to obtain flows flᵢⱼ through the layered network *}
    for all (i, j) ∈ E do fᵢⱼ ← fᵢⱼ + flᵢⱼ; {* cumulative flows *}
    call the LAYER procedure
end {* while *}
```

To illustrate the complete algorithm, let us continue with the example given in Fig. 3-17. The layered network (first time around) was given in Fig. 3-18. A saturating flow through the layered network of Fig. 3-18 is

$$fl_{sa} = fl_{ac} = fl_{ct} = 20$$

and 0 in all other edges.

The network of Fig. 3-17 with the flow pattern obtained so far is shown in Fig. 3-19. All usable edges and their capacities are shown. Note that the usable edges in reverse are shown with dashes. The layered network extracted from Fig. 3-19 is shown in Fig. 3-20.

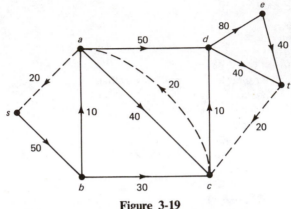

Figure 3-19

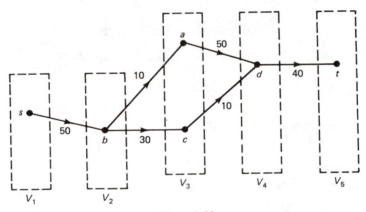

Figure 3-20

The saturating flow through this layered network is $fl_{sb} = fl_{dt} = 20,$ and $fl_{ba} = fl_{ad} = fl_{bc} = fl_{cd} = 10.$

Next, the cumulative flow pattern obtained so far in the network of Fig. 3-17 is shown in Fig. 3-21.

Once again we extract a layered network (from Fig. 3-21). It turns out to be a single path, as shown in Fig. 3-22. The saturating flow through this path is obviously 20 units (through each edge).

Once again, the usable edges in the network (with the current cumulative flow pattern) are as shown in Fig. 3-23.

When the LAYER procedure is applied to Fig. 3-23, it returns the Boolean variable *stpath* = *false*, implying that there is no usable path from s to t in Fig. 3-23.

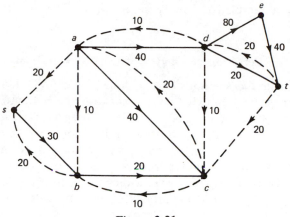

Figure 3-21

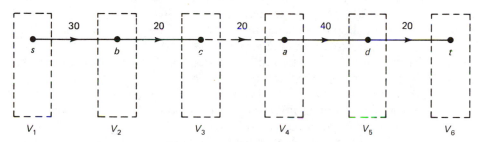

Figure 3-22

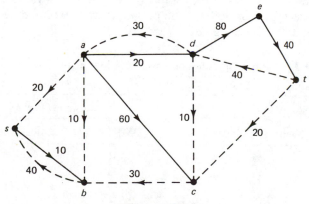

Figure 3-23

Hence the current cumulative flow is the max-flow. It yields the maximum flow value.

Performance Analysis of DMKM Algorithm

Let us first estimate the amount of work done in the SATURATE procedure (in finding a saturating flow through a layered network of n nodes and m edges). Since at least one node (the reference node) is deleted in each interation, this procedure will have at most n iterations. In the ith iteration, the amount of effort in distributing the flow is $O(n + m_i)$, when m_i is the number of edges deleted in the ith iteration. The value n comes from the fact that at each node (other than the reference node) we use at most one outgoing edge without saturating it or one incoming edge without saturating it. Therefore, the total effort is of the order of

$$O\left(\sum_i (n + m_i) \right) = O(n^2 + m) = O(n^2)$$

It should be noted that the recomputation of potential at each node requires no extra effort. Identification of the reference node requires finding the minimum of at most n numbers per iteration. Therefore, this effort is also bounded by $O(n^2)$.

Thus each stage (of saturating a layered network) requires $O(n^2)$ effort, and there are at most $n - 1$ such steps in the entire procedure (because the number of layers in each successive layered network grows by at least one, and the path length from s to t is bounded by $n - 1$). Moreover, the time required for extracting a layered network (consisting of initialization, a Dijkstra's-shortest-path-type forward traversal and labeling, and a backward traversal and pruning) is also bounded by $O(n^2)$. Hence the entire algorithm can be executed in time $O(n^3)$.

Table 3-3 gives the run times of the two most efficient max-flow algorithms—the DMKM algorithm as implemented here in the MAXFLOW procedure, and Karzanov's algorithm [1974], which we did not discuss in this section. The input networks are randomly generated, complete digraphs (i.e., with exactly one directed edge between each pair of nodes). The programs were executed on an Amdahl 470 V/8 computer, and times are given in seconds.

Table 3-3. Computing Times of Two Max-Flow Algorithms on Complete Networks

	Run Time (sec)	
Number of Nodes	DMKM	Karzanov
40	0.364	0.165
60	0.743	0.600
80	1.699	1.206
100	2.534	2.372

The purpose of Table 3-3 is just to give a feeling for the execution times for and their growth with the size of the network. These should not be taken as expected time complexities. That would require an extensive empirical study on networks of varying sizes and densities, taking averages over several networks of a given size and density, as well as varying the $s - t$ pair for each network. Such studies have been conducted and reported in the literature. It may be observed from this table, however, that as n grows, Karzanov's and DMKM algorithms are quite comparable in their performances, at least for the input used here.

Computer Implementation of the Maximum-Flow Algorithm

The MAXFLOW procedure finds a flow pattern that yields the maximum flow value from a specified node S to another specified note T in a given weighted network with N nodes and M edges. As mentioned earlier, this algorithm is due to Dinic [1970] with modification suggested by Malhotra, Kumar, and Maheshwari [1978] (hence the name DMKM). The procedure follows closely the method outlined in this section.

The MAXFLOW procedure calls the LAYER and SATURATE procedures alternately until layering is no longer possible (i.e., no usable path remains from S to T). A Boolean variable STPATH is set FALSE when that happens.

The SATURATE procedure, in turn, calls REFNODE, PULL, and PUSH. Two elementary procedures ADD and DELETE are required by both the LAYER and SATURATE procedures. The ADD(A, I, X) procedure adds a given node X to a set whose elements are pointed by A[I], the Ith entry in array A. Similarly, the DELETE(A, I, X) procedure deletes node X from the set.

To keep the code somewhat shorter and simpler, we have used matrix CAPA to represent the given network. The reader can convert the code for other forms of representation. In fact, if the input network is sparse, it would be more efficient to use another data structure, say the forward star, instead of a matrix (as discussed earlier in Section 3.1). Note that a nonexistent edge has zero capacity [i.e., CAPA [I, J] = 0 if there is no edge (I, J) in the network].

At each stage there is a layered network. The layered network is represented by listing the set of immediate successors and immediate predecessors (this duplication is for ease of processing) for every node in the network. Two linked lists for each node are used for this purpose. The arrays (each of length N) PRED and SUCC are the heads of these linked lists. Whenever an element is to be added, a node pointed by AV with two fields DATA and LINK is obtained from a common storage pool in which all available nodes are linked together.

Global Constants

N   number of nodes in the given network

MAX   total number of elements for lists LAYERS, PRED, and SUCC; it
        should be at least 3M, where M is the number of edges in the network

Data Types

```
TYPE  ARRNN  = ARRAY[1..N,1..N] OF INTEGER;
      ARRN   = ARRAY[1..N] OF INTEGER;
      ARRMAX = ARRAY[1..MAX] OF INTEGER;
```

Procedure Parameters

```
PROCEDURE MAXFLOW(
    N,S,T,MAX      :INTEGER;
    VAR CAPA,FLOW:ARRNN);
```

Input

N	number of nodes in the given network
S	source node
T	sink node
MAX	total number of elements for lists LAYERS, PRED, and SUCC; MAX $\geq 3 \cdot$ the number of edges in the network
CAPA[1..N,1..N]	array of capacities of the edge

Output

FLOW[1..N,1..N]	array of max-flow pattern

Arrays and Variables (Local to the Procedure MAXFLOW)

LCAPA[1..N,1..N]	array of capacities of the edges in the layered network
LFLOW[1..N,1..N]	array of saturating flow in the layered network
EL[1..N,1..N]	array of edge set of the layered network such that if an edge (I, J) is useful, $EL[I, J] = 1$; otherwise, $EL[I, J] = 0$
LABELS[1..N]	array for layer number of each node; initially, every element is set to -1
LAYERS[1..N]	array of pointers for nodes in each layer; nodes in a layer are linked together
PRED[1..N]	array of pointers for predecessors for each node; all predecessors of a node are linked together
SUCC[1..N]	array of pointers for successors for each node; all successors of a node are linked together
INPOT[1..N]	array of total additional flow that can go into each node

OUTPOT[1..N]	array of total additional flow that can possibly flow out of each node
POTEN[1..N]	array of maximum additional flow that can be forced through each node
INFLOW[1..N]	array of inflow of each node in the layered network
OUTFLOW[1..N]	array of outflow of each node in the layered network
L	layer number of the sink node in the layered network
R	reference node
RLAYER	layer number of the reference node
STPATH	Boolean variable such that if there is no path from S to T, it is set FALSE
FLAG	Boolean variable such that if the layered network is not saturated, it is set TRUE
AV	pointer to a linked list of available storage pool
DATA[1..MAX]	array to implement a linked list (data)
LINK[1..MAX]	array to implement a linked list (links)

Note: MAX should be a least 3M, where M is the number of edges.

Example

For the network shown in Fig. 3-17, we have N = 7, S = 1, T = 7, and the capacity array is as follows (nodes are ordered in s, a, b, c, d, e, t):

$$
CAPA[1..7,1..7] = \begin{bmatrix}
0 & 20 & 50 & 0 & 0 & 0 & 0 \\
0 & 0 & 0 & 60 & 50 & 0 & 0 \\
0 & 10 & 0 & 30 & 0 & 0 & 0 \\
0 & 0 & 0 & 0 & 10 & 0 & 20 \\
0 & 0 & 0 & 0 & 0 & 80 & 40 \\
0 & 0 & 0 & 0 & 0 & 0 & 40 \\
0 & 0 & 0 & 0 & 0 & 0 & 0
\end{bmatrix}
$$

The capacity array of the first layered network is

$$
LCAPA[1..7,1..7] = \begin{bmatrix}
0 & 20 & 50 & 0 & 0 & 0 & 0 \\
0 & 0 & 0 & 60 & 50 & 0 & 0 \\
0 & 0 & 0 & 30 & 0 & 0 & 0 \\
0 & 0 & 0 & 0 & 0 & 0 & 20 \\
0 & 0 & 0 & 0 & 0 & 0 & 40 \\
0 & 0 & 0 & 0 & 0 & 0 & 0 \\
0 & 0 & 0 & 0 & 0 & 0 & 0
\end{bmatrix}
$$

The values of other arrays for the first layered network are as follows:

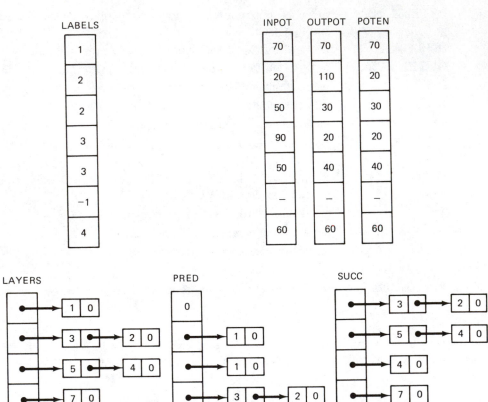

The final max-flow pattern obtained is

$$
\text{FLOW}[1..7,1..7] =
\begin{bmatrix}
0 & 20 & 40 & 0 & 0 & 0 & 0 \\
0 & 0 & 0 & 0 & 30 & 0 & 0 \\
0 & 10 & 0 & 30 & 0 & 0 & 0 \\
0 & 0 & 0 & 0 & 10 & 0 & 20 \\
0 & 0 & 0 & 0 & 0 & 0 & 40 \\
0 & 0 & 0 & 0 & 0 & 0 & 0 \\
0 & 0 & 0 & 0 & 0 & 0 & 0
\end{bmatrix}
$$

Pascal Procedure MAXFLOW

```
PROCEDURE MAXFLOW(
   N,S,T,MAX      : INTEGER;
   VAR CAPA,FLOW : ARRNN);

VAR    LCAPA,LFLOW,EL           : ARRNN;
       LABELS,LAYERS,PRED,SUCC  : ARRN;
       DATA,LINK                : ARRMAX;
       STPATH                   : BOOLEAN;
       AV,I,J,L,DIF             : INTEGER;


PROCEDURE ADD(
   VAR A : ARRN;
   I,X   : INTEGER);

VAR   P: INTEGER;
      FOUND: BOOLEAN;
BEGIN
   FOUND:= FALSE;
   P:= A[I];
   WHILE (NOT FOUND) AND (P <> 0) DO
      IF DATA[P] = X THEN FOUND:= TRUE
      ELSE P:= LINK[P];
   IF NOT FOUND THEN
   BEGIN
      P:= AV; AV:= LINK[AV]; (* GET A NODE *)
      DATA[P]:= X;
      LINK[P]:= A[I];
      A[I]:= P
   END
END; (* ADD *)


PROCEDURE DELETE(
   VAR A : ARRN;
   I,X   : INTEGER);

VAR   P,Q: INTEGER;
BEGIN
   P:= A[I];
   IF DATA[P] = X THEN A[I]:= LINK[P]
   ELSE
   BEGIN
      WHILE DATA[P] <> X DO
      BEGIN
         Q:= P;
         P:= LINK[P]
      END;
      LINK[Q]:= LINK[P]
   END
END; (* DELETE *)
```

```
PROCEDURE LAYER(
    VAR L         : INTEGER;
    VAR LCAPA,EL  : ARRNN;
    VAR STPATH    : BOOLEAN);


VAR I,J,P,Q,X,Y,V,W : INTEGER;
BEGIN
    STPATH:= TRUE;
    FOR V:= 1 TO N DO
    BEGIN
        LABELS[V]:= -1;
        SUCC[V]   := 0;
        PRED[V]   := 0
    END;
    P:= AV; AV:= LINK[AV]; (* GET A NODE *)
    DATA[P]:= S; LINK[P]:= 0;
    LAYERS[1]:= P;
    LABELS[S]:= 1;
    FOR X:= 1 TO N DO
        FOR Y:= 1 TO N DO
            LCAPA[X,Y]:= CAPA[X,Y];
    FOR X:= 1 TO N DO
        FOR Y:= 1 TO N DO
        BEGIN
            EL[X,Y]:= 0;
            IF FLOW[X,Y] > 0 THEN
            BEGIN
                LCAPA[X,Y]:= CAPA[X,Y] - FLOW[X,Y];
                LCAPA[Y,X]:= FLOW[X,Y] + LCAPA[Y,X]
            END
        END;
    I:= 1;
(* INITIALIZATION OVER *)


(* FORWARD TRAVERSAL AND LABELING *)
    WHILE (LAYERS[I] <> 0) AND (LABELS[T] = -1) DO
    BEGIN
        LAYERS[I+1]:= 0;
        P:= LAYERS[I];
        WHILE P <> 0 DO
        BEGIN
            X:= DATA[P];
            FOR Y:= 1 TO N DO
                IF ((LABELS[Y] = -1) OR (LABELS[Y] = I+1))
                    AND (LCAPA[X,Y] > 0) THEN
                BEGIN
                    ADD(LAYERS,I+1,Y);
                    LABELS[Y]:= I + 1;
                    ADD(SUCC,X,Y);
                    ADD(PRED,Y,X);
                    EL[X,Y]:= 1
                END
```

```
                ELSE LCAPA[X,Y]:= 0;
              P:= LINK[P]
          END;
          I:= I + 1
      END; (* WHILE *)
      L:= I;

  (* BACKWARD TRAVERSAL AND PRUNING *)
      IF LABELS[T] = -1 THEN STPATH:= FALSE (* NO USEFUL PATH *)
      ELSE
      BEGIN
          J:= I;
          WHILE J <> 1 DO
          BEGIN
              P:= LAYERS[J];
              WHILE P <> 0 DO
              BEGIN
                  W:= DATA[P];
                  IF (SUCC[W] = 0) AND (W <> T) THEN
                  BEGIN
                      Q:= PRED[W];
                      WHILE Q <> 0 DO
                      BEGIN
                          X:= DATA[Q];
                          EL[X,W]:= 0;
                          DELETE(SUCC,X,W);
                          LCAPA[X,W]:= 0;
                          Q:= LINK[Q]
                      END;
                      DELETE(LAYERS,J,W);
                      PRED[W]:= 0;
                      LABELS[W]:= -1
                  END;
                  P:= LINK[P]
              END;
              J:= J - 1
          END
      END (* ELSE *)
  END; (* LAYER *)

  PROCEDURE SATURATE(
      VAR L          : INTEGER;
      VAR LFLOW,EL : ARRNN);

  VAR INPOT,OUTPOT,POTEN,INFLOW,OUTFLOW : ARRN;
      V,X,Y,I,J,K,P,R,RLAYER              : INTEGER;
      FLAG                               : BOOLEAN;


  PROCEDURE REFNODE(
      VAR L,R,RLAYER : INTEGER);

  VAR   V: INTEGER;
```

```
BEGIN
   POTEN[S]:= OUTPOT[S];
   POTEN[T]:= INPOT[T];
   IF POTEN[S] < POTEN[T] THEN
   BEGIN
      R:= S;
      RLAYER:= 1
   END
   ELSE
   BEGIN
      R:= T;
      RLAYER:= L
   END;
(* INITIALIZATION OVER *)

   FOR V:= 1 TO N DO
      IF (LABELS[V] <> -1) AND (V <> S) AND (V <> T) THEN
      BEGIN
         IF INPOT[V] < OUTPOT[V] THEN POTEN[V]:= INPOT[V]
         ELSE POTEN[V]:= OUTPOT[V];
         IF POTEN[V] < POTEN[R] THEN
         BEGIN
            R:= V;
            RLAYER:= LABELS[V]
         END
      END
END; (* REFNODE *)

PROCEDURE PUSH(
   VAR I : INTEGER);

VAR   P,Q,U,V,AVACAP: INTEGER;
BEGIN
   P:= LAYERS[I];
   WHILE P <> 0 DO
   BEGIN
      U:= DATA[P];
      IF OUTFLOW[U] > 0 THEN
      BEGIN
         Q:= SUCC[U];
         WHILE (OUTFLOW[U] > 0) AND (Q <> 0) DO
         BEGIN
            V:= DATA[Q];
            IF LABELS[V] <> -1 THEN
            BEGIN
               AVACAP:= LCAPA[U,V] - LFLOW[U,V];
               IF AVACAP > 0 THEN
               BEGIN (* SEND MIN{AVACAP,OUTFLOW[U]} THROUGH (U,V) *)
                  IF AVACAP > OUTFLOW[U] THEN
                     AVACAP:= OUTFLOW[U];
```

```
                    LFLOW[U,V]:= LFLOW[U,V] + AVACAP;
                    OUTFLOW[U]:= OUTFLOW[U] - AVACAP;
                    OUTFLOW[V]:= OUTFLOW[V] + AVACAP;
                    OUTPOT[U]:= OUTPOT[U] - AVACAP;
                    INPOT[V]:= INPOT[V] - AVACAP
                  END
                END;
                Q:= LINK[Q]
            END (* NODE V *)
        END;
        P:= LINK[P]
    END (* NODE U *)
END (* PUSH *);

PROCEDURE PULL(
    VAR J : INTEGER);

VAR    P,Q,U,V,AVACAP: INTEGER;
BEGIN
    P:= LAYERS[J];
    WHILE P <> 0 DO
    BEGIN
        U:= DATA[P];
        IF INFLOW[U] > 0 THEN
        BEGIN
            Q:= PRED[U];
            WHILE (INFLOW[U] > 0) AND (Q <> 0) DO
            BEGIN
                V:= DATA[Q];
                IF LABELS[V] <> -1 THEN
                BEGIN
                    AVACAP:= LCAPA[V,U] - LFLOW[V,U];
                    IF AVACAP > 0 THEN
                    BEGIN (* SEND MIN{AVACAP,INFLOW[U]} THROUGH (V,U) *)
                        IF AVACAP > INFLOW[U] THEN
                            AVACAP:= INFLOW[U];
                        LFLOW[V,U]:= LFLOW[V,U] + AVACAP;
                        INFLOW[U]:= INFLOW[U] - AVACAP;
                        INFLOW[V]:= INFLOW[V] + AVACAP;
                        OUTPOT[V]:= OUTPOT[V] - AVACAP;
                        INPOT[U]:= INPOT[U] - AVACAP
                    END
                END;
                Q:= LINK[Q]
            END (* NODE V *)
        END;
        P:= LINK[P]
    END (* NODE U *)
END; (* PULL *)

BEGIN (* SATURATE *)
    FOR V:= 1 TO N DO
```

```
    BEGIN
       INPOT[V]:= O;
       OUTPOT[V]:= O
    END;
    FOR I:= 1 TO N DO
       FOR J:= 1 TO N DO
          IF EL[I,J] = 1 THEN
          BEGIN
             INPOT[J]:= INPOT[J] + LCAPA[I,J];
             OUTPOT[I]:= OUTPOT[I] + LCAPA[I,J]
          END;
    FOR I:= 1 TO N DO
       FOR J:= 1 TO N DO
          LFLOW[I,J]:= O;
    FLAG:= TRUE; (* GL IS UNSATURATED *)
(* INITIALIZATION OVER *)


    WHILE FLAG DO (* WHILE GL IS UNSATURATED *)
    BEGIN
       REFNODE(L,R,RLAYER);
       IF POTEN[R] <> O THEN
       BEGIN
          INFLOW[R]:=POTEN[R];
          OUTFLOW[R]:= POTEN[R];
          FOR V:= 1 TO N DO
             IF (LABELS[V] <> -1) AND (V <> R) THEN
             BEGIN
                INFLOW[V]:=O;
                OUTFLOW[V]:= O
             END;
          FOR K:= RLAYER TO L-1 DO
             PUSH(K);
          FOR J:= RLAYER DOWNTO 2 DO
             PULL(J)
       END;
       IF (POTEN[R] <> O) OR ((R <> S) AND (R <> T)) THEN
       BEGIN
          LABELS[R]:= -1;
          P:= SUCC[R];
          WHILE P <> O DO
          BEGIN
             Y:= DATA[P];
             INPOT[Y]:= INPOT[Y] - (LCAPA[R,Y] - LFLOW[R,Y]);
             EL[R,Y]:= O;
             DELETE(SUCC,R,Y);
             DELETE(PRED,Y,R);
             P:= LINK[P]
          END;
          P:= PRED[R];
          WHILE P <> O DO
```

```
          BEGIN
             X:= DATA[P];
             OUTPOT[X]:= OUTPOT[X] - (LCAPA[X,R] - LFLOW[X,R]);
             EL[X,R]:= 0;
             DELETE(SUCC,X,R);
             DELETE(PRED,R,X);
             P:= LINK[P]
          END
      END
      ELSE  FLAG:= FALSE (* GL IS SATURATED *)
   END (* WHILE *)
END; (* SATURATE *)


PROCEDURE INITIALIZE(
   MAX : INTEGER);

VAR I : INTEGER;
BEGIN
   FOR I:= 1 TO MAX DO
      LINK[I]:= I + 1;
   AV:= 1
END; (* INITIALIZE *)


BEGIN (* MAXFLOW *)
   FOR I:= 1 TO N DO
      FOR J:= 1 TO N DO
         FLOW[I,J]:= 0;
   INITIALIZE(MAX); (* MAKE A LINKED LIST *)
(* INITIALIZATION OVER *)

   LAYER(L,LCAPA,EL,STPATH);
   WHILE STPATH DO
   BEGIN
      SATURATE(L,LFLOW,EL);
      FOR I:= 1 TO N DO
         FOR J:= 1 TO N DO
            IF LFLOW[I,J] > 0 THEN
            BEGIN
               DIF:= LFLOW[I,J] - FLOW[J,I];
               IF DIF > 0 THEN
               BEGIN
                  FLOW[I,J]:= FLOW[I,J] + DIF;
                  FLOW[J,I]:= 0
               END
               ELSE FLOW[J,I]:= -DIF
            END;
      INITIALIZE(MAX); (* MAKE A LINKED LIST *)
      LAYER(L,LCAPA,EL,STPATH)
   END
END; (* MAXFLOW *)
```

PROBLEMS

3-38. Conceptually, the simplest approach for constructing a maximum flow from s to t is by repeatedly finding a path from s to t and increasing the flow along the path by the maximum amount allowed by the edge capacities. This, indeed, was the strategy behind the labeling procedure discussed in Section 1.3.

   (a) Show that with this approach the flow pattern may not converge to a correct solution if the edge capacities are reals rather than integers (see Ford and Fulkerson [1962] for an example).

   (b) Show that even when the edge capacities are integers, the time complexity may not be a polynomial in n or m but a function of the value of the maximum flow $\mathscr{F}$ (see Section 1.3).

   (c) If in the procedure above we replace the statement "finding a path from s to t" with "finding a shortest path from s to t," difficulties (a) and (b) both disappear.

3-39. Show that in a flow network if all edge capacities are integers, there is a flow pattern such that the value of the maximum flow from s to t is integral and the flow through every edge is also an integer.

3-40. In some applications it is also necessary to impose capacity restrictions on nodes (e.g., relay stations in communication networks, or transshipment points in transportation networks). Show how you will convert this problem into the original form of the max-flow problem given in this section.

3-41. Suppose that the flow network has several source nodes and several destination nodes and the object is to obtain the maximum possible value of the total flow from sources to destinations. Show how to convert this problem into a one-source, one-destination flow problem as discussed in this section.

3-42. Show how you will solve the maximum flow problem in a network containing both directed and undirected edges. (Reduce this version to the original form of the max-flow problem.)

3-43. Combining Problems 3-40 and 3-42, show how a flow network with some edges undirected, as well as with some nodes having capacity restrictions, can be reduced to the standard flow network discussed in this section.

3-44. Verify that each conversion in Problems 3-40 through 3-43 can be achieved in time which is bounded by a polynomial in the size of the network.

Comment: Problems 3-45 through 3-48 are difficult and may require a literature search.

3-45. In the flow network in this section (as well as in Section 1.3) we assumed that there was only an upper limit on the flow through each edge (edge capacity). In some applications, a lower limit is also specified for each edge (to prevent pipes from freezing in Alaska, for example). Show how you would:

(a) Determine if there is a feasible flow pattern at all.
(b) Determine a maximum flow pattern.
(c) Determine a minimum flow pattern.

3-46. In this section it was assumed that the amount of flow did not vary along an edge. In many practical networks, the flow suffers a significant loss during transmission, due to leakage, evaporation, and so forth. How will you handle such *lossy networks*?

3-47. In some practical situations it becomes necessary to deal with several distinct commodities flowing simultaneously (e.g., cars, bicycles, and horse carts sharing a common highway). How will you handle such a *multicommodity flow* problem?

3-48. Suppose, after solving the maximum flow problem for a given network, you discover that the capacity of one of the edges in the network was incorrect. Is there a way to salvage the results, or will you have to solve the entire problem again?

3-49. Let every edge in a flow network G be of unit capacity. Then show that:

(a) The value of the maximum flow from s to t equals the maximum number of edge-disjoint directed paths from s to t in G.
(b) This value equals the minimum number of edges whose removal destroys all paths from s to t in G.

3-50. Show that the maximum number of node-disjoint paths from a node x to node y in a directed graph is equal to the minimum number of nodes whose removal destroys all paths from x to y. (*Hint:* Use Problems 3-39 and 3-40. Problems 3-49 and 3-50 are versions of Menger's theorem.)

3-51. Let us define a *line of a matrix* as either a row or a column. Consider a $p \times q$ binary matrix M. A set of lines is said to cover M if all 1's in M lie on one of these lines. A pair of 1's are called *independent* if they are not in the same line. The König-Egerváry theorem says that the minimum number of lines that cover M equals the maximum number of independent 1's in M. Prove the König-Egerváry theorem by formulating it as a max-flow problem.

3-52. Given two vectors $P = (p_1, p_2, \ldots, p_r)$ and $Q = (q_1, q_2, \ldots, q_s)$ of positive integers. Devise a procedure for determining if there exists an $r \times s$ binary matrix whose row sums are p_i's and the column sums are q_i's. (*Hint:* Formulate this as a complete bipartite flow network with r sources and s destinations.)

3-53. The MAXFLOW procedure presented in this section uses matrices to represent the capacities in the input network, as well as the flow pattern. In a sparse network we would be better off using adjacency lists (or forward stars) instead of matrices. Rewrite the MAXFLOW procedure using adjacency lists. Compare empirically the performance of the two for large sparse networks.

3-54. In the MAXFLOW procedure we have avoided using recursive procedures, for the sake of efficiency but at the cost of making the program longer. Rewrite the procedure using recursion, when needed to make it more compact.

3-55. Identify the min-cuts (cuts of minimum capacities, as discussed in Section 1.3) for the networks in Figs. 3-13, 3-15, and 3-17. Verify that the max-flow min-cut theorem (Theorem 1-5) holds for these networks.

3-56. A part of the procedure given in Section 1.3 solves the max-flow problem for a very special class of networks (complete bipartite networks called transportation networks). Compare the efficiency of the procedure given in Section 1.3 with that of this section, by running on a number of randomly generated complete bipartite flow networks.

REFERENCES AND REMARKS

The maximum-flow problem was first posed and solved by Ford and Fulkerson in

FORD, L. R., Jr., and D. R. FULKERSON, Maximal Flow through a Network, *Canad. J. Math.* 8(1956), 399–404

and independently by Elias et al. in

ELIAS, P., A. FEINSTEIN, and C. E. SHANNON, Note on Maximum Flow through a Network, *IRE Trans. Inform. Theory*, IT-2(1956), 117–119

Both of these papers proved the "max-flow min-cut" theorem, which is a remarkable result but does not directly lead to an efficient algorithm for finding a maximum flow. This theorem has been stated and discussed in Section 1.3.

The classic reference on network-flow theory is

FORD, L. R., JR., and D. R. FULKERSON, *Flows in Networks*, Princeton University Press, Princeton, N.J., 1962

which contains the Ford–Fulkerson algorithm for maximum flow as well as many other combinatorial problems that can be solved through the max-flow problem.

The labeling algorithm proposed by Ford and Fulkerson, which is based on the max-flow min-cut theorem, is also explained in Section 1.3. In spite of its poor performance in pathological cases (see Section 1.3 and Problem 3-38), the Ford–Fulkerson algorithm remained in practical use for over 15 years, with various refinements, modifications, and implementations. An ALGOL listing of Ford–Fulkerson algorithm is given in

BAYER, G., Algorithm 324, MAXFLOW, *Comm. ACM* 11(1968), 117–118; 16(1973), 309

The variations of Ford–Fulkerson algorithm have been primarily in labeling and scanning of nodes (in the process of finding augmenting paths and flows along them).

Edmonds and Karp, in the following paper, pointed out that the order in which the nodes are scanned may be crucial for efficiency of the Ford–Fulkerson algorithm:

EDMONDS, J., and R. M. KARP, Theoretical Improvement in Algorithmic Efficiency for Network Flow Problems, *J. ACM* 19(1972), 248–264

They showed that in the Ford–Fulkerson algorithm if we were always to augment along the shortest paths (having fewest edges from) s to t, then the lengths of the augmenting paths would be nondecreasing. By imposing this simple order the computational time bound was reduced to $O(nm^2)$ in an n-node m-edge network, regardless of the values of the edge capacities. Thus formally, Edmonds and Karp gave the first polynominal-time (in the network size) algorithm for the max-flow problem, and it works even when the capacities are irrational numbers. Furthermore, they showed that the time bound of the original Ford–Fulkerson algorithm cannot be improved over $O(m\mathcal{F})$, where $\mathcal{F}$ is the flow value through the network from s to t, and m is the number of edges. E. A. Dinic made the next contribution. Independently of Edmonds and Karp, Dinic also proposed in

DINIC, E. A., Algorithm for Solution of a Problem of Maximal Flow in a Network with Power Estimation, *Soviet Math. Dokl.* 11(1970), 1277–1280

the idea of using the shortest augmenting paths first (rather than any augmenting path, as in the original Ford–Fulkerson algorithm) together with the added fundamental improvement of dividing the problem into phases. The concept of layered network is due to Dinic. His algorithm requires $O(n^2m)$ computational steps. A good description of Dinic's algorithm is given in

EVEN, S., *Graph Algorithms*, Computer Science Press, Potomac, Md., 1979, Chapter 5

All subsequent max-flow algorithms use Dinic's idea of phases and layered networks, but use better methods of finding saturating flow through a layered network. The max-flow algorithm presented here uses Dinic's idea of phases and layered network and Malhotra, Kumar, and Maheshwari's method of saturating a layered network, which appeared in

MALHOTRA, V. M., M. PRAMODH KUMAR, and N. MAHESHWARI, An $O(|V|^3)$ Algorithm for Finding the Maximum Flows in Networks, *Inform. Process. Lett.* 7(1978), 227–278

A fundamentally different approach for the max-flow problem was suggested by Karzanov in

KARZANOV, A. V., Determining the Maximal Flow in a Network by the Method of Pre-flows, *Soviet Math. Dokl.* 15(1974), 434–437

For a good treatment of Karzanov's algorithm and its performance, see

BARTZ, A. E., Construction and Analysis of Network Flow Problem Which Forces Karzanov Algorithm to $O(n^3)$ Running Time, TN-83, MIT, LCS, Apr. 1977

HAMACHER, H., Numerical Investigations on the Maximum Flow Algorithm of Karzanov, Report 78-7, Köln University, Mathematics Institute, Cologne, W. Germany, Apr. 1978

A FORTRAN coding of Karzanov's algorithm which runs in $O(n^3)$ time is given in

NIJENHUIS, A., and H. S. WILF, *Combinatorial Algorithms*, 2nd ed., Academic Press, New York, 1978, pp. 196–216

A number of excellent survey papers on relative performance of various max-flow algorithms have appeared. Some of these are

CHEUNG, T. Y., Computational Comparison of Eight Methods for the Maximum Network Flow Problem, *ACM Trans. Math. Software* 6(1980), 1–16

GALIL, Z., On the Theoretical Efficiency of Various Network Flow Algorithms, *Theoret. Comput. Sci.* 14(1981), 103–111

GLOVER, F., D. KLINGMAN, J. MOTE, and D. WHITMAN, Comprehensive Computer Evaluation and Enhancement of Maximum Flow Algorithm, Research Report 356, Center for Cybernetic Studies, University of Texas, Austin, Oct. 1979

Another interesting and recent algorithm for the max-flow problem can be found in

SLEATOR, D., An $O(mn \log n)$ Algorithm for Maximum Network Flow, Ph.D. thesis, Department of Computer Science, Stanford University, 1980

3.6. MINIMUM-COST FLOW PROBLEM

In the preceding section we considered the problem of sending the maximum amount of flow value from source s to sink t in a network with specified edge capacities. There was no cost involved. We will now consider the problem of sending a specified amount of flow value θ (which we will call *target flow*) from s to t in a network G in which every edge (i, j) has a capacity c_{ij} as well as a nonnegative cost d_{ij} associated with it. The "cost" d_{ij} of sending a unit flow through edge (i, j) may represent the driving time, gas consumption, or any other measure depending on the application.

Obviously, if θ is greater than the maximum flow value $\mathcal{F}$ from s to t, no solution is possible. If θ is smaller than or equal to $\mathcal{F}$, then, in general, many different flow patterns will result in establishing a flow value θ from s to t in G. Our purpose is to find that flow pattern which minimizes the total cost. This problem is called the *minimum-cost flow problem* (or the *min-cost flow problem*) and may be stated more formally as follows:

To minimize

$$\sum_{(i, j)} d_{ij} f_{ij} \tag{3-5}$$

under the following constraints:

1. For every edge (i, j) in G,

$$0 \leqslant f_{ij} \leqslant c_{ij} \tag{3-6}$$

2. There is a specified node s in G, called the *source*, for which

$$\sum_{i} f_{si} - \sum_{i} f_{is} = \theta \tag{3-7}$$

where the summation $\sum_i f_{ij}$ is over all incoming edges to node j and $\sum_j f_{ij}$ is over all outgoing edges from node i.

3. There is another specified node t in G, called the *sink* or *destination*, for which

$$\sum_{i} f_{ti} - \sum_{i} f_{it} = -\theta \tag{3-8}$$

4. For each of the remaining nodes j, called *intermediate* nodes,

$$\sum_{i} f_{ji} - \sum_{i} f_{ij} = 0 \tag{3-9}$$

Observe that if all $c_{ij} = 1$, and $\theta = 1$, that is, if there is no edge-capacity constraint, the problem is reduced to one of finding a *cheapest* (shortest) path from s to t with d_{ij} as the cost (distance or weight). In fact, even if the edge capacity constraints are imposed, the method of shipping θ amount of commodity along the cheapest path will work if the smallest edge capacity along this path is not less than θ. On the other hand, if the target flow value θ is greater than the smallest edge capacity along the cheapest path, the rest of flow has to be shipped through more expensive paths. This observation suggests that we can solve the min-cost flow problem by repeated application of a shortest-path algorithm, as follows.

First, we find the shortest-path (least-cost path) from s to t using a shortest-path algorithm (from Section 3.3). Then we send as many units of flow as possible along this path. If we have achieved the target flow value θ, we are through; otherwise, we modify (to be discussed shortly) the network, taking into account the flow pattern obtained so far. In this modified network, once again we find the shortest path from s to t and send as many units of flow as possible. These two basic steps of finding the shortest path (in the modified network) and modifying the network (due to the added flow) are repeated alternately until either the specified target flow value θ is obtained or there is no path left from s to t (i.e., θ exceeds the maximum possible flow value $\mathcal{F}$ through G).

Network Modification

The *modified network* (also called the *incremental network*) with respect to a flow pattern in a network G is the network G^* with the same structure as G but with *modified capacities* c_{ij}^* and *modified costs* d_{ij}^* defined as follows:

If there is a nonzero flow f_{ji} through an existing edge (j, i), we have a (fictitious) edge (i, j) "usable" in the reverse direction with capacity f_{ji}; that is,

$$c_{ij}^* = f_{ji} \qquad \text{if} \quad f_{ji} > 0$$

The cost associated with this edge is $-d_{ji}$, because the edge (i, j) is usable only by virtue of a flow reduction (and hence cost reduction); that is,

$$d_{ij}^* = -d_{ji} \qquad \text{if} \quad f_{ji} > 0$$

Also, if there is a flow f_{ij} through an existing edge (i, j) which is less than its capacity c_{ij}, we can send additional flow through this unsaturated edge (i, j) in forward direction (provided, of course, that there is no flow f_{ji}). Thus

$$c_{ij}^* = c_{ij} - f_{ij} \qquad \text{if} \quad f_{ji} = 0 \quad \text{and} \quad c_{ij} > f_{ij}$$

The cost of sending 1 unit of additional flow through this unsaturated edge is the same as the original cost. That is,

$$d_{ij}^* = d_{ij} \qquad \text{if} \quad f_{ji} = 0 \quad \text{and} \quad c_{ij} > f_{ij}$$

Finally, since a saturated edge is not usable in the forward direction, we have

$$c_{ij}^* = c_{ij} - f_{ij} = 0 \qquad \text{if} \quad f_{ij} = c_{ij} \quad \text{and} \quad f_{ji} = 0$$

and

$$d_{ij}^* = \infty \qquad \text{if} \quad f_{ij} = c_{ij} \quad \text{and} \quad f_{ji} = 0$$

Note that we maintain throughout the constraint $f_{ij}f_{ji} = 0$, because we cannot have flows in opposing directions through any edge at the same time. All edges with no flows remain unchanged in the modified network G^*.

Busacker–Gowen's Algorithm

Now, having defined the modified network, we can express the *min-cost path-flow* algorithm as follows:

Algorithm 3-7(a):   Outline of Busacker and Gowen's Min-Cost Flow Algorithm

```
begin
    G* ← G;
    while (flow value < θ) and (a path exists from s to t in G*) do
    begin
        find a shortest path P from s to t in G*;
        if no such path exists then exit;
        send additional flow along P until either P is saturated or the flow reaches θ;
        if P saturates then obtain a modified network G*
    end
end
```

This "incremental approach" of achieving larger and larger flows through paths of increasing costs was originally proposed by Busacker and Gowen [1961]. It is also called the *dual method* of solving the min-cost flow problem, because the first feasible solution obtained is an optimal solution. The following small example illustrates further details of this algorithm.

The five-node seven-edge network G shown in Fig. 3-24, has two labels on each edge. The first number is the cost of sending a unit flow and the second number is the capacity. We wish to find a minimum-cost flow of 25 units from s to t.

Iteration 1

The shortest (least-cost) path from s to t in Fig. 3-24 is $P = (s, a, c, t)$. The cost of sending 1 unit of flow through this path is $3 + 4 + 3 = 10$. The maximum flow possible through this path is $\min\{18, 15, 17\} = 15$ which is less than 25, the desired amount. Therefore, we establish a flow of 15 units through this path. Thus the first

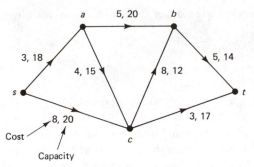

Figure 3-24   *Flow network with edge costs and capacities.*

flow pattern obtained is

$$f_{sa} = f_{ac} = f_{ct} = 15$$

We modify the network to take this flow pattern into account. The modified network is shown in Fig. 3-25. We go into the second iteration.

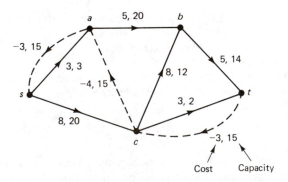

Figure 3-25   *Network after first modification.*

Iteration 2

In Fig. 3-25 the least-cost path from s to t is $P' = (s, c, t)$. The unit cost through this path is $8 + 3 = 11$. The maximum flow through this path is $\min\{20, 2\} = 2$. We ship 2 units through this path, thus achieving so far a total flow value of $15 + 2 = 17$ which is less than 25. The cumulative flow pattern obtained thus far is

$$f_{sa} = f_{ac} = 15, \qquad f_{ct} = 17, \qquad f_{sc} = 2$$

The modified network with respect to this flow pattern is given in Fig. 3-26. Next, we go into the third iteration.

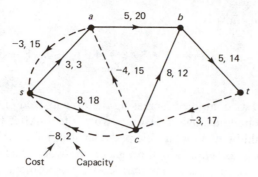

Figure 3-26   *Network after second modification.*

Iteration 3

In Fig. 3-26 the shortest path from s to t is $P'' = (s, a, b, t)$. The unit cost in P'' is $3 + 5 + 5 = 13$ and the maximum flow possible through P'' is $\min\{3, 20, 14\} = 3$. Sending 3 units through P'' results into a total flow value of $17 + 3 = 20$ and into the cumulative flow pattern of

$$f_{sa} = 18, \qquad f_{ac} = 15 \qquad f_{sc} = 2, \qquad f_{ct} = 17 \qquad f_{ab} = 3, \qquad f_{bt} = 3$$

The modified network with respect to this flow pattern is given in Fig. 3-27.

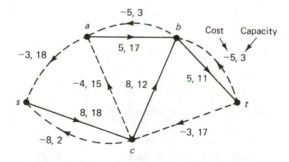

Figure 3-27   *Network after third modification.*

Iteration 4

The shortest path from s to t in Fig. 3-27 is $P''' = (s, c, a, b, t)$, with unit flow cost $8 - 4 + 5 + 5 = 14$ and maximum capacity $\min\{18, 15, 17, 11\} = 11$. Since we have achieved a flow value of 20 units, we require only 5 additional units of flow, which we can send through this path. This results into a total flow value of 25 units from s to t, and the final flow pattern is

$$f_{sa} = 18, \qquad f_{sc} = 7 \qquad f_{ab} = 8, \qquad f_{ac} = 10 \qquad f_{bt} = 8, \qquad f_{ct} = 17$$

The total cost of this flow is

$$\sum_{(i,\,j)} d_{ij} f_{ij} = 3 \cdot 18 + 8 \cdot 7 + 5 \cdot 8 + 4 \cdot 10 + 5 \cdot 8 + 3 \cdot 17 = 281$$

Note that had we continued increasing the flow instead of stopping at 25 units, we would have obtained a flow of 31 units from s to t. After that, in the modified network, there would be no more usable paths left from s to t. Hence 31 units would be the value of the maximum flow through the network. Thus the maximum flow problem is a special case of the minimum-cost flow problem. However, as we will discuss later, this may be an inefficient way of solving the max-flow problem.

Now we will describe somewhat more formally Busacker and Gowen's min-cost flow algorithm just illustrated. We are given network G in which two vertices s and t are specified and the value of the target flow θ is also specified. Initially, we set the current value of the flow from s to t as 0. We also set the *modified network* G^* as G itself, to begin with. Moreover, we maintain a Boolean variable *stPath* to indicate the existence of a usable path from s to t. Initially, we set *stPath* to *true*.

The algorithm may now be described compactly as follows [the formal proof of the algorithm is left as an exercise (Problem 3-60)]:

Algorithm 3-7(b):   Busacker and Gowen's Min-Cost Flow Algorithm

```
INITIALIZATION:
    CurrentFlow ← 0;
    G* ← G; ⟨* modified network is G itself *⟩
    stPath ← true;


ITERATION:
    while (CurrentFlow < θ) and stPath do
    begin
        find shortest path P from s to t in G*;
        if no path exists then stPath ← false;
        if stPath then
        begin
            delta ← smallest usable edge capacity in P;
                ⟨* maximum feasible flow through P *⟩
            if CurrentFlow + delta > θ then delta ← θ − CurrentFlow;
            for each edge (i, j) on P do ⟨* establish a flow of amount delta
                through P by suitably increasing or decreasing flows through
                edges in P, and adjust costs *⟩
            begin
                if (i, j) is a forward edge ⟨* i.e., capacity cᵢⱼ* > 0 *⟩ then
```

```
    begin
        increase flow f_ij through (i, j);
        if (i, j) is unsaturated then cost*(i, j) stays the same
        else cost*(i, j) ← ∞;
        if (i, j) is usable in backward direction then cost*(j, i) ← − cost(i, j)
    end
    else {* if (i, j) is a backward edge *}
    begin
        decrease flow f_ji through (j, i);
        adjust cost of edge (j, i) as in the forward case
    end;
        CurrentFlow ← CurrrentFlow + delta
    end {* for *}
end {* stPath *}
end {* while *}
```

Shortest-Path Subprocedure

In implementing the min-cost flow algorithm just described, a crucial decision that we must make is in the choice of the shortest-path algorithm. Three shortest-path algorithms were described in Section 3.3, of which Floyd's algorithm is not suitable for our purpose, because it computes shortest paths between all pairs of nodes (at added computational cost), whereas we only need a shortest path from s to t. If the choice were between the Bellman–Moore algorithm and the Dijkstra's, we could select either. Both are fast. The following are some important considerations:

1. *Negative-weight edges.* In this min-cost flow algorithm, as we proceed with modified networks, many edge costs become negative. Dijkstra's algorithm does not normally work for networks with negative weights, as discussed in Section 3.3. However, the negative weights can be eliminated by an artificial device of adding suitable numbers to various edge costs (or weights) to keep them all nonnegative, without changing their relative costs. Thus the shortest path remains unaffected. This device was suggested independently by Iri [1969] and Edmonds and Karp [1972], and it enables one to use Dijkstra's algorithm for the min-cost flow problem. This does require some additional computations, thus making Dijkstra's algorithm more expensive than it would otherwise be. The Bellman–Moore algorithm, of course, has no difficulty in handling negative edges.

2. *One-to-one and one-to-all paths.* Recall that Dijkstra's algorithm could be terminated as soon as the destination node t received a permanent label, that is, as soon as the shortest path from s to t was obtained. In the Bellman–Moore algorithm, on the other hand, we must wait until the shortest paths from s to all other nodes have been obtained. Therefore, the latter is more suitable for

one-to-all shortest-path problems. It may, therefore, appear that Dijkstra's algorithm would have an advantage in the min-cost flow algorithm. However, due to negativity of edge weights, it turns out that in using Dijkstra's algorithm also we must wait until all nodes have been permanently labeled.

3. *Edge density.*   It was mentioned in Section 3.3 that the PDM implementation of the Bellman–Moore algorithm outperformed Dijkstra's algorithm only when the network was sparse. Because of the added computational work in modifying Dijkstra's algorithm, our experience shows that the use of PDM procedure leads to overall shorter time for a typical min-cost flow problem, even when the network is complete (i.e., there is an edge between every pair of nodes, see Table 3.2).

4. *Worst-case input versus average-case input.*   Although in a typical network the PDM implementation of the Bellman–Moore algorithm is faster (and more so for sparse networks), for a pathologically worst-case input (see Problem 3-15), it could take an exponential amount of time. Of course, a network such as the one shown in Problem 3-15 is not encountered in practice.

In light of these four considerations, we will implement the Busacker–Gowen min-cost flow algorithm [i.e., Algorithm 3-7(b)] using the (slightly modified) PDM procedure (Algorithm 3-2) for shortest paths. We will leave the use of the modified Dijkstra's procedure as an exercise.

Data Structure

The choice of data structure will also have a significant influence on the computation time. Since new edges are created as the network gets modified, it is not sufficient to have just a fixed number of edges (say m, in the forward-star form). We must keep a space for representing $2m$ edges, if the network originally has m edges. Because of this, and to be consistent with the MAXFLOW algorithm in the preceding section, we will use matrices to represent edge capacities (CAPA) and costs (COST). The reader is encouraged to make changes in order to use other data structures, such as the list of edges on forward star, which would be more efficient for very sparse and large networks.

Computer Implementation of Busacker– Gowen's Min-cost Flow Algorithm

The BUSACKER procedure finds a minimum-cost flow from a specified node S to another specified node T in an N-node network, represented by a cost matrix COST and a capacity matrix CAPA. The procedure closely follows the method outlined in Algorithm 3-7(b).

In the initialization step the value of current flow is set to zero, and the cost matrix COST is copied into the modified cost matrix MCOST. Then the PDM procedure is called to find a cheapest path P from S to T, using MCOST. If there is no path from S to T, the current flow value is the maximum possible flow. Otherwise, find DELTA, which is the maximum incremental flow through P. Then adjust each flow in P and the modified edge costs and capacities. This process continues until CURRENTFLOW, the value of current flow, equals TARGET-FLOW, the desired value of the flow, or until no more flow increase is possible (i.e., no S-T path exists).

Global Constant

N   number of nodes in the given network

Data Types

```
TYPE   ARRNN = ARRAY[1..N,1..N] OF INTEGER;
       ARRN  = ARRAY[1..N] OF INTEGER;
```

Procedure Parameters

```
PROCEDURE BUSACKER(
    N,S,T,INF,TARGETFLOW :INTEGER;
    VAR COST,CAPA,FLOW  :ARRNN;
    VAR TOTALCOST        :INTEGER);
```

Input

N	number of nodes in the given network
S	source node
T	sink node
INF	an integer much larger than largest edge cost
TARGETFLOW	flow value to be established from S to T
COST[1..N,1..N]	cost matrix of the given network
CAPA[1..N,1..N]	capacity matrix of the given network

Output

FLOW[1..N,1..N]	minimum-cost flow pattern for the given network
TOTALCOST	total cost

Example

For the five-node seven-edge network shown in Fig. 3-24, the cost and the capacity matrices are as follows (nodes are in order s, a, b, c, t):

$$\text{COST}[1..5,1..5] = \begin{bmatrix} \infty & 3 & \infty & 8 & \infty \\ \infty & \infty & 5 & 4 & \infty \\ \infty & \infty & \infty & \infty & 5 \\ \infty & \infty & 8 & \infty & 3 \\ \infty & \infty & \infty & \infty & \infty \end{bmatrix}$$

$$\text{CAPA}[1..5,1..5] = \begin{bmatrix} 0 & 18 & 0 & 20 & 0 \\ 0 & 0 & 20 & 15 & 0 \\ 0 & 0 & 0 & 0 & 14 \\ 0 & 0 & 12 & 0 & 17 \\ 0 & 0 & 0 & 0 & 0 \end{bmatrix}$$

The other inputs are N = 5, S = 1, T = 5, INF = 999999, and TARGETFLOW = 25. The output flow pattern, FLOW, is

$$\text{FLOW}[1..5,1..5] = \begin{bmatrix} 0 & 18 & 0 & 7 & 0 \\ 0 & 0 & 8 & 10 & 0 \\ 0 & 0 & 0 & 0 & 8 \\ 0 & 0 & 0 & 0 & 17 \\ 0 & 0 & 0 & 0 & 0 \end{bmatrix}$$

and the total cost is

$$\text{TOTALCOST} = 281$$

Pascal Procedure BUSACKER

```
PROCEDURE BUSACKER(
    N,S,T,INF,TARGETFLOW : INTEGER;
    VAR COST,CAPA,FLOW    : ARRNN;
    VAR TOTALCOST         : INTEGER);

VAR MCOST                        : ARRNN;
    DIST,PRED                    : ARRN;
    STPATH                       : BOOLEAN;
    I,J,CURRENTFLOW,DELTA,D : INTEGER;

PROCEDURE PDM(
    N,S,T,INF     : INTEGER;
    VAR A         : ARRNN;
    VAR STPATH    : BOOLEAN;
    VAR DIST,PRED : ARRN);

VAR QUEUE                         : ARRN;
    U,V,HEAD,NEXT,TEMP,NEWLABEL : INTEGER;
```

```
BEGIN
   FOR V:= 1 TO N DO
   BEGIN
      DIST[V]:= INF;
      PRED[V]:= -1;
      QUEUE[V]:= -1
   END;
   DIST[S]:= 0; U:= S;
   HEAD:= S; QUEUE[HEAD]:= INF;
(* INITIALIZATION OVER *)

   WHILE U <> INF DO              (* QUEUE IS EMPTY IF U = INF *)
   BEGIN
      NEXT:= QUEUE[U];
      QUEUE[U]:= 0;
      FOR V:= 1 TO N DO
         IF A[U,V] <> INF THEN
         BEGIN
            NEWLABEL:= DIST[U] + A[U,V];
            IF DIST[V] > NEWLABEL THEN
            BEGIN
               DIST[V]:= NEWLABEL;
               PRED[V]:= U;
               TEMP:= NEXT;
               IF QUEUE[V] < 0 THEN  (* V WAS NEVER IN QUEUE *)
               BEGIN
                  QUEUE[HEAD]:= V;
                  HEAD:= V; QUEUE[HEAD]:= INF;
                  IF TEMP = INF THEN TEMP:= V;
                  NEXT:= TEMP
               END
               ELSE
               IF QUEUE[V] = 0 THEN   (* V WAS IN QUEUE, BUT *)
               BEGIN                   (* NOT IN QUEUE NOW    *)
                  QUEUE[V]:= TEMP;
                  NEXT:= V
               END
               ELSE NEXT:= TEMP             (* V IS IN QUEUE *)
            END
         END;
      U:= NEXT
   END; (* WHILE *)
   IF PRED[T] = -1 THEN STPATH:= FALSE
END; (* PDM *)

BEGIN  (* BUSACKER *)
   FOR I:= 1 TO N DO
      FOR J:= 1 TO N DO
      BEGIN
         FLOW[I,J]:= 0;
         MCOST[I,J]:= COST[I,J]
      END;
```

```
      CURRENTFLOW:= 0;
      STPATH:= TRUE;
(* INITIALIZATION OVER *)

   WHILE (CURRENTFLOW < TARGETFLOW) AND (STPATH) DO
   BEGIN
      PDM(N,S,T,INF,MCOST,STPATH,DIST,PRED);
      IF STPATH THEN
      BEGIN
         DELTA:= INF;
         I:= T;
         REPEAT
            J:= I;
            I:= PRED[J];
            IF CAPA[I,J] > 0 THEN              (* FORWARD EDGE *)
            BEGIN
               D:= CAPA[I,J] - FLOW[I,J];
               IF D < DELTA THEN DELTA:= D
            END
            ELSE                               (* BACKWARD EDGE *)
            IF FLOW[J,I] < DELTA THEN DELTA:= FLOW[J,I]
         UNTIL I = S;
         IF CURRENTFLOW + DELTA > TARGETFLOW THEN
            DELTA:= TARGETFLOW - CURRENTFLOW;
         I:= T;
         REPEAT
            J:= I;
            I:= PRED[J];
            IF CAPA[I,J] > 0 THEN              (* FORWARD EDGE *)
            BEGIN
               FLOW[I,J]:= FLOW[I,J] + DELTA;
               IF FLOW[I,J] = CAPA[I,J] THEN MCOST[I,J]:= INF;
               MCOST[J,I]:= -COST[I,J]
            END
            ELSE                               (* BACKWARD EDGE *)
            BEGIN
               FLOW[J,I]:= FLOW[J,I] - DELTA;
               MCOST[J,I]:= COST[J,I];
               IF FLOW[J,I] = 0 THEN MCOST[I,J]:= COST[I,J]
            END
         UNTIL I = S;
         CURRENTFLOW:= CURRENTFLOW + DELTA
      END (* STPATH *)
   END; (* WHILE *)
   TOTALCOST:= 0;
   FOR I:= 1 TO N DO
      FOR J:= 1 TO N DO
         IF FLOW[I,J] > 0 THEN
            TOTALCOST:= TOTALCOST + COST[I,J] * FLOW[I,J]
END; (* BUSACKER *)
```

Performance Analysis

In each execution of the outer loop of the min-cost flow algorithm we obtain a (next-shortest) path with nonzero capacity. Since the target flow value is θ units, it is clear that the outer loop will not be executed more than θ times. If we were to use Dijkstra's shortest-path algorithm, each execution of the loop would require $O(n^2)$ number of steps. Thus the computational complexity would be bounded by $O(n^2\theta)$. This bound, however, is unsatisfactory, because θ could be very large even for a network with a few nodes and edges. It is desirable to have a time bound, which is a function of n and m only.

In obtaining $O(n^2\theta)$ bound we are being overly pessimistic because in each iteration we expect to send more than just 1 unit flow through the new path (if the edge capacities are large). In fact, it has been shown that in a flow network of n nodes and m edges there exists (in theory) a sequence of no more than m minimum-cost flow-augmenting paths which are sufficient to yield a min-cost flow of any specified value θ (see Lawler [1976, p. 133]). Had this theoretical bound been achieved by the Busacker–Gowen algorithm, we would get a time bound of $O(n^2m)$, provided that Dijkstra's algorithm was used for finding shortest paths.

From a practical point of view, it appears that the shortest-path subprocedure is usually called only a few times—rarely m times in an m-edge network. In view of this large gap between the theoretical bound and the practical observations, as well as the fact that we are using the PDM procedure (not DIJKSTRA), we have to rely on empirical results for relative efficiencies of various min-cost flow algorithms (see Papadimitriou and Steiglitz [1982, pp. 153–154]).

Table 3-4 shows the computational times for solving four randomly generated min-cost flow problems, in complete networks of 40, 60, 80, and 100 nodes. The execution times are measured in seconds and the computer used is Amdahl 470 V/8. The table is intended to provide an idea of the execution time and also to show the rate of growth with the number of nodes in the network.

Of the three columns, PDM represents the computational time of the BUSACKER procedure (exactly as given in this section) using the PDM procedure for obtaining shortest paths. The second column is for a modified procedure which

Table 3-4.   Computing Times for the Busacker–Gowen Algorithm on Complete Networks

Number of Nodes	Run Time (sec)		
	PDM	BM	Dijkstra
40	0.105	0.179	0.156
60	0.190	0.368	0.255
80	0.330	0.753	0.506
100	0.349	0.775	0.547

Table 3-5. Computing Times for the BUSACKER Procedure Modified to Produce a Min-Cost Max-Flow Pattern

Number of Nodes	Run Time (sec)
40	0.96
60	3.96
80	9.52
100	18.12

uses the Bellman–Moore algorithm with only single-ended queue (see Problem 3-66). The third column is for the case when Dijkstra's algorithm is used for finding the shortest paths (see Problem 3-67).

The timings in Table 3-4 are for the cases when the target flow θ was specified at some arbitrary value (significantly smaller than the maximum flow value, but much larger than the total capacities of the first few shortest paths from s to t). If θ is specified to be the maximum value, that is, if the program were allowed to run until no further flow augmentation is possible, the execution times would naturally be longer (see Problem 3-62). Table 3-5 shows the run times of the BUSACKER procedure, slightly modified to run continuously until the flow value achieved is the maximum. The inputs are the same complete networks as used in Table 3-4, and so is the computer.

Compare this table with the PDM column in Table 3-4 and observe that the rate of growth of the former is much greater (with n), as expected. It is also instructive to compare Table 3-5 with Table 3-3 (both use the same computer and the same complete networks as inputs). The min-cost max-flow pattern takes longer to compute than does the max-flow pattern. Furthermore, the rate of growth of the execution time (with number of nodes) of the BUSACKER procedure is greater than the MAXFLOW procedure (which is also to be expected).

PROBLEMS

3-57. In the four-node five-edge network shown below, you are to ship 4 units of flow from s to t as cheaply as possible. Obtain the solution (flow pattern) by hand, using the Busacker–Gowen algorithm given in this section. Observe the curious phenomenon that no flow is taking place through the cheapest path (s, a, b, t) in the final flow pattern. Explain this anomaly.

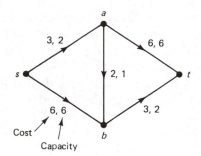

3-58.   A flow network may have nodes (other than s and t) which have only incoming edges or only outgoing edges. Such nodes can be deleted because they cannot serve as intermediate nodes. Similarly, a node x with only one incoming edge and only one outgoing edge can also be eliminated by replacing the two edges *in series* with a single edge whose capacity is the smaller of the two capacities and whose cost is the sum of the two costs. In the network given below you are to ship 50 units from s to t at a minimum cost, but first reduce the network by eliminating as many nodes as you can without affecting the flow, of course. (*Hint:* Eliminate nodes 4, 8, and then 5.) Write a general algorithm for preprocessing a large, sparse network

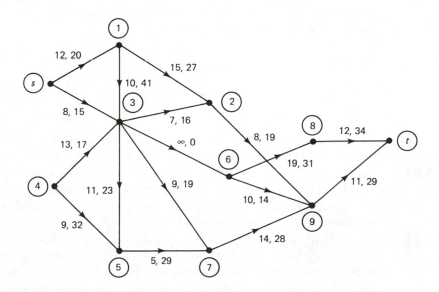

which will eliminate such "useless" nodes and edges. (Be careful in handling any *parallel edges* that may result from reduction.)

3-59. Caterer problem. A caterer is required to supply $x_1, x_2, \ldots, x_r$ fresh napkins on r successive days. He can buy new napkins at α cents each or have them cleaned in a fast laundry (in 1 day) at β cents per napkin or have them cleaned in a slow laundry (in 2 days) at $\gamma(<\beta<\alpha)$ cents each. The problem is to select a purchase-laundry policy which meets demand x_i at a minimum cost. Formulate this as a min-cost flow problem. (This problem arises in scheduling military equipment maintenance, among other practical situations.)

3-60. Give a formal proof for the following result, which intuitively seems apparent and is the basis of the Busacker–Gowen algorithm for the min-cost flow. Let F_1 be a minimum-cost flow pattern of value θ_1 in network G. Let G_1^* be the modified network with respect to F_1. Let P be a minimum-cost path from s to t in G_1^*. Prove that a flow pattern F_2 obtained by sending additional units from s to t along P is also a min-cost flow. (*Hint:* See Ford and Fulkerson [1962, p. 121]—referred to in Section 3.5.)

3-61. In some flow applications there may be a cost associated with each node as a unit flow passes through the node. Show how you will convert this problem into the problem with edge costs only.

3-62. Instead of solving the min-cost flow problem for a specified value θ, we wish to find the cheapest flow for sending the maximum amount possible from s to t. Make necessary changes in Algorithm 3-7(b) to solve this problem.

3-63. Instead of specifying a θ and then minimizing the total cost, suppose that we wish to achieve as large a value of flow from s to t as possible for a specified budget b. That is, instead of Eq. (3-5), we

maximize $\qquad\qquad\qquad\qquad \theta$

subject to $\qquad\qquad\qquad \sum_{(i,\,j)} d_{ij} f_{ij} \leqslant b$

and constraints (3-6), (3-7), (3-8), and (3-9). Modify Algorithm 3-7(b) so that we stop the flow whenever the total cost reaches the budget limitation b.

3-64. Instead of having only upper limits on the edge capacities, suppose that your network has both lower and upper limits on the permissible flow f_{ij} through each edge (i, j). Will this have any serious effect on the BUSACKER procedure, or will a simple modification take care of this situation? Explain.

3-65. Modify the BUSACKER procedure so that it employs the forward-star form for edge capacities and costs (instead of two matrices). Compare the execution time of the original procedure and the modified procedure by running them on your computer for several large sparse networks (e.g., a grid network).

3-66. The PDM implementation of the Bellman–Moore algorithm uses a double-ended queue (as explained in Section 3.3). This tends to speed up the execution in a typical network, but it can take an extremely long time for a worst-case network (an example is given in Problem 3-15). Modify the BUSACKER procedure so that it employs only one-ended queue (in implementing the Bellman–Moore algorithm). Compare the performance of this modification empirically.

3-67. Modify the BUSACKER procedure so that it uses Dijkstra's shortest-path algorithm (with appropriate alterations to take care of the negative edges) instead of the Bellman–Moore procedure. Now conduct a timing study on your computer using randomly generated networks to compare the performance of the modified procedure against the procedure given in this section.

3-68. In this section it was assumed that the cost of sending a unit of flow through an edge (i, j) was a constant d_{ij}—independent of the flow f_{ij}. In some real-life situations this cost d_{ij} may be flow dependent. For example, as the traffic increases on the road, the travel time also increases. Will the algorithm given in this section still work? Give reasons (see Klein [1967]).

3-69. Let G be a given network with a flow pattern $F = [f_{ij}]$ with a flow value θ from s to t. Let G^* be the corresponding modified network. Show that F is a minimum-cost flow pattern (among all flow patterns of flow value θ from s to t) if and only if G^* has no cycle with a net negative cost. [Klein's primal method of solving the min-cost flow problem is based on this property (see T. C. Hu, *Integer Programming and Network Flows*, Addison-Wesley, Reading, Mass., 1969).]

3-70. In the context of the min-cost flow problem, discuss

(a) The multisource, multidestination network (see Problem 3-41).
(b) Lossy networks (see Problem 3-46).
(c) Multicommodity flows (see Problem 3-47).

REFERENCES AND REMARKS

The min-cost flow problem is one of the oldest and the best known network optimization problems. The transportation problem defined in Section 1.3 can be viewed as a min-cost flow problem defined on a complete bipartite network, if all the supply nodes are connected to a *super source node* and all the demand nodes are connected to a *super sink node*. In the form of a transportation problem, the min-cost flow problem was posed and solved independently by F. L. Hitchcock, L. Kantorovich, and T. C. Koopmans between 1939 and 1941.

As in the case of the transportation problem (in Section 1.3), the min-cost flow problem can obviously be formulated as an LP problem and solved using the simplex method—in primal form or dual form. However, the most efficient algorithms are based on the graph-theoretic approach. The algorithms based on the graph-theoretic approach can be classified into three groups.

1.   Start with an arbitrary flow pattern with the desired flow value θ from s to t, and then continuously improve the pattern until a minimum-cost flow pattern is achieved. An algorithm based on this "primal" approach was given by Klein in the following paper:

KLEIN, M., A Primal Method for Minimum Cost Flows, *Management Sci.* 14(1967), 205–220

2.   A "dual" approach would be to start with a minimum-cost flow pattern of value less than θ and continuously increase the flow until it reaches the desired flow value θ—always maintaining the minimality of the cost. This indeed is the approach behind the min-cost path algorithm of Busacker and Gowen, which appeared in

BUSACKER, R. G., and P. J. GOWEN, A Procedure for Determining a Family of Minimum-Cost Flow Patterns, Operations Research Office Technical Report 15, Johns Hopkins University, Baltimore, 1961

The BUSACKER procedure described in this section is based on their method.

3.   A third approach is the so-called out-of-kilter method and was first suggested by Fulkerson in

FULKERSON, D. R., An Out-of-Kilter Method for Minimal Cost Flow Problems, *SIAM J. Appl. Math.* 9(1961), 18–27

Figure 3-28 shows this broad classification of various min-cost flow algorithms.
The following report contains an excellent overview of various algorithms and their relationships for solving the min-cost flow problem:

ZADEH, N., Near-Equivalence of Network Flow Algorithms, Technical Report 26, Dept. of Operations Research, Stanford University, Stanford, Calif., Dec. 1, 1979

Of all the procedures for solving the min-cost flow problem, probably out-of-kilter is the most popular. A good description of it can be found in

LAWLER, E. L., *Combinatorial Optimization. Networks and Matroids*, Holt, Rinehart and Winston, New York, 1976, Chapter 4

MINIEKA, E., *Optimization Algorithms for Networks and Graphs*, Dekker, New York, 1978, Chapter 4

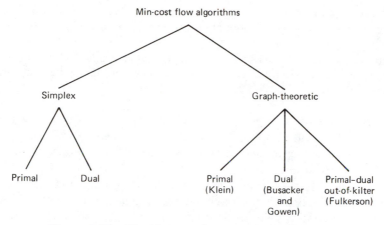

Figure 3-28 *Classification of min-cost flow algorithms.*

A FORTRAN code for the out-of-kilter algorithm is available as SHARE Program SDA 3536, Out-of-Kilter Network Routine, OKF3 (1967) and is also given in

PHILLIPS, D. T., and A. GARCIA - DIAZ, *Fundamentals of Network Analysis*, Prentice-Hall, Englewood Cliffs, N.J. 1982

The last reference not only gives a complete description of the out-of-kilter method as a general tool for solving network problems, but also shows how it can be used to solve more specific problems, such as the transportation problem, the assignment problem, the max-flow problem, the min-cost flow problem, the transshipment problem, and so forth.

An ALGOL code for the out-of-kilter algorithm can be found in

BRAY, T. A., and C. WITZGALL, Algorithm 336, NETFLOW, *Comm. ACM* 11(1968), 631–632 (with corrections in 1969, vol. 12)

All these computer codes of the out-of-kilter algorithm solve the min-cost flow problem in a network where edge capacities have both upper as well as lower bounds.

Another FORTRAN code, called NETFLOW, for solving the min-cost flow problem with both upper and lower bounds on the capacities is given in

KENNINGTON, J. L., and R. V. HELGASON, *Algorithms for Network Programming*, Wiley-Interscience, New York, 1980 (Appendix F)

NETFLOW uses a specialized primal simplex, with data structures and initial solution tailored for computational efficiency.

A number of empirical studies have been conducted to determine the comparative speeds of various min-cost flow algorithms. One such study is reported in

GLOVER, F., D. KARNEY, and D. KLINGMAN, Implementation and Computational Comparisons of Primal, Dual, and Primal–Dual Computer Codes for Minimum Cost Network Flow Problems, *Networks* 4(1974), 191–212

The execution speed is greatly dependent on the size, density, structure, and cost matrix of the input network. The data structures being used, the choice of the initial solution, and other implementational details are also of critical importance in determining the execution time.

Our computational experience and those of many others indicate that for most real-life networks the Busacker–Gowen algorithm (implemented in this section) is at least as fast as the out-of-kilter method and conceptually a great deal simpler.

For a good treatment of the computational complexities of various min-cost flow algorithms, from a theoretical point of view, see Lawler [1976] and

PAPADIMITRIOU, C. H., and K. STEIGLITZ, *Combinatorial Optimization: Algorithms and Complexity*, Prentice-Hall, Englewood Cliffs, N.J. 1982

Adaptation of Dijkstra's method for handling negative-weight edges arising in the Busacker–Gowen algorithm can be found (together with several other important results) in

EDMONDS, J., and R. M. KARP, Theoretical Improvement in Algorithmic Efficiency for Network Flow Problems, *J. ACM* 19(1972) 248–64

IRI, M., *Network Flow, Transportation and Scheduling: Theory and Algorithms*, Academic Press, New York, 1969

A good reference for finding how various network flow problems relate to the problems of planning and utilization of road networks is

POTTS, R. B., AND R. M. OLIVER, *Flows in Transportation Networks*, Academic Press, New York, 1972

3.7   MAXIMUM-CARDINALITY MATCHING

A set of edges M in an undirected graph $G = (V, E)$ is called a *matching* if no two edges in M have a node in common. For example, in Fig. 3-29 the edge set $\{(v_1, v_6), (v_2, v_4)\}$, shown in heavy lines, is a matching, because the two edges are not incident on a common node. The set $\{(v_1, v_6), (v_2, v_5), (v_3, v_4)\}$ is another matching. But

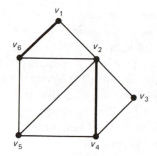

Figure 3-29 *Graph and a matching.*

$\langle (v_2, v_3), (v_3, v_4) \rangle$ is *not* a matching because the edges share a node, namely v_3. A single edge in a graph is obviously a matching.

Applications

A well-known optimization problem encountered in numerous practical situations is: In a given graph G, find a maximum matching, that is, a matching with as many edges as possible. This problem is called the *matching problem* or the *cardinality matching problem*. Clearly, a maximum matching in an n-node graph cannot have more than $\lfloor n/2 \rfloor$ edges. It may have fewer edges. For example, the matching with two edges shown in Fig. 3-30 is a maximum matching in a graph of eight nodes.

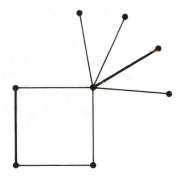

Figure 3-30 *Maximum matching.*

The matching problem was first posed and solved by Claude Berge in 1957. Since then it has attracted a great deal of research interest because of its intuitive appeal, its many applications, and several interesting variations. Consider, for example, the following situation: Suppose that four applicants a_1, a_2, a_3, and a_4 are available to fill six vacant positions p_1, p_2, p_3, p_4, p_5, and p_6. Applicant a_1 is qualified to fill position p_2 or p_5. Applicant a_2 can fill p_2 or p_5. Applicant a_3 is qualified for p_1, p_2, p_3, p_4, or p_6. Applicant a_4 can fill jobs p_2 or p_5. This situation is represented by the graph in Fig. 3-31. The vacant positions and applicants are represented by nodes. The edges represent the qualifications of each applicant for filling different positions. The problem is to fill as many positions as possible from the given set of applicants. This is a matching problem in a bipartite graph. A graph $G = (V, E)$, such as the one in Fig. 3-31, is called a *bipartite graph* if all its nodes can be partitioned into two sets, say V_1 and V_2, such that every edge in the graph joins some node from V_1 to some node in V_2. That is, no edge joins two nodes within the same set. Solving the matching problem in a bipartite graph turns out to be considerably easier than solving in a general graph.

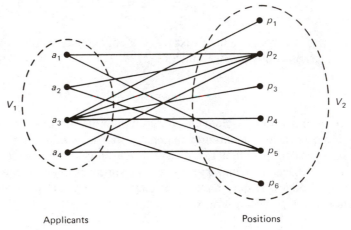

Applicants   Positions

Figure 3-31 *Matching in a bipartite graph.*

A generalization of the matching problem is called the *weighted matching problem*. Here the given graph has weights associated with its edges. Consider the following example of the weighted matching problem in a bipartite graph.

There are four workers w_1, w_2, w_3, w_4 and four machines m_1, m_2, m_3, m_4. The productivity of each worker varies from machine to machine, and is given by the following productivity matrix:

$$
\begin{array}{c c c c c}
 & m_1 & m_2 & m_3 & m_4 \\
w_1 & 9 & 8 & 10 & 14 \\
w_2 & 12 & 11 & 13 & 13 \\
w_3 & 7 & 5 & 22 & 18 \\
w_4 & 8 & 10 & 14 & 9 \\
\end{array}
$$

Find an assignment for maximizing the total productivity. (In general, the number of workers and machines need not be equal. In case the ith worker is not qualified to operate the jth machine, the corresponding entry is set to $-\infty$.)

The *assignment problem* defined in Problem 1-51 is similar to this, except that there the cost of the assignment was being minimized and here the productivity of the assignment is being maximized. One problem can easily be converted into the other.

Some related problems (the *stable marriage problem*, the *problem of distinct representatives*, etc.) will be discussed as exercises. Apart from these operations research applications, the matching problem has applications in computer code optimization problems, in deadlock-free systems, and in a variety of other situations.

Augmenting Path Algorithm

Before we can discuss an algorithm for solving the matching problem, we need to define a few terms. With respect to a given matching M in a graph G, an edge is said to be *matched* if it is in M and *unmatched* if it is not in M. Similarly, a node x is said to be *matched* or *saturated* if it is an end node of some matched edge, say (x, y). The two end nodes x and y of a matched edge (x, y) are said to be the *mates* of each other. A node that is not matched is called an *exposed* or *free* node (with respect to the specified matching M, of course). In Fig. 3-29, nodes v_3 and v_5 are exposed and the rest are saturated.

A path $P = (v_1, v_2, \ldots, v_k)$ is called an *alternating path* with respect to a given matching M if edges of P are alternately in M and not in M. For example, in Fig. 3-32 path (a, b, c, d, e) is alternating with respect to matching $M = \{(a, b), (c, d), (e, j), (h, i)\}$. So are the paths (e, h, i, g), and (g, c, d, j, e, h, i). But (e, h, i, g, a) is *not* an alternating path.

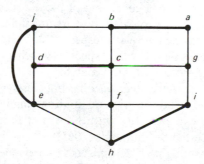

Figure 3-32   *Graph G with matching $M = \{(a, b), (c, d), (e, j), (h, i)\}$.*

An alternating path that begins at an exposed node and ends at (another) exposed node is called an *augmenting path*. For example, (g, a, b, c, d, j, e, f) in Fig. 3-32 is an augmenting path. The importance of an augmenting path lies in the following theorem due to Berge.

Berge's Theorem

A matching M in graph G is maximum if and only if G has no augmenting path with respect to M. ∎

Let P be an augmenting path with respect to matching M in G. If we reverse the role of the edges in augmenting path P (i.e., the edges of P currently in matching M are removed from the matching and those edges of P not in M are put into the matching), we will get another matching M' which has one more edge than M did. Therefore, M cannot be a maximum matching.

Conversely, suppose that M is not a maximum matching, and let M' be a maximum matching in G. Let subgraph $H = M \cup M' - M \cap M'$ denote the symmetric difference $M \oplus M'$ of M and M'. That is, H is the set of edges that are in M or in M' but not in both. Since a node in H can be incident with (1) either one edge of M, or (2) one edge of M', or (3) one edge of both M and M', each component of H is either a cycle of even length, alternating in M and M', or else a path with edges alternately in M and M'. Since M' is larger that M, H contains more edges from M' than from M. Therefore, some component of H must consist of a path P, which starts and ends with edges in M'. This path P is an augmenting path with respect to M, because it starts and ends with nodes exposed with respect to M. Thus, if M is not a maximum matching, there must exist an augmenting path in G with respect to M.

This important theorem, discovered by Berge in 1957, forms the basis of all algorithms for the matching problem. It suggests the following method of finding a maximum matching in a given graph G. Start with an arbitrary matching M (which may be empty). Find an augmenting path P with respect to M. Then construct another matching $M' = M \oplus P$, consisting of those edges of M or P that are not in both M and P. Clearly, the size of M' is one greater than that of M. Now, start with M', then find an augmenting path with respect to M', and proceed as before. Repeat this until a matching is obtained with respect to which there is no augmenting path. Then this final matching is a maximum matching.

The method just outlined may become hopelessly time consuming unless we have a systematic and efficient way of finding augmenting paths, because the number of paths grow exponentially with the size of the graph. The following is an elegant approach suggested by Edmonds [1965].

To find an augmenting path with respect to a given matching M, we must start at an exposed node. Let us start at an arbitrary exposed node r. If there is another exposed node s adjacent to r, add the edge (r, s) to the current matching M. If, on the other hand, all nodes adjacent to r are matched, we grow a tree rooted at r. Let root r be placed at level 0 and all nodes adjacent to it, say, $a_1, a_2, \ldots, a_q$, at level 1 (see Fig. 3-33). (Nodes $a_1, a_2, \ldots, a_q$ are all matched.) Now include in the tree all matched edges incident on $a_1, a_2, \ldots, a_q$, thus putting the mates of these nodes, say, $b_1, b_2, \ldots, b_q$, at level 2. Next, all nodes adjacent to b_i's (which are not yet in the tree) are placed at level 3. And so on. A tree so constructed is called an *alternating tree*. More formally, an alternating tree with respect to a matching M is a tree rooted at an exposed node and has the property that all paths emanating from the root are alternating paths. (Recall that there is exactly one path between every pair of nodes in a tree.)

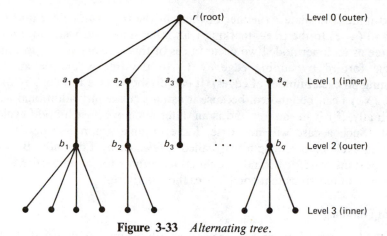

Figure 3-33 *Alternating tree.*

Let the nodes placed at odd levels be called *inner nodes* and those at even levels be called *outer nodes*. While growing an alternating tree, if we are at an inner node, all we do is to add the corresponding matched edge to the tree. Therefore, every inner node in the tree is of degree 2. But at an outer node x several distinct cases can arise, as we examine an edge (x, y) incident on x. If y is an exposed node and not yet in the tree, an augmenting path has been found. On the other hand, if y is a

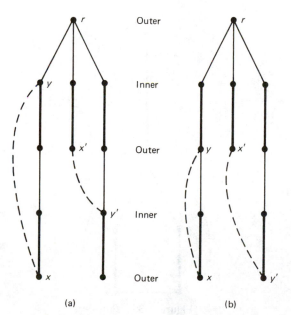

Figure 3-34 *Cycles formed with alternating tree:* (*a*) *Even cycles;* (*b*) *Odd cycles.*

matched node with mate z (neither of them in the tree, so far), we add both edges (x, y) and (y, z) to the tree—marking y an inner node and z an outer node. If y is in the tree as an inner node [two occurrences of this case are shown in Fig. 3-34(a)], the cycle formed by adding edge (x, y) to the tree would be an even cycle (containing an even number of edges). It can be shown (see Problem 3-72) that such an edge (x, y) can be ignored, because it cannot create any additional augmenting path. Finally, if y is in the tree and is an outer node, we have an odd cycle [see Fig. 3-34(b)]. (Such a case will not arise if the original graph is bipartite.) Such an alternating cycle of odd length was called a *blossom* by Edmonds. Blossom is the most important concept behind matching algorithms for nonbipartite graphs. It is the presence of blossoms that complicates the algorithm.

Blossom Handling

In Edmonds's original algorithm, as soon as a blossom was detected, it was shrunk to a single node (called a *pseudonode*), and a record was kept for future expansion. The pseudonode replacing a blossom was kept in the alternating tree as an outer node, and then tree growing proceeded as before. Although the shrinking of a blossom appears simple conceptually, its computer implementation is quite involved. Edmonds showed that using Berge's theorem and his concept of alternating trees, together with shrinking and expansion of blossoms, a maximum matching could be obtained in $O(n^4)$ time.

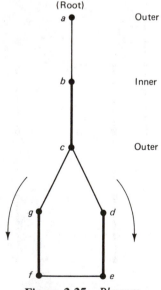

Figure 3-35 *Blossom.*

Subsequently, Gabow [1972] refined Edmonds's algorithm to avoid explicit shrinking and expansion of blossoms, and thus obtained an $O(n^3)$ procedure. Lawler [1976] also obtained a modified version of Edmonds's algorithm which worked in $O(n^3)$ time, in which explicit contraction and expansion of blossoms was not required. An efficient labeling technique was used for a systematic method of backtracking through blossoms.

We will employ a much simpler labeling technique developed by Pape and Conradt [1980]. They develop each odd cycle in two alternating paths (rather than shrinking it). It is based on the realization that a node v will be at an even distance (an outer node) along one path from the root and at an odd distance (an inner node) along another path from the root if and only if v is on an odd cycle. For example, in Fig. 3-35, along the alternating path (a, b, c, d, e, f, g), node f is an inner node, whereas the same node f along path (a, b, c, g, f, e) is an outer node. Thus all five nodes on the odd cycle in Fig. 3-35 can get two labels—outer and inner both.

Growing Alternating Tree

The alternating tree will be grown in a breadth-first manner (see Problem 3-7) starting at an arbitrary exposed node r. As mentioned earlier, the growing is done primarily from outer nodes. (At an inner node all we do is to add the corresponding matched edge.) For conducting a breadth-first search, we maintain a queue Q containing the outer nodes currently in the tree but which have not been explored from. We take the node x from the head of the queue, delete it from the queue and fan out from x, examining all edges adjacent to x (see Fig. 3-33).

We keep a Boolean array *nontree* of size n (n being the number of nodes in the graph), to keep track of the nodes that are in the tree as a root or as inner nodes. Initially, before we start growing a fresh tree from root r we set

> *nontree(r)* ← *false*;
> **for** all $v \in V$, $v \neq r$ **do** *nontree(v)* ← *true*

Each time a new node v comes into the tree as an inner node we set *nontree(v)* ← *false*. Once again, recall that whenever an inner node enters the tree, its mate, an outer node, comes with it automatically.

Another array, called *grandfather*, of size n is maintained in order to trace the path back up to the root when an augmenting path is discovered and the matching is to be enlarged. For an outer node w, *grandfather(w)* = u if there is a path (u, v, w) in the alternating tree from u to w, and (v, w) is a matched edge.

Initial Matching

Although one can start with any initial matching, including the empty one, it would be more efficient to start with as large a matching as possible. A reasonably large initial matching can be obtained rapidly by considering each exposed node v in the graph and matching it with the first exposed node adjacent to it.

Let the variable *expo* denote the number of exposed nodes in the graph, and let an array (of size n) *mate* denote the matching in the graph. That is, if (x, y) is an edge in the matching, then $mate(x) = y$ and $mate(y) = x$. For an exposed node i, $mate(i) = 0$. The following procedure gives this initial matching:

Algorithm 3-8(a):   Initial Matching

```
begin
    for all v ∈ V do mate(v) ← 0;
    expo ← n;  {* number of exposed nodes = n *}
    for all u ∈ V do
    begin
        if mate(u) = 0 then
        begin
            v ← an exposed node adjacent to u;
            mate(u) ← v;
            mate(v) ← u;
            expo ← expo − 2
        end
    end
end
```

Maximum Matching

We will start with the initial matching just obtained and given by array *mate*. We also have the number of exposed nodes denoted by variable *expo*. We need to continue the search for a larger matching only if $expo \geq 2$. In that case, we start with the first exposed node r we encounter and begin growing an alternating tree from r in a breadth-first manner.

As we try to expand from an outer node x (taken from the head of the queue, Q) and look at its adjacent node y, there will be several different cases. If $nontree(y) = false$, we ignore edge (x, y) (and proceed with the next node adjacent to x). This is because $nontree(y) = false$ implies that y is an inner node in the tree, and it could simply be the mate of x or (x, y) forming an even cycle with the tree. In either case we can ignore (x, y). Thus we need to consider only those nodes adjacent to x whose *nontree* values are *true*.

Now, for the case $nontree(y) = true$, node y may be matched or may be exposed. If it is exposed [i.e., if $mate(y) = 0$], we have found an augmenting path. We will enlarge the matching by tracing the augmenting path up to the root and by interchanging the matched and unmatched edges in this path. We then abandon the current tree but continue the main loop for growing an alternating tree from the next exposed node.

If, on the other hand, y is not exposed, we must first check to make sure that y is not an ancestor of x in the tree; otherwise, a false augmenting path may be reported (see Problem 3-73). This checking for an ancestor is done by tracing upward from x to the root using the array *grandfather*. In case y is an ancestor of x, we do nothing. If y is not an ancestor of x, and y is matched with node z, we incorporate them into the current tree [i.e., insert outer node z into the queue, set *grandfather*$(z) \leftarrow x$ and *nontree*$(y) \leftarrow false$].

We exit from the tree-construction loop as soon as an alternating path is found (i.e., Boolean variable *found* = *true*) or when there are no more outer nodes left to expand from (i.e., when the queue is empty).

The matching procedure itself is terminated (i.e., the matching obtained is a maximum) when either the number of exposed nodes is less than 2 or when every existing exposed node has been tried as the roots of an alternating tree without yielding an augmenting path.

This maximum matching algorithm can be described more formally and compactly as follows:

Algorithm 3-8(b):   Maximum Matching

```
begin
     start with an initial matching;
     for all r ∈ V do
          if (mate(r) = 0) and (expo ⩾ 2) then
          begin  (* grow alternating tree rooted at r *)
               for all v ∈ V do nontree(v) ← true;
               nontree(r) ← false;
               initialize Q to contain r only;
               found ← false;  (* augmenting path not found *)
               repeat  (* until either Q = ∅ or found = true *)
                    delete head node x from Q;
                    while not found do
                    begin
                         for every y adjacent to x do
                              if nontree(y) = true then
                              begin
                                   enlarge matching;
                                   expo ← expo − 2;
                                   found ← true
                              end
                              else if mate(y) ≠ x then
                              begin
                                   z ← mate(y);
                                   if (x ≠ r) or (z not an ancestor of x) then
                                   begin  (* add edges (x, y) and (y, z) to tree *)
                                        nontree(y) ← false:
                                        grandfather(z) ← x;
                                        insert z at tail of Q
                                   end
```

```
                               end
                         end {* while *}
                     until found or (Q = ∅)
                 end {* for *}
         end
```

The matching is enlarged when we have found an exposed node y adjacent to the current node x. The procedure for enlarging traces the alternating path from y to r and converts every unmatched edge into a matched edge and vice versa, as follows:

Algorithm 3-8(c):   Enlarge Matching

```
mate(y) ← x;
repeat
    next ← mate(x)  {* next is the old mate of x *}
    mate(x) ← y  {* y is the new mate of x *}
    if next ≠ 0 then  {* if x is not the root *}
    begin
        x ← grandfather(x);
        mate(next) ← x;
        y ← next
    end
until next = 0
```

Finally, the procedure for determining if node z (which is the mate of y and is grandson of current node x) is an ancestor of x in the tree already constructed is as follows:

Algorithm 3-8(d):   Checking for Odd Cycle Formation

```
if x ≠ r then {* root has no ancestor *}
begin
    u ← grandfather(x);
    while (u ≠ r) and (u ≠ y) do
        u ← grandfather(u);  {* back up the tree *}
    if u = r then z is not an ancestor of x
end
```

Observe that the edges (x, y) and (y, z) are incorporated into the tree in Algorithm 3-8(b) only when z is not an ancestor of x (i.e., no odd cycle has formed).

An Example

Let us now work out a small example to see how this deceptively simple algorithm handles blossoms. Readers are encouraged to solve by hand a few additional problems of their own in order to truly understand the subtleties of this algorithm.

We wish to find a maximum matching in the 11-node 12-edge graph G given in Fig. 3-36(a), in which the initial matching produced by Algorithm 3-8(a) is shown by the heavy lines. (Readers are encouraged to verify the initial matching.)

The first exposed node encountered is 9, and therefore we start growing an alternating tree with node 9 as its root, according to Algorithm 3-8(b). Edges (9, 1) and (1, 2) are added to the tree. The next growth begins at outer node 2. After completion of the **for** loop (for all nodes adjacent to 2), the alternating tree, the two arrays, and the Q are as shown in Fig. 3-36(b).

Next, node 4 is deleted from head of Q and its adjacent nodes $\{3, 6, 8\}$ are examined in that order. Node 3 is ignored because *nontree*(3) = *false*. Node 6 is considered. Since *mate*(6) = 5 $\neq$ root, a check for ancestorship of 4, by Algorithm 3-8(d) reveals that 5 is not an ancestor of 4 in the tree constructed so far. Therefore, edges (4, 6) and (6, 5) are added to the tree; similarly, edges (4, 8) and (8, 7). Next, we explore from node 6 at the head of Q, and look at edges (6, 4) and (6, 7). The tree is grown through both. The situation at this point is as depicted in Fig. 3-36(c).

Now, we explore from node 5 and need to consider only edge (5, 2). Since *mate*(2) = 1, and 5 $\neq$ root, we check to see if 1 is an ancestor of 5, by Algorithm 3-8(d), as follows:

```
u ← grandfather(5) = 4
Since u ≠ r and u ≠ 2, we set
    u ← grandfather(u) = grandfather(4) = 2.
Now u = 2 = y.
Since u ≠ r, 2 is an ancestor of 5.
Therefore, we do nothing.
```

Next, we take node 7, and find that for both its neighbors 6 and 8, *nontree*(6) = *nontree*(8) = *false*. Hence, we do nothing from node 7. The next node in the Q is 3. Nodes adjacent to 3 are 2 and 4. Since *nontree*(2) = *true*, we examine edge (3, 2). Since *mate*(2) = 1 $\neq$ root, we check to see if 1 is an ancestor of 3, and find that it is. Hence nothing is done from 3.

Finally, we explore from 8, the only remaining node in Q. Of the three nodes adjacent to 8, we see that *nontree*(4) = *nontree*(7) = *false*. But *nontree*(10) = *true* and *mate*(10) = 0. Therefore, we have found an alternating path and hence enlarge matching by Algorithm 3-8(c).

After enlarging the matching size by 1, we set *expo* ← *expo* − 2 = 3 − 2 = 1, and *found* ← *true*. Since *found* = *true*, we get out of the **repeat**... **until** loop in Algorithm 3-8(b), abandoning the current tree (and the associated Q and arrays—*nontree*, *grandfather*, *mate*). We would have continued in the outermost loop, starting fresh from an exposed node, but since *expo* = 1, we are through.

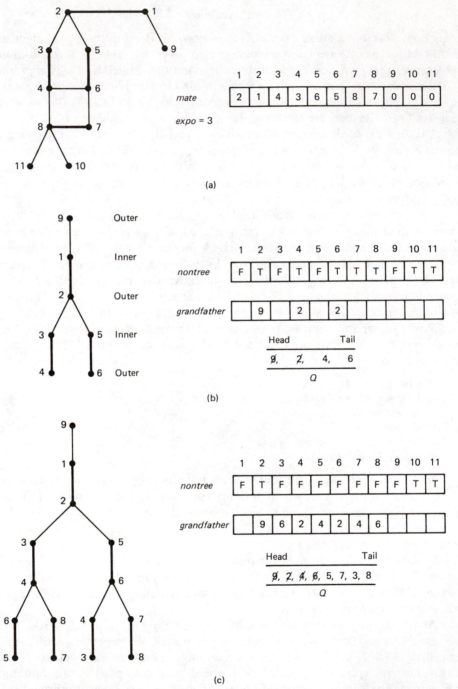

Figure 3-36 (*a*) *Initial matching in graph* G; (*b*) *Alternating tree*; (*c*) *Alternating tree.*

Complexity Analysis

The initial matching is obtained by Algorithm 3-8(a), which executes the two **for** loops exactly n times each. In the second loop, in each iteration, all the adjacent nodes are examined. Thus, in Algorithm 3-8(a) each edge of the given graph is examined twice, and hence the time required for getting an initial matching is $O(m)$.

In Algorithm 3-8(b), the outermost loop may be executed at most n times, since an alternating tree is attempted no more than once with each node as the root. (Typically, this loop will be executed only a few times, because generally there will only be a few exposed nodes with respect to an initial matching.) For each value of root r, the **repeat . . . until** loop is executed at most $n - 1$ times, because a node may be inserted into the queue at most once (why?), and each iteration of this loop deletes one node from the queue.

Now, for each node x deleted from the queue the **while** loop is executed once for each neighbor of x. Therefore, the condition *nontree*(y) = *true* is tested only nm times during the execution of the entire algorithm, in the worst case. Note, however, that this condition can be true at most once per node y, per value of root r, since once a node y has been found not to be in the alternating tree, y is added to the tree [setting *nontree*(y) ← *false*]. Hence in each tree-growing iteration, the condition *nontree*(y) = *true* will be satisfied no more than n times. It is only after this condition is satisfied that we enlarge the matching [Algorithm 3-8(c)] or add two edges (x, y) and (y, z) to the tree [after checking for ancestry, using Algorithm 3-8(d)]. Both of these operations require at most $O(n)$ time. Hence the entire algorithm requires $O(n^3)$ time in the worst case. A complete analysis of the Pascal MATCH procedure is left as an exercise (Problem 3-74).

It turns out that in a randomly generated, undirected, connected graph, the performance of this algorithm is not nearly as bad as $O(n^3)$. The primary reason for this good performance is that the initial matching is often close to being a maximum matching, and therefore only a few iterations are performed. Table 3-6 gives the execution times of the MATCH procedure for five randomly generated graphs each with an average degree of 10 for its nodes. The computer used was Amdahl 470 V/8 and time is in seconds. This table, as in previous sections, is intended only to provide

Table 3-6. Computing Times for the Pape–Conradt Matching Algorithm

Number of Nodes	Number of Edges	Run Time (sec)
100	500	0.005
200	1000	0.009
300	1500	0.028
400	2000	0.036
500	2500	0.041

a feeling for the run times. It should not be construed as the average-case complexity for this algorithm.

Extensive timing studies performed by Derigs [1980] show that this algorithm is indeed very fast, probably close to being the fastest among all known matching algorithms for all practical purposes.

Computer Implementation of Matching Algorithm

The MATCH Procedure implements the Pape–Conradt algorithm for finding a maximum-cardinality matching in an undirected graph with no isolated nodes. The procedure follows the outline given in Algorithm 3-8(b).

First, the MAKEFIRSTMATCHING subprocedure obtains an initial matching as described in Algorithm 3-8(a). Then from every exposed node ROOT, the building of an alternating tree is attempted. If an augmenting path is found, the current matching is enlarged, as outlined in Algorithm 3-8(c). If, on the other hand, it is discovered after checking for ancestry [i.e., odd cycle formation according to Algorithm 3-8(d)] that the node under consideration (NBHR) and its mate (MATWI) are to be included in the current tree, the Boolean variable ADDTOTREE is set to TRUE. In that case MATWI is added to the Q.

The input network in this procedure is assumed to be in the forward-star form (see Section 3.1) and has N nodes and M edges. The array POINTER points to the starting place in the array ENDV, which contains the list of adjacent nodes. Since the given graph is undirected, every edge is represented twice—once from each end node.

Global Constants

N   number of nodes in the given network

N1   N + 1

M2   twice the number of edges in the given network

Data Types

```
TYPE   ARRM2 = ARRAY[1..M2] OF INTEGER;
       ARRN  = ARRAY[1..N] OF INTEGER;
       ARRNB = ARRAY[1..N] OF BOOLEAN;
       ARRN1 = ARRAY[1..N1] OF INTEGER;
```

Procedure Parameters

```
PROCEDURE MATCH(
      N             :INTEGER;
      VAR POINTER :ARRN1;
      VAR ENDV    :ARRM2;
      VAR MATE    :ARRN;
      VAR EXPO     :INTEGER);
```

Input

N	number of nodes in the given network
POINTER[1..N1]	array of pointers in forward-star form
ENDV[1..M2]	array of edges in forward-star form

Output

MATE[1..N]	indicates matched nodes
EXPO	number of exposed (unmatched) nodes

Example

For the graph shown in Fig. 3-36(a), the data representation in forward-star form is as follows:

$$N = 11$$
$$POINTER[1..12] = [1, 3, 6, 8, 11, 13, 16, 18, 22, 23, 24, 25]$$
$$ENDV[1..24] = [2, 9, 1, 3, 5, 2, 4, 3, 6, 8, 2, 6, 4, 5, 7, 6, 8, 4, 7, 10, 11, 1, 8, 8]$$

The solution obtained is

$$MATE[1..11] = [9, 5, 4, 3, 2, 7, 6, 10, 1, 8, 0]$$
$$EXPO = 1$$

Pascal Procedure MATCH

```
PROCEDURE MATCH(
     N          : INTEGER;
     VAR POINTER: ARRN1;
     VAR ENDV   : ARRM2;
     VAR MATE   : ARRN;
     VAR EXPO   : INTEGER);

VAR ADDTOTREE, FOUND                                  : BOOLEAN;
    GRANDFATHER, Q                                    : ARRN;
    HEAD, LAST, MATWI, NBHR, NEXT, ROOT, TAIL, VTX, X, Y: INTEGER;
    NONTREE                                           : ARRNB;

PROCEDURE MAKEFIRSTMATCHING;

VAR LAST, X, Y: INTEGER;

BEGIN
   EXPO := N;                     (* ALL NODES ARE INITIALLY UNMATCHED *)
   FOR X := 1 TO N DO MATE[X] := 0;
   FOR X := 1 TO N DO
      IF MATE[X] = 0 THEN
```

```
            BEGIN
                Y := POINTER[X];
                LAST := POINTER[X+1] - 1;
                WHILE (MATE[ENDV[Y]] <> 0) AND (Y < LAST) DO Y := Y + 1;
                IF MATE[ENDV[Y]] = 0 THEN
                BEGIN
                    MATE[ENDV[Y]] := X;               (* A PAIR OF NODES IS MATCHED *)
                    MATE[X] := ENDV[Y];
                    EXPO := EXPO - 2
                END
            END
        END;

BEGIN (* MATCH *)
    MAKEFIRSTMATCHING;
    FOR ROOT := 1 TO N DO
        IF (EXPO >= 2) AND (MATE[ROOT] = 0) THEN
        BEGIN                  (* BUILD TREE ONLY IF ROOT IS AN EXPOSED NODE *)
            FOR X := 1 TO N DO NONTREE[X] := TRUE;       (* EMPTY TREE... *)
            NONTREE[ROOT] := FALSE;                  (* ...EXCEPT FOR ROOT *)
            Q[1] := ROOT;                       (* THE ROOT IS IN THE QUEUE *)
            HEAD := 1;
            TAIL := 1;
            FOUND := FALSE;                 (* NO AUGMENTING PATH FOUND YET *)
            REPEAT
                VTX := Q[HEAD];
                HEAD := HEAD + 1;
                X := POINTER[VTX];
                LAST := POINTER[VTX+1] - 1;
                WHILE (NOT FOUND) AND (X <= LAST) DO
                BEGIN                         (* CONSIDER NEXT VTX NEIGHBOR *)
                    IF NONTREE[ENDV[X]] THEN
                    BEGIN
                        NBHR := ENDV[X];
                        MATWI := MATE[NBHR];
                        IF MATWI = 0 THEN
                        BEGIN                         (* AUGMENTING PATH FOUND *)
                            MATE[NBHR] := VTX;
                            REPEAT
                                NEXT := MATE[VTX];
                                MATE[VTX] := NBHR;
                                IF NEXT <> 0 THEN
                                BEGIN
                                    VTX := GRANDFATHER[VTX];
                                    MATE[NEXT] := VTX;
                                    NBHR := NEXT
                                END
                            UNTIL NEXT = 0;
                            EXPO := EXPO - 2;
                            FOUND := TRUE
                        END
                        ELSE IF MATWI <> VTX THEN
```

```
          BEGIN
            IF VTX = ROOT THEN ADDTOTREE := TRUE
            ELSE
            BEGIN
              Y := GRANDFATHER[VTX];
              WHILE (Y <> ROOT) AND (Y <> NBHR) DO
                  Y := GRANDFATHER[Y];
              IF Y = ROOT THEN ADDTOTREE := TRUE
              ELSE ADDTOTREE := FALSE
            END;
            IF ADDTOTREE THEN
            BEGIN          (* ANCESTRY NOT THROUGH BLOSSOM BASE *)
              NONTREE[NBHR] := FALSE;    (* MARK INNER NODE *)
              GRANDFATHER[MATWI] := VTX;       (* TREE LINK *)
              TAIL := TAIL + 1;      (* PUT OUTER NODE IN Q *)
              Q[TAIL] := MATWI
            END
          END
        END; (* IF *)
        X := X + 1
      END (* WHILE *)
    UNTIL FOUND OR (HEAD > TAIL)
  END (* IF *)
END; (* MATCH *)
```

PROBLEMS

3-71. Show that a graph has an odd-length cycle if and only if, at some juncture in Algorithm 3-8(b), there exists a nontree edge going from an outer node to another outer node. (*Hint:* An outer node is by definition at even distance from the root, and no nontree edges emanate from an inner node.)

3-72. Show that a nontree edge going from an outer node to an inner node does not contribute to any additional augmenting path and can therefore be ignored.

3-73. Using the following graph (with initial matching shown in heavy lines), show that in Algorithm 3-8(b), if we do not check for node z to be an ancestor of x in the tree, the algorithm would incorrectly report a nonexisting augmenting path $(3, 1, 2, 4, 6, 2, 1, 5)$.

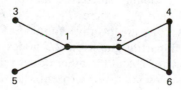

3-74. Make a complete analysis of the maximum matching algorithm, and show that its worst-case time complexity is bounded by $O(n^3)$—n being the number of nodes in the graph.

3-75. Suppose that in a graph G there is no augmenting path from root r with respect to matching M. Let P be an augmenting path between two other exposed nodes u and v. Then show that there is no augmenting path from r with respect to the enlarged matching $M' = M \cup P - M \cap P$.

3-76. Use the result of Problem 3-75 and induction on the number of augmentations to prove that if there is no augmenting path from root r at any time [in Algorithm 3-8(b)], there will never be an augmenting path from r (after any number of matching enlargements).

3-77. Devise an $O(m + n)$ algorithm to determine whether a given graph is bipartite.

3-78. Formulate the matching problem for a bipartite graph as a max-flow problem. What would be the computational complexity of such an algorithm, in view of the discussions in Section 3.5? Why can we not generalize this approach to solve the matching problem in a nonbipartite graph?

3-79. A matching M in a graph G is called a *perfect matching* or a *complete matching* if every node in G is incident on some edge in M. For example, in Fig. 3-29, $M = \{(v_1, v_6), (v_2, v_5), (v_3, v_4)\}$ is a perfect matching. Investigate conditions under which a maximum matching is perfect.

3-80. **P. Hall's theorem.**   In a bipartite graph $G = (V, E)$, let V_1 and V_2 be the two disjoint sets of nodes (i.e., $V_1 \cap V_2 = \varnothing$ and $V_1 \cup V_2 = V$), which are of equal cardinalities (i.e., $|V_1| = |V_2| = n/2$). Show that G has a perfect matching if and only if for each subset $S \subseteq V_1$ there are at least $|S|$ nodes in V_2 which are collectively adjacent to nodes in S. (*Hint:* One way to prove this is to convert the matching problem into the max-flow problem, as in Problem 3-78.)

3-81. **Distinct representatives.**   A set of p senators $\{s_1, s_2, \ldots, s_p\}$ are members of q different committees $\{c_1, c_2, \ldots, c_q\}$, $q \leqslant p$. The membership is described by a bipartite graph in which the two node sets are $V_1 = \{c_1, c_2, \ldots, c_q\}$ and $V_2 = \{s_1, s_2, \ldots, s_p\}$, such that committee c_i has senator s_j as its member if and only if there is an edge (c_i, s_j) in the graph. When is it possible to form a delegation of q senators which has a distinct representative from each committee? Give a necessary and sufficient condition. (*Hint:* See Problem 3-80.)

3-82. Show that a $k \times k$ doubly stochastic matrix D can be expressed as $D = c_1 P_1 + c_2 P_2 + \ldots + c_q P_q$, where $P_1, P_2, \ldots, P_q$ are $k \times k$ permutation matrices; $c_1, c_2, \ldots, c_q$ are positive numbers; and $c_1 + c_2 + \ldots + c_q = 1$. (*Hint:* Define a bipartite graph on $2k$ nodes in which one node set, V_1, represents the rows of D and the other nodes set, V_2, represents the columns of D. An edge represents a nonzero entry in D. Using P. Hall's theorem, show the existence of a perfect matching. This result is called the Birkhoff–von Neumann theorem.)

3-83. **Stable marriage problem.** A set of q men and a set of q women are to be paired in a mass marriage. Each man has ranked the women from 1 to q, and each woman has ranked the men from 1 to q. A marriage is called *stable* if for each two men m_i and m_j and their brides w_i and w_j both of the following conditions are satisfied:

1. Either m_i ranks w_i higher than w_j, or w_j ranks m_j higher than m_i
2. Either m_j ranks w_j higher than w_i, or w_i ranks m_i higher than m_j

Show that a stable-marriage pairing of men and women always exists and devise an algorithm to find such a pairing.

3-84. Let $G = (V, E)$ be a bipartite graph with V_1 and V_2 as the two disjoint sets of nodes. Let $|V_1| \leqslant |V_2|$ and let M be a complete matching of V_1 into V_2 (i.e., every node in V_1 is matched). Show that there exists a node x in V_1 such that for every edge e incident on x there is a complete matching containing e. Devise an algorithm to locate such a node in V_1.

3-85. Develop an algorithm for finding a minimum-weight perfect matching in a weighted, complete, bipartite graph. [This is called the *linear sum assignment problem* or simply the *assignment problem*. It may be solved as an LP problem (see Problem 1-51). It may also be solved as a minimum-cost flow problem (see Section 3-6).]

3-86. In a weighted graph G the weight of a matching M is defined as the sum of the weights of edges in M. Consider the following three problems in a general weighted graph G (not necessarily bipartite):

  (a) To find a matching with maximum weight (not necessarily of maximum cardinality).

  (b) To find that maximum-cardinality matching which has maximum weight.

  (c) To find that maximum-cardinality matching which has minimum weight.

Show how all three problems can be transformed into a minimum-weight perfect-matching problem on a complete graph with an even number of vertices. (*Hint:* See Burkard and Derigs [1980, pp. 38–40].)

3-87. In a graph $G = (V, E)$, a subset of edges $E' \subseteq E$ is said to be an *edge cover* if every node in G is incident on at least one edge in E'. Show that the number of edges in a maximum matching plus the number of edges in a smallest edge cover in G equals the number of nodes in G, provided that G has no isolated nodes (see Problems 2-36 and 2-37).

3-88. A *node cover* in a graph $G = (V, E)$ is a subset of nodes $C \subseteq V$ if every edge in E is incident on at least one node in C. Show that in a bipartite graph the number of edges in a maximum matching equals the number of nodes in a minimum node cover (see also Problems 2-34 and 2-35).

3-89. In the bin-packing problem (see Problem 4-87 for definition), suppose that the weight of each item to be packed is greater than one-third the bin capacity. Show that this special case of the bin-packing problem can be formulated as a maximum-matching problem (and thereby solved in polynomial time).

3-90. Two-processor scheduling. There are two identical processors and a set of n jobs $\mathcal{J} = \{J_1, J_2, \ldots, J_n\}$ to be executed. Each job requires 1 unit of processing time. A partial ordering relation " $\leqslant$ " is also given, in the form of an acyclic digraph $G = (\mathcal{J}, E)$, so that if job J_i must be executed before J_k can be started, there is a directed edge $(J_i, J_k) \in E$. Devise an algorithm to schedule the jobs on the two processors so that all the jobs can be completed at the earliest possible time. [*Hint:* Construct an undirected "compatibility graph" $G' = (\mathcal{J}, E')$, such that an edge $(J_p, J_q) \in E'$ if and only if there is no directed path from J_p to J_q or from J_q to J_p (i.e., jobs J_p and J_q can be processed at the same time). Let M be a maximum matching in G'. Then show that the total processing time $T \geqslant n - |M|$. Also show that there is always a schedule with $T = n - |M|$. Give an efficient algorithm for finding such a schedule (see Fujii et al. [1969]). See Problem 4-39 for another scheduling problem that can be solved using a matching algorithm.]

3-91. Show that in a graph G there always exists a maximum matching which saturates a specified node v in G (as long as v is not an isolated node).

3-92. Mendelsohn – Dulmage theorem. Let G be a bipartite graph with V_1 and V_2 as the two node sets. Let M_1 and M_2 be two arbitrary matchings in G. Show that there exists a matching which saturates all nodes saturated by M_1 in V_1 and by M_2 in V_2.

REFERENCES AND REMARKS

Matchings in bipartite graphs have been investigated for a long time in both classical combinatorial analysis (e.g., system of distinct representatives) and operations research (e.g., assignment problem). However, Berge was the first to use the concept of alternating paths. He showed in the paper

BERGE, C., Two Theorems in Graph Theory, *Proc. Nat. Acad. Sci. USA* 43(1957), 842–844

that a matching is maximum if and only if it does not admit an augmenting path. This classic result has been the backbone of all matching algorithms for nonbipartite graphs.

Berge's original suggestion was to trace out an alternating path from an exposed node until it can go no further, and then if not augmenting, backtrack and try again. This exhaustive tracing of all paths requires an exponential time. A PL/1 code of this very inefficient, exponential-time algorithm can be found in

DÖRFLER, W., and J. MÜHLBACHER, Bestimmung eines maximalen Matchings in beliebigen Graphen, *Computing* 9(1972), 251–257

The first polynomial-bounded algorithm for solving the matching problem based on Berge's theorem was developed by Edmonds in

EDMONDS, J., Path, Trees and Flowers, *Canad. J. Math*. 17(1965), 449–467

He defined blossoms (in nonbipartite graphs) and showed that his method would construct a maximum matching in $O(n^4)$ time. A closer examination reveals that Edmonds's method is of complexity $O(mn^2)$. In fact, Witzgall and Zahn gave an algorithm requiring $O(mn^2)$, in

WITZGALL, C., and C. T. ZAHN, JR., Modification of Edmonds' Maximum Matching Algorithm, *J. Res. Nat. Bur. Standards* 69B (1965), 91–98

A number of subsequent improvements were suggested based on labeling techniques for alternating tree construction and blossom handling. The two best known among these are

GABOW, H., An Efficient Implementation of Edmonds's Algorithm for Maximum Matchings on Graphs, *J. ACM* 23(1975), 221–234

LAWLER, E. L., *Combinatorial Optimization*: *Networks and Matroids*, Holt, Rinehart and Winston, New York, 1976, Chapter 6

Both of these are of $O(n^3)$ time complexity.

The following two algorithms are somewhat of an improvement over Gabow's and Lawler's algorithms, because they require $O(mn)$ time rather than $O(n^3)$ (improvement for sparse graphs).

HESKE, A., Verfahrensvergleiche bei der Bestimmung maximaler Matchings in Allgemeinen Graphen, Diploma thesis, Mathematisches Institut Universität zu Köln, 1978

KAMEDA, T., and I. MUNRO, A $O(|V| \cdot |E|)$ Algorithm for Maximum Matching of Graphs, *Computing* 12(1974), 91–98

A FORTRAN code of Gabow's algorithm modified by Heske to work in $O(mn)$ time is available in

DERIGS, U., and A. HESKE, A Computational Study on Some Methods for Solving the Cardinality Matching Problem, *Angew. Inform. Jg.* 22(1980), 249–254

A FORTRAN code of Lawler's algorithm is given in

BURKARD, R. E., and U. DERIGS, *Assignment and Matching Problems*: *Methods with FORTRAN-Programs*, Lecture Notes in Economics and Mathematical Systems 184, Springer-Verlag, New York, 1980

Algorithm 3-8(b) and the MATCH procedure presented in this section are due to Pape and Conradt, described in

PAPE, U., and D. CONRADT, Maximales Matching in Graphen, in H. Späth (ed.), *Ausgewählte Operations Research Software in FORTRAN*, Oldenburg, Munich, 1980, pp. 103–114

A FORTRAN code of Algorithm 3-8(b) is given in the paper cited above.

Extensive empirical studies reported in Derigs and Heske [1980] and in the following two papers show that the Pape–Conradt implementation is not only the simplest to understand but is also one of the fastest (much faster than Gabow's algorithm and its variations, for example):

CONRADT, D., Eine effiziente Implementierung des Verfahrens von Edmonds zur Bestimmung eines maximalen Matchings, TU Berlin, FB 20, Bericht 78-18, 1978

DERIGS, U., Methods for Solving the Cardinality Matching Problem (A State of the Art Study), Industrieseminar Universität zu Köln, July 1980

In all the matching algorithms mentioned so far, the length of augmenting paths chosen is arbitrary. Hopcroft and Karp, in the following paper, proposed a fundamentally new idea:

HOPCROFT, J. E., and R. M. KARP, A $n^{5/2}$ Algorithm for Maximum Matching in Bipartite Graphs, *SIAM J. Comput.* 2(1973), 225–31

They proposed to augment along a shortest augmenting path—in fact, along a maximal set of node-disjoint shortest augmenting paths in each phase. (Note the similarity of this with Dinic's max-flow algorithm discussed in the preceding section.) Using this approach, they gave an $O(n^{2.5})$ algorithm for maximum matching in bipartite graphs. A clever extension of this method for general graphs was made in

EVEN, S., and O. KARIV, An $O(n^{2.5})$ Algorithm for Maximum Matching in General Graphs, *Proc. 16th Annual Symp. on Foundations of Computer Science*, IEEE, Berkeley, Calif., 1975, pp. 100–112

which solves the general matching problem in $O(n^{2.5})$ time. However, the Even–Kariv algorithm is extremely involved, and its storage requirement is very high. According to Derigs [1980], the Even–Kariv algorithm is not practical; it is primarily of theoretical interest.

Theoretically, the fastest algorithm for maximum matching in a nonbipartite graph has been proposed in

MICALI, S., and V. V. VAZIRANI, An $O(\sqrt{|V|} \cdot |E|)$ Algorithm for Finding Maximum Matching in General Graphs, *Proc. 21st Annual Symp. on Foundations of Computer Science*, IEEE, Long Beach, Calif. 1980, pp. 17–27

The literature on the subject of matching is extensive. Matching problems are closely related to flow problems (see Section 3.5 and Problem 3-78), covering problems (see Problems 3-78 and 3-88), and they can, of course, be formulated as LP problems (see Section 2.2). There are numerous variations—matching in bipartite graphs, matching in weighted graphs, matching in directed graphs, and so on. We have confined ourselves to one problem, namely the cardinality-matching problem and have presented one algorithm—which is believed to be the best overall algorithm from a practical point of view, available at present. For discussions of matching-related problems, other algorithms, and applications, we recommend relevant chapters in the following books: Burkard and Derigs [1980], Lawler [1976], Minieka [1978], and Papadimitriou and Steiglitz [1982]. (The last three have been referred to in Section 3.6.)

Problem 3-86 is solved in Burkard and Derigs [1980]. A backtrack algorithm and a Pascal code for the stable marriage problem (Problem 3-83) is given in

WIRTH, N., *Algorithms + Data Structures = Programs*, Prentice Hall, Englewood Cliffs, N.J., 1976, pp. 148–154

Problem 3-89 is from Papadimitriou and Steiglitz [1982] (see the References and Remarks of Section 3.6) and Problem 3-90 is from

FUJII, M., T. KASAMI, and K. NINAMIYA, Optimal Sequencing of Two Equivalent Processors, *SIAM J. Appl. Math.* 17(1969), 784–89; Erratum, *ibid.* 20(1971), 141

3.8. TRAVELING SALESMAN PROBLEM

All the network problems considered so far in this chapter were of the type for which we could find an optimal solution in polynomial time (i.e., computation time was some polynomial in n, the size of the input). Now, in this section we will consider a network problem belonging to another class—the class of those problems for which no polynomial-time algorithm has been found. Such computationally hard problems (called NP-hard problems) require different solution methods. General techniques to solve such hard problems were discussed in Chapter 1 in connection with zero–one programming and in Chapter 2 while solving the knapsack problem.

The problem we are going to consider is called the *traveling salesman problem* (TSP) or the traveling salesperson problem (now that traveling saleswomen are becoming almost as frequent as traveling salesmen). It is one of the oldest problems in network optimization and has attracted a great deal of attention in the past 40 years or so. The statement of the traveling salesman problem is simple.

A traveling salesman must visit every city in his territory exactly once and then return to his starting point. Given the cost of travel between all pairs of cities, how should he plan his itinerary so that he visits each city exactly once and so that the total cost of his entire tour is minimum? In network theory terms, the problem is to find a minimum-weight cycle of length n in a given weighted, complete graph of n nodes. (Recall that a complete graph is one in which there is an edge between every pair of nodes.) Consider for example, a five-city traveling salesman problem, shown as a complete graph in Fig. 3-37. The weight of cycle (cost of the route) (a, b, c, d, e, a) is 206, whereas the minimum-weight cycle is (a, b, e, c, d, a) with weight 89.

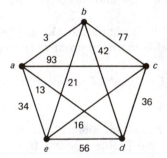

Figure 3-37   *Traveling salesman problem for five cities.*

The traveling salesman problem is a classic problem in combinatorial optimization. It arises in numerous applications. Consider, for example, the following situation. The arm of an automatic riveting machine, putting rivets, say, on an aircraft wing has to move from spot to spot. After putting n rivets on n different specified spots, it returns to the starting position. Finding an optimal route for the arm is the traveling salesman problem. Or consider another situation: A machine shop is to produce n different items, requiring retooling whenever the production item is changed. The cost of retooling between items is given. Finding an optimal production sequence is simply the traveling salesman problem. Another application of the traveling salesman problem is in the no-wait flow-shop problem, described in Section 4.2.3.

There are several variants of the traveling salesman problem, as defined above. Let $W = [w_{ij}]$ denote the weight matrix of a graph. Then, (1) the edge weights may not be symmetric (i.e., $w_{ij} \neq w_{ji}$), in which case we are dealing with a directed graph

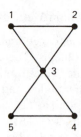

Figure 3-38 *Graph without a cycle of length 5.*

and solving an *asymmetric traveling salesman problem*. (2) Sometimes the given graph is not complete. That is, there are pairs of nodes which are not directly connected with edges. Although every complete graph always has cycles of length n, a graph that is not complete may not have a cycle of length n. For example, the graph in Fig. 3-38 has no cycle of length 5. In such a graph the traveling salesman problem has no solution. The problem of determining if a given graph with n nodes has a cycle of length n is called the *Hamiltonian cycle problem*, after the Irish mathematician Sir William Rowan Hamilton, who first posed this problem in 1859. A cycle passing through every node in the graph is called a *Hamiltonian cycle*. (3) If the traveling salesman does not wish to return to the starting node, after visiting all nodes in the graph (exactly once), we have a different problem. We need to find a minimum-weight *path* of length $(n - 1)$ and not a cycle of length n.

It should be noted that a very drastic change to the solution of this problem may take place if the traveling salesman is allowed to visit a city more than once (if it is cheaper to do that). For example, in Fig. 3-39, if he is allowed to visit a node more than once, then the optimal route (a, b, c, d, e, c, a) (which is not a cycle) has a weight of 60, but if he were not allowed to visit a node more than once, then route (a, b, c, d, e, a) is the only possible solution, with weight 140. The condition under which at least one of the Hamiltonian cycles is always a minimum-weight cycle is that the weights satisfy the *triangle inequality* $w_{ij} \leqslant w_{ik} + w_{kj}$ (see Problem 3-96).

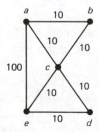

Figure 3-39 *Minimum-weight route is not Hamiltonian.*

Solution Methods

As mentioned in Chapter 2, an exhaustive enumeration is one of the methods used for solving computationally hard problems. There are $n!$ permutations of n nodes. But only $(n - 1)!$ of these are distinct Hamiltonian cycles (in a complete directed graph), because we can fix any of the n nodes, and thereafter there are $(n - 1)!$ ways to arrange the remaining $(n - 1)$ nodes in the cycle. In case the given network is undirected, there are $(n - 1)!/2$ distinct cycles.

Theoretically, the problem may be solved by generating all $(n - 1)!$ cycles and comparing their weights. Such an approach, however, is extremely inefficient. For example, even a graph of 20 nodes has $19! > 10^{17}$ cycles, and would require years of continuous computing for such an exhaustive-enumeration algorithm.

As mentioned earlier, the traveling salesman problem is one of a number of combinatorial optimization problems for which no efficient (polynomial-time) algorithms are known. All known algorithms require a computation time that is exponential in the number of cities, n.

When faced with such a computationally difficult problem, there are two approaches that one may take:

1. Use refined techniques such as branch-and-bound or dynamic programming, which drastically reduce the effect in exhaustive enumeration. Such refined enumerative techniques are certain to produce an optimal solution, but for worst-case inputs they may require an exponential number of calculations and thus become prohibitively expensive for some inputs.
2. Employ approximate but fast (polynomial-time) solution methods, which may not produce an optimal solution, but suboptimal solutions that are acceptably close to optimal for most inputs.

Both of these methods are useful, and we will discuss specific algorithms based on both these approaches to solve the traveling salesman problem. Let us consider the branch-and-bound method first.

3.8.1. A BRANCH-AND-BOUND ALGORITHM

Branch-and-bound algorithms are based on a tree search where at each step all possible solutions of the current problem are partitioned into two or more subsets, each represented by nodes in a decision tree. In this case, we will partition the solutions at each step into two subsets: those that contain a specific edge (i, j) and those that do not.

This *branching* (partitioning or splitting) is performed according to some heuristic (to be described shortly) which reduces the amount of search to be conducted for the optimal solution. After branching, lower bounds are computed on the cost for each of the two subsets. The next solution space to be searched is chosen to be the one with the smaller of the two lower-cost bounds.

This process is continued until a Hamiltonian cycle is obtained. Then, only those subsets of solutions need be searched whose lower bounds are smaller than the value of the current solution. This branching and bounding of solution space allows us to discard large subsets of solutions, thereby saving a great deal of fruitless search. The basic process in computation of lower bounds is that of reduction, which we will describe next, and then with the help of an example show how this branch-and-bound method can be used to solve the TSP.

Reduction

Let us observe that a Hamiltonian cycle (cycle of length n) will contain exactly one element from each row of cost matrix (or weight matrix) W and exactly one element from each column of W. If a constant q is subtracted from any row or from any column of W, the cost of all Hamiltonian cycles (TS tours) is reduced by q. Therefore, the relative costs of different cycles remain the same. Thus the optimal tour remains optimal. If such a subtraction is done from rows and columns, such that each row and each column contains at least one zero while keeping remaining w_{ij}'s nonnegative, the total amount subtracted will be a lower bound on the cost of any solution. This process of subtracting constants from rows and columns is called *reduction*.

	1	2	3	4	5	6
1	∞	3	93	13	33	9
2	4	∞	77	42	21	16
3	45	17	∞	36	16	28
4	39	90	80	∞	56	7
5	28	46	88	33	∞	25
6	3	88	18	46	92	∞

Let us now consider an example which is adapted from Reingold et al. [1977]. The cost matrix given above can be reduced by subtracting 3, 4, 16, 7, 25, and 3 from rows 1 through 6, respectively, and then subtracting 15 and 8 from columns 3 and 4, respectively, leaving the reduced matrix. Since a total of 81 was subtracted, 81 is a lower bound on the cost of all solutions for this problem.

	1	2	3	4	5	6
1	∞	0	75	2	30	6
2	0	∞	58	30	17	12
3	29	1	∞	12	0	12
4	32	83	58	∞	49	0
5	3	21	48	0	∞	0
6	0	85	0	35	89	∞

Now, how do we split the set of all solutions into two classes? Suppose that we choose the edge $(6, 3)$ on which to split the solution space. One set will contain all solutions that exclude the edge $(6, 3)$; knowing that $(6, 3)$ is excluded, we can change the corresponding entry in the cost matrix, $w_{63} \leftarrow \infty$. The resulting matrix can then have 48 subtracted from the third column (and nothing from the sixth row), thus making the lower bound of $81 + 48 = 129$, for all solutions that exclude edge $(6, 3)$. The other set contains all solutions that include edge $(6, 3)$, so the sixth row and the third column must be deleted from the reduced cost matrix, because now we can never go from 6 to anywhere else or arrive at 3 from anywhere else. The result is a cost matrix one less in dimension. Furthermore, since all solutions in this subset use edge $(6, 3)$, the edge $(3, 6)$ is no longer usable, and we must set $w_{36} \leftarrow \infty$. The binary search tree at this stage is as shown in Fig. 3-40.

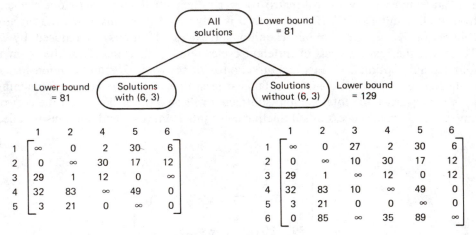

Figure 3-40   *Splitting of solutions.*

The edge $(6, 3)$ was used for the solutions because, of all edges, it caused the greatest increase in the lower bound of the right subtree (in the search tree of Fig. 3-40). This rule is used to split each node of the search tree because we prefer to look for optimal solution by following left branches rather than right branches; the left branches reduce the dimension of the problem, whereas the right branches only add another ∞ and perhaps a few new zeros without changing the dimension. It is not difficult to determine which edge gives the largest increase in the lower bound of the right subtree: Choose the zero that, when changed to infinity, allows the most to be subtracted from its row and column.

Thus, in the 5×5 matrix representing the problem for the left subtree in Fig. 3-40, the zero entry of $(4, 6)$, when changed to ∞, allows 32 to be subtracted from the fourth row (and nothing from the sixth column). This is the most of all zeros in this matrix. Therefore, the next splitting is to be done around edge $(4, 6)$. It increases the lower bound of all those solutions that include edge $(6, 3)$ and exclude edge $(4, 6)$ to $81 + 32 = 113$. It decreases the dimension of the matrix on the left branch to a 4×4, since row four and column six are deleted. This situation is shown in Fig. 3-41. Note that since edges $(4, 6)$ and $(6, 3)$ are included in the solutions, edge $(3, 4)$ is

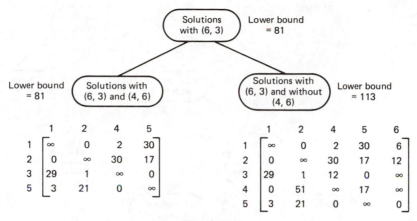

Figure 3-41

no longer usable. This is enforced by setting entry $(3, 4)$ to ∞. In general, then, if the edge added to the partial tour is from i_u to j_1 and the partial tour contains paths $(i_1, i_2, \ldots, i_u)$ and $(j_1, j_2, \ldots, j_k)$, the edge whose use is to be prevented is (j_k, i_1).

For splitting the leftmost node further, edge $(2, 1)$ is the best; for when the zero in $(2, 1)$ position is replaced with ∞, it allows 17 to be subtracted from the second row and 3 to be subtracted from the first column. This quantity is more than allowed by any other zero entry.

After splitting on edge $(2, 1)$, the cost matrix on the left side is 3×3. Since we have included edge $(2, 1)$ in the leftmost solutions, the edge $(1, 2)$ is forbidden by setting $(1, 2)$ entry to ∞. This matrix,

$$
\begin{array}{c c c c}
 & 2 & 4 & 5 \\
1 & \infty & 2 & 30 \\
3 & 1 & \infty & 0 \\
5 & 21 & 0 & \infty
\end{array}
$$

can be reduced by subtracting 1 from column two and 2 from row one. This produces the cost matrix

$$
\begin{array}{c c c c}
 & 2 & 4 & 5 \\
1 & \infty & 0 & 28 \\
3 & 0 & \infty & 0 \\
5 & 20 & 0 & \infty
\end{array}
$$

with the associated lower bound on the solution subset as $81 + 1 + 2 = 84$. Application of this method to the sample problem yields the binary tree shown in Fig. 3-42.

Note that after $n - 2$ edges have been chosen, the remaining cost matrix is 2×2 in dimension. At this point all of these $n - 2$ edges either form a single path

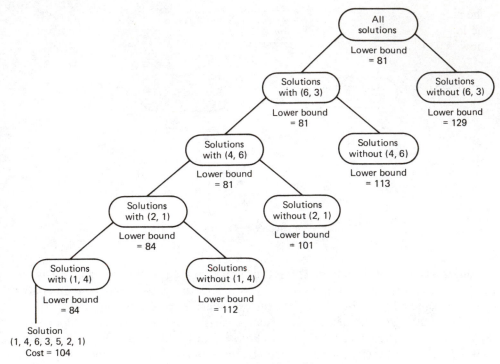

Figure 3-42   *Branch-and-bound binary tree.*

or form two separate paths. In either case, if the original network is directed (i.e., the distance matrix is asymmetric), we have no choice left in selection of the two remaining edges that complete the traveling salesman's tour. For instance, in the six-city illustrative problem that we have been solving, after edges (6, 3), (4, 6), (2, 1), and (1, 4) are selected, we have no choice but to add edges (3, 5) and (5, 2) to complete the traveling salesman's route.

Now we have obtained one TS route, namely 1, 4, 6, 3, 5, 2, 1, which costs 104. All nodes in the search tree (Fig. 3-42) with lower bound greater than 104 can be rejected, as they will not lead to a cheaper route. In Fig. 3-42 only one node has lower bound less than 104, and it must be expanded further. The node with lower bound of 101 includes edges (6, 3) and (4, 6) but excludes edge (2, 1). The cost matrix associated with this node is

$$
\begin{array}{c@{\quad}cccc}
 & 1 & 2 & 4 & 5 \\
\begin{array}{c}1\\2\\3\\5\end{array} &
\left[\begin{array}{cccc}
\infty & 0 & 2 & 30 \\
\infty & \infty & 13 & 0 \\
26 & 1 & \infty & 0 \\
0 & 21 & 0 & \infty
\end{array}\right]
\end{array}
$$

The next splitting must be on edge (5, 1), because exclusion of edge (5, 1) adds 26 to the lower bound, which is the maximum. The rest of the subtree from this node is shown in Fig. 3-43.

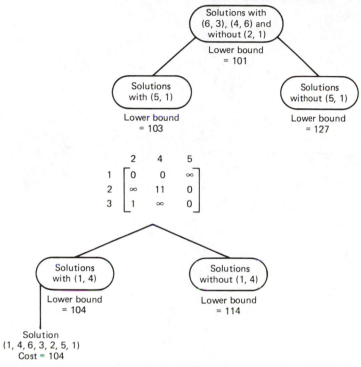

Figure 3-43

Thus we have found two optimal solutions, each costing 104. All other solutions cost more. As shown in this example, a traveling salesman problem may have more than one optimal solution. Also notice that the first solution found is optimal; we cannot expect this to happen in general. For this particular example, only 13 nodes were examined, far fewer than all 120 distinct cycles in a six-city (directed) TS problem.

Algorithm Development

For this branch-and-bound algorithm it is assumed that the given network is directed (i.e., the $n \times n$ cost or weight matrix W is not necessarily symmetric). The algorithm can easily be modified to take care of the undirected case (see Problem 3-97).

As illustrated in the foregoing example, with every node in the search tree there is an *associated cost matrix A*. At the beginning, the cost matrix (associated with the root) A is the given weight matrix W; then, as we move down from the root following the left branches, the size of the cost matrices gets smaller and smaller. Finally, when A is a 2×2 matrix, no further branching can be done. We simply add the last two edges (about which we have no choice) and get a Hamiltonian cycle.

Observe that the reduced matrix at this junction would be one of the following two forms:

$$
\begin{array}{c}
\begin{array}{cc} w & x \end{array} \\
\begin{array}{c} u \\ v \end{array}
\begin{bmatrix} 0 & \infty \\ \infty & 0 \end{bmatrix}
\end{array}
\qquad
\begin{array}{c}
\begin{array}{cc} w & x \end{array} \\
\begin{array}{c} u \\ v \end{array}
\begin{bmatrix} \infty & 0 \\ 0 & \infty \end{bmatrix}
\end{array}
$$

where nodes u, v, w, and x may be four distinct nodes or only three distinct nodes. In any case we need to look at only one entry (say, in the first row and the first column) to determine which edges are to be included. This can be achieved as follows:

```
if A(1, 1) = ∞ then
        add edges (u, x) and (v, w) to the current route
else
        add edges (u, w) and (v, x) to the current route
```

As we continue downward along the search tree, at each node we must reduce the cost matrix A, that is, subtract from each row and each column appropriate amounts such that there is at least one zero entry in every row and every column of A (keeping all entries nonnegative). This total reduction value is then added to the cost of the current partial solution. The reduction of matrix A may be described as follows:

Algorithm 3-9(a):   Reduction of Matrix A

```
function REDUCE(A);
begin
    rvalue ← 0;  {* reduction value *}
    for i ← 1 to size do  {* size of A *}
    begin
        rowred(i) ← smallest element in ith row;
        if rowred(i) > 0 then
        begin
            subtract rowred(i) from every finite element in ith row;
            rvalue ← rvalue + rowred(i)
        end
    end;
    for j ← 1 to size do
```

```
          begin
               colred(j) ← smallest element in jth column;
               if colred(j) > 0 then
               begin
                    subtract colred(j) from every finite element in jth column;
                    rvalue ← rvalue + colred(j)
               end
          end;
          REDUCE ← rvalue
     end
```

After reducing A, we increase the cost associated with the node by *rvalue*. The cost associated with a node x in the solution tree represents the lower bound on the cost of the partial solution (up to node x). If this cost is less than *tweight*, the cost of the best complete solution obtained so far, we branch downward (searching for a better solution); otherwise, we abandon the current node and go up the tree.

In order to branch down, we must determine the best edge on which to split the solutions. That is, find that zero entry in the reduced matrix A which when set to ∞ would maximize the amount to be subtracted from that row and the column.

The best edge can be determined by the following procedure, which takes the reduced cost matrix A of size *size* and computes entry (r, c) corresponding to the best edge and the value, *most*.

Algorithm 3-9(b):   Selecting the Best Edge (r, c)

```
procedure BESTEDGE(A, size r, c, most);
begin
     most ← − ∞;
     for i ← 1 to size do {* row *}
          for j ← 1 to size do {* column *}
               if aij = 0 then
               begin
                    minr ← smallest entry in ith row, other than aij;
                    minc ← smallest entry in jth column, other than aij;
                    total ← minr + minc;
                    if total > most then
                    begin
                         most ← total;
                         r ← i; {* row index of best edge *}
                         c ← j { *column index of best edge *}
                    end
               end
     end
```

Cycle Prevention

Suppose that we have just branched on edge (u_i, v_1), which is the included edge (u_i, v_1) to the left subtree and the excluded edge (u_i, v_1) from the right subtree. Clearly, u_i must be the terminal node of some maximal path $(u_1, u_2, \ldots, u_i)$ and v_1 be starting node of some other path $(v_1, v_2, \ldots, v_k)$. (Either or both of these two paths may be empty.) To prevent formation of a short cycle (of length less than n) we must preclude edge (v_k, u_1) from ever appearing on the left subtree (see Fig. 3-44).

Nodes u_1 and v_k can be identified by tracing backward from node u_i and tracing forward from node v_1.

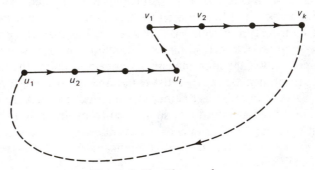

Figure 3-44 *Short cycle.*

Now we will describe the heart of the branch-and-bound TSP algorithm by means of a recursive procedure EXPLORE. The EXPLORE procedure considers a given partial solution and searches for a better solution. Let *edges* denote the number of edges included in the partial solution. The variable *cost* is the cost of this partial solution, and A is the associated matrix [of size $(n - edges) \times (n - edges)$]. The EXPLORE procedure maintains a global copy of the best solution obtained so far together with its weight *tweight*.

Algorithm 3-9(c): Exploring Search Tree in Depth-First Fashion

```
procedure EXPLORE(edges, cost, A);
begin
    cost ← cost + REDUCE(A);
    if cost < tweight then
        if edges = n − 2 then
        begin
            add the last two edges;
            tweight ← cost;
            record the new solution
        end
```

    else
    begin
        apply procedure BESTEDGE to find (r, c), the best edge to split
          the solutions on;
        let *most* be amount subtracted from row r and column c;
        lowerbound ← *cost* + *most*;
        prevent cycle;
        newA ← A − column c − row r; {∗ delete row r and column c ∗}
        EXPLORE(*edges* + 1, *cost*, *newA*); {∗ left subtree ∗}
        restore A by adding column c and row r;
        if *lowerbound* < *tweight* **then**
        begin {∗ explore right subtree ∗}
            a_{rc} ← ∞;
            EXPLORE(*edges*, *cost*, A);
            a_{rc} ← 0
        end
    end;
    unreduce A {∗ add what was originally subtracted ∗}
end

Complexity Analysis

As mentioned at the beginning of this section, no polynomial-time algorithm has been found for the traveling salesman problem (producing an exact solution), in spite of 30 years of intensive search. In fact, it is highly unlikely that such an algorithm will ever be found. The branch-and-bound algorithm is basically a method of exhaustive search, and for a worst-case input, one may end up examining all possible solutions. For an n-city asymmetric (directed) TSP there are $(n - 1)!$ distinct Hamiltonian cycles, and thus the worst-case time complexity of this branch-and-bound algorithm could be as bad as $O(n!)$. In a typical case, however, the situation is not as bad. For example, in the particular six-city problem considered in this section, we examined only 13 nodes in the solution tree; that is, 13 sets of reductions, splitting, and so on, were performed. This is far fewer than $120 = 5!$, the number of distinct Hamiltonian cycles.

The execution time is strongly dependent on the problem instance. Table 3-7 shows the minimum, maximum, and median run times for the BABTSP procedure

Table 3-7. Computing Times (sec) for the BABTSP Procedure

Number of Nodes	Minimum	Maximum	Median
10	0.030	0.16	0.072
20	0.243	4.99	1.360
30	0.776	79.90	12.100
40	11.000	991.00	401.000

(to be described shortly) for randomly generated (with uniform distribution) weight matrices of size 10, 20, 30, and 40. The times are in virtual CPU seconds on an Amdahl 470 V/8. Ten different weight matrices were generated for each value of n.

Table 3-7 shows how rapidly the execution time grows with the size of the network. It also shows the enormous variation between two networks of the same size, and thus the median figure is not very meaningful. Finally, it shows that it would be quite expensive to obtain an exact solution of the TSP problem (using the branch-and-bound algorithm) for networks with nodes much greater than 40.

Computer Implementation of the Branch-and-Bound TSP Algorithm

The BABTSP procedure finds an exact solution of the traveling salesman problem in a network with N nodes, given as an N × N weight matrix W of integers.

Since the input is a complete network, it is economical to represent it as an N × N weight matrix W. We assume that W may be asymmetric and that the edge weights are integers. (In case they are reals, they should be so declared or converted into integers by multiplying every entry with some suitable number.) The final answer, which is a minimum-weight Hamiltonian cycle (i.e., a TS route), will be given by a linear array ROUTE of length N, such that ROUTE[1] is node 1, ROUTE[2] is successor of 1, ROUTE[3] is the successor of ROUTE[2], and so on. Thus array ROUTE is simply a permutation of integers $\{1, 2, \ldots, N\}$, with the first element always as 1. The total weight TWEIGHT of this cycle is also given as an output.

The current best solution (i.e., the minimum-weight Hamiltonian cycle obtained so far) is saved by another (similar) array BEST, such that (I, BEST[I]) is an edge in the solution, for every I = 1, 2, ..., N.

We also need to keep the partial solution (the list of edges included in the solution). It is convenient to keep this list of edges in the form of two arrays—forward pointers (FWDPTR) and backward pointers (BACKPTR). If (I, J) is an edge in the partial solution, FWDPTR[I] = J and BACKPTR[J] = I. Both of these arrays are initialized to 0 in the code.

At all times we must also keep a record of which rows and columns of the original matrix W are in the current matrix A. This can be achieved by keeping two linear arrays ROW and COL, such that ROW[I] = K implies that the Kth row of W (i.e., node K) now occupies Ith row in A; similarly for COL. For example, in Fig. 3-41 the 4 × 4 matrix A on the left side is represented by the two arrays as ROW = $[1, 2, 3, 5, -, -]$ and COL = $[1, 2, 4, 5, -, -]$. Initially, we set ROW[I] ← COL[I] ← I, for all I = 1 to N. Then at all times the following holds: a_{IJ} = W[ROW[I], COL[J]]—all reduction from that row and column.

The BABTSP procedure calls the EXPLORE recursive procedure, which explores the entire solution space in a depth-first fashion. The EXPLORE procedure, in turn, calls the function REDUCE(ROW,COL,ROWRED,COLRED), which reduces the associated matrix (defined by arrays ROW and COL) and computes amounts ROWRED[I] subtracted from its Ith row and COLRED[J] subtracted from its Jth column. It also calls the BESTEDGE(R,C,MOST) procedure, which finds the best edge on which to split the solution space and the value, MOST, by which the lower bounds of the two differ.

The subprocedures follow closely the outlines given in Algorithms 3-9.

Global Constant

N   number of nodes in the given network

Data Types

```
TYPE   ARRN  = ARRAY[1..N] OF INTEGER;
       ARRNN = ARRAY[1..N, 1..N] OF INTEGER;
```

Procedure Parameters

```
PROCEDURE BABTSP(
     N,INF          :INTEGER;
     VAR W          :ARRNN;
     VAR ROUTE      :ARRN;
     VAR TWEIGHT :INTEGER);
```

Input

N   number of nodes in the given network
INF   integer much larger than the largest edge weight
W[1..N,1..N]   weight matrix of the given network

Output

ROUTE[1..N]   array giving the optimal route
TWEIGHT   total weight of the route

Example

The input

$$N = 5 \qquad W[1..5,1..5] = \begin{bmatrix} \infty & 38 & 31 & 44 & 8 \\ 45 & \infty & 15 & 48 & 38 \\ 43 & 32 & \infty & 26 & 47 \\ 38 & 19 & 14 & \infty & 44 \\ 4 & 25 & 33 & 15 & \infty \end{bmatrix},$$

produced output as follows:

$$ROUTE[1..5] = [1, 5, 4, 2, 3]$$
$$TWEIGHT = 100$$

Pascal Procedure BABTSP

```
PROCEDURE BABTSP(
   N, INF    : INTEGER;
   VAR W     : ARRNN;
   VAR ROUTE : ARRN;
   VAR TWEIGHT: INTEGER);

VAR BACKPTR, BEST, COL, FWDPTR, ROW: ARRN;
    I, INDEX                        : INTEGER;

PROCEDURE EXPLORE(
   EDGES, COST : INTEGER;
   VAR ROW, COL: ARRN);

VAR AVOID, C, COLROWVAL, FIRST, I, J,
          LAST, LOWERBOUND, MOST, R, SIZE: INTEGER;
    COLRED, NEWCOL, NEWROW, ROWRED       : ARRN;

FUNCTION MIN(
   I, J: INTEGER): INTEGER;

BEGIN
   IF I <= J THEN MIN := I
   ELSE MIN := J
END;

FUNCTION REDUCE(
   VAR ROW, COL, ROWRED, COLRED: ARRN): INTEGER;

VAR I, J, RVALUE, TEMP: INTEGER;

BEGIN
   RVALUE := 0;                                (* REDUCE ROWS *)
   FOR I := 1 TO SIZE DO
   BEGIN
      TEMP := INF;
      FOR J := 1 TO SIZE DO
         TEMP := MIN(TEMP,W[ROW[I],COL[J]]);
      IF TEMP > 0 THEN
      BEGIN
         FOR J := 1 TO SIZE DO
            IF W[ROW[I],COL[J]] < INF THEN
               W[ROW[I],COL[J]] := W[ROW[I],COL[J]] - TEMP;
         RVALUE := RVALUE + TEMP
      END;
      ROWRED[I] := TEMP
   END;
   FOR J := 1 TO SIZE DO                        (* REDUCE COLUMNS *)
   BEGIN
      TEMP := INF;
      FOR I := 1 TO SIZE DO
         TEMP := MIN(TEMP,W[ROW[I],COL[J]]);
```

```
        IF TEMP > 0 THEN
        BEGIN
            FOR I := 1 TO SIZE DO
                IF W[ROW[I],COL[J]] < INF THEN
                    W[ROW[I],COL[J]] := W[ROW[I],COL[J]] - TEMP;
            RVALUE := RVALUE + TEMP
        END;
        COLRED[J] := TEMP
    END;
    REDUCE := RVALUE
END;

PROCEDURE BESTEDGE(
    VAR R, C, MOST: INTEGER);

VAR I, J, K, MINCOLELT, MINROWELT, ZEROES: INTEGER;

BEGIN
    MOST := -INF;
    FOR I := 1 TO SIZE DO
        FOR J := 1 TO SIZE DO
            IF W[ROW[I],COL[J]] = 0 THEN
            BEGIN
                MINROWELT := INF;
                ZEROES := 0;
                FOR K := 1 TO SIZE DO
                    IF W[ROW[I],COL[K]] = 0 THEN ZEROES := ZEROES + 1
                    ELSE MINROWELT := MIN(MINROWELT,W[ROW[I],COL[K]]);
                IF ZEROES > 1 THEN MINROWELT := 0;
                MINCOLELT := INF;
                ZEROES := 0;
                FOR K := 1 TO SIZE DO
                    IF W[ROW[K],COL[J]] = 0 THEN ZEROES := ZEROES + 1
                    ELSE MINCOLELT := MIN(MINCOLELT,W[ROW[K],COL[J]]);
                IF ZEROES > 1 THEN MINCOLELT := 0;
                IF (MINROWELT + MINCOLELT) > MOST THEN
                BEGIN                       (* A BETTER EDGE HAS BEEN FOUND *)
                    MOST := MINROWELT + MINCOLELT;
                    R := I;                 (* ROW INDEX OF BETTER EDGE *)
                    C := J                  (* COLUMN INDEX OF BETTER EDGE *)
                END
            END
END;

BEGIN (* EXPLORE *)
    SIZE := N - EDGES;          (* NUMBER OF ROWS, COLS LEFT IN MATRIX *)
    COST := COST + REDUCE(ROW,COL,ROWRED,COLRED);
    IF COST < TWEIGHT THEN
        IF EDGES = (N-2) THEN
```

```
         BEGIN                              (* LAST TWO EDGES ARE FORCED *)
            FOR I := 1 TO N DO BEST[I] := FWDPTR[I];
            IF W[ROW[1],COL[1]] = INF THEN AVOID := 1
            ELSE AVOID := 2;
            BEST[ROW[1]] := COL[3-AVOID];
            BEST[ROW[2]] := COL[AVOID];
            TWEIGHT := COST
         END
         ELSE
         BEGIN
            BESTEDGE(R,C,MOST);
            LOWERBOUND := COST + MOST;
            FWDPTR[ROW[R]] := COL[C];                (* RECORD CHOSEN EDGE *)
            BACKPTR[COL[C]] := ROW[R];
            LAST := COL[C];                              (* PREVENT CYCLES *)
            WHILE FWDPTR[LAST] <> 0 DO LAST := FWDPTR[LAST];
            FIRST := ROW[R];
            WHILE BACKPTR[FIRST] <> 0 DO FIRST := BACKPTR[FIRST];
            COLROWVAL := W[LAST,FIRST];
            W[LAST,FIRST] := INF;
            FOR I := 1 TO R-1 DO NEWROW[I] := ROW[I];     (* REMOVE ROW *)
            FOR I := R TO SIZE-1 DO NEWROW[I] := ROW[I+1];
            FOR I := 1 TO C-1 DO NEWCOL[I] := COL[I];     (* REMOVE COL *)
            FOR I := C TO SIZE-1 DO NEWCOL[I] := COL[I+1];
            EXPLORE(EDGES+1,COST,NEWROW,NEWCOL);
            W[LAST,FIRST] := COLROWVAL;        (* RESTORE PREVIOUS VALUES *)
            BACKPTR[COL[C]] := 0;
            FWDPTR[ROW[R]] := 0;
            IF LOWERBOUND < TWEIGHT THEN
            BEGIN
               W[ROW[R],COL[C]] := INF;  (* EXCLUDE EDGE ALREADY CHOSEN *)
               EXPLORE(EDGES,COST,ROW,COL);
               W[ROW[R],COL[C]] := 0          (* RESTORE EXCLUDED EDGE *)
            END
         END;
      FOR I := 1 TO SIZE DO                       (* UNREDUCE MATRIX *)
         FOR J := 1 TO SIZE DO
            W[ROW[I],COL[J]] := W[ROW[I],COL[J]] + ROWRED[I] + COLRED[J]
END;

BEGIN (* BABTSP *)
   FOR I := 1 TO N DO
   BEGIN
      ROW[I] := I;
      COL[I] := I;
      FWDPTR[I] := 0;
      BACKPTR[I] := 0
   END;
   TWEIGHT := INF;
   EXPLORE(0,0,ROW,COL);
   INDEX := 1;
   FOR I := 1 TO N DO
```

```
BEGIN
   ROUTE[I] := INDEX;
   INDEX := BEST[INDEX]
END
END;
```

3.8.2. APPROXIMATE ALGORITHMS

As mentioned earlier, the traveling salesman problem is one of those combinatorial optimization problems for which no efficient algorithms (i.e., with computation time bounded by a polynomial in n, the number of nodes) are known. The branch-and-bound method just presented requires exhaustive searching of a solution space whose size is exponential in the number of nodes. For a worst-case input the computation time of the branch-and-bound algorithm could become exponential in n.

When faced with such computationally hard (NP-hard) problems, one practical approach is to relax the requirement that an algorithm produce an optimal result, and be satisfied with a suboptimal solution reasonably close to the optimal. Such relaxation of the optimality constraint often reduces the computation time from an exponential to a polynomial function (of the size of the input). Such approximate algorithms are the only realistic methods of solving computationally hard problems of large sizes. It is hoped that there will be only a moderate sacrifice in the cost of the suboptimal solutions thus obtained.

For the traveling salesman problem, a number of such approximate algorithms have been proposed and studied. Here we will consider two of the most effective, each representative of a class of approximate algorithms. The first is called an insertion algorithm, and the second one is called a local-search algorithm.

Insertion Algorithm

An insertion algorithm arbitrarily picks a city as the starting node, say s, of the tour. From among the remaining $(n - 1)$ cities it selects another city (according to a criterion to be discussed shortly), say p. Now we have a subtour or cycle of two nodes (s, p, s). Then a third node, q, out of the $(n - 2)$ unvisited nodes is selected, and is inserted into the current cycle to produce cycle (s, p, q, s) or (s, q, p, s)—whichever is cheaper. This enlargement is continued. At any stage let V_T denote the set of nodes included in the subtour, and, as usual, let V be the entire node set for this problem.

The kth iteration $(1 \leqslant k \leqslant n - 1)$ enlarges the cycle of size k to one of size $k + 1$ by means of the following two steps:

Selection step.   In the set $V - V_T$ of unvisited nodes, determine which node is to be added to the cycle next.

Insertion step.   Determine where (i.e., between which two nodes of the cycle) the newly selected node is to be inserted to enlarge the current subtour.

A number of heuristics have been suggested and experimented with for the selection step. Some of these are: (1) pick any unvisited node at random—*arbitrary insertion*; (2) pick that unvisited node which is nearest to the current subtour—*nearest insertion*; (3) compare the costs of insertions of all unvisited nodes (in all insertion positions) and pick the one with smallest cost—*cheapest insertion*; and (4) pick that unvisited node which is farthest from the current subtour—*farthest insertion*.

Of the four heuristics mentioned, *farthest insertion* appears to be the best overall strategy, according to a number of extensive and independent empirical studies (see Adrabiński and Sysło [1980] and Golden et al. [1980]). The underlying intuition behind this approach is that if a rough outline of the tour can be constructed through the widely-separated nodes, then the finer details of the tour resulting from the inclusion of the nearer nodes can be filled in without greatly increasing the total length of the tour. We will, therefore, present the farthest insertion algorithm and leave the other three as exercises.

In order to find efficiently that unvisited node which is farthest from V_T, the nodes in the current cycle, let us maintain a distance array *dist* (of length n) such that for every node v not in the cycle, $dist(v)$ is the distance to v from that node in the current cycle from which v is closest. The farthest node f is the one with the largest value in the *dist* array, and it is the node to be inserted next. Each time a new node f is added to the cycle, the *dist* array is updated such that its entries are the minimum of the current entries in the *dist* array and the fth row in W. Observe the similarity between this procedure and the Prim–Dijkstra minimum spanning tree procedure (Algorithm 3-5).

Having settled on the selection step, let us now look at the insertion step. Assume that there are k nodes in the current cycle, and the next (farthest) node to be inserted is f. We examine every edge (i, j) in the current tour to determine the insertion cost of f between node i and j, which is

$$c_{ij} = w_{if} + w_{fj} - w_{ij}$$

Among all k edges in the cycle we select edge (t, h)—with tail t and head h—for which c_{th} has the smallest value (c_{ij} could be negative). Then insert node f between t and h. The weight of the cycle is updated. We also update the *dist* array.

The following FITSP (farthest insertion traveling salesman procedure) algorithm describes this heuristic more precisely and compactly.

Algorithm 3-10: Farthest-Insertion Approximate TSP Algorithm

```
INITIALIZATION:
    V_T ← {s}; {* Initial cycle of one node s and fictitious cycle of
                    zero weight at s *}
    E_T ← {(s, s)};
```

$w_{ss} \leftarrow 0$;
$tweight \leftarrow 0$; {* total weight of the subtour *}
for every node $u \in V - V_T$ **do** $dist(u) \leftarrow w_{s\,u}$;

ITERATION:
    while $|V_T| < n$ **do**
    begin
        $f \leftarrow$ node in $V - V_T$ with largest value of $dist(f)$;
        for every edge $(i, j) \in E_T$ **do**
            $c_{ij} \leftarrow w_{if} + w_{fj} - w_{ij}$; {* insertion cost *}
        $(t, h) \leftarrow$ edge in E_T with smallest value of c_{th};
        $E_T \leftarrow E_T \cup \langle (t, f), (f, h) \rangle - \langle (t, h) \rangle$;
        $V_T \leftarrow V_T \cup \{f\}$;
        $tweight \leftarrow tweight + c_{th}$;
        for all $x \in (V - V_T)$ **do** $dist(x) \leftarrow \min\langle dist(x), w_{fx} \rangle$
    end

To keep track of V_T, the nodes in the current cycle, as well as E_T, the edges in the current cycle, we will maintain an array, *cycle*, of length n, defined as follows; $cycle(i) = 0$ if and only if node i is not in the current cycle; and $cycle(i) = j$ if and only if (i, j) is an edge in the current cycle.

An Example

As an example, let us solve the same six-city, traveling salesman problem which was used to illustrate the branch-and-bound method in the preceding subsection. Let us arbitrarily pick node 1 as the starting node s. The *dist* array at this juncture will be

$$dist = (-, 3, 93, 13, 33, 9)$$

which is row 1 of weight matrix W, except $dist(1)$, which is immaterial. The other array is

$$cycle = (1, 0, 0, 0, 0, 0)$$

The largest entry in *dist* array is 93, corresponding to node 3. Therefore, the subtour is enlarged to $(1, 3, 1)$ and the total distance traveled (*tweight*) is $w_{13} + w_{31} = 93 + 45 = 138$.

The *dist* array is now modified to have entries that are the smaller of *dist* and row 3 of W. That is

$$dist = (-, 3, -, 13, 16, 9)$$

and

$$cycle = (3, 0, 1, 0, 0, 0)$$

This completes the first iteration. Now in the second iteration the farthest node from the current subtour is 5, corresponding to the largest value, 16, in the *dist* array. Node 5 can be inserted in two different ways. The insertion costs are

$$c_{13} = w_{15} + w_{53} - w_{13} = 33 + 88 - 93 = 28$$

$$c_{31} = w_{35} + w_{51} - w_{31} = 16 + 28 - 45 = -1$$

Performing the insertion with lower cost, we obtain *tweight* $= 138 - 1 = 137$, and the two arrays are

$$cycle = (3, 0, 5, 0, 1, 0)$$

$$dist = (-, 3, -, 13, -, 9)$$

In the third iteration, node 4 is the farthest. The three insertion costs of node 4 are

$$c_{13} = w_{14} + w_{43} - w_{13} = 0$$

$$c_{35} = w_{34} + w_{45} - w_{35} = 76$$

$$c_{51} = w_{54} + w_{41} - w_{51} = 44$$

We, therefore, perform the lowest-cost insertion (at zero cost). The new subtour is $(1, 4, 3, 5, 1)$, with a value *tweight* of 137. The updated arrays are

$$cycle = (4, 0, 5, 3, 1, 0)$$

$$dist = (-, 3, -, -, -, 7)$$

For the fourth iteration we observe that the farthest node is 6. The insertion costs of node 6 are

$$c_{14} = w_{16} + w_{64} - w_{14} = 42$$

$$c_{43} = w_{46} + w_{63} - w_{43} = -55$$

$$c_{35} = w_{36} + w_{65} - w_{35} = 104$$

$$c_{51} = w_{56} + w_{61} - w_{51} = 0$$

Since the second value is the smallest, we therefore insert node 6 between nodes 4 and 3. This gives us the new tour as $(1, 4, 6, 3, 5, 1)$, with a total distance *tweight* $= 137 - 55 = 82$.

The corresponding updated arrays are

$$cycle = (4, 0, 5, 6, 1, 3)$$
$$dist = (-, 3, -, -, -, -)$$

In the fifth and the last iteration, we must insert node 2. Its five insertion costs are $c_{14} = 32$, $c_{46} = 99$, $c_{63} = 147$, $c_{35} = 22$, and $c_{51} = 22$.

There are two minimum values; we could pick either. Let us choose c_{35}. Then we obtain the final solution as $(1, 4, 6, 3, 2, 5, 1)$, with the total distance traveled *tweight* = 82 + 22 = 104.

The solution just obtained happens to be one of the two optimal TS routes. In general, we cannot expect to be so lucky and get an optimal solution by this approximate method.

This example illustrates how a "good" solution is finally obtained as we continue to insert nodes at points of least-cost insertions. The reader is encouraged to work out several small examples by hand to gain further insight.

It should be observed that, in general, different starting nodes will produce different final solutions and with different costs. Sometimes, when a better solution is required it is advisable to execute this algorithm n times—starting once from each node—and then keep the best of the n solutions.

Time Complexity of FITSP

The outer loop in Algorithm 3-10 is executed $(n - 1)$ times, as a new node is inserted in the cycle in each iteration. During the kth iteration, k different values for c_{ij} are computed and compared. Also in each iteration two arrays are updated, each requiring at most n time units. Thus the entire algorithm is of time complexity $O(n^2)$. If the FITSP was to be run n times, starting from each node once, we would need $O(n^3)$ time.

It is also evident that (unlike in the case of the BABTSP procedure), the execution time for the FITSP is not dependent on the distribution of the entries in the weight matrix, W. The run time depends entirely on the size of the network. Table 3-8 shows the run times for the Pascal FITSP procedure (listing to be given

Table 3-8. Computing Times (sec) for the FITSP Procedure

Number of Nodes	One Call of FITSP	n Calls of FITSP
10	0.0025	0.025
20	0.0065	0.129
40	0.0212	0.848
60	0.0453	2.720
80	0.0775	6.200

shortly) for randomly generated problems with 10 to 80 nodes. As usual, the times are in virtual CPU seconds on an Amdahl 470 V/8.

Table 3-8 shows that the time for one call of FITSP grows as n^2, and for n calls it grows as n^3. Compare Tables 3-7 and 3-8 and observe the enormous difference between the computational costs of approximate and exact solutions.

Computer Implementation of Farthest Insertion TSP Algorithm

The FITSP procedure computes an approximate tour (Hamiltonian cycle) for the traveling salesman problem, in a complete directed network of N nodes, given in the form of an N × N weight (distance matrix or cost matrix) matrix W. The starting node S is also specified. All entries in W are nonnegative integers, except the diagonal entries, which could be set to any arbitrary value.

Global Constant

N   number of nodes in the given network

Data Types

```
TYPE   ARRN  = ARRAY[1..N] OF INTEGER;
       ARRNB = ARRAY[1..N] OF BOOLEAN;
       ARRNN = ARRAY[1..N,1..N] OF INTEGER;
```

Procedure Parameters

```
PROCEDURE FITSP(
    N,S,INF        :INTEGER;
    VAR W          :ARRNN;
    VAR ROUTE      :ARRN;
    VAR TWEIGHT    :INTEGER);
```

Input

N   number of nodes in the given network
S   starting node
INF   infinity, an integer much larger than the largest entry in W
W[1..N,1..N]   weight matrix of the network

Output

ROUTE[1..N]   traveling salesman route; ROUTE [I] is the Ith node visited and ROUTE [1] is the starting node S
TWEIGHT   total weight or cost of the tour

Example

We use the same six-city problem which was considered in the text:

$$N = 6, \quad S = 1, \quad W[1..6,1..6] = \begin{bmatrix} \infty & 3 & 93 & 13 & 33 & 9 \\ 4 & \infty & 77 & 42 & 21 & 16 \\ 45 & 17 & \infty & 36 & 16 & 28 \\ 39 & 90 & 80 & \infty & 56 & 7 \\ 28 & 46 & 88 & 33 & \infty & 25 \\ 3 & 88 & 18 & 46 & 92 & \infty \end{bmatrix}$$

The solution obtained is

$$ROUTE[1..6] = [1, 4, 6, 3, 2, 5]$$

$$TWEIGHT = 104$$

Pascal Procedure FITSP

```
PROCEDURE FITSP(
    N, S, INF  : INTEGER;
    VAR W      : ARRNN;
    VAR ROUTE  : ARRN;
    VAR TWEIGHT: INTEGER);

VAR END1, END2, FARTHEST, I, INDEX,
        INSCOST, J, MAXDIST, NEWCOST, NEXTINDEX: INTEGER;
    CYCLE, DIST                             : ARRN;

BEGIN
    FOR I := 1 TO N DO CYCLE[I] := 0;
    CYCLE[S] := S;
    FOR I := 1 TO N DO DIST[I] := W[S,I];
    TWEIGHT := 0;
    FOR I := 1 TO N-1 DO
    BEGIN
        MAXDIST := -INF;
        FOR J := 1 TO N DO
            IF CYCLE[J] = 0 THEN
                IF DIST[J] > MAXDIST THEN
                BEGIN
                    MAXDIST := DIST[J];
                    FARTHEST := J
                END;
        INSCOST := INF;
        INDEX := S;
        FOR J := 1 TO I DO
        BEGIN
            NEXTINDEX := CYCLE[INDEX];
            NEWCOST := W[INDEX,FARTHEST] + W[FARTHEST,NEXTINDEX] -
                        W[INDEX, NEXTINDEX];
```

```
        IF NEWCOST < INSCOST THEN
        BEGIN
            INSCOST := NEWCOST;
            END1 := INDEX;
            END2 := NEXTINDEX
        END;
        INDEX := NEXTINDEX
    END;
    CYCLE[FARTHEST] := END2;
    CYCLE[END1] := FARTHEST;
    TWEIGHT := TWEIGHT + INSCOST;
    FOR J := 1 TO N DO
        IF CYCLE[J] = 0 THEN
            IF W[FARTHEST,J] < DIST[J] THEN
                DIST[J] := W[FARTHEST,J]
    END;
    INDEX := S;
    FOR I := 1 TO N DO
    BEGIN
        ROUTE[I] := INDEX;
        INDEX := CYCLE[INDEX]
    END
END;
```

3.8.3 LOCAL SEARCH HEURISTICS

One of the best known and most successful heuristics for obtaining a nearly optimal solution of the traveling salesman problem is the following. Start with an arbitrary Hamiltonian cycle (an initial TS tour) H; delete r edges from H, thus producing r disconnected paths (some of which may be isolated nodes). Reconnect these r paths in such a way as to produce another TS tour H' using edges different from those that were removed from H. Thus H and H' differ from each other by exactly r edges; the remaining $(n - r)$ edges are common. Figures 3-45 and 3-46 illustrate 2- and 3-exchanges, respectively. Compute the total weight $w(H')$ of tour H'. If $w(H') < w(H)$, replace H with H' and repeat the process; otherwise, pick another set of r edges from H to exchange. Such exchanges (of sets of r edges) continue until no additional improvement can be made by exchanging r edges. The final solution, which cannot be improved by exchanging any of its r edges, is called an *r-optimal* (or *r-opt*, for short) solution.

Clearly, the edge-exchange procedure will terminate at a local optimum (not necessarily a global optimum), thus producing an approximate solution. This approximate algorithm for TSP illustrates a general approach to solving many combinatorial optimization problems, known as *local search* or *neighborhood search*.

This method of successive improvement in the TS route by exchanging r edges can be used for symmetric as well as asymmetric problems. Since in the two

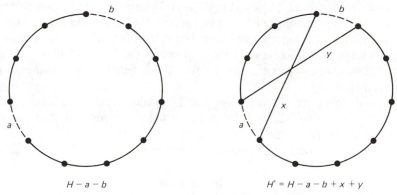

$$H - a - b$$ $$H' = H - a - b + x + y$$

Figure 3-45   *A 2-exchange.*

preceding sections we have used asymmetric TSPs, let us consider the symmetric TS problem for developing and illustrating the local-search heuristic. The asymmetric case is left as an exercise.

It is easy to see that for a symmetric (undirected) network with numbers of nodes $n \geqslant 5$, the value of r can vary from 2 to n. [Observe that for an asymmetric (directed) network r cannot be less than 3.]

In general, the higher the value of r in the r-exchange procedure, the better will be the solution. But the computational expense also increases rapidly with the value of r. There are $\binom{n}{r}$ subsets of r edges in a cycle with n edges. Exchanging each of these even once requires $O(n^r)$ time. One must, therefore, make a careful trade-off between the accuracy of the solution and computational expense. There is very little theory developed, and one has to depend primarily on intuition and empirical studies for guidance. For example, a number of extensive, independent empirical studies conducted by Lin [1965], Lin and Kernighan [1973], Adrabiński and Sysło

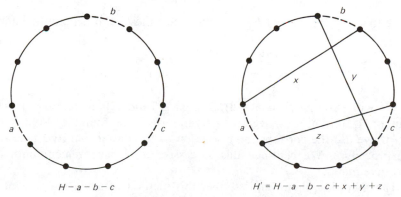

$$H - a - b - c$$ $$H' = H - a - b - c + x + y + z$$

Figure 3-46   *A 3-exchange.*

[1980], and Golden et al. [1980], suggest that 3-opt solutions are much better than 2-opt, but that 4-opt solutions are not sufficiently better than 3-opt solutions to justify the increased computational cost. The reason for this strange phenomenon is not clear, except that for relatively small networks (which is all that has been tried) the 3-opt search is powerful enough to cover a sufficiently large number of points in the solution space.

We will, therefore, discuss 2-opt and 3-opt algorithms: the former, because of its simplicity and the fact that for many smaller networks it gives a very good solution—particularly if the initial cycle is already good. We will also develop the 3-opt algorithm, because of its undisputed superiority.

Initial Tour

The choice of the initial Hamiltonian cycle can have the most dramatic impact on the final solution. Although pathological cases can be constructed where a better initial tour does not lead to a better r-optimal solution, in most cases a better initial solution does produce a better final solution, as observed by Adrabiński and Sysło [1980] and Golden et al. [1980], among others. We will therefore start with a good initial tour (say, one produced by the FITSP procedure given in the preceding section) rather than a random cycle.

2-Optimal Algorithm

Let the initial cycle consist of the following set of (undirected) edges $H = \{x_1, x_2, \ldots, x_n\}$ in the order $x_1, x_2, \ldots, x_n$. Let $X = \{x_i, x_j\}$ be a set of two edges in H which we delete and replace with edges $Y = \{y_p, y_q\}$, if there is an improvement. That is, $H' = (H - X) \cup Y$ is a new and improved tour.

Observe that: (1) the two edges x_i, x_j in X cannot be adjacent and that (2) once set X has been chosen, set Y is determined. Thus there are

$$\frac{n(n-1)}{2} - n = \frac{n(n-3)}{2}$$

possible tours H' for a given H. For any one of these let δ denote the improvement

$$\delta = w(H) - w(H')$$
$$= w(x_i) + w(x_j) - w(y_p) - w(y_q)$$

We will examine all $n(n-3)/2$ tours H' and retain the one for which δ is maximum. If this δ_{max} is negative or zero, we have found a 2-opt solution. If $\delta_{max} > 0$, we use the corresponding solution as the initial tour and then repeat the entire procedure. We continue this successive tour improvement until δ_{max} is a nonpositive number.

The following algorithm describes this method of achieving a 2-opt approximate solution of the TSP, by successive replacement of two edges.

Algorithm 3-11:   2-Opt Approximate TSP Algorithm

begin
    let $H = (x_1, x_2, \ldots, x_n)$ be the current tour;
    repeat
        $\delta_{max} \leftarrow 0$;
        for $i \leftarrow 1$ **to** $(n - 2)$ **do**
            for $j \leftarrow (i + 2)$ **to** n or $(n - 1)$ **do**
                ⟨∗ last case is only when $i = 1$ ∗⟩
                if $(w(x_i) + w(x_j)) - (w(y_p) + w(y_q)) > \delta_{max}$ **then**
                    begin
                        $\delta_{max} \leftarrow (w(x_i) + w(x_j)) - (w(y_p) + w(y_q))$;
                        save i and j
                end;
        if $\delta_{max} > 0$ **then** $H \leftarrow H - \{x_i, x_j\} \cup \{y_p, y_q\}$
    until $\delta_{max} = 0$
end

Computer Implementation of the 2-Opt TSP Algorithm

The TWOOPT procedure finds an approximate solution of the symmetric traveling salesman problem with N nodes, given as an N × N weight matrix W of integers. It starts with the initial tour (specified by array ROUTE) and improves the tour repeatedly by exchanging two edges at a time. The procedure closely follows Algorithm 3-11. The edge exchange continues until no better solution can be found by exchanging a pair of edges of the current solution. The final solution, which is a local optimum, is given by the array ROUTE (as in the case of the BABTSP and FITSP procedures).

Global Constant

N   number of nodes in the given network

Data Types

```
TYPE   ARRN  = ARRAY[1..N] OF INTEGER;
       ARRNN = ARRAY[1..N,1..N] OF INTEGER;
```

Procedure Parameters

```
PROCEDURE TWOOPT(
    N            :INTEGER;
    VAR W        :ARRNN;
    VAR ROUTE    :ARRN;
    VAR TWEIGHT :INTEGER);
```

Input

N	number of nodes in the given network
W[1..N,1..N]	weight matrix of the given network
ROUTE[1..N]	array specifying the initial TS route

Output

ROUTE[1..N]	array giving the final route
TWEIGHT	total weight of the route

Example

The following input

$$N = 10$$

$$W[1..10,1..10] = \begin{bmatrix} — & 10 & 55 & 78 & 93 & 79 & 48 & 49 & 76 & 11 \\ 10 & — & 77 & 46 & 64 & 86 & 37 & 8 & 29 & 35 \\ 55 & 77 & — & 79 & 9 & 82 & 13 & 54 & 80 & 84 \\ 78 & 46 & 79 & — & 34 & 29 & 9 & 35 & 98 & 91 \\ 93 & 64 & 9 & 34 & — & 74 & 83 & 44 & 67 & 68 \\ 79 & 86 & 82 & 29 & 74 & — & 42 & 98 & 85 & 8 \\ 48 & 37 & 13 & 9 & 83 & 42 & — & 84 & 5 & 2 \\ 49 & 8 & 54 & 35 & 44 & 98 & 84 & — & 98 & 28 \\ 76 & 29 & 80 & 98 & 67 & 85 & 5 & 98 & — & 48 \\ 11 & 35 & 84 & 91 & 68 & 8 & 2 & 28 & 48 & — \end{bmatrix}$$

(Elements of the main diagonal are not used in the procedure.)

$$ROUTE[1..10] = [1, 2, 3, 4, 5, 6, 7, 8, 9, 10]$$

produced the following output:

$$ROUTE[1..10] = [1, 3, 5, 4, 8, 2, 9, 7, 6, 10]$$

$$TWEIGHT = 236$$

Pascal Procedure TWOOPT

```
PROCEDURE TWOOPT(
   N          : INTEGER;
   VAR W      : ARRNN;
   VAR ROUTE  : ARRN;
   VAR TWEIGHT: INTEGER);

VAR AHEAD, I, I1, I2, INDEX, J, J1, J2, LAST,
     LIMIT, MAX, MAX1, NEXT, S1, S2, T1, T2: INTEGER;
   PTR                                     : ARRN;
```

```
BEGIN
    FOR I := 1 TO N-1 DO PTR[ROUTE[I]] := ROUTE[I+1];
    PTR[ROUTE[N]] := ROUTE[1];
    REPEAT
        MAX := 0;
        I1 := 1;
        FOR I := 1 TO N-2 DO
        BEGIN
            IF I = 1 THEN LIMIT := N-1
            ELSE LIMIT := N;
            I2 := PTR[I1];
            J1 := PTR[I2];
            FOR J := I+2 TO LIMIT DO
            BEGIN
                J2 := PTR[J1];
                MAX1 := W[I1,I2] + W[J1,J2] - (W[I1,J1] + W[I2,J2]);
                IF MAX1 > MAX THEN
                BEGIN                   (* BETTER PAIR OF EDGES HAS BEEN FOUND *)
                    S1 := I1;
                    S2 := I2;
                    T1 := J1;
                    T2 := J2;
                    MAX := MAX1
                END;
                J1 := J2
            END;
            I1 := I2
        END;
        IF MAX > 0 THEN
        BEGIN                                       (* SWAP PAIR OF EDGES *)
            PTR[S1] := T1;
            NEXT := S2;
            LAST := T2;
            REPEAT                          (* REVERSE APPROPRIATE LINKS *)
                AHEAD := PTR[NEXT];
                PTR[NEXT] := LAST;
                LAST := NEXT;
                NEXT := AHEAD
            UNTIL NEXT = T2;
            TWEIGHT := TWEIGHT - MAX         (* ROUTE IS NOW SHORTER *)
        END
    UNTIL MAX = 0;
    INDEX := 1;
    FOR I := 1 TO N DO
    BEGIN
        ROUTE[I] := INDEX;
        INDEX := PTR[INDEX]
    END
END; (* TWOOPT *)
```

3-Optimal Algorithm

Removal of three edges from a Hamiltonian cycle H results into three paths which can be joined with three edges to form a new cycle H' in eight possible ways, as shown in Fig. 3-47. Similarly, four paths can be joined in 48 possible ways. For a general expression of different ways of joining r paths, see Problem 3-105. Of these eight cycles in Fig. 3-47, observe that part (a) is the initial tour H itself and parts (b), (c), and (d) are obtained by exchanging only two edges. Thus, if H is already 2-optimal, then parts (b), (c), and (d) need not be considered. Also observe that of the remaining four, parts (e), (f), and (g) are simply rotations of each other, and therefore need not be distinguished. In other words, parts (f) and (g) can be obtained from part (e) by simply relabeling the nodes. Finally, part (h) is distinct from all others. Thus there are only two distinct ways of replacing three edges in a Hamiltonian cycle. These are: replace edge set $\{(u_1, u_2), (v_1, v_2), (z_1, z_2)\}$ with set $\{(v_1, u_1), (z_1, u_2), (z_2, v_2)\}$ or with set $\{(u_1, v_2), (z_1, u_2), (v_1, z_2)\}$, as shown in Figs. 3-47(e) and (h), respectively.

Observe that in the second case [i.e., in Fig. 3-47(h)] the paths are traversed along the same direction in the new Hamiltonian cycle as they were in the old, whereas in the first case [i.e., in Fig. 3-47(e)] the direction of one of the paths—z_2 to u_1— is reversed. While exchanging edges this must be taken into account, even though the network is undirected. As we will see shortly, this reversal is carried out by the REVERSE subprocedure, which is invoked in one case but not in the other.

First, we select which three edges we might delete, and then compare the gains in each of the two possible exchanges. We pick the more advantageous of the two (by invoking the SWAPCHECK subprocedure). This computation is carried out for all potential (of order n^3) triplets of edges in the current solution. When all possible triplets have been considered, we pick the one whose deletion (and replacement) would yield the maximum gain. If this gain is positive, we actually carry out the exchange of edges and repeat the whole process, starting with the new TS route. Otherwise (if this gain is not positive), the execution is terminated, because we have obtained a 3-optimum solution.

The only detail that remains to be worked out is an orderly and simple fashion in which we select edges such that all potential 3-exchanges have been considered without duplications or omissions. The first of the three edges can be any one of the n edges in the current tour, unlike in the 2-opt algorithm. This is because in the 2-opt case there is only one way of rejoining the two paths (if a different cycle is to be obtained), while in 3-opt there are several. Since the three cases in Fig. 3-47(e), (f), and (g) are being collapsed into one, we cannot set the bounds on the first edge as we did in the 2-opt case; similarly, for the other two edges. The details are left as an exercise (Problem 3-117).

A great deal of sophistication is possible in choosing the triplets. These have been discussed in Lin [1965], Christofides and Eilon [1972], and Lin and Kernighan [1973]. The THREEOPT procedure presented here strikes a reasonable balance between simplicity and efficiency.

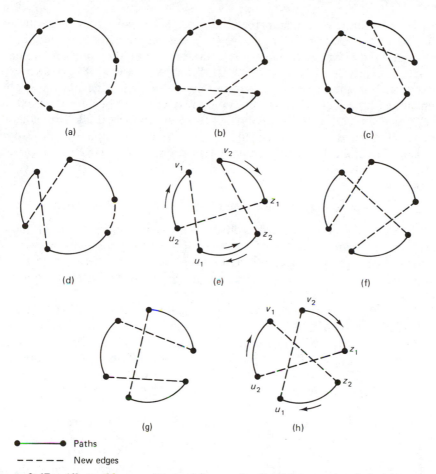

Paths

New edges

Figure 3-47 *All possible connections of three paths. Solid lines, paths; dashed lines, new edges.*

Clearly, one iteration (actual swapping of an edge triplet) requires $O(n^3)$ operations. It turns out that the total number of iterations is usually small—much less than n. For instance, in the 10-node example used for the THREEOPT procedure, the following sets of five exchanges took place, starting with tour $(1, 2, 3, 4, 5, 6, 7, 8, 9, 10, 1)$:

edges $(7, 8), (8, 9), (2, 3)$	replaced with	$(7, 9), (8, 3), (2, 8)$
edges $(8, 3), (3, 4), (5, 6)$	replaced with	$(3, 8), (6, 4), (5, 3)$
edges $(7, 6), (4, 5), (10, 9)$	replaced with	$(4, 7), (9, 5), (10, 6)$
edges $(2, 1), (9, 5), (3, 8)$	replaced with	$(9, 2), (8, 5), (3, 1)$
edges $(4, 7), (2, 8), (3, 1)$	replaced with	$(4, 8), (2, 1), (3, 7)$

Observe from this example that the same edge may go in and out of the solution several times. For instance, edge $(1, 2)$ was originally in the solution. It was thrown out in the fourth exchange. It came back into the solution in the fifth exchange; similarly for edges $(4, 7)$, $(3, 1)$, $(8, 3)$, $(9, 5)$, and $(2, 8)$. Heuristics have been suggested in the literature to reduce this source of inefficiency. Also observe that in some iterations (e.g., iteration 2) only two edges are being exchanged rather than all three. In fact, it can be easily seen from the code, that all 2-exchanges of a given Hamiltonian cycle also appear in our implementation of the 3-optimal algorithm. Hence, the 3-optimal solutions generated by the THREEOPT procedure are also 2-optimal.

Computer Implementation of the 3-Opt TSP Algorithm

The THREEOPT procedure starts with the initial solution specified by the array ROUTE in an undirected N-node network given as the weight matrix W. It exchanges three edges at a time, and continues to do so until there is no set of three edges in the current TS route whose exchange would yield a better route (i.e., one with smaller weight).

The THREEOPT procedure invokes two subprocedures: SWAPCHECK and REVERSE. The SWAPCHECK procedure compares the two distinct ways of exchanging the current triplet of edges (u_1, u_2), (v_1, v_2), (z_1, z_2), as illustrated in Fig. 3-47(e) and (h). The first of these is called ASYMMETRIC and the second, called SYMMETRIC, selects the better of the two (i.e., the one with larger gains). The REVERSE procedure reverses the path when the two end nodes [say, z_2 and u_1 in Fig. 3-47(e)] are specified.

As in the case of the 2-opt procedure, here also the edges in the current solution are kept in the array PTR. That is, (I, PTR[I]) is an edge in the current solution, incident on node I.

Initially, from the array ROUTE (which specifies the order in which the nodes are visited), the array PTR is created. Then for every possible triplets of edges in the current solution PTR, the better (of the two) candidate for swapping is determined by invoking the SWAPCHECK procedure. The best triplet (i.e., the one producing the maximum reduction in weight) among all of these is determined. Then an actual exchange of edges is made if this triplet swapping results in a positive gain (i.e., causes a reduction in the weight). The process is repeated until the gain is no longer positive.

Global Constant

N number of nodes in the given network

Data Types

```
TYPE  ARRN  = ARRAY[1..N] OF INTEGER;
      ARRNN = ARRAY[1..N,1..N] OF INTEGER;
```

Procedure Parameters

```
PROCEDURE THREEOPT(
    N              :INTEGER;
    VAR W          :ARRNN;
    VAR ROUTE    :ARRN;
    VAR TWEIGHT :INTEGER);
```

Input

N   number of nodes in the given network
W[1..N,1..N]   weight matrix of the given network
ROUTE[1..N]   array specifying the initial TS route

Output

ROUTE[1..N]   array giving the final route
TWEIGHT   total weight of the route

Example

We use the same 10×10 example as the one used for procedure TWOOPT. The output given by the THREEOPT procedure is

$$ROUTE = [1, 10, 6, 4, 8, 5, 3, 7, 9, 2]$$
$$TWEIGHT = 193$$

Pascal Procedure THREEOPT

```
PROCEDURE THREEOPT(
    N          : INTEGER;
    VAR W        : ARRNN;
    VAR ROUTE   : ARRN;
    VAR TWEIGHT: INTEGER);

TYPE SWAPTYPE    = (ASYMMETRIC, SYMMETRIC);
     SWAPRECORD = RECORD
                     X1, X2, Y1, Y2, Z1, Z2, GAIN: INTEGER;
                     CHOICE                       : SWAPTYPE
                  END;

VAR BESTSWAP, SWAP: SWAPRECORD;
    I, INDEX, J, K: INTEGER;
    PTR           : ARRN;
```

```
PROCEDURE SWAPCHECK(
   VAR SWAP: SWAPRECORD);

VAR DELWEIGHT, MAX: INTEGER;

BEGIN
   WITH SWAP DO
   BEGIN
      GAIN := 0;                        (* AMOUNT TO BE GAINED BY SWAPPING *)
      DELWEIGHT := W[X1,X2] + W[Y1,Y2] + W[Z1,Z2];
      MAX := DELWEIGHT - (W[Y1,X1] + W[Z1,X2] + W[Z2,Y2]);
      IF MAX > GAIN THEN
      BEGIN
         GAIN := MAX;
         CHOICE := ASYMMETRIC
      END;
      MAX := DELWEIGHT - (W[X1,Y2] + W[Z1,X2] + W[Y1,Z2]);
      IF MAX > GAIN THEN
      BEGIN
         GAIN := MAX;
         CHOICE := SYMMETRIC
      END
   END
END; (* SWAPCHECK *)

PROCEDURE REVERSE(
   START, FINISH: INTEGER);

VAR AHEAD, LAST, NEXT: INTEGER;

BEGIN
   IF START <> FINISH THEN
   BEGIN
      LAST := START;
      NEXT := PTR[LAST];
      REPEAT   (* REVERSE POINTERS FROM ONE PAST START THROUGH FINISH *)
         AHEAD := PTR[NEXT];
         PTR[NEXT] := LAST;
         LAST := NEXT;
         NEXT := AHEAD;
      UNTIL LAST = FINISH
   END
END; (* REVERSE *)


BEGIN (* THREEOPT *)
   FOR I := 1 TO N-1 DO               (* GIVEN ROUTE, ESTABLISH POINTERS *)
      PTR[ROUTE[I]] := ROUTE[I+1];
```

```
    PTR[ROUTE[N]] := ROUTE[1];
    REPEAT                      (* REPEAT UNTIL NO MORE GAIN FROM SWAPPING *)
        BESTSWAP.GAIN := 0;
        SWAP.X1 := 1;
        FOR I := 1 TO N DO
        BEGIN
            SWAP.X2 := PTR[SWAP.X1];
            SWAP.Y1 := SWAP.X2;
            FOR J := 2 TO N-3 DO
            BEGIN
                SWAP.Y2 := PTR[SWAP.Y1];
                SWAP.Z1 := PTR[SWAP.Y2];
                FOR K := J+2 TO N-1 DO
                BEGIN
                    SWAP.Z2 := PTR[SWAP.Z1];
                    SWAPCHECK(SWAP);  (* FOR THESE 3 EDGES, FIND BEST SWAP *)
                    IF SWAP.GAIN > BESTSWAP.GAIN THEN BESTSWAP := SWAP;
                    SWAP.Z1 := SWAP.Z2
                END;
                SWAP.Y1 := SWAP.Y2
            END;
            SWAP.X1 := SWAP.X2
        END;
        IF BESTSWAP.GAIN > 0 THEN
            WITH BESTSWAP DO
            BEGIN
                IF CHOICE = ASYMMETRIC THEN
                BEGIN
                    REVERSE(Z2, X1);
                    PTR[Y1] := X1;
                    PTR[Z2] := Y2
                END
                ELSE
                BEGIN
                    PTR[X1] := Y2;
                    PTR[Y1] := Z2
                END;
                PTR[Z1] := X2;
                TWEIGHT := TWEIGHT - GAIN
            END
    UNTIL BESTSWAP.GAIN = 0;
    INDEX := 1;                    (* GIVEN POINTERS, DETERMINE NEW ROUTE *)
    FOR I := 1 TO N DO
    BEGIN
        ROUTE[I] := INDEX;
        INDEX := PTR[INDEX]
    END
END; (* THREEOPT *)
```

Performance of Heuristic Algorithms for TSP

The performance of an approximate algorithm is measured not only by determining how fast it runs but also how close the solution it produces comes to the optimal. The empirical data shown in Table 3-9 below, gives some idea of the performance of the TWOOPT and THREEOPT procedures on both counts—the execution times and the length of the traveling salesman route produced by them. Complete, random networks of 10, 20, 30, and 40 nodes were employed as inputs. In each of the four cases, 20 different random networks were generated, and both the execution times and lengths of the tours produced were averaged over these 20 inputs. The edge-weights were chosen to be symmetric and satisfied the triangle inequality. All execution times are in virtual CPU-seconds on an Amdahl 470 V/8.

Table 3-9.   Average Tour Lengths and Execution Times of the TWOOPT and THREEOPT Procedures

Number of Nodes	Ave. Length of Initial Tours	TWOOPT		THREEOPT	
		Ave. Length of Final Tours	Ave. Run Times (sec.)	Ave. Length of Final Tours	Ave. Run Times (sec.)
10	270.4	142.7	0.002	141.9	0.021
20	528.5	186.1	0.019	180.8	0.591
30	793.2	226.5	0.073	217.7	3.749
40	1040.5	254.2	0.182	243.2	13.074

3.8.4   MODIFICATIONS

Heuristics with Guaranteed Bounds

Although in practice the two approximate TSP methods discussed so far (farthest-insertion algorithm, and 2- or 3-opt algorithm) usually produce results that are not too far off from the optimal, they do not guarantee any bound on the ratio

$$\frac{\text{weight of the approximate solution}}{\text{weight of the optimal solution}}$$

for an arbitrary n-node TSP. In fact, it has been shown by Sahni and Gonzales [1976] that finding any polynomial-time approximate algorithm that would guarantee a constant upper bound for this ratio is NP-hard (i.e., is as hard as finding an exact TSP algorithm).

However, if the edge weights satisfy the triangle inequality, that is, if

$$w_{ij} + w_{jk} \geq w_{ik}$$

for all triplets i, j, k (or equivalently, if the traveling salesman is allowed to visit nodes more than once), and if the weight matrix is symmetric, that is, if

$$w_{ij} = w_{ji}$$

for all pairs i, j, then there are approximate algorithms which guarantee a fixed bound on the approximate to optimal tour ratio. Rosenkrantz et al. [1977] showed that for the nearest-insertion heuristic, this ratio can be no worse than 2. That is, if the weight matrix is symmetric and satisfies the triangle inequality the nearest insertion algorithm will never produce a TS route with weight more than twice that of the optimal. The proof of this interesting result is left as an exercise (Problem 3-115).

An entirely different heuristic has been proposed by Christofides [1976]. It produces tours with weight no more than 50% above optimal (for symmetric matrices satisfying the triangle inequality). Christofides's algorithm involves first obtaining a minimum spanning tree and then finding a minimum weight matching on the odd nodes of this tree. The details are left as an exercise (Problem 3-116). It has been shown by Cornuéjols and Nemhauser [1978] that there exist weight matrices for which this ratio of $3 : 2$ is best possible.

It has been observed consistently and independently by many people that the average performance of these heuristics is far better than the worst-case performance indicated by these ratios.

Solvable Cases of the TSP

There are a number of special types of weight matrices for which the traveling salesman problem can be solved exactly by polynomial-time algorithms. These are called *solvable cases* of the TSP. The first nontrivial solvable case of the TSP was found by Gilmore and Gomory in 1964. Some of the other special cases were found by Lawler in 1971 and Sysło in 1973. A lucid survey of this aspect of the TSP is given in Lenstra et al. [1983].

Hamiltonian Path Problem

When the traveling salesman visits every node exactly once, but does not return to the starting node, he traces a path (rather than a cycle) with $n - 1$ edges. Such a path may be called a *Hamiltonian path*. Since this path is also a spanning tree, it is also referred to as a *Hamiltonian tree*. It has also been referred to as a *traveling salesman curve*.

There are routing problems in practice when one is required to find a minimum Hamiltonian path rather than a minimum Hamiltonian cycle. The following is another application (called the *back-plane wiring problem*). Given a set of n pins on the back plane (of an electronic board or panel into which modules are plugged in front), we are required to connect them together (electrically, with wires) using the smallest total length of wire such that no more than two wire ends are wrapped on each pin (because of the limited length of the pins). Thus we are

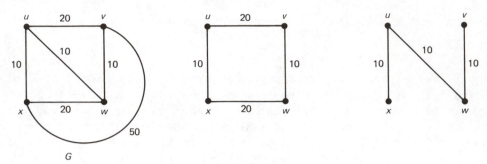

Figure 3-48   *Minimum Hamiltonian cycle and path in G.*

required to find a minimum spanning tree in which the degree of every node is 2 or less. This is nothing but the minimum Hamiltonian path problem.

It is interesting to observe that deleting the largest edge from a minimum Hamiltonian cycle does not necessarily yield a minimum Hamiltonian path in the network. For example, in Fig. 3-48 the minimum Hamiltonian cycle is (u, v, w, x, u) and deleting the largest edge from it does not give the minimum Hamiltonian path.

There is yet another way to view the minimum Hamiltonian path problem, namely, the problem of finding a shortest path from among all shortest paths between all pairs of nodes such that the rest of the nodes are intermediate nodes.

Like the traveling salesman problem, the minimum Hamiltonian path problem has been shown to be NP-hard. The branch-and-bound algorithm can, of course, be modified to solve this problem exactly. A number of approximate algorithms have also been suggested in the literature.

PROBLEMS

3-93.   Write down all 12 Hamiltonian cycles in the five-node undirected, complete network shown in Fig. 3-37. Calculate their weights and list the cycles in ascending order of their weights.

3-94.   Write a (quick-and-dirty) exact TSP program for very small problems which generates all $(n - 1)!$ Hamiltonian cycles in an n-node directed network and finds the one with smallest weight. For a few small randomly generated complete networks, compare the execution times of this algorithm with that of the branch-and-bound algorithm given in this section (the BABTSP procedure).

3-95. Using the branch-and-bound algorithm described in this section, find a minimum-weight TS tour, by hand, for the following six-node network:

$$W = \begin{bmatrix} \infty & 38 & 31 & 6 & 44 & 8 \\ 45 & \infty & 15 & 9 & 48 & 38 \\ 43 & 32 & \infty & 15 & 26 & 47 \\ 10 & 11 & 19 & \infty & 8 & 44 \\ 38 & 19 & 14 & 3 & \infty & 27 \\ 4 & 25 & 33 & 15 & 9 & \infty \end{bmatrix}$$

3-96. Show that in an undirected network in which the edge weights satisfy the triangle inequality, no (closed) tour that visits every node can have a weight smaller than the smallest-weight Hamiltonian cycle.

3-97. Modify the BABTSP procedure so that it is particularly suitable for undirected networks.

3-98. Suppose that the traveling salesman has an additional constraint that he must visit city y immediately after he has visited city x. Show how you will modify the branch-and-bound algorithm (Algorithm 3-9) to take care of this constraint.

3-99. Suppose that the constraint in Problem 3-98 is relaxed such that the traveling salesman must visit city y any time (not necessarily immediately) after he has visited city x. Now modify the branch-and-bound algorithm to incorporate this constraint.

3-100. Solve the six-node TSP given in Problem 3-95 using the FITSP algorithm, by hand. Compare your result with the one obtained by branch and bound in Problem 3-95.

3-101. For the network given in Problem 3-95, start with the tour $(1, 2, 3, 4, 5, 6, 1)$, and use the 3-opt algorithm to find an improved solution, by hand.

3-102. Show that in exchanging three edges for the 3-opt TSP algorithm, any two edges (of the three being replaced) may form a path of length 2 by being adjacent. But all three edges must not form a single path if the three are to be replaced.

3-103. Show that the number of all distinct triplets of edges that may be replaced is

$$\frac{n(n + 1)(n - 4)}{6}$$

(*Hint:* Use Problem 3-102).

3-104. Construct an example of a directed network to show that starting from a better initial tour does not necessarily lead to a better final solution by the 3-opt TSP algorithm. That is, construct a complete network G which has two Hamiltonian cycles H_1 and H_2 (together with others) such that

$w(H_1) > w(H_2)$ and such that the final tours H_1' and H_2' corresponding to the initial tours H_1 and H_2 satisfy the inequality $w(H_1') < w(H_2')$.

3-105. Show that there are $f(r)$ distinct ways of obtaining a Hamiltonian cycle by joining r different undirected paths, where

$$f(r) = f(r-1) \cdot 2(r-1)$$
$$f(1) = 1$$

3-106. Construct an undirected graph in which every node is of degree $\geqslant 2$ and which has no Hamiltonian cycle but does have a Hamiltonian path.

3-107. Figure 3-48 gives an example of a symmetric network in which the minimum Hamiltonian cycle minus its largest edge is not a minimum Hamiltonian path.

(a) Construct a similar (symmetric) example with five nodes.

(b) Construct another (symmetric) example with four nodes, in which the edge weights satisfy the triangle inequality.

(c) Construct an asymmetric example showing the same phenomenon.

3-108. Let G be any complete undirected network in which the edge weights satisfy the triangle inequality. Show that the optimal TS route in G has a weight at most twice the weight of a minimum-weight spanning tree in G. (*Hint:* See Rosenkrantz et al. [1977].)

3-109. For each of the following statements, either give a counterexample to show that it is incorrect or give a proof to show that it is correct: In every complete undirected network in which the edge weights satisfy the triangle inequality:

(a) There exists an optimum TS route which does not contain the largest edge in G (therefore, the largest edge can be deleted).

(b) There exists an optimum TS route which contains the smallest edge in G (therefore, the smallest edge can always be included).

3-110. Discuss why it is not possible to solve the general, undirected traveling salesman problem as an LP assignment model (see Problem 1-51).

3-111. For the six-node TSP employed to illustrate the branch-and-bound algorithm is this section, calculate the lower bounds to the optimal solution using:

(a) The minimum spanning tree problem.

(b) The assignment problem.

3-112. An ice cream manufacturer produces six different flavors of ice cream on the same machine each day. There is a (widely varying) setup time t_{ij} spent in cleaning and preparing the machinery when a change is made from flavor i to flavor j. The following matrix $T = [t_{ij}]$ gives the setup times. Find a schedule that minimizes the total amount of setup time on the machine. Assume that at the end of the day the machine is set up for the next day's manufacturing.

$$T = \begin{bmatrix} - & 7 & 20 & 21 & 12 & 23 \\ 27 & - & 13 & 16 & 46 & 5 \\ 53 & 15 & - & 25 & 27 & 6 \\ 16 & 2 & 35 & - & 47 & 10 \\ 31 & 29 & 5 & 18 & - & 4 \\ 28 & 24 & 1 & 17 & 5 & - \end{bmatrix}$$

(*Hint:* Formulate the problem as an asymmetric TSP, and solve.)

3-113. Investigate the problem of scheduling of the reading of records from the magnetic disk or magnetic drum in your computing facility. Show how it can be formulated as a TSP.

3-114. (*Difficult problem*)   Explore the necessary and sufficient conditions under which a minimum Hamiltonian cycle minus its largest edge always yields a minimum Hamiltonian path.

3-115. Analogous to the farthest-insertion algorithm (FITSP) developed and coded in the text, develop the nearest-insertion algorithm for obtaining an approximate solution of the TSP. Show that if the given network is undirected and the edge weights satisfy the triangle inequality, the weight of the approximate solution obtained with the nearest-insertion algorithm will never be more than twice that of the optimum. (*Hint:* For a proof, see Rosenkrantz et al. [1977].)

3-116. Describe and implement the heuristic developed by Christofides for an approximate solution of the TSP. Show that in the case of undirected networks satisfying the triangle inequality, the TS tours produced by this algorithm will never be of weights greater than 150% of the optimum.

3-117. Study the THREEOPT procedure and Fig. 3-47 carefully. Then, analogous to Algorithm 3-11 (which describes the 2-Opt TSP heuristic), give an algorithm in a Pascal-like language for the 3-optimal heuristic. Show which of the cases cause exchange of only two edges. Justify your selection of edge triplets (see also Problems 3-102 and 3-103).

3-118. *k*-person TSP.   In this section we considered only the one-person TSP, in which there is one traveling salesman who is required to visit all *n* nodes. In many real-life vehicle-routing problems and scheduling problems, there are *k* salesmen who must collectively visit all *n* nodes at a minimum total traveling expense. Fuel delivery by a fleet of vehicles is one such example. Perform a literature search and devise approximate algorithms for solving the *k*-person traveling salesman problem.

3-119. Transform the problem of finding a shortest TS route (which can visit a node more than once) to the problem of finding a shortest Hamiltonian cycle (which visits every node exactly once) in a certain modified network.

(*Hint:* Consider the network on the same set of nodes with the weights $[w_{ij}^*]$ being the lengths of the shortest paths in the original network.)

3-120. Show how the problems of (a) finding a shortest Hamiltonian cycle, and (b) finding a shortest Hamiltonian path in a network can be computationally transformed to each other.

REFERENCES AND REMARKS

It is widely believed that it was Karl Menger who made the first documented statement of the traveling salesman problem in

MENGER, K., *Ergebnisse eines Kolloquiums* 3(1930), 11–12

Independently, Hassler Whitney also mentioned the traveling salesman problem in 1934 in a lecture at Princeton University. In 1937, while Merrill Flood was working on a New Jersey school bus routing problem, the traveling salesman problem was pointed out to him by A. W. Tucker. This seems to be the first mention of the traveling salesman problem as a practical problem.

By the early 1950s, the traveling salesman problem had become a popular optimization problem, partly because of its simple statement and its reputation for computational difficulty. It has fascinated professionals and laypersons alike. Some 500 technical papers, book chapters, and research reports have been published in the past 30 years on the traveling salesman problem. For an early review of the TSP, see

FLOOD, M. M., The Traveling-Salesman Problem, *Oper. Res.* 4(1956), 61–75

A later survey appears in

BELLMORE, M., and G. L. NEMHAUSER, The Traveling Salesman Problem: A Survey, *Oper. Res.* 16(1968), 538–58

More recent surveys with emphasis on computational aspects are

BURKARD, R. E., Traveling Salesman and Assignment Problems: A Survey, *Ann. Discrete Math.* 4(1979), 193–215

CHRISTOFIDES, N., The Traveling Salesman Problem, in N. Christofides, A. Mingozzi, P. Toth, and C. Sandi (eds), *Combinatorial Optimization*, Wiley, New York, 1979, pp. 131–149

HOFFMAN, A. J., E. L. JOHNSON, P. WOLFE, and M. HELD, Aspects of the Traveling Salesman Problem, IBM Research Report RC-8787, IBM Thomas J. Watson Research Center, Yorktown Heights, N.Y., 1981

An up-to-date and comprehensive treatment of various aspects of the TSP can be found in the following collection of recent papers:

LENSTRA, J. K., E. L. LAWLER, and A. H. G. RINNOOY KAN, (eds.), *The Travelling Salesman Problem*, Wiley, New York, 1983

The first TSP of serious size was solved in 1954 by George Dantzig, Ray Fulkerson, and Selmer Johnson at the Rand Corporation. Using a hybrid method—a repetitive use of LP and of direct human intervention—they solved a symmetric TSP with 42 nodes consisting of road distances between 42 major cities in the United States. This work was reported in

DANTZIG, G. B., D. R. FULKERSON, and S. M. JOHNSON, Solution of a Large-Scale Travelling-Salesman Problem, *Oper. Res.* 2(1954), 393–410

Much of the fascination with the TSP is because of the fact that several other combinatorial problems (such as the assignment problem, minimum spanning tree problem) sounding much like TSP can be solved so much faster (in polynomial time).

Branch-and-bound techniques have been used successfully in optimization problems since the late 1950s. Several different branch-and-bound algorithms are possible for the TS problem. A survey of these is given in Bellmore and Nemhauser [1968]. See also

HOROWITZ, E., and S. SAHNI, *Fundamentals of Computer Algorithms*, Computer Science Press, Potomac, Md., 1978, pp. 403–416

The clever approach of reducing matrix to obtain the lower bounds was proposed in

LITTLE, J. D. C., K. G. MURTY, D. W. SWEENEY, and C. KAREL, An Algorithm for Traveling Salesman Problem, *Oper. Res.* 11(1963), 972–989

See also

REINGOLD, E. M., J. NIEVERGELT, and N. DEO, *Combinatorial Algorithms: Theory and Practice*, Prentice-Hall, Englewood Cliffs, N.J., 1977, pp. 121–126

An ALGOL implementation of the branch-and-bound algorithm for the TSP (asymmetric) is available in

KNÖDEL, W., Algorithm 7: Travelling Salesman Problem, *Computing* 3(1968), 151–156

A FORTRAN listing is given in

PHILLIPS, D. T., and A. GARCIA - DIAZ, *Fundamentals of Network Analysis*, Prentice-Hall, Englewood Cliffs, N.J., 1981, pp. 97–109, 444–447

An excellent survey of the branch-and-bound approach to the TSP and other early approaches to the TSP are given in

GOMORY, R. E., The Traveling Salesman Problem, *Proc. IBM Scientific Computing Symp. on Combinatorial Problems*, Mar. 1964, IBM Data Processing Division, White Plains, N.Y., 1966, pp. 93–121

Another branch-and-bound approach for the TSP and its expected complexity has been studied in depth in

SMITH, D. R., On the Computational Complexity of Branch and Bound Search Strategies, Ph.D. dissertation, Duke University, 1979 (available as Report NPS 52-70-004, Computer Science, Naval Post Graduate School, Monterey, Calif.)

Evidence is presented in this report that asymmetric TSP can be solved exactly in time $O(n^3 \log n)$ on the average.

Held and Karp developed a branch-and-bound algorithm for the TSP using the bound from a minimum spanning tree, which is reported in

HELD, M., and R. M. KARP, The Travelling Salesman and Minimum Spanning Trees, Part I, *Oper. Res.* 18(1970), 1138–1162; Part II, *Math. Programming* 1(1971), 6–26

Bounds from the solutions of the assignment problem, the matching problem, and the shortest n-path problem have also been suggested and explored. For a brief survey and references, see Christofides [1979].

Over the past three decades the TSP has remained the prototype of a "hard" combinatorial problem. Since the introduction of NP-completeness in 1971 and subsequent inclusion of the TSP in the NP-complete class, some of the mystery has gone out of the TSP. For a complete treatment of NP-completeness and its relationship to the TSP, see

GAREY, M. R., and D. S. JOHNSON, *Computers and Intractability: A Guide to the Theory of NP-Completeness*, W. H. Freeman, San Francisco, 1979

The NP-complete results have given a new impetus to heuristic methods for solving "hard" combinatorial problems. There seems to be little use at present in looking for a polynomially bounded algorithm for the general TSP. Much work has been done on fast approximate algorithms for the TSP. There is a trade-off between the speed of an algorithm and its ability to yield tours which are close to the optimal. The following two studies (among others) have compared the performances of a number of heuristics:

ADRABIŃSKI, A., and M. M. SYSŁO, Computational Experiments with Some Heuristic Algorithms for the Travelling Salesman Problem, TR.N-78, Institute of Computer Science, Wrocław University, Wrocław, Poland, 1980; also in *Zastos. Mat.* 18(1983), 91–95

GOLDEN, B., L. BODIN, T. DOYLE, and W. STEWART, JR., Approximate Traveling Salesman Algorithms, *Oper. Res.* 28(1980), 694–711

There are two broad classes of heuristics that have been found to be effective in obtaining approximate solutions to the traveling salesman problem: (1) incremental or insertion heuristics, in which the route is built by inserting one edge at a time chosen with some "greedy" (shortsighted) approach, and (2) edge-exchange or local search heuristics. In this chapter we have discussed both.

In the following paper Sahni and Gonzales showed that if the triangle inequality is not satisfied, the problem of finding an approximate solution for the TSP within any fixed bound ratio of the optimum is as hard as finding an exact solution:

SAHNI, S., and T. GONZALES, P-Complete Approximation Problems, *J. ACM* 23(1976) 555–565

However, Rosenkrantz et al. have applied worst-case analysis to several insertion heuristics for symmetric networks which satisfy the triangle inequality. They show that none of the insertion algorithms examined have a worst-case bound on approximate to optimum tour ratio less than 2. This bound is the same as that yielded by doubling up the edges in a minimum spanning tree. Their result is reported in

ROSENKRANTZ, D. J., R. E. STEARNS, and P. M. LEWIS, An Analysis of Several Heuristics for the Traveling Salesman Problem, *SIAM J. Comput.* 6(1977), 563–581

This bound of 2 is shown for the closest insertion algorithm. For the same class of networks, using an altogether different heuristic, Christofides developed an algorithm with a worst-case bound on approximate-to-optimum tour ratio of 3 : 2, in

CHRISTOFIDES, N., Worst-Case Analysis of a New Heuristic for the Traveling Salesman Problem, Technical Report TR-GS1A, Carnegie-Mellon University, Pittsburgh, Pa., 1976; also to appear in *Math. Programming*

Furthermore, it was shown in

CORNUÉJOLS, G., and G. L. NEMHAUSER, Tight Bounds for Christofides Travelling Salesman Heuristic, *Math. Programming* 14(1978), 116–121

that there are networks for which this worst-case ratio is achieved by Christofides's algorithm. Thus, the ratio of 3 : 2 is a tight bound.

The edge exchange strategy was first applied to the traveling salesman problem by Croes in the following paper:

CROES, A., A Method for Solving Traveling-Salesman Problems, *Oper. Res.* 5(1958), 791–812

He suggested the 2-optimal algorithm for the symmetric TSP. About the same time and independently, a 3-edge strategy was suggested by Bock in

BOCK, F., An Algorithm for Solving "Traveling-Salesman" and Related Network Optimization Problems, 14th National Meeting of the ORSA, St. Louis, Mo., Oct. 24, 1958

However, it was Lin who truly established through extensive empirical study that the 3-exchange algorithm was indeed an excellent approximation algorithm for the (symmetric) TSP. This study is reported in

LIN, S., Computer Solutions of the Traveling Salesman Problem, *Bell System Tech. J.* 44(1965), 2245–2269

The 2-optimal and 3-optimal algorithms presented in this section are slightly modified and somewhat simplified versions of the original algorithm given by Lin. A substantial improvement in implementation of the 3-optimal (in general, *r*-optimal) algorithm was given by Christofides and Eilon in

CHRISTOFIDES, N., and S. EILON, Algorithms for Large-Scale Traveling Salesman Problems, *Operational Res. Quart.* 23(1972) 511–518

Christofides and Eilon, with their faster 2-optimal and 3-optimal algorithms, were able to solve TS problems with 500 nodes in less than 3 minutes on the CDC-6600 computer.

Lin and Kernighan added another level of sophistication to the *r*-optimal algorithm. Instead of having a fixed value of 2 or 3, *r* was allowed to vary. The following paper of theirs is now a classic paper on the local-search heuristics for the symmetric TSP:

LIN, S., and B. W. KERNIGHAN, An Effective Heuristic Algorithm for the Travelling-Salesman Problem, *Oper. Res.* 21(1973), 498–516

Extension of the *r*-optimal approach to directed networks is explored in

KANELLAKIS, P., and C. H. PAPADIMITRIOU, Local Search for the Asymmetric Traveling Salesman Problem, *Oper. Res* 28(1980), 1086–1099

The complexity of the edge-exchange (or local-search) heuristic has been explored in

PAPADIMITRIOU, C. H., and K. STEIGLITZ, On the Complexity of Local Search for the Traveling Salesman Problem, *SIAM J. Comput.* 6(1977), 76–83

See also their book

PAPADIMITRIOU, C. H., and K. STEIGLITZ, *Combinatorial Optimization: Algorithms and Complexity*, Prentice-Hall, Englewood Cliffs, N.J., 1982, Chapter 19

In the following paper, Jenkyns has obtained bounds for solutions to the TSP using the theory of independence system:

JENKYNS, T. A., The Greedy Traveling Salesman Problem, *Networks* 9(1979), 363–373

For asymmetric TSP for which the triangle inequality is satisfied, the worst-case performances of various heuristics is derived in

FRIEZE, A. M., G. GALBIATI, and F. MAFFIOLI, On the Worst-Case Performance of Some Algorithms for the Asymmetric Traveling Salesman Problem, *Networks* 12(1982), 23–39

The largest symmetric TSP known to have been solved exactly is of size 318, using a combination of methods, including the Lin–Kernighan heuristic to find a good starting tour, and the 0–1 LP procedure with degree 2 constraint. The details are given in

CROWDER, H., and M. W. PADBERG, Solving Large-Scale Symmetric Traveling Salesman Problems to Optimality, *Management Sci.* 26(1980), 495–509

PADBERG, M. W., and S. HONG, On the Symmetric Travelling Salesman Problem: A Computational Study, *Math. Programming Studies* 12(1980), 78–107

The Hamiltonian path problem, as a back-plane wiring problem, was first posed and viewed as a degree-constrained minimum spanning tree problem in

DEO, N., and S. L. HAKIMI, The Shortest Generalized Hamiltonian Tree, in M. E. Van Valkenburg (ed.), *Proc. 3rd Annual Allerton Conf. on Circuit and System Theory*, Monticello, Ill., Oct. 20–22, 1965, pp. 879–888

An LP formulation of the problem was also given.

Several different types of the k-person TSP can be transformed to the standard TSP, see for instance

BELLMORE, M., and S. HONG, Transformation of the Multisalesmen Problem to the Standard Traveling Salesman Problem, *J. ACM* 21(1974), 500–504

Polynomial time approximation algorithms for some variants of the k-person TSP together with their evaluation appeared in

FREDERICKSON, G. N., M. S. HECHT, and C. E. KIM, Approximation Algorithms for some Routing Problems, *SIAM J. Comput.* 7(1978), 178–193

FRIEZE, A. M., An Extension of Christofides Heuristic to the k-person Travelling Salesman Problem, Dept. Comput. Science, Queen Mary College, University of London, 1980; submitted to *Discrete Appl. Math.*

Approximation algorithms for the traveling salesman curve problem have been proposed and evaluated in

JA'JA', J., and V. K. PRASANNA KUMAR, Approximation Algorithms for Several Variations of the Traveling Salesman Problem, TR CS-81-3, Dept. Comput. Science, The Pennsylvania State University, 1981

The Hamiltonian path problem as a wiring problem (together with three other interesting applications of the TSP) is also discussed in

LENSTRA, J. K., and A. H. G. RINNOOY KAN, Some Simple Applications of the Travelling Salesman Problem, *Operational Res. Quart.* 26(1975), 717–733

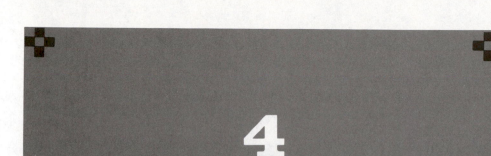

4

Coloring and Scheduling

The problems discussed in this chapter are concerned with assigning (scheduling) one class of objects to another class, provided that the assignment is feasible and that a certain objective is reached or at least approached.

Section 4.1 deals with the graph-coloring problem, which is a pure graph-theoretic scheduling model and computationally is one of the most intractable discrete problems. Not only is this problem NP-complete but also finding coloring close to optimal is NP-complete and most of the approximation algorithms behave very badly in the worst case. It is usually difficult to solve problems with 50 to 70 vertices using an implicit enumeration algorithm. We discuss two basic approaches to vertex coloring of a graph—independent set and sequential—and present implementations of three sequential algorithms, two of which are approximate. The results of extensive computational experiments with the codes presented and with several other coloring algorithms are also included.

Scheduling problems, discussed in Section 4.2, are much more complex models than the coloring problem. Roughly, we have to assign a number of tasks to a number of machines. There are usually several restrictions put on these two types of objects and on the relations between them. In particular, the tasks may be partially ordered by technological constraints. The goal is to find a schedule of tasks that optimizes a given performance measure of the task–machine system. Scheduling theory is one of the most rapidly developing branches of discrete optimization and applied combinatorics. We restrict our considerations to very simple classical models, and basic ideas and approaches. We illustrate that several very complex scheduling problems can be solved by relatively simple procedures and algorithms. An additional goal is to indicate some algorithmic relations between scheduling problems and other topics presented in the book.

4.1. GRAPH COLORING

In this section we consider the problem of coloring the vertices of a graph using a minimum number of colors, subject to the restriction that no two adjacent vertices get the same color. Therefore, in a colored graph, the vertices of the same color are mutually nonadjacent. Such a subset of vertices of a graph is called *independent* or *compatible*. In other words, the coloring problem of a graph G calls for a partition of the vertex set of G into a minimum number of independent subsets.

The coloring problem was originally formulated in the last century by cartographers who asked for a minimum number of colors to color a political map in such a way that no two neighboring countries (i.e., countries sharing a common boundary line which is not a single point) get the same color. They predicted that four colors always suffice (see the References and Remarks at the end of this section).

A contemporary application of the coloring problem is the scheduling problem of school examinations. Let $x_1, x_2, \ldots, x_n$ be the set of school courses that can be chosen by any student. In a schedule of examinations it is required that no two examinations be assigned to the same time period if there is a student registered for both of the corresponding courses. Although there always exists a schedule without conflicts in which each examination is assigned to a different time period, it is often desirable to determine a schedule involving either the minimum number or a given number (say, required by the school staff) of time periods.

This scheduling problem can be represented by a graph G in which vertices correspond to the school courses and two vertices are joined by an edge if the corresponding courses are taken by at least one of the students. Now the problem of scheduling the examinations with the fewest time periods or a given number k of periods is equivalent to the partitioning of the vertices of G into either the least or k independent subsets, respectively.

Thus the problem of scheduling examinations is equivalent to the graph-coloring problem stated in its pure form. However, in practical problems there are usually more restrictions generated by student and staff requests which have to be taken into consideration in finding a satisfactory timetable of examinations. For instance, because of space limitations, we may want a partition of examinations in which the number of examinations taking place in any one period is limited. Owing to such restrictions, the mathematical models are no longer simple coloring problems. It should be noted that for more general problems, the aspect of coloring of the graph vertices (i.e., unrestricted partition of the vertex set into independent subsets) becomes less significant.

Other scheduling problems and loading problems can also be defined as the graph coloring problem. We refer the reader to the pertinent literature and to Section 4.2, which is devoted entirely to scheduling problems.

4.1.1. DEFINITIONS AND BASIC PROPERTIES

Definitions

We refer the reader to Section 3.1 for basic graph-theoretic terms. Here we define only notions related to the coloring problem.

An assignment of colors (or elements of some set) to the vertices of a graph G, one color to each vertex, is called a *coloring* of G if adjacent vertices are colored by different colors. Sometimes we refer to this coloring of a graph G as a *complete* coloring of G, in contrast to a *partial* coloring of G, which is an assignment of colors not necessarily to all vertices of G. Coloring with k colors is called *k-coloring*. A graph G is *k-colorable* if there exists an *l*-coloring of G, where $l \leq k$. It is evident that every graph with n vertices is n-colorable.

G is assumed to be a simple graph that is a symmetric graph without multiple edges and self-loops, for the tuples of multiple edges may be replaced by single edges and a vertex with self-loops cannot be colored.

The minimum k for which a graph G is k-colorable is called the *chromatic number* of G and is denoted by $\chi(G)$. A graph G is *k-chromatic* if $\chi(G) = k$.

It is sometimes very useful to define a coloring of a graph $G = (V, E)$ using a *coloring function*. A function f determines a k-coloring of G if

$$f: V \overset{\text{on}}{\to} \{1, 2, \ldots, k\}$$

and if $(i, j) \in E$, then $f(i) \neq f(j)$ for all $(i, j) \in E$. A function that defines k-coloring is called the *k-coloring function*.

In coloring G, a set of vertices of the same color is called a *color class*. Such subsets of vertices of a graph in which no two vertices are joined by an edge are called *independent*. A subset W of vertices is a *maximal* independent set (MIS) if there is no independent set in which W is contained properly. A *maximum* independent set (MmIS) is a maximum cardinality independent set.

Every k-coloring function of a graph G defines a partition of the vertex set of G into k independent subsets, and vice versa. To see this, let f be a k-coloring function of G. Then the corresponding partition is defined as

$$V_i = \{v : f(v) = i\} \quad \text{for } i = 1, 2, \ldots, k$$

On the other hand, if $V = V_1 \cup V_2 \cup \ldots \cup V_k$ is a partition of V into subsets of mutually nonadjacent vertices, a k-coloring function of G can be defined as

$$f(v) = i$$

where i is such that $v \in V_i$.

Although these two definitions of a coloring are equivalent, it is sometimes more convenient to use one than the other. In this section we shall use k-coloring functions discussing sequential methods of coloring and set partitioning when some properties of independent sets of a graph are used.

Note that different coloring functions may generate the same color partition of the vertex set; such colorings are called *equivalent*. In what follows we show how to avoid producing equivalent colorings of a graph and colorings of subsets of vertices which may lead to equivalent colorings. Such partial colorings are called *redundant*.

Our main goal is to present basic computational methods for finding the chromatic number of a graph and a corresponding coloring of vertices together with the fundamental theoretic background.

For practical reasons we are interested in efficient methods for evaluating the chromatic number of a graph and for constructing a chromatic coloring (i.e., an optimal coloring). Unfortunately, in general, both problems are very difficult. There has been no algorithm proposed to solve the optimal coloring problem which runs in time bounded by a polynomial function in the number of vertices of a graph. It was proved that the problem of k-coloring of a graph belongs to the class of NP-complete problems. This is a collection of combinatorial problems that are equivalent in the sense that either each of them can be solved by a polynomial-time algorithm or none of them can.

The k-coloring problem remains difficult even in the presence of some strong restrictions put on a graph and integer k. For instance, the problem of finding whether a planar graph having no more than four edges incident with every vertex can be colored with three colors also has no polynomial-time algorithm (compare Problem 4-14).

In view of the complexity results, it is very unlikely that an exact method (i.e., one that always produces an optimal solution) can be constructed which could solve in a reasonable time the coloring problem for real-world graphs with hundreds of vertices. With this conclusion in mind, there have been proposed many simple heuristic coloring algorithms which produce colorings relatively quickly, although do not guarantee optimal solutions. While discussing approximation algorithms, we shall consider the proximity of their solutions to the optimal solution. More specifically, let $\chi_A(G)$ denote the number of colors used by algorithm A to color a graph G, and $\chi_A(n) = \max\{\chi_A(G)/\chi(G) : G$ has no more than n vertices$\}$. The function $\chi_A(n)$ is called the *goodness function* of A since it indicates how badly the algorithm A can behave. It is evident that $\chi_A(n) \leqslant n$ for every coloring algorithm A, but it is perhaps surprising that for many heuristic coloring algorithms, function $\chi_A(n)$ is linear in the number of vertices of a graph.

It is quite easy to determine the chromatic number for several classes of graphs. For example,

$$\chi(C_{2n}) = 2 \quad \text{and} \quad \chi(C_{2n+1}) = 3, \qquad n = 1, 2, \ldots$$

where C_i is the cycle with i vertices, and

$$\chi(K_n) = n$$

where K_n denotes the complete n-vertex graph. For every tree T with at least two vertices, $\chi(T) = 2$ [see Problem 4-4(a)].

A graph G is *k-partite* $(k \geqslant 1)$ if the set of vertices of G can be partitioned into k subsets $V_1, V_2, \ldots, V_k$ such that every edge of G joins vertices from different subsets. For $k = 2$, such graphs are called *bipartite*. It is easy to see that if G is k-partite, then $\chi(G) \leqslant k$. Therefore, every bipartite graph with at least one edge is 2-chromatic. On the other hand, if G is 2-chromatic, then the color classes define the 2-partition of the vertex set of G. D. König proved in 1936 (see Biggs et al. [1976]) that a graph G is bipartite if and only if it contains no cycle of odd length. Hence we can conclude that a nonempty graph G is 2-chromatic if and only if it contains no cycle of odd length [see Problem 4-4(b)]. Using this fact, it is easy to design an efficient algorithm for testing whether a graph is 2-chromatic (see Problem 4-6). Unfortunately, König's theorem has no generalization for $k > 2$.

Two Very Simple Bounds

Although the number of papers on the coloring problem exceeds that on any other graph problem, no formula has been found for the chromatic number of an arbitrary graph and we must thus be satisfied with bound estimates.

A very simple lower bound for the chromatic number of a graph G can be obtained by considering the largest subset of mutually adjacent vertices of G. Such a subset of vertices is called a *clique* of G. The *clique number* of G, denoted $\omega(G)$, is the number of vertices in the largest clique of G. Since in any coloring of G, all vertices of any clique of G must have different colors, we get

$$\omega(B) \leqslant \chi(G) \qquad (4\text{-}1)$$

Two comments are in order relative to this bound. First, no polynomial-time algorithm has been proposed for calculating the clique number of an arbitrary graph. Moreover, the problems of testing whether $\chi(G) \leqslant k$ and $\omega(G) \geqslant k$ hold are equivalent in the sense that a polynomial-time algorithm for one of them would provide a polynomial-time algorithm for solving the other. Second, it can be shown that for some graphs this bound can be very poor. Such graphs have been found by many authors, and the following construction is due to Mycielski (see Ore [1967]).

We construct a sequence of graphs $M_1, M_2, M_3, \ldots$ such that $\chi(M_i) = i$ and M_i contains no triangle; that is, $\omega(M_i) = 2$. Therefore, (4-1) can yield arbitrarily bad estimates for the chromatic number.

First, $M_1 = K_1$, $M_2 = K_2$ and for $k \geqslant 2$, M_{k+1} is constructed from M_k in the following way. Let us assume that M_k is a graph with p vertices, contains no triangle, and that $\chi(M_k) = k$. Graph M_{k+1} has $2p + 1$ vertices $v_1, v_2, \ldots, v_p, u_1, \ldots, u_p, w_{k+1}$, where $v_1, v_2, \ldots, v_p$ are the vertices of M_k and $u_1, u_2, \ldots, u_p$ are the copies of v_i's. The set of edges of M_{k+1} contains: (1) all the edges of M_k; (2) the edges that join u_i with neighbors of v_i for $i = 1, 2, \ldots, p$; and (3) (w_{k+1}, u_i) for $i = 1, 2, \ldots, p$. Figure 4-1 illustrates the construction of M_3 from M_2, and M_4 from M_3. It can be easily proved by induction on k that M_k is triangle-free and k-chromatic [Problem 4-10(a)]. The computational results presented in Section 4.1.5 show that the graphs constructed by Mycielski are very hard instances for several coloring algorithms.

Now we shall derive a very simple upper bound to the chromatic number of an arbitrary graph. Let $\Delta(G)$ denote the maximum degree of a vertex of G. We have

$$\chi(G) \leqslant \Delta(G) + 1 \qquad (4\text{-}2)$$

This inequality follows from the observation that if $\Delta(G) + 1$ colors are available, then at each vertex v of the graph G at least one of the colors can be used, since at most $\Delta(G)$ colors are used to color neighbors of v.

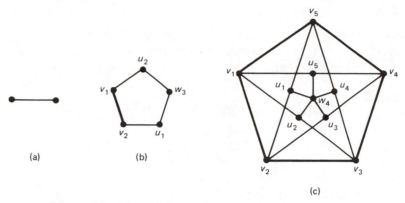

Figure 4-1 *Mycielski's graphs: (a) M_2; (b) M_3; (c) M_4.*

It was proved by Brooks (see Ore [1967]) that there are only two classes of graphs for which (4-2) holds with equality: odd cycles and complete graphs.

As in the case of the previous bound, there are also several classes of graphs for which bound (4-2) is not particularly good. For instance, every star $K_{1,n}$ (i.e., a graph consisting of n edges incident at one vertex) is 2-chromatic as a bipartite graph and $\Delta(K_{1,n}) + 1 = n + 1$. Thus the difference between $\chi(G)$ and $\Delta(G) + 1$ can be arbitrarily large. However, bound (4-2) differs from (4-1) in that it can be easily calculated in time bounded by $O(m)$, where m is the number of edges of a graph.

While presenting sequential algorithms we shall improve bound (4-2) by using certain theoretic and algorithmic tools.

4.1.2. INDEPENDENT SET APPROACH

As we have already pointed out, a k-coloring of a graph G is equivalent to a partition of the set of vertices of G into k independent sets $V_1, V_2, \ldots, V_k$ such that $V_i \cap V_j = \varnothing$ for $i \neq j, i, j = 1, 2, \ldots, k$, and $\cup_{i=1}^{k} V_i = V$. Such a partition is called a k-coloring partition of V.

An *independent set approach* to find a k-coloring of a graph is a method which first colors vertices with color 1, that is, finds set V_1, then finds set V_2, and so on. Such a method can evidently produce an optimal coloring for every graph G. It suffices to assume that the vertices of G are given in such an order that vertices of V_1' appear first, then those of V_2', and so on, where $V_1', V_2', \ldots$ are the color classes of an optimal coloring of G.

One may ask what are the properties of independent sets in k-coloring partitions of V in general and in an optimal partition, in particular. The following theorem has been proved by several authors.

Theorem 4-1

For every k-coloring partition $\{V_1, V_2, \ldots, V_k\}$ of a graph G there exists an l-coloring partition $\{V_1', V_2', \ldots, V_l'\}$ of G such that $l \leqslant k$ and at least one of the sets in the latter partition is a maximal independent set. ∎

The proof of this theorem is very simple. Let $\{V_1, V_2, \ldots, V_k\}$ be a k-coloring partition of V. If V_1 is not a maximal independent set of G, it can be augmented by vertices of the other subsets to become such a set V_1'. Now, evidently, every set $V_i - V_1'$ for $i = 2, 3, \ldots, k$ is independent, however, some of them may be empty. Therefore, it is clear that V_1' and $\{V_i' = V_i - V_1' : V_i' \neq \varnothing, i = 2, 3, \ldots, k\}$ form an l-coloring partition of G, $l \leqslant k$ and that V_1' is a maximal independent set.

Let $G_{|W}$ denote the subgraph of G induced by W, where $W \subseteq V$, that is, $G_{|W} = (W, E')$ and E' contains those edges of G which have both ends in W. By Theorem 4-1, in every optimal coloring of G, either one color class is maximal or can be augmented to a maximal set. Therefore, there exists a maximal independent subset U of the vertices of G such that

$$\chi(G) = \chi(G_{|V-U}) + 1$$

There is a finite number of maximal independent sets W in G. Therefore, minimizing over all such subsets, we obtain

$$\chi(G) = \min_{W \subseteq V} \chi(G_{|V-W}) + 1 \tag{4-3}$$

where we assume that $\chi(\varnothing) = 0$. Equation (4-3) is a basis for various algorithms for finding an optimal coloring of a graph, exact and approximate.

An Exact Method

An exact procedure for finding a chromatic partition of a graph by using the independent set approach is a backtracking method, as are most of the exact methods for coloring a graph. First, all maximal independent subsets of G are generated. Then the same step is repeated for $G_{|V-W}$ for every maximal independent subset W of G, and so on.

This method can be restated as follows. If a graph G is k-chromatic, then it can be colored with k colors, coloring first with color 1 a maximal independent subset W_1 of G, next coloring with another color a maximal independent set of $G_{|V-W_1}$, and so on, until all the vertices of G are colored. However, this method does not determine which MIS should be colored at each step; therefore, every MIS of a current subgraph must be considered.

Let $\mathcal{G}_l$, $l = 1, 2, \ldots$, denote the family of all maximal (in the set-theoretic sense) l-chromatic subgraphs of G. Notice that $\mathcal{G}_1$ is the family of all MISs of G and that $\chi(G)$ is the smallest l for which $G \in \mathcal{G}_l$. To find such l, we start with $\mathcal{G}_0 = \{\varnothing\}$ and apply the following algorithm due to Christofides [1975]. (For the sake of simplicity we assume that the empty set in $\mathcal{G}_0$ denotes the graph with the empty vertex set.)

Algorithm 4-1:   Christofides's Algorithm

```
begin
    k ← 0;
    𝒢₀ ← {∅};
    while there exists no graph H in 𝒢ₖ such that V(H) = V(G) do
    begin
        F ← ∅;
        for every H ∈ 𝒢ₖ do
            for every MIS W of G|V(G) − V(H) do F ← F ∪ ⟨G|V(H) ∪ W⟩;
        k ← k + 1;
        𝒢ₖ ← maximal graphs of F
    end
    {∗ k is the chromatic number of G ∗}
end
```

This algorithm can be improved in many ways; for instance, all nonmaximal elements of F can be excluded when they are created and the process of generation of new members of F can be stopped when a subgraph of G is found which is isomorphic to G. Other improvements are discussed in the references.

Example 4-1

Let us apply Christofides's algorithm to the graph G shown in Fig. 4-2. The vertex sets of the maximal l-chromatic subgraphs of G for $l = 1, 2$ are as follows:

$$\mathcal{G}_1 = \{\{1,3\}, \{1,6\}, \{2,4\}, \{3,4\}, \{3,5\}, \{4,6\}\}$$
$$\mathcal{G}_2 = \{\{1,3\} \cup \{2,4\}, \{1,3\} \cup \{4,6\}, \{1,6\} \cup \{2,4\}, \{1,6\} \cup \{3,4\},$$
$$\{1,6\} \cup \{3,5\}, \{2,4\} \cup \{3,5\}, \{3,5\} \cup \{4,6\}\}$$

Now using 2-chromatic subgraph $G_{|\{1,3\} \cup \{2,4\}}$, we generate 3-chromatic subgraphs on $\{1,3\} \cup \{2,4\} \cup \{5\}$ and $\{1,3\} \cup \{2,4\} \cup \{6\}$. Using the subgraph on $\{1,3\} \cup \{4,6\}$, we generate the subgraphs on $\{1,3\} \cup \{4,6\} \cup \{2\}$ and $\{1,3\} \cup \{4,6\} \cup \{5\}$. The subgraph on $\{1,6\} \cup \{2,4\}$ produces the subgraph on $\{1,6\} \cup \{2,4\} \cup \{3,5\}$, which contains all vertices of G; therefore, $\chi(G) = 3$ and $\{1,6\}$, $\{2,4\}$, and $\{3,5\}$ is one of the 3-chromatic partitions of G. ∎

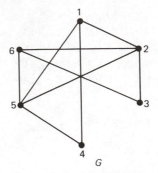

Figure 4-2 *Sample graph used to illustrate coloring algorithms.*

We now estimate the running time which is needed to solve (4-3) recursively in the worst case. Suppose that for a fixed subgraph G' of G, we have already found $\chi(G'')$ for all proper subgraphs G'' of G'. Therefore, the time required to compute $\chi(G')$ depends on the number of MIS's in G' and the time required to generate them. Let $r = |V(G')|$. We make use of the following results: The number of MISs in a graph on r vertices is not greater then $3^{r/3}$, and there exists an algorithm for generating all MISs of an r-vertex graph G in time $O(mrk)$, where $m = |E(G)|$ and k is the number of MISs in G. Thus the time to compute $\chi(G')$ is bounded by $O(mr3^{r/3})$. Since all subgraphs G' of G may be encountered, the overall running time required to solve (4-3) is bounded by a function of order

$$\sum_{r=0}^{n} \binom{n}{r} mr3^{r/3} < mn \sum_{r=0}^{n} \binom{n}{r} 3^{r/3} = mn(1 + 3^{1/3})^n \simeq mn2.445^n$$

Thus the coloring problem can be solved by an algorithm with the worst-case running time bounded by $O(mn2.445^n)$, where n and m are the number of vertices and the number of edges of a graph to be colored. The same approach can be used to derive time bounds for the problems of the existence of 3- and 4-coloring of a graph (see Problem 4-16).

Approximate Algorithms

An idea to generate a set of vertices of one color at a time yielded several approximate algorithms. We present here two algorithms which generate a sequence of independent subsets of a graph that form a color partition. The maximum independent set algorithm (MmIS algorithm) generates a sequence of color classes which are maximum cardinality independent subsets in subgraphs of uncolored vertices.

Algorithm 4-2: Maximum Independent Set Algorithm (MmIS Algorithm)

```
        begin
            U ← V;
            k ← 0;
            while U ≠ ∅ do
            begin
                k ← k + 1;
 *              find a maximum independent subset W in G|U;
                for each u ∈ W do f(u) ← k;
                U ← U − W
            end
            ⟨* k is the number of colors used by this algorithm to color the graph
                G = (V, E) *⟩
        end
```

Algorithm MmIS has a very serious drawback. The subproblem of finding a maximum cardinality independent subset in a graph, which must be solved to generate a new color class has also no polynomial-time algorithm. Therefore, the same arguments of inefficiency that forced us to abandon the search for an optimal coloring now apply to the MmIS algorithm.

Instead of using maximum independent subsets as color classes we may generate efficiently approximately maximum independent subsets. Let U be a subset of vertices of a graph G. The following procedure returns an approximately MmIS W contained in U.

Algorithm 4-3: Finding an Approximately Maximum Independent Subset of a Graph

```
    procedure AMmISet(U, W);
        begin
            U₁ ← U;
            W ← ∅;
            while U₁ ≠ ∅ do
            begin
                find vertex u of minimum degree in G|U₁;
                W ← W ∪ {u};
                U₁ ← U₁ − {u} − {v ∈ U₁ : (v, u) ∈ E(G)}
            end
        end
```

Starting with $W = \emptyset$, at each step, the set W is augmented by a minimum-degree vertex of the subgraph generated by the nonneighbors of W.

The approximately maximum independent set algorithm (AMmIS) is the MmIS algorithm with the line ($*$) replaced by AMmISet(U, W). It is easy to implement the AMmISet procedure to run in polynomial time (Problem 4-17); therefore, the AMmIS algorithm runs also in polynomial time.

Both algorithms appear to have a better worst-case behavior than any other heuristic method for graph coloring. It was proved that $\chi_{\text{MmIS}}(n)$ is $O(\log n)$ and $\chi_{\text{AMmIS}}(n)$ is $O(n/\log n)$. Although the values of both goodness functions can be arbitrarily large, they are much smaller bounds than that for other approximation algorithms.

Example 4-2

For the graph shown in Fig. 4-2, MmIS and AMmIS algorithms give the same 4-partition $\{1, 3\}$, $\{2, 4\}$, $\{5\}$, $\{6\}$. (In both algorithms, vertices and their subsets are considered in the order defined by the vertex labels.) ∎

Leighton [1979] proposed the *recursive-largest-first* algorithm (RLF), which combines the strategy of the LFS algorithm (described in Section 4.1.3) with the structure of the AMmIS algorithms. The RLF algorithm appears to be one of the best heuristic algorithms for coloring graphs.

The AMmIS and RLF algorithms have been implemented and compared with the other algorithms presented in the sequel (see Section 4.1.5).

4.1.3. APPROXIMATE SEQUENTIAL ALGORITHMS

Simple Sequential Algorithms

Let $v_1, v_2, \ldots, v_n$ be an ordering of vertices of a graph G. In a *sequential method* for coloring G, vertex v_i is added to the subgraph induced by already colored vertices $v_1, v_2, \ldots, v_{i-1}$ and a new coloring of $v_1, v_2, \ldots, v_{i-1}, v_i$ is determined. This step is repeated for $i = 1, 2, \ldots, n$, where for $i = 1$ the subgraph is empty. At each step, an attempt is made to use relatively small number of colors. In a basic sequential algorithm, the vertex v_i is assigned a color with the smallest number. Formally, this procedure can be described as follows. Suppose that the vertices of a graph $G = (V, E)$ have been ordered $v_1, v_2, \ldots, v_n$.

Algorithm 4-4:   Simple Sequential Algorithm(S)

```
begin
    f(v₁) ← 1;
    for i ← 2, 3, ..., n do
        f(vᵢ) ← min{k: k ⩾ 1 and f(vⱼ) ≠ k for every vⱼ (1 ⩽ j < i) adjacent to vᵢ}
end
```

Note that in the simple sequential algorithm while coloring v_i, the vertices colored before v_i retain their colors.

It is easy to determine an upper bound $u_S(G; v_1, v_2, \ldots, v_n)$ for the number of colors $\chi_S(G)$ used by the sequential algorithm applied to G and the ordering of its vertices $v_1, v_2, \ldots, v_n$. Every vertex v_i can be colored by color i, therefore $f(v_i) \leqslant i$. On the other hand, at least one of the first $\deg(v_i) + 1$ colors can be assigned to v_i. Hence

$$f(v_i) \leqslant \min\{i, \deg(v_i) + 1\}$$

for every $i = 1, 2, \ldots, n$, and thus

$$\chi_S(G) \leqslant u_S(G; v_1, v_2, \ldots, v_n) = \max_{1 \leqslant i \leqslant n} \min\{i, \deg(v_i) + 1\}. \qquad (4\text{-}4)$$

Note that this inequality was obtained without any assumption about the ordering of vertices of G.*

Example 4-3

The sequential algorithm applied to the graph G of Fig. 4-2 and the ordering of its vertices $1, 2, 3, 4, 5, 6$ produces the following coloring $f(1) = 1$, $f(2) = 2$, $f(3) = 1$, $f(4) = 2$, $f(5) = 3$, and $f(6) = 4$. In this case, $u_S(G; 1, 2, \ldots, 6) = \max\{1, 2, 3, 3, 5, 4\} = 5$. ∎

This example illustrates a typical behavior of the sequential algorithm. Although the bound (4-4) was derived from this algorithm, the actual number of colors used to color the vertices of G in a given order $v_1, v_2, \ldots, v_n$ is almost always less than $u_S(G; v_1, v_2, \ldots, v_n)$ (see Section 4.1.5).

One of the first versions of the sequential algorithm was proposed by Welsh and Powell [1967], who first ordered the vertices according to nonincreasing degree, $\deg(v_1) \geqslant \deg(v_2) \geqslant \ldots \geqslant \deg(v_n)$. Such an ordering is called the *largest-first* (LF) ordering and the sequential algorithm applied to a graph with such an ordering of vertices is called the *largest-first sequential algorithm* (LFS algorithm). The following theorem shows the superiority of an LF ordering over the other orderings when they are used in the sequential algorithm.

Theorem 4-2

Let $u_1, u_2, \ldots, u_n$ be an LF ordering of the vertices of a graph G and let us denote $u_{LF}(G) = u_S(G; u_1, u_2, \ldots, u_n)$. Then

$$u_{LF}(G) = \min u_S(G; v_1, v_2, \ldots, v_n)$$

where the minimum is taken over all orderings $v_1, v_2, \ldots, v_n$ of the vertices of G. ∎

---

*We warn the reader that inequality (4-4), in most of the references, is proved for the ordering of vertices according to nonincreasing degree. This assumption, however, is superfluous.

The theorem follows from the following observation. If for an ordering $v_1, v_2, \ldots, v_n$ there exists i ($1 \leqslant i \leqslant n - 1$) such that $\deg(v_{i+1}) > \deg(v_i)$, then $u_S(G; v_1, v_2, \ldots, v_i, v_{i+1}, \ldots, v_n) \geqslant u_S(G; v_1, v_2, \ldots, v_{i+1}, v_i, \ldots, v_n)$. The proof of this fact is left to the reader (Problem 4-18).

Example 4-4

The sequential algorithm applied to the graph of Fig. 4-2 and the LF ordering of its vertices $2, 5, 1, 6, 3, 4$ produces 3-coloring $f(2) = 1$, $f(5) = 2$, $f(1) = 3$, $f(6) = 3$, $f(3) = 2, f(4) = 1$.   ∎

The sequential algorithm may behave very badly in the worst case. Consider the family of graphs $G_{2n} = (V_{2n}, E_{2n})$, $n \geqslant 3$, shown in Fig. 4-3 and defined as follows:

$$V_{2n} = \{u_i, v_i : i = 1, 2, \ldots, n\}, \; E_{2n} = \{(u_i, v_j) : i, j = 1, 2, \ldots, n, i \neq j\}.$$

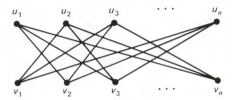

Figure 4-3   *Graph G_{2n}.*

Each G_{2n}, as a bipartite graph, is 2-chromatic. However, the sequential algorithm applied to G_{2n} and the ordering $u_1, v_1, u_2, v_2, \ldots, u_n, v_n$, which in fact is an LF ordering, defines $f(u_i) = f(v_i) = i$, for every $i = 1, 2, \ldots, n$. Thus $\chi_S(2n) \geqslant n$ and therefore the sequential algorithm has the worst growth rate possible.

As we have already seen, there is evidence that the sequential algorithm performs considerably better than the bound (4-4) suggests. A closer inspection of the algorithm and the proof of (4-4) reveals that for a given ordering $v_1, v_2, \ldots, v_n$ of the vertices of a graph G, instead of $f(v_i) \leqslant 1 + \deg(v_i)$ we have in fact $f(v_i) \leqslant 1 + \deg_i(v_i)$, where $\deg_i(v_i)$ denotes the degree of vertex v_i in the subgraph of G induced by $v_1, v_2, \ldots, v_i$. Therefore, the algorithm never requires more than $\max\{1 + \deg_i(v_i) : 1 \leqslant i \leqslant n\}$ colors; hence

$$\chi_S(G) \leqslant u_S'(G; v_1, v_2, \ldots, v_n) = 1 + \max_{1 \leqslant i \leqslant n} \deg_i(v_i)$$

The following procedure finds a vertex ordering which minimizes $u_S'(G; v_1, v_2, \ldots, v_n)$ (see Matula et al. [1972]):

1. v_n is a minimum degree vertex of G.
2.   For $i = n - 1, n - 2, \ldots, 2, 1, v_i$ is a minimum degree vertex in the subgraph of G induced by $V - \{v_n, v_{n-1}, \ldots, v_{i+1}\}$.

Such an ordering is called *smallest-last* (SL) since by the construction we have

$$\deg_i(v_i) = \min_{1 \leqslant j \leqslant i} \deg_i(v_j)$$

Let us denote

$$u_{\mathrm{SL}}(G) = 1 + \max_{1 \leqslant i \leqslant n} \min_{1 \leqslant j \leqslant i} \deg_i(v_j)$$

where $v_1, v_2, \ldots, v_n$ is an SL ordering of the vertices of G. We have

$$\chi_{\mathrm{S}}(G) \leqslant u_{\mathrm{SL}}(G) \tag{4-5}$$

It is left to the reader [Problem 4-20(a)] to show that the bound (4-5) implies the following bound, known as the Szekeres–Wilf inequality:

$$\chi(G) \leqslant 1 + \max_{H \subseteq G} \min_{v \in V(H)} \deg_H(v) \tag{4-6}$$

where the maximum is over all subgraphs H of G and $\deg_H(v)$ denotes the degree of v in H.

It is also easy to prove that

$$u'_{\mathrm{S}}(G; v_1, v_2, \ldots, v_n) \leqslant u_{\mathrm{S}}(G; v_1, v_2, \ldots, v_n)$$

for every ordering $v_1, v_2, \ldots, v_n$ of the vertices of G and that

$$u_{\mathrm{SL}}(G) \leqslant u_{\mathrm{LF}}(G) \tag{4-7}$$

Despite the last inequality, which may suggest that the SLS algorithm is superior to the LFS algorithm, the former shares several properties with the latter. In both cases, the number of colors used is much smaller than the value of the corresponding bound (see Section 4.1.5) and the goodness function is a linear function in the number of vertices of a graph.

Inequality (4-6) is very useful for bounding the chromatic number of graphs which contain subgraphs of bounded minimum degree. For instance, it is well known that every planar graph has a vertex of degree not exceeding 5. Since all subgraphs of a planar graph are planar, bounds (4-5) and (4-6) show that the SLS algorithm colors every planar graph with no more than six colors (see Problem 4-27).

There are several other heuristic algorithms for graph coloring which are sequential in nature. One of them, called *saturation-largest-first* algorithm, due to Brélaz [1979] is the subject of Problem 4-24. Some computational experiments with sequential algorithms are reported in Section 4.1.5.

After presenting implementations of ordering and simple sequential algorithms, we shall describe a modification of the sequential algorithm which at each step of coloring, if a new color is to be used, tries to change a partial coloring to free a color for currently colored vertex.

**Computer Implementation
of Ordering Algorithms**

The following integer function ORDERING finds either a largest-first (if BOOL = FALSE) or a smallest-last ordering (if BOOL = TRUE) of vertices of a graph and its value is equal to the bound resulting from the obtained ordering, that is, is equal either to (4-4) or to (4-5), respectively.

The algorithms applied in function ORDERING to find both orderings guarantee the optimal time complexity of the function within a constant factor. A radix sort algorithm is used to find the former ordering. Since the vertex degrees are from the range 0 and $N - 1$, where N is the number of vertices of a graph, a radix sort algorithm can be implemented to run in time $O(N)$. We leave to the reader the details of the implementation, which can easily be seen from the Pascal code.

The second part of the algorithm is executed when BOOL = TRUE and generates a smallest-last ordering of vertices starting with the largest-first one found in the first part of the function. In the beginning, the last element of the former ordering is also the last one of the latter. Then we remove this element from the graph and modify the sequence of the other vertices so that they still are in a nonincreasing order. This is done by changing the positions (if necessary) of the neighbors of the vertex which has been removed most recently. Therefore, at each step, the number of operations performed is proportional to $\deg_i(v_i)$ and it results in the time complexity $O(M)$ of the entire algorithm when BOOL = TRUE.

In the function ORDERING, the graph is represented in the linked adjacency list form (see the example below).

Global Constant

N   number of vertices of a graph to be colored

Data Types

```
TYPE  ARRN      = ARRAY[1..N] OF INTEGER;
      ARR0N     = ARRAY[0..N] OF INTEGER;
      VERTPOINT = @VERTLIST;
      GRAPH     = ARRAY[1..N] OF
                     RECORD
                        DEGREE,COLOR :INTEGER;
                        ADJLIST         :VERTPOINT
                     END;
      VERTLIST  = RECORD
                     VERTEX :INTEGER;
                     NEXT   :VERTPOINT
                  END;
```

Comment: In our Pascal codes, character @ stands for the pointer arrow ↑.

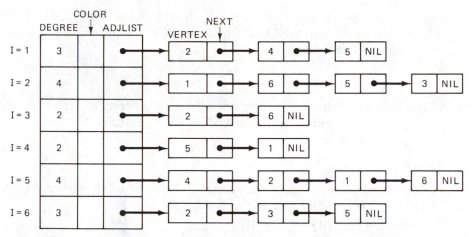

Figure 4-4 *Representation of the graph in Fig. 4-2 in the functions* ORDERING *and* SEQCOLORING.

The structure of a variable of type **GRAPH** is illustrated in Fig. 4-4 for the graph of Fig. 4-2. Three fields are assigned to each vertex I:

DEGREE   contains the vertex degree.
COLOR   is the color number of I.
ADJLIST   is a linked list of neighbors of I.

Field COLOR is not used in function ORDERING, and is used in function SEQCOLORING for which type GRAPH is identical.

Function Parameters

```
FUNCTION ORDERING(
    N       :INTEGER;
    BOOL    :BOOLEAN;
    VAR GR  :GRAPH;
    VAR SEQ :ARRN):INTEGER;
```

Input

N   number of vertices of a graph to be colored
BOOL   Boolean variable such that if BOOL = TRUE, then the function finds a smallest-last ordering and a largest-first ordering, otherwise

GR   variable of type GRAPH which represents the graph; fields DEGREE and ADJLIST must be set before calling the function (see the example below)

Output

SEQ[1..N]   array which contains a corresponding ordering of vertices of the graph, either largest-first if BOOL = FALSE or smallest-last, otherwise

Example

For the graph shown in Fig. 4-2 we have N = 6 and the graph is represented in function ORDERING as illustrated in Fig. 4-4.

The solution obtained is as follows:

BOOL	Value of ORDERING	SEQ[1..6]
FALSE	4	[5, 2, 6, 1, 4, 3]
TRUE	3	[2, 5, 1, 6, 4, 3]

Pascal Function ORDERING

```
FUNCTION ORDERING(
    N       :INTEGER;
    BOOL    :BOOLEAN;
    VAR GR :GRAPH;
    VAR SEQ:ARRN):INTEGER;

VAR BOUND,I,I1,I2,J1,J2,J3,J4,J5:INTEGER;
    INN,D                       :ARRN;
    RAD,RADIX                   :ARRON;
    P                           :VERTPOINT;
BEGIN
  (* FUNCTION ORDERING ORDERS VERTICES OF THE GRAPH.
     IF BOOL=TRUE THEN SEQ[1], SEQ[2],...,SEQ[N] IS A SMALLEST
     LAST ORDERING AND IF BOOL=FALSE THEN IT IS A LARGEST-FIRST
     ORDERING. *)
  (* RADIX METHOD IS APPLIED TO THE DEGREE SEQUENCE
     OF THE GRAPH TO FIND ITS LARGEST-FIRST ORDERING *)
    FOR I:=0 TO N DO RADIX[I]:=0;
    FOR I:=1 TO N DO
    BEGIN
        I1:=GR[I].DEGREE;
        RADIX[I1]:=RADIX[I1]+1
    END;
    I1:=0;
    I:=N;
    WHILE I1 < N DO
```

```
BEGIN
    I:=I-1;
    I1:=RADIX[I]+I1;
    RADIX[I]:=I1;
    RAD[I]:=I1
END;
FOR I:=1 TO N DO
BEGIN
    I1:=GR[I].DEGREE;
    I2:=RADIX[I1];
    INN[I]:=I2;
    SEQ[I2]:=I;
    RADIX[I1]:=I2-1
END;
(*   SEQ[1],SEQ[2],...,SEQ[N]  IS THE SEQUENCE OF VERTICES
     ORDERED BY NON-INCREASING DEGREE   *)
BOUND:=1;
IF BOOL THEN
BEGIN   (*  FINDING A SMALLEST-LAST ORDER *)
    FOR I:=1 TO N DO
        D[I]:=GR[I].DEGREE;
    FOR I:=N DOWNTO 3 DO
    BEGIN
        I1:=SEQ[I];
        J1:=D[I1];
        IF J1 > BOUND THEN BOUND:=J1;
        RAD[J1]:=I-1;
        P:=GR[I1].ADJLIST;
        WHILE P <> NIL DO
        BEGIN
            J1:=P@.VERTEX;
            J2:=INN[J1];
            IF J2 < I THEN
            BEGIN
                J3:=D[J1];
                J4:=RAD[J3];
                J5:=SEQ[J4];
                SEQ[J2]:=J5;
                SEQ[J4]:=J1;
                INN[J5]:=J2;
                INN[J1]:=J4;
                RAD[J3]:=J4-1;
                D[J1]:=J3-1
            END (* J2 < I *);
            P:=P@.NEXT
        END (* LIST OF NEIGHBORS OF I1 *)
    END (* I=N,N-1,...,3 *);
    BOUND:=BOUND+1
END (* BOOL = TRUE *)
ELSE
    FOR I:=2 TO N DO
```

```
    BEGIN
        J1:=GR[SEQ[I]].DEGREE+1;
        IF J1 > I THEN J1:=I;
        IF J1 > BOUND THEN BOUND:=J1
    END ;
  ORDERING:=BOUND
END (* ORDERING *);
```

**Computer Implementation of the Simple
Sequential Algorithm**

The integer function SEQCOLORING is a straightforward implementation of the simple sequential algorithm. The value of this function is the number of colors used by the algorithm, and additionally the corresponding vertex colors are stored in fields COLOR of each vertex record of array GR.

A graph is represented in the linked adjacency list form.

It is easy to check that function SEQCOLORING colors a graph with M edges in $O(M)$ time.

Global Constant and Data Types

The same as for function ORDERING.

Function Parameters

```
FUNCTION SEQCOLORING(
     N        :INTEGER;
     VAR GR   :GRAPH;
     VAR SEQ :ARRN):INTEGER;
```

Input

N	number of vertices of a graph to be colored
GR	variable of type GRAPH which represents the graph; fields DEGREE and ADJLIST must be set before calling the function
SEQ[1..N]	array which contains an ordering of vertices of the graph in which they are to be colored

Output

GR[I].COLOR   color of vertex I, for $I = 1, 2, \ldots, N$

Example

For the graph in Fig. 4-2 we have N = 6, and the representation of the graph and values of SEQ elements are given in the Example for function ORDERING.

The solution obtained is as follows:

SEQ[1..6]	Value of SEQCOLORING	COLORs of Vertices $1, 2, \ldots, 6$
$[5, 2, 6, 1, 4, 3]$	3	$[1, 2, 3, 3, 2, 1]$
$[2, 5, 1, 6, 4, 3]$	3	$[1, 2, 3, 3, 1, 2]$

Pascal Function SEQCOLORING

```
FUNCTION SEQCOLORING(
   N     :INTEGER;
   VAR GR :GRAPH;
   VAR SEQ:ARRN):INTEGER;


VAR I,I1,I2,K,L:INTEGER;
    COL         :ARRN;
    P           :VERTPOINT;
BEGIN
  (* COLORING BY THE SEQUENTIAL ALGORITHM. VERTICES ARE
     PROCESSED IN THE ORDER GIVEN IN ARRAY SEQ. *)
  FOR I:=1 TO N DO
     GR[I].COLOR:=0;
  K:=1;
  GR[SEQ[1]].COLOR:=1;
  FOR I:=2 TO N DO
  BEGIN
     I1:=SEQ[I];
     I2:=GR[I1].DEGREE+1;
     IF I < I2 THEN I2:=I;
     FOR L:=1 TO I2 DO COL[L]:=0;
     P:=GR[I1].ADJLIST;
     WHILE P <> NIL DO
     BEGIN
        L:=GR[P@.VERTEX].COLOR;
        IF L <> 0 THEN COL[L]:=1;
        P:=P@.NEXT
     END;
     L:=1;
     WHILE COL[L] <> 0 DO L:=L+1;
     GR[I1].COLOR:=L;
     IF K < L THEN K:=L
  END (* I=2,3,...,N *);
  SEQCOLORING:=K
END; (* SEQCOLORING *)
```

Sequential with Interchange Algorithm

The sequential algorithm can be improved by rearranging already used colors when a new color must be added. Let us assume that $v_1, v_2, \ldots, v_{i-1}$ have been colored by k colors and the simple sequential algorithm colors v_i with color $k + 1$. It means that among the vertices $v_1, \ldots, v_{i-1}$ there are neighbors of v_i colored with $1, 2, \ldots, k$. If there exists a complete subgraph on k vertices in the subgraph generated by the neighbors of v_i, then the new color is necessary. Otherwise, it may be possible to interchange the colors of some neighbors of v_i, preserve the k-coloring of the subgraph generated by $v_1, v_2, \ldots, v_{i-1}$, and free one of the first k colors for v_i.

Let G_{pq} be the subgraph of G induced by the vertices colored with color p or q, $1 \leqslant p, q \leqslant k$. The rearrangement of colors which can free a color depends on interchanging colors p and q, for some vertices in G_{pq} and for some p and q. First note that if G_{pq} has two vertices u and v of different colors, both adjacent to v_i and both belonging to the same connected component of G_{pq}, then no interchange of p and q can free one of these two colors for v_i. Therefore, if this holds for every p and q $(1 \leqslant p, q \leqslant k, p \neq q)$, we must set $f(v_i) = k + 1$. On the other hand, if there exist p and q such that in every connected component of G_{pq} the vertices adjacent to v_i are of at most one color, then the following procedure, called the (p, q)-*interchange*, can be applied.

Let U_p be the set of vertices of those connected components of G_{pq} which contain a neighbor v of vertex v_i such that $f(v) = p$. Now for every $u \in U_p$, if $f(u) = p$, set $f(u) = q$, and if $f(u) = q$, set $f(u) = p$.

It is easy to see that the resulting function f is still a coloring function of the vertices $v_1, v_2, \ldots, v_{i-1}$ and that there is no vertex of color p adjacent to v_i. Therefore, we can set $f(v_i) = p$. The whole procedure can be described as follows. Suppose that the vertices of a graph G are ordered as $v_1, v_2, \ldots, v_n$.

Algorithm 4-5:   Sequential with Interchange Algorithm (SI)

```
begin
    f(v₁) ← 1;
    k ← 1;
    for i ← 2, 3, …, n do
    begin
        g ← min{h: h ⩾ 1 and f(vⱼ) ≠ h for every vⱼ (1 ⩽ j ⩽ i − 1) adjacent to vᵢ};
        if g ⩽ k then f(vᵢ) ← g
        else
            if (p, q)-interchange of colors can be performed for
                    some p and q such that 1 ⩽ p, q ⩽ k, p < q then
            begin rearrange colors of G_pq; f(vᵢ) ← p end
            else begin f(vᵢ) ← g; k ← k + 1 end
    end
end
```

Example 4-5

Let us now apply the sequential with interchange algorithm to the graph shown in Fig. 4-3 and ordering $u_1, v_1, u_2, v_2, \ldots, u_n, v_n$. The first four vertices are colored similarly as in the simple sequential algorithm [see also Problem 4-23(b)]. In the next step, that is, while coloring u_3, we have $g = 3$ and since $k = 2$ we check whether the $(1, 2)$-interchange of colors is possible. The subgraph $G_{1,2}$ consists of two connected components and in both the neighbors of u_3 are monochromatic (see Fig. 4-6). It is therefore possible to free color 1 or 2 for u_3. Finally, we obtain $f(u_1) = f(u_2) = f(u_3) = 1$ and $f(v_1) = f(v_2) = 2$, which leads to the following 2-coloring of the graph $f(u_i) = 1$ and $f(v_i) = 2$, for every $i = 1, 2, \ldots, n$, without any other recoloring of the vertices in the next steps. ∎

It is obvious that the sequential with interchange algorithm can be applied to any ordering of vertices of a graph, in particular, for a LF and SL orderings (see the computational results in Section 4.1.5).

We leave it to the reader to verify that the SI algorithm can never use more than two colors on bipartite graphs [Problem 4-22(b)]; however, for graphs with chromatic number greater than 2 there exists a family of graphs for which the goodness function of this algorithm is linear in n.

Although the SI algorithm was designed to improve the S algorithm, there exists no proof that SI always uses at most as many colors as S. In fact, there exist graphs and orderings of their vertices for which an interchange of two colors at an early step (the reduction of the number of colors used in a subgraph) can result in algorithm SI using more colors than algorithm S. (Such a graph with 45 vertices and their ordering where found by the authors while generating random graphs and random orderings!)

As pointed out in the preceding section, the SLS algorithm colors every planar graph with at most six colors. It can be proved (see Matula et al. [1972]) that the SLSI algorithm colors every such graph with at most five colors. The proof of this fact reminds us of that of the 5-colorability of planar graphs.

Computer Implementation of the Sequential with Interchange Algorithm

The value of integer function INTERSEQCOLORING is the number of colors used by the sequential with interchange algorithm applied to a graph represented by variable GR of type GRAPH and an ordering of its vertices given in array SEQ.

The sequential with interchange algorithm works similarly to the sequential algorithm except when a new color is introduced. In this case, it is checked if there exists a subgraph G_{pq} induced by the vertices of colors p and q such that no connected component of G_{pq} has more than one color vertex adjacent to the vertex that is to be colored.

First, procedure MATES is called which links together the two instances of each edge; that is, each neighbor j of vertex i is linked to neighbor i of vertex j. Such a structure, when an edge is deleted, also allows us to delete the other one without additional searching. We leave to the reader to verify that procedure MATES works correctly and needs $O(M)$ time [see Problem 4-23(a)].

Then a sequential algorithm is applied to a given in array SEQ sequence of vertices $v_1, v_2, \ldots, v_n$. If a new color $g = k + 1$ is to be assigned to vertex v_i [see Problem 4-23(b)], it is checked whether there exists a subgraph G_{pq} $(1 \leqslant p, q < k$, $p < q)$ suitable for the (p, q)-interchange of colors. The data structures used in this function are such that the time spent on identifying G_{pq} is proportional to the number of edges in G_{pq}. Since each edge of the graph may belong to at most one subgraph G_{pq}, the total time of verifying whether there exists an (p, q)-interchange of colors is bounded by $O(M)$. If such a subgraph G_{pq} exists, the corresponding interchange of colors and modification of the entries of fields COLPOINT, COL-MATE1, and COLMATE2 for the vertices in G_{pq} and their neighbors is also done in $O(M)$ time. Therefore, the time complexity of the INTERSEQCOLORING function is $O(LM)$, where L is the number of colors used or $O(NM)$, since $L \leqslant N$.

Global Constants

N, M, HI   see the input below

Data Types

```
TYPE  ARRN        = ARRAY[1..N] OF INTEGER;
      ARR0N       = ARRAY[0..N] OF INTEGER;
      ARRM        = ARRAY[1..M] OF INTEGER;
      VERTPOINT   = @VERTLIST;
      ARRNPOINT   = ARRAY[1..N] OF VERTPOINT;
      ARRMPOINT   = ARRAY[1..M] OF VERTPOINT;
      GRAPH       = ARRAY[1..N] OF
                      RECORD
                          DEGREE,COLOR :INTEGER;
                          ADJLIST           :VERTPOINT;
                          COLPOINT          :ARRAY[1..HI] OF VERTPOINT
                      END;
      VERTLIST    = RECORD
                          VERTEX                             :INTEGER;
                          NEXT,COLMATE1,COLMATE2,MATE :VERTPOINT
                      END;
```

Comment: In our Pascal codes, character @ stands for the pointer arrow ↑.

The structure of a variable of type GRAPH is illustrated in Fig. 4-5 for graph G_6 of

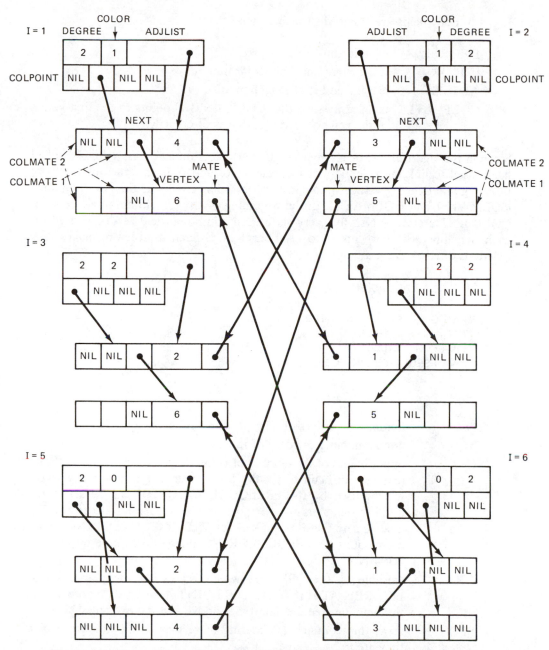

Figure 4-5 *Representation of G_6 in the* INTERSEQCOLORING *function after coloring the first four vertices.*

Fig. 4-3. Four fields are assigned to each vertex I:

DEGREE   contains the vertex degree.

COLOR   is a current and final color number of I.

ADJLIST   is a linked list of neighbors of I.

COLPOINT   is an array such that COLPOINT[K] points to the first vertex in the neighbor list of I that is of color K.

Since in the algorithm vertices may change their colors many times, a doubly linked list is used to link the neighbors of I of each color and to enable the operations of inserting and deleting vertices in a constant time. The entries of this list are named COLMATE1 and COLMATE2. In the beginning of the algorithm, all elements of COLPOINT are set to NIL for every vertex. An additional field MATE is used for each neighbor J of I to link it to neighbor I of J. Such a structure makes easy deletions of edges from a graph.

Function Parameters

```
FUNCTION INTERSEQCOLORING(
    N,M,HI   :INTEGER;
    VAR GR   :GRAPH;
    VAR SEQ  :ARRN):INTEGER;
```

Input

N   number of vertices of a graph to be colored

M   double number of edges of the graph

HI   upper bound of the number of colors used by the algorithm (Since it is very difficult to predict for a given graph how many colors will be used by INTERSEQCOLORING, HI can be set to the number of colors used by the simple sequential algorithm. We assume here that it is unlikely for INTERSEQCOLORING to use more colors than SEQCOLORING does. In any case we may set HI = N.)

GR   variable of type GRAPH which represents the graph. Fields DEGREE, ADJLIST, and COLPOINT must be set before calling the function (see the data types above and the example below)

SEQ[1..N]   array that contains an ordering of vertices of the graph in which they are to be colored

Output

GR[I].COLOR   color of vertex I, for I = 1, 2, ..., N

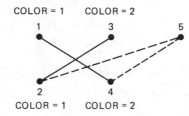

Figure 4-6 *Subgraph $G_{1,2}$ of the graph G_6 after coloring the first four vertices.*

Example

Let us consider the graph G_{2n} of Fig. 4-3 for $n = 3$. Assume that vertices u_1, v_1, u_2, v_2, u_3, and v_3 are numbered 1, 2, 3, 4, 5, and 6. In this case N = 6, M = 18, and we may assume that HI = 4, since $\Delta(G_6) = 3$. Figure 4-5 shows the representation of G_6 in the INTERSEQCOLORING function after vertices 1, 2, 3, and 4 have been colored.

The solution obtained: When vertex 5 is to be colored, the simple sequential approach finds that it has to get color 3. Since it is a new color, the sequential with interchange algorithm checks whether (1, 2)-interchange of colors is possible. The graph $G_{1,2}$ is shown in Fig. 4-6. Since no connected component of $G_{1,2}$ has more than one color vertex adjacent to 5, (1, 2)-interchange can be applied. Therefore, vertex 2 gets color 2 and vertex 3 gets color 1. Now we may assign color 1 to vertex 5 and then vertex 6 will get color 2.

Pascal Function INTERSEQCOLORING

```
FUNCTION INTERSEQCOLORING(
   N,M,HI :INTEGER;
   VAR GR :GRAPH;
   VAR SEQ:ARRN):INTEGER;



TYPE PQUEUE=@QUEUE;
    QUEUE =RECORD
               VERTEX:INTEGER;
               NEXT  :PQUEUE
            END;
VAR H,H1,H2,I,I1,I2,J,J1,K,KK,K1,L,LL:INTEGER;
    EX                          :BOOLEAN;
    SGN                         :ARRN;
    HQ,PQ,PQ1                   :PQUEUE;
    P,P1,P2,P3                  :VERTPOINT;
PROCEDURE MATES;
(* THIS PROCEDURE MATCHES INTO PAIRS ELEMENTS
   OF ADJACENCY LISTS ( VERTLIST ) WHICH CORRESPOND
   TO THE SAME EDGE OF THE GRAPH *)
```

```
VAR DEGG:ARRN;
    ARR1:ARRM;
    ARR2:ARRMPOINT;
    ARR3:ARRNPOINT;
BEGIN
    SGN[1]:=1;
    I1:=GR[1].DEGREE;
    DEGG[1]:=I1;
    FOR I:=2 TO N DO
    BEGIN
        SGN[I]:=I1+1;
        I1:=I1+GR[I].DEGREE;
        DEGG[I]:=I1
    END;
    I1:=1;
    FOR I:=1 TO N DO
    BEGIN
        I2:=DEGG[I];
        P:=GR[I].ADJLIST;
        FOR J:=I1 TO I2 DO
        BEGIN
            L:=P@.VERTEX;
            K:=SGN[L];
            ARR1[K]:=I;
            ARR2[K]:=P;
            SGN[L]:=K+1;
            P:=P@.NEXT
        END;
        I1:=I2+1
    END (*  I  *);
    I1:=1;
    FOR I:=1 TO N DO
    BEGIN
        I2:=DEGG[I];
        FOR J:=I1 TO I2 DO
            ARR3[ARR1[J]]:=ARR2[J];
        P:=GR[I].ADJLIST;
        FOR J:=I1 TO I2 DO
        BEGIN
            P@.MATE:=ARR3[P@.VERTEX];
            P:=P@.NEXT
        END;
        I1:=I2+1
    END (*  I  *)
END (*  MATES  *);
PROCEDURE INSERT(I:INTEGER);
(* THIS PROCEDURE INSERTS VERTEX P1@.VERTEX IN
   LIST OF I-COLOR NEIGHBORS OF VERTEX J *)
```

```
BEGIN
   P2:=GR[J].COLPOINT[I];
   GR[J].COLPOINT[I]:=P1;
   P1@.COLMATE1:=P2;
   P1@.COLMATE2:=NIL;
   IF P2 <> NIL THEN
      P2@.COLMATE2:=P1
END   (*  INSERT  *);
BEGIN
(* BODY OF PROCEDURE INTERSEQCOLORING *)
   MATES;
   FOR I:=1 TO N DO
   BEGIN  (* INITIALIZATION *)
      GR[I].COLOR:=0;
      FOR J:=1 TO HI DO
      GR[I].COLPOINT[J]:=NIL
   END;
   K:=1;
   K1:=1;
   FOR I:=1 TO N DO
   BEGIN
      I1:=SEQ[I];
(* I1-VERTEX TO BE COLORED *)
      IF I > 1 THEN
      BEGIN
         I2:=GR[I1].DEGREE+1;
         IF I < I2 THEN I2:=I;
         FOR L:=1 TO I2 DO SGN[L]:=0;
         P:=GR[I1].ADJLIST;
         WHILE P <> NIL DO
         BEGIN
            L:=GR[P@.VERTEX].COLOR;
            IF L <> 0 THEN SGN[L]:=1;
            P:=P@.NEXT
         END;
         K1:=1;
         WHILE SGN[K1] <> 0 DO K1:=K1+1;
     (*   K1 IS THE SMALLEST COLOR NUMBER WHICH
          CAN BE USED TO COLOR VERTEX I1 *)
         IF (K1 > K) AND ((I > 4) OR ((I = 4) AND (K = 2))) THEN
         BEGIN
         (* SEARCH FOR INTERCHANGE OF COLORS *)
            EX:=TRUE;
            KK:=0;
            WHILE EX AND (KK < K) DO
            BEGIN
               KK:=KK+1;
               HQ:=NIL;
               P:=GR[I1].COLPOINT[KK];
               WHILE P <> NIL DO
```

```
              BEGIN
                  NEW(PQ1);
                  PQ1@.NEXT:=HQ;
                  HQ:=PQ1;
                  PQ1@.VERTEX:=P@.VERTEX;
                  P:=P@.COLMATE1
              END;
              LL:=KK;
              WHILE EX AND (LL < K) DO
              BEGIN
                  FOR J:=1 TO N DO SGN[J]:=0;
                  PQ:=HQ;
                  WHILE PQ <> NIL DO
                  BEGIN
                      SGN[PQ@.VERTEX]:=1;
                      PQ:=PQ@.NEXT
                  END;
                  LL:=LL+1;
                  PQ:=HQ;
                  WHILE PQ <> NIL DO
                  BEGIN
                  (* CONSTRUCTION OF (KK,LL)-SUBGRAPH *)
                      H:=PQ@.VERTEX;
                      H1:=SGN[H];
                      IF GR[H].COLOR = KK THEN H2:=LL ELSE H2:=KK;
                      P:=GR[H].COLPOINT[H2];
                      WHILE P <> NIL DO
                      BEGIN
                      (* SEARCH FOR H2-COLOR NEIGHBORS OF VERTEX H *)
                          I2:=P@.VERTEX;
                          IF SGN[I2] = 0 THEN
                          BEGIN
                              SGN[I2]:=-H1;
                              NEW(PQ1);
                              PQ1@.NEXT:=PQ@.NEXT;
                              PQ@.NEXT:=PQ1;
                              PQ1@.VERTEX:=I2
                          END;
                          P:=P@.COLMATE1
                      END;
                      PQ:=PQ@.NEXT
                  END;
                  EX:=FALSE;
          (* SEARCH IF I1 IS NOT ADJACENT (EX=FALSE) TO
             ANY LL-COLOR VERTEX OF (KK,LL)-SUBGRAPH *)
                  P:=GR[I1].COLPOINT[LL];
                  WHILE (P <> NIL) AND (NOT EX) DO
                  BEGIN
                      EX:=SGN[P@.VERTEX] <> 0;
                      P:=P@.COLMATE1
                  END
              END (* SEARCH FOR COLOR LL *)
          END (* SEARCH FOR COLOR KK *);
```

```
(* IF EX=FALSE THEN (KK,LL)-SUBGRAPH FEASIBLE FOR
   INTERCHANGE OF COLORS HAS BEEN FOUND *)
      IF NOT EX THEN
      BEGIN
        PQ:=HQ;
        WHILE PQ <> NIL DO
        BEGIN
          I2:=PQ@.VERTEX;
          IF GR[I2].COLOR = KK THEN
             GR[I2].COLOR:=LL
          ELSE GR[I2].COLOR:=KK;
          P:=GR[I2].COLPOINT[LL];
          GR[I2].COLPOINT[LL]:=GR[I2].COLPOINT[KK];
          GR[I2].COLPOINT[KK]:=P;
          PQ:=PQ@.NEXT
        END;
  (* INTERCHANGE OF POINTERS TO COLORS IN NEIGHBORHOOD
     OF VERTICES OF (KK,LL)-SUBGRAPH *)
        PQ:=HQ;
        WHILE PQ <> NIL DO
        BEGIN
          I2:=PQ@.VERTEX;
          H:=GR[I2].COLOR;
          IF H = LL THEN H1:=KK ELSE H1:=LL;
          P:=GR[I2].ADJLIST;
          WHILE P <> NIL DO
          BEGIN
            J:=P@.VERTEX;
            IF GR[J].COLOR <> H1 THEN
            BEGIN
              P1:=P@.MATE;
              P2:=P1@.COLMATE1;
              P3:=P1@.COLMATE2;
              IF P3 = NIL THEN
                 GR[J].COLPOINT[H1]:=P2
              ELSE P3@.COLMATE1:=P2;
              IF P2 <> NIL THEN
                 P2@.COLMATE2:=P3;
              P1@.COLMATE2:=NIL;
              INSERT(H)
            END;
            P:=P@.NEXT
          END (* NEIGHBORS OF I2 *);
          PQ:=PQ@.NEXT
        END (* VERTICES IN (KK,LL)-SUBGRAPH *);
        K1:=KK
      END (* NOT EX *)
    END (* SEARCH FOR INTERCHANGE OF COLORS *)
END (* I1>1 *);
```

```
(* VERTEX I1 MAY HAVE COLOR K1 *)
   GR[I1].COLOR:=K1;
   IF K1 > K THEN K:=K1;
   P:=GR[I1].ADJLIST;
   WHILE P <> NIL DO
   BEGIN
       J:=P@.VERTEX;
       P1:=P@.MATE;
       INSERT(K1);
       P:=P@.NEXT
   END
END (* I *);
INTERSEQCOLORING:=K
END (* INTERSEQCOLORING *);
```

4.1.4. BACKTRACKING SEQUENTIAL ALGORITHM

In this section we show how an idea of the sequential coloring can be extended to give rise to a backtracking method which will color every graph optimally, that is, with the minimum number of colors.

Let $v_1, v_2, \ldots, v_n$ be an arbitrary ordering of vertices of a graph G. Initially, color 1 is assigned to vertex v_1 and then the remaining vertices are colored consecutively applying the following rule:

> Assume that vertices $v_1, v_2, \ldots, v_{i-1}$ are already colored, then find the set U_i of feasible colors for v_i and assign $\min\{U_i\}$ to vertex v_i.

First, we determine the set U_i of feasible colors for vertex v_i. Since the method we present here is of the complete search type, we have to take into account all possible assignments of colors for v_i except those which can lead to either nonoptimal or to equivalent colorings. The following theorem restricts the range of color numbers for v_i.

Theorem 4-3

Let $f: \{v_1, v_2, \ldots, v_{i-1}\} \rightarrow \{1, 2, \ldots, l\}$ be a coloring of $i - 1$ vertices of G by using exactly l colors. If only nonredundant colorings are to be generated, then a color for vertex v_i should satisfy $1 \leqslant f(v_i) \leqslant l + 1$.   ∎

This theorem can be shown by induction. Assume that the collection of partial and complete colorings found so far during the enumeration process of different colorings contains no redundant coloring. We now use the partial coloring f: $\{v_1, v_2, \ldots, v_{i-1}\} \rightarrow \{1, 2, \ldots, l\}$ to produce colorings of $v_1, v_2, \ldots, v_{i-1}, v_i$. Any coloring f' of vertices $v_1, v_2, \ldots, v_i$ cannot be equivalent to any coloring produced up to this point since f is not redundant. Therefore, we have to check for redundancy only among colorings produced from f. A coloring such that $f(v_i) \leqslant l$ is not redundant because colors not greater than l have been already used to color vertices $v_1, v_1, \ldots, v_{i-1}$. Otherwise, if more than one unused color is utilized to color v_i, then the colorings produced would be redundant with respect to each other. Therefore, $1 \leqslant f(v_i) \leqslant l + 1$.

Let l_{i-1} denote the maximum color number used to color vertices $v_1, v_2, \ldots, v_{i-1}$. Every color j that belongs to U_i must satisfy the following conditions:

$j \leqslant l_{i-1} + 1$,   by Theorem 4-3

$j \leqslant \min\{i, \deg(v_i) + 1\}$,   by (4-4)

$\quad j$ is not a color of any vertex v_h $(1 \leqslant h \leqslant i - 1)$ which is a neighbor of v_i.

If a complete q-coloring of G has been already found, then additionally

$$j \leqslant q - 1$$

Now we describe the backtracking steps of the algorithm. There are two situations when we cannot proceed any further: if the vertex v_i to be currently colored satisfies $U_i = \varnothing$, or if the last colored vertex is v_n. In the former case we backtrack to vertex v_{i-1}. Because of the last bound on j, the latter case occurs when a new better complete coloring of the graph has been found. Assume that l colors have been used in this solution. We set q equal l and again attempt to find a coloring with at most $q - 1$ colors. To accomplish this we have to recolor at least those vertices that have got color q. Therefore, we have to backtrack to vertex v_{i-1} such that v_i is the vertex with the lowest index colored with q. In both cases of backtracking we remove from U_{i-1} the color used in the current solution and assign the smallest one from among the other colors in U_{i-1}.

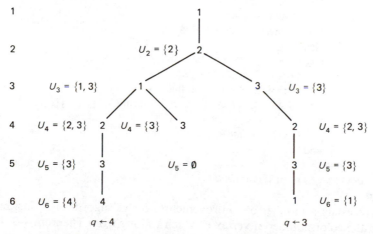

Figure 4-7 *Backtracking tree for the graph of Fig. 4-2.*

Example 4-6

We apply this method to the graph shown in Fig. 4-2. Figure 4-7 shows the search tree of the method, where the numbers in vertices denote the colors assigned to the vertices listed in the column on the left-hand side. ∎

The simplest version of the backtracking sequential algorithm can be described as follows.

Algorithm 4-6:   Backtracking Sequential Algorithm

```
begin
    {* k is the number of a vertex being considered, q is the number of colors used in the
       best solution found so far, l is the number of colors used in the current partial
       coloring, l_k is l for vertex v_k, and lowerbound is the value of a lower bound for the
       chromatic number of the graph which is colored *}
    l ← 1; k ← 1; f(v_1) ← 1; q ← n + 1;
    increase ← true;                          {* increase has value true when the forward *}
    repeat                                    {* step is to be performed and false for *}
        if increase then                      {* the backward step *}
        begin {* forward step *}
            l_k ← l; k ← k + 1;
            find U_k
        end;
        if U_k = ∅ then
        begin      {* backward step *}
            k ← k – 1; l ← l_k;
            increase ← false
        end
        else
        begin
            j ← the smallest number in U_k;
            U_k ← U_k – {j};
            f(v_k) ← j;
            if j > l then l ← l + 1;
            if k < n then increase ← true
            else
            begin  {* a new, better solution has been found *}
                store the present solution;
                find the smallest i such that f(v_i) = l;
                delete colors l, l + 1,..., q – 1 from U_1, U_2,..., U_{l–1};
                q ← l; l ← q – 1; k ← i – 1;
                increase ← false
            end
        end {* U_k ≠ ∅ *}
    until (k + 1) or (q = lowerbound)
end
```

There exists a number of improvements to this algorithm. Consider first the bound on the color number of vertex v_i which follows from Theorem 4-3. If vertex v_i

cannot be colored with any of the colors used for the previous $i - 1$ vertices, then there is only one feasible color for v_i, namely $l_{i-1} + 1$. In particular, the only color that can be assigned to vertex v_1 is color 1. Similarly, v_2 can be colored by 1 and 2, and only by 2 if v_2 is adjacent to v_1. Repeating this argument, it is obvious that if $v_1, v_2, \ldots, v_h$ constitute a clique in G, then these vertices can be colored only by the colors $1, 2, \ldots, h$, respectively. Therefore, we can terminate the algorithm in the case when all backtrackings to v_{h+1} have been considered.

The ordering of vertices was assumed to be arbitrary. However, taking as the first vertices those which constitute a largest clique in G will improve the efficiency of the algorithm. The remaining vertices can be ordered so that for every i, vertex v_i is adjacent to more of the vertices $v_1, v_2, \ldots, v_{i-1}$ than any other vertex v_j, $j > i$. Since the problem of finding the vertices of a largest clique in G is of the same difficulty as the coloring problem, we apply this method starting with a vertex of maximum degree in G (ties may be broken by choosing the vertex of greater degree). The resulting ordering of vertices of G is referred to as a *greedy* LF (GLF) ordering (see Brown [1972]).

The algorithm can be improved also by using a look-ahead procedure in an effort to reduce the number of forward steps which do not lead to better complete colorings. For instance, if for a given assignment of color j to vertex v_i, there exists its neighbor v_k such that $k > i$ and $U_k = \{j\}$, then we can cancel the assignment $f(v_i) \leftarrow j$, remove j from U_i, and proceed to the next feasible color for v_i. Thus we cut off all colorings (partial or complete) that agree with the current coloring on the first i vertices. Such a look-ahead procedure is implemented in our Pascal version of the backtracking sequential algorithm.

To decrease the number of forward steps one may augment the look-ahead procedure with a dynamic reordering of the uncolored vertices, so that the vertex to be colored next at a given stage of the algorithm is the one with the smallest number of feasible color numbers available for its coloring (see Brown [1972] and Korman [1979]).

There are several other modifications and improvements of the simple backtracking sequential algorithm which attempt to limit the search (see Korman [1979]). The reader should realize, however, that there always exists a substantial trade-off between the number of forward steps of an algorithm and its computational time. It happens very often that a considerable saving in the number of steps is obtained by increasing the computational time.

Computer Implementation
of the Backtracking Algorithm

The value of the integer function BACKTRACKSEQCOLORING is the number of colors used by the backtracking sequential algorithm with the simple look-ahead procedure. A graph to be colored is represented by a variable GR of type GRAPH and vertices are considered in the order given in array SEQ.

The forward steps combined with look-ahead and backward steps are controlled by using vertex fields COLBOUND, NUMCOL of type integer, and COLLIST, which is an N-element array of integers. For a given vertex I, field COLBOUND contains an upper bound for the color number which can be used to color vertex I and NUMCOL is the number of feasible colors for coloring I available at a given stage of the algorithm. Initially, if I is the Kth vertex in SEQ (i.e., SEQ[K] = I), then GR[I].COLBOUND and GR[I].NUMCOL are set to min{GR[I].DEGREE + 1, K} [see inequality (4-3)]. During the course of the algorithm, values of COLBOUND are changed when a new complete coloring is found. If the number of colors of the best coloring found so far is Q, then for every vertex, COLBOUND is set to Q − 1 if it was greater than this number, since in the next steps only colorings by at most Q − 1 colors are of interest. Additionally it is checked if a color assigned to a vertex I is not greater than L + 1, where L is the maximum color number used to color the vertices preceding I in the ordering SEQ.

The arrays COLLISTs are used to control the smallest color assignments and the look-ahead procedure. For a vertex K1 = SEQ[K] and J = 1, 2, ..., N, we assume that

$$
GR[K1].COLLIST[J] = \begin{cases} 0, & \text{if color J is greater than GR[K1].COLBOUND} \\ N, & \text{if color J is available for vertex K1 (i.e., J has} \\ & \text{not been used by any neighbor of K1 which} \\ & \text{precedes K1 in SEQ)} \\ I, & \text{if vertex K1 has a neighbor colored with J, where} \\ & \text{I < K and SEQ[I] is the first such vertex in} \\ & \text{the ordering SEQ} \end{cases}
$$

Initially, for every vertex, we assume that COLLIST[J] = N for J = 1, 2, ..., COLBOUND and COLLIST[J] = 0 for J = COLBOUND + 1, ..., N.

With such a data structure it is quite obvious how to implement the look-ahead procedure. When vertex K1 = SEQ[K] is to be colored and COL is the smallest number that can be used (i.e., COL is not greater than L + 1 and not greater than COLBOUND for K1), we look ahead to find whether there exists a neighbor SEQ[H] of K1 such that H > K and for this vertex NUMCOL = 1 and COLLIST[COL] = N. If this is the case, then no complete coloring for a current partial one (i.e., up to vertex K1) can be reached since after coloring K1 with COL there will be no free color for SEQ[H]. Therefore, we either have to try the next feasible color for K1, if one exists, or to backtrack to the predecessor of K1 in SEQ.

Evidently, the values of fields NUMCOL and COLLIST are subject to some changes during the forward and backward steps. This is quite an easy task, however, and can be deduced from the function code, so we leave it for the reader to figure out.

An ordering of vertices that are to be colored may have a great influence on efficiency of the algorithm for particular graphs. For instance, while coloring

Mycielski's graph M_5, the number of forward steps (COUNT) for the largest-first ordering was seven times smaller than that for the smallest-last ordering (see Section 4.1.5 for computational results).

Global Constants

N, LOWERBOUND   see the input below

Data Types

```
TYPE  ARRN       = ARRAY[1..N] OF INTEGER;
      ARR0N      = ARRAY[0..N] OF INTEGER;
      VERTPOINT = @VERTLIST;
      GRAPH      = ARRAY[1..N] OF
                     RECORD
                        DEGREE,COLBOUND,COLOR :INTEGER;
                        CURCOL,LK,NUMCOL          :INTEGER;
                        COLLIST                   :ARRAY[1..N] OF INTEGER;
                        ADJLIST                   :VERTPOINT
                     END;
      VERTLIST   = RECORD
                     VERTEX :INTEGER;
                     NEXT    :VERTPOINT
                   END;
```

Comment: In our Pascal codes, character @ stands for the pointer arrow ↑.

The structure of a variable of type GRAPH is similar to that shown in Fig. 4-4 except that some additional fields are added to keep track of forward and backward steps of the algorithm. Eight fields are assigned to each vertex I:

DEGREE	contains the vertex degree.
COLBOUND	contains the bound on the color number which can be used to color I.
COLOR	contains the vertex color of the best coloring found so far.
CURCOL	contains the vertex color of coloring which is constructed.
LK	equals the maximum color number used by vertices preceding I in the ordering given in SEQ.
NUMCOL	equals the number of feasible colors available for vertex I at a given stage of the algorithm.
COLLIST	contains the information about the colors available for vertex I.
ADJLIST	is a linked list of neighbors of I.

Function Parameters

```
FUNCTION BACKTRACKSEQCOLORING(
    N,LOWERBOUND :INTEGER;
    VAR COUNT      :INTEGER;
    VAR GR         :GRAPH;
    VAR SEQ        :ARRN):INTEGER;
```

Input

N	number of vertices of a graph to be colored
LOWERBOUND	lower bound to the chromatic number of the graph. It can be equal to the clique number of the graph, if it is known, or to the size of another complete subgraph of the graph. In particular, we may set LOWERBOUND to 3 if the graph has a triangle, or to 2 otherwise. (We assume that a graph has at least one edge.)
GR	variable of type GRAPH which represents the graph. Only fields DEGREE and ADJLIST must be set before calling the function
SEQ[1..N]	array that contains an ordering of vertices of the graph in which they are to be colored

Output

GR[I].COLOR	color of vertex I for I = 1, 2, ..., N in the optimal coloring of the graph
COUNT	number of forward steps performed by the algorithm to find the optimal coloring

Example

For the graph of Fig. 4-2, fields DEGREE and ADJLIST are displayed in Fig. 4-4, and the complete backtracking tree of the algorithm when vertices are colored in the order 1, 2, 3, 4, 5, 6 is shown in Fig. 4-7. ∎

Pascal Function BACKTRACKSEQCOLORING

```
FUNCTION BACKTRACKSEQCOLORING(
   N,LOWERBOUND:INTEGER;
   VAR COUNT    :INTEGER;
   VAR GR       :GRAPH;
   VAR SEQ      :ARRN):INTEGER;
```

```
VAR COL,H,I,J,K,K1,L,Q:INTEGER;
    INCREASE            :BOOLEAN;
    P                   :VERTPOINT;
    SEQ1                :ARRN;

PROCEDURE BACKSTEP;
(*  BACKTRACKING FROM VERTEX K1 TO ITS PREDECESSOR IN SEQ *)
BEGIN
    K:=K-1;
    K1:=SEQ[K];
    COL:=GR[K1].CURCOL;
    L:=GR[K1].LK;
    P:=GR[K1].ADJLIST;
    WHILE P <> NIL DO
    BEGIN
        H:=P@.VERTEX;
        IF SEQ1[H] > K THEN
            WITH GR[H] DO
            BEGIN
                IF COLLIST[COL] = K THEN
                BEGIN
                    COLLIST[COL]:=N;
                    NUMCOL:=NUMCOL+1
                END
            END;
        P:=P@.NEXT
    END;
    COL:=COL+1;
    INCREASE:=FALSE
END (* BACKSTEP *);

BEGIN    (*  MAIN BODY  *)
    FOR I:=1 TO N DO
        SEQ1[SEQ[I]]:=I;
    FOR I:=1 TO N DO
        WITH GR[SEQ[I]] DO
        BEGIN
            NUMCOL:=DEGREE+1;
            IF NUMCOL > I THEN NUMCOL:=I;
            COLBOUND:=NUMCOL;
            FOR J:=1 TO NUMCOL DO
                COLLIST[J]:=N;
            FOR J:=NUMCOL+1 TO N DO
                COLLIST[J]:=0
        END;
    COUNT:=0;
    K:=1;
    K1:=SEQ[1];  (* K1 - VERTEX TO BE COLORED *)
    COL:=1;
    Q:=N;
    L:=0;
    INCREASE:=TRUE;
```

```
REPEAT
   IF INCREASE THEN   (*  FORWARD STEP  *)
   BEGIN
      I:=GR[K1].COLBOUND;
      IF I > L+1 THEN I:=L+1;
      WHILE (GR[K1].COLLIST[COL] < K) AND (COL <= I) DO
         COL:=COL+1;
   (* COL IS THE COLOR FOR VERTEX K1 *)
      IF COL = I+1 THEN INCREASE:=FALSE
      ELSE
      BEGIN
         IF K = N THEN
         BEGIN
         (* NEW COMPLETE COLORING HAS BEEN FOUND  *)
            COUNT:=COUNT+1;
            GR[K1].CURCOL:=COL;
            FOR I:=1 TO N DO
               WITH GR[I] DO COLOR:=CURCOL;
            IF COL > L THEN L:=L+1;
            Q:=L;
            IF Q > LOWERBOUND THEN
            BEGIN
            (* BACKTRACKING TO FIRST VERTEX OF COLOR Q *)
               I:=1;
               WHILE GR[SEQ[I]].COLOR <> Q DO I:=I+1;
               FOR J:=N DOWNTO I DO
                  BACKSTEP;
               L:=Q-1;
               FOR I:=1 TO N DO
                  WITH GR[I] DO
                     IF COLBOUND > L THEN
                     BEGIN
                        FOR J:=L+1 TO COLBOUND DO
                           IF COLLIST[J] = N THEN
                              NUMCOL:=NUMCOL-1;
                        COLBOUND:=L
                     END
         END
         END (* K = N *)
         ELSE (* K < N *)
         BEGIN
            P:=GR[K1].ADJLIST;
            WHILE (P <> NIL) AND INCREASE DO
            BEGIN
               H:=P@.VERTEX;
               IF SEQ1[H] > K THEN
                  WITH GR[H] DO
                     INCREASE:= NOT ((NUMCOL = 1)
                              AND (COLLIST[COL] >= K));
               P:=P@.NEXT
            END;
```

```
                   IF INCREASE THEN
                   BEGIN
                       COUNT:=COUNT+1;
                       GR[K1].CURCOL:=COL;
                       GR[K1].LK:=L;
                       IF COL > L THEN L:=L+1;
                       P:=GR[K1].ADJLIST;
                       WHILE P <> NIL DO
                       BEGIN
                           H:=P@.VERTEX;
                           IF SEQ1[H] > K THEN
                               WITH GR[H] DO
                                   IF COLLIST[COL] >= K THEN
                                   BEGIN
                                       COLLIST[COL]:=K;
                                       NUMCOL:=NUMCOL-1
                                   END;
                           P:=P@.NEXT
                       END;
                       K:=K+1;
                       K1:=SEQ[K];
                       COL:=1
                   END (* INCREASE *)
                   ELSE COL:=COL+1
                END (* K < N *)
             END (* COL <= I *)
          END (* INCREASE = TRUE *)
          ELSE
          BEGIN  (* BACKWARD STEP *)
             INCREASE:=TRUE;
             IF (COL > GR[K1].COLBOUND) OR (COL > L+1) THEN
                 BACKSTEP
          END (* INCREASE = FALSE *)
      UNTIL (K = 1) OR (Q = LOWERBOUND);
      BACKTRACKSEQCOLORING:=Q
END; (* BACKTRACKSEQCOLORING *)
```

4.1.5. COMPUTATIONAL RESULTS

The functions SEQCOLORING, INTERSEQCOLORING, and BACKTRACK-SEQCOLORING have been tested on many graphs, in particular, we used the following graphs:

1. Mycielski's graphs M_3, M_4, and M_5 with 5, 11, and 23 vertices, respectively.

2. The graph from Matula et al. [1972] with 34 vertices.

3. Random graphs G_{np}, for $n = 15(15)90$ and $p = 0.1(0.2)0.9$ such that $\Pr((u, v) \in E(G_{np})) = p$ and these probabilities for each of the $\binom{n}{2}$ possible edges are mutually independent.

The following Pascal functions have been also tested and compared with the codes enclosed here:

AMMIS   implementation of the approximately maximum independent set algorithm (see Algorithms 4-2 and 4-3)

DSATUR   implementation of the algorithm due to Brélaz [1979] and described in Problem 4-24

RLF   Pascal version of the RLF algorithm published in Leighton [1979]

BACKTRACK   implementation of the simple backtracking sequential algorithm (see Algorithm 4-6)

Table 4-1.   Computational Results from the Application of Some Heuristic and Backtracking Algorithms to Color Mycielski's Graphs M_3, M_4, M_5 and the Graph from Matula et al. [1972]

	Graph G			
	M_3	M_4	M_5	Matula et al. [1972]
$\chi(G)$	3	4	5	8
Upper Bounds				
$\Delta(G) + 1$, (4-4), (4-5)	3, 3, 3	6, 5, 4	12, 7, 6	21, 14, 10
AMMIS algorithm				
Number of colors/time[a]	3/< 1	4/2	5/5	9/17
SEQCOLORING				
Number of colors				
LF and SL orderings	3	4	5	8
Five random orderings	3, 3, 3, 3, 3	4, 4, 4, 4, 4	6, 5, 5, 5, 5	9, 8, 9, 10, 9
Average time[a]	< 1	< 1	1	2
INTERSEQCOLORING				
Number of colors	3	4	5	8
Average time[a]	1	3	9	25
BACKTRACK				
BACKTRACKSEQCOLORING				
(COUNT/time[a])				
LF ordering	4/< 1	80/4	31917/1923	182/19
	5/< 1	22/3	909/122	65/16
SL ordering	4/< 1	57/3	27140/1668	108/11
	5/< 1	36/4	6959/860	43/11
GLF ordering	5/< 1	56/3	30813/1925	33/3
	5/< 1	40/4	7644/961	34/8

[a] Time in milliseconds on Amdahl 470 V/8.

Table 4-2. Computational Results from the Application of Some Heuristic Algorithms to Color Random Graphs

n	Density, p	Δ+1	(4-4)	(4-5)	χ	AMMIS Number of Colors/Time[a]	SEQCOLORING/INTERSEQCOLORING LF, SL, GLF Orderings	Three Random Orderings	Average Time[a]	DSATUR	RLF
30	0.1	9	7	4	4	4/6	4,4,4/4,4,4	5,4,4/4,4,4	1/6	4/6	4/5
	0.3	15	11	7	6	7/11	6,6,6/6,6,6	7,6,7/6,6,6	2/15	6/7	6/9
	0.5	18	14	11	6	8/14	8,7,7/7,7,7	8,8,7/7,7,7	3/26	7/9	7/13
	0.7	25	20	18	10	11/24	12,11,12/10,10,11	11,12,11/11,10,10	3/48	10/10	10/21
	0.9	30	26	24	17	19/47	17,17/17,17,17	20,20,20/17,18,17	4/102	17/13	17/41
60	0.1	14	10	5	4	5/22	5,5,4/4,4,4	6,5,5/4,4,4	3/26	4/20	4/16
	0.3	29	20	14	7	9/44	9,10,10/9,8,8	10,10,11/9,9,9	6/106	9/27	8/36
	0.5	41	30	24		15/83	13,14,13/12,12,12	14,16,14/13,13,12	9/187	14/35	11/65
	0.7	50	41	35		20/145	20,19,18/18,18,18	21,20,22/18,18,19	13/339	20/44	18/117
	0.9	59	51	49		30/265	28,28/28,28,27,26	31,34,32/28,29,29	16/733	26/51	26/209
90	0.1	16	13	7		7/53	6,6,6/5,5,5	7,7,7/6,5,6	5/67	5/43	5/35
	0.3	40	30	22		12/120	12,13,13/11,11,10	14,13,12/11,12,12	13/281	12/61	11/89
	0.5	57	45	37		18/228	19,18,18/16,15,15	19,19,20/16,17,17	21/533	17/78	15/173
	0.7	73	62	55		27/435	28,27,27/24,23,23	27,27,29/26,24,24	29/1225	26/96	23/323
	0.9	88	78	74		43/834	41,41,39/38,38,38	44,42,42/39,39,38	35/2753	39/115	38/615

[a] Time in milliseconds on Amdahl 470 V/8.

Table 4-3. Computational Results from the Application of Backtracking Algorithms to Color Random Graphs of Type 3

Time COUNT

n	Density, p	χ	BACKTRACK[a]						BACKTRACKSEQCOLORING[b]					
			Ordering						Ordering					
			LF		SL		GLF		LF		SL		GLF	
			Time	COUNT	Time	COUNT	Time	COUNT	Time	COUNT	Time	COUNT	Time	COUNT
15	0.1	3	1	14	2	14	1	14	1	15	1	15	1	15
	0.3	4	2	23	2	14	2	14	2	16	2	15	2	15
	0.5	4	3	26	3	21	1	14	3	21	4	23	2	15
	0.7	7	3	18	2	14	2	14	4	16	3	15	3	15
	0.9	8	5	30	3	15	3	14	5	21	4	15	4	15
30	0.1	4	23	280	4	43	3	38	10	76	5	41	4	36
	0.3	6	25	193	5	35	4	29	8	41	6	33	6	30
	0.5	6	971	6328	53	339	16	98	148	607	33	135	15	62
	0.7	10	1139	5324	289	1348	682	3213	281	845	117	337	292	849
	0.9	17	44	156	9	30	8	29	44	103	14	31	14	30
45	0.1	3	284	3201	17	196	18	199	24	181	15	121	19	164
	0.3	6	6566	41638	3146	20101	30	191	1345	5762	748	2952	31	124
	0.5	9					8577	38259	2560	6383			1081	2875
	0.7	13												
	0.9	21												

[a] Time in milliseconds obtained on AMDAHL 470 V/6, which is 25% slower than V/8.
[b] Time in milliseconds obtained on AMDAHL 470 V/8.

Table 4-4. Computational Results from the Application of the Function BACKTRACKSEQCOLORING to Color Some Random Graphs of Type 3 in at Most 1000 Forward Steps

n	Density, p	χ	LF Ordering χ'[a]	LF Ordering Time[b]	LF Ordering COUNT	SL Ordering χ'	SL Ordering Time[b]	SL Ordering COUNT	GLF Ordering χ'	GLF Ordering Time[b]	GLF Ordering COUNT
45	0.1	3	3	24	181	3	15	121	3	19	164
	0.3	6	8	167	1000	7	258	1000	6	31	124
	0.5	9	10	370	1000	10	355	1000	9	349	1000
	0.7	13	14	476	1000	13	484	1000	14	464	1000
	0.9	21	22	594	1000	23	548	1000	21	480	857
60	0.1	4	5	160	1000	4	26	159	4	14	76
	0.3	7	8	338	1000	10	210	1000	8	288	1000
	0.5		13	431	1000	12	415	1000	12	451	1000
	0.7		20	598	1000	18	599	1000	17	673	1000
	0.9		27	699	1000	28	635	1000	27	710	1000

[a] χ' is the number of colors used by function BACKTRACKSEQCOLORING in the COUNT number of forward steps.
[b] Time in milliseconds on Amdahl 470 V/8.

Table 4-1 contains the computational results of approximate and backtracking algorithms applied to graphs 1 and 2 and Tables 4-2, 4-3, and 4-4 contain the computational results obtained for graphs 3.

As mentioned before, the simple sequential algorithm (function SEQCOLORING) colors almost all graphs with a number of colors less than or equal to the best upper bound even for random orderings of vertices.

The results show also that adding the interchange step to the simple sequential algorithm (function INTERSEQCOLORING) is significantly beneficial.

The functions INTERSEQCOLORING, DSATUR, and RLF generate solutions of almost the same quality; however, the last two are much faster.

Tables 4-1 and 4-3 show that the look-ahead procedure produces great improvements when incorporated into the backtracking sequential algorithm.

Mycielski's graphs turn out to be very hard instances for backtracking algorithms. The reader may wonder why the look-ahead procedure results in such different algorithm efficiencies when applied to different vertex orderings. At least in the case of graph M_5, the explanation of this phenomenon is quite simple. For the LF ordering, the first vertex that got color 5 was the seventh from the bottom of the search tree, and for the SL ordering it was the third one. Therefore, after finding the first complete coloring (which in fact is optimal), the algorithm backtracks to the fifteenth level of the search tree in the former case, and only to the nineteenth level in the latter.

Table 4-4 contains the computational results of the application of the function BACKTRACKSEQCOLORING with a fixed number of forward steps (COUNT = 1000) to color some random graphs of type 3.

The results of experiments reported in Tables 4-1, 4-3, and 4-4 support the claim that backtracking algorithms spend less time on vertex orderings which force the use of the biggest unavoidable color as soon as possible. This also explains why the GLF ordering turns out to be superior to LF and SL ones when combined with backtracking algorithms.

PROBLEMS

4-1.   Show that every k-chromatic graph with n vertices is l-colorable, where $k \leqslant l \leqslant n$.

4-2.   Reductions.

    (a) Prove that if a graph G has two vertices u and v such that the set of neighbors of u is contained in that of v, then the problem of optimal coloring of G can be reduced to that of coloring the subgraph of G induced by $V - \{u\}$.

    (b) A subset $U \subset V$ is called a *separation clique* of G if the subgraph induced by U is a clique and $V - U$ induces a disconnected subgraph.

Show that to color G it suffices to optimally color all components of $G|_{V-U}$.

4-3. Assume that we are given an *update algorithm* A for optimal coloring of graphs; that is, A can be used to correct the chromatic number of a graph after the addition or deletion of an edge. Is it possible that A is a polynomial-time algorithm? (*Hint:* Design an algorithm for optimal coloring of a graph which will use A as a subroutine.)

4-4. **(a)** Show that every tree with at least two vertices is 2-colorable.
(b) (König's Theorem, 1930) Apply this result to prove that a nonempty graph is 2-chromatic if and only if it has no cycle of odd length.

4-5. Prove that G is 2-colorable if and only if there exists an orientation of its edges such that the resulting digraph $\vec{G}$ contains no path of length 2, where the length of a path is defined as the number of its oriented edges.

4-6. Design an algorithm which in time $O(n + m)$ verifies whether a graph with n vertices and m edges is bipartite (i.e., 2-colorable).

4-7. Show that two colors are sufficient for coloring the regions generated by the intersections of straight lines in the plane, so that no two regions sharing a common boundary line (not a single point) are of the same color.

4-8. Show that the regions of a plane graph G can be colored with two colors if and only if every vertex in G is of even degree.

4-9. (Whitney: see Ore [1967]) Prove that the regions of a Hamiltonian plane graph can be colored with four colors.

4-10. **(a)** Prove that M_k, the kth Mycielski's graph, is triangle-free and $\chi(M_k) = k$.
(b) Prove that M_k has $3 \cdot 2^{k-2} - 1$ vertices. Evaluate the degrees of vertices and the number of edges of M_k. Design a procedure for automatic generation of the sequence of graphs $\{M_k\}$.
(c) Prove that every subgraph $M_k - \{u\}$ is $(k-1)$-colorable; that is, M_k is *critically k-chromatic*.

4-11. **(a)** (Gallai: see Ore [1967]) Prove that if a graph can be oriented in such a way that no direct path has more than k vertices, then $\chi(G) \leqslant k$.
(b) Show that $\chi(G) \leqslant d^*(G) + 1$, where $d^*(G)$ is the length of the longest path in G.

4-12. A *labeling* (or *evaluation*) Π of the vertices of a graph $G = (V, E)$ is a one-to-one mapping from V to the positive integers $N = \{1, 2, \ldots, n\}$, where $n = |V|$. The *bandwidth of Π* is defined to be

$$B_\Pi(G) = \max\{|\Pi(u) - \Pi(v)|: \{u, v\} \in E\}$$

and the *bandwidth of G* is defined as follows:

$$B(G) = \min\{B_\Pi(G): \Pi \text{ is a labeling of } G\}$$

Prove that

$$\chi(G) \leqslant B(G) + 1$$

4-13. Let $\beta_0(G)$ denote the *vertex independence number* of G; that is, $\beta_0(G)$ is the maximum number of nonadjacent vertices in G. Verify the inequalities

$$\frac{n}{\beta_0(G)} \leqslant \chi(G) \leqslant n - \beta_0(G) + 1$$

and find graphs for which these inequalities become equalities and strict inequalities.

4-14. Design a polynomial-time algorithm for finding the chromatic number of a graph having no vertex of degree exceeding 3.

4-15. Find an infinite family of graphs such that the ith graph has $3^{i/3}$ maximal independent subsets.

4-16. (Lawler [1976])   Let G be a graph with n vertices and m edges.
 (a) Show how to test whether G is 3-colorable in $O(mn3^{n/3})$ time.
 (b) Show how to test whether G is 4-colorable in $O((m + n)2^n)$ time.

4-17. Implement the AMmIS algorithm to run in $O(n^3)$ time.

4-18. Prove Theorem 4-2.

4-19. (Matula et al. [1972])   Let $G = (V, E)$ be a *complete k-partite* graph; that is, vertex set V can be partitioned into k subsets $V_1, V_2, \ldots, V_k$ and for every $u \in V_i$ and $v \in V_j$ $(i \neq j)$, $(u, v) \in E$. Prove that the sequential algorithm applied to G produces k-coloring of G for every ordering of its vertices.

4-20. **(a)** Prove inequality (4-6).
 (b) Prove inequality (4-7).

4-21. **(a)** Find the LF and SL orderings of the vertices of graph G shown in Fig. 4-8 (break ties by choosing the vertex of lexicographically smaller label).
 (b) Find the LF and SL bounds to $\chi(G)$ and color G by applying the sequential algorithm to the LF and SL orderings.

4-22. **(a)** Show that the SL sequential algorithm colors Mycielski's graphs $\{M_k\}$ optimally.
 (b) Prove that the sequential with interchange algorithm colors bipartite graphs optimally.

4-23. **(a)** Show that the MATES procedure works correctly and that its time complexity is $O(M)$.

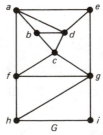

Figure 4-8 *Graph for Problems* 4-21, 4-24, *and* 4-25.

(b) Show that in the sequential with interchange algorithm (see Algorithm 4-5), a search for an interchange of colors may be restricted to the case when the following expression is satisfied $g > 2$ *and* $g > k$ *and* ($i > 4$ *or* ($i = 4$ *and* $k = 2$)).

4-24. (Brélaz [1979]) Let G be a partially colored graph. A *saturation degree* of an uncolored vertex v is defined as the number of different colors to which v is adjacent. The following algorithm (DSATUR) colors a graph and finds a *saturation-largest-first* (SLF) ordering of the vertices of G:

(a) Color a vertex of maximum degree with color 1.

(b) As v_i take an uncolored vertex of maximum saturation degree (break ties by choosing the vertex of greater degree) and color v_i with the least possible color ($i = 2, 3, \ldots, n$).

Implement this algorithm and apply to the graphs of Figs. 4-2 and 4-8.

4-25. Apply the simple backtracking sequential algorithm and the backtracking sequential with look-ahead procedure to the graph of Fig. 4-8 using LF, SL, and SLF orderings of vertices.

4-26. (Zykov: see Corneil and Graham [1973]) Let u and v be two nonadjacent vertices of a graph G and define two graphs $G'_{uv} = G \cup (u, v)$ and G''_{uv}, where G'_{uv} is obtained from G by adding the edge (u, v) and G''_{uv} by merging u and v into one vertex which is joined to all neighbors of u and v.

(a) Show that

$$\chi(G) = \min\{\chi(G'_{uv}), \chi(G''_{uv})\} \tag{4-8}$$

and design an algorithm for finding $\chi(G)$ based on this relation. [*Hint:* Note that (4-8) cannot be applied only to complete graphs.] Apply your algorithm to the graph in Fig. 4-2.

(b) (Dutton and Brigham [1981]) Design and implement an approximation algorithm which is based on the relation (4-8) and merges at each step a pair of nonadjacent vertices with the maximum number of common adjacent vertices. Compare this algorithm with other approximation algorithms on a set of random graphs.

4-27. Prove that every nonbipartite outerplanar graph is 3-chromatic. Design a linear time algorithm for coloring outerplanar graphs.

4-28. An *edge coloring* of a graph G is an assignment of a color to each edge of G so that the edges incident to any vertex have distinct colors. The *chromatic index*, of a graph G, $\chi'(G)$, is the least number of colors needed to color the edges of G.

(a) Show how the following scheduling problem can be solved by coloring edges of a corresponding graph: Assume that each student must be examined orally by each of the professors who taught him; how many examination periods are needed?

(b) V. G. Vizing (1963; see Fiorini and Wilson [1977]) proved that the chromatic index of a graph G satisfies the inequalities

$$\Delta(G) \leqslant \chi'(G) \leqslant \Delta(G) + 1$$

where $\Delta(G)$ is the maximum degree of G. Adopt a sequential vertex coloring approach to coloring edges of a graph and show that the number of colors used, $\chi'_S(G)$ satisfies the inequality

$$\chi'_S(G)/\chi'(G) < 2$$

REFERENCES AND REMARKS

The coloring problem is perhaps one of the most famous problems in graph theory and is well known in its special version dealing only with planar graphs which are graph-theoretic models of planar political maps. There was a claim, known as the four-color problem (FCP), that such graphs (or equivalently, political maps in the plane or on the sphere) can be always colored by at most four colors.

A great portion of the history of graph theory is related to the FCP, which has stimulated research in this field during the last century. The reader may consult the following book for a collection of the most important discoveries and papers in graph theory which appeared between the first paper published by L. Euler and the first book by D. König:

BIGGS, N. L., E. K. LLOYD, and R. J. WILSON, *Graph Theory 1736–1936*, Oxford University Press, London, 1976

All attempts made to solve the FCP and the results obtained before 1966 are presented in

ORE, O., *The Four Color Problem*, Academic Press, New York, 1967

For more than 100 years, from 1852 to 1976, the FCP was known as the four-color conjecture, and finally, K. Appel, W. Haken, and J. A. Koch, with the essential help of a computer, proved this conjecture in 1976. The solution has been published in very technical papers which appeared in the *Illinois Journal of Mathematics* (in 1976 and 1977). The reader who wants to understand only the main points of the proof without going into details is referred to

HAKEN, W., An Attempt to Understand the Four Color Problem, *J. Graph Theory* 1(1977), 193–206

BERNHART, F. R., A Digest of the Four Color Theorem, *J. Graph Theory* 1(1977), 207–225

The most recent reference for the equivalent formulations of the FCP and the solution method is

SAATY, T., and P. KAINEN, *The Four Color Problem: Assaults and Conquest*, McGraw-Hill, New York, 1977

A linear-time algorithm for 5-coloring of planar graphs was presented in

CHIBA, N., T. NISHIZEKI, and N. SAITO, A Linear Algorithm for Five-coloring a Planar Graph, in N. Saito and T. Nishizeki (eds.), *Graph Theory and Algorithms*, Lecture Notes in Computer Science, vol. 108, Springer-Verlag, Berlin 1981, pp. 9–19

The applications of coloring to some scheduling problems with several additional real-world constraints have been presented in many papers, see, for instance,

DE WERRA, D., How to Color a Graph, and A Few Remarks on Chromatic Scheduling, in B. Roy (ed.), *Combinatorial Programming: Methods and Applications*, D. Reidel, New York, 1975, pp. 305–325, 337–342

OPSUT, R., and F. S. ROBERTS, On the Fleet Maintenance, Mobile Radio Frequency, Task Assignment, and Traffic Phasing Problem, in G. Chartrand, Y. Alavi, D. L. Goldsmith, L. Lesniak-Foster, and D. R. Lick (eds.), *The Theory and Applications of Graphs*, Wiley, New York, 1981, pp. 479–492

The main reference for general results in the computational complexity of combinatorial problems is

GAREY, M. R., and D. S. JOHNSON, *Computers and Intractability: A Guide to the Theory of NP-Completeness*, W. H. Freeman, San Francisco, 1979

The following papers deal only with the complexity of the coloring problem and the behavior of its algorithms:

JOHNSON, D. S., Worst Case Behavior of Graph Coloring Algorithms, in *Proc. 5th Southeastern Conf. on Combinatorics, Graph Theory, and Computing, Congressus Numerantium X*, Utilitas Mathematica, Winnipeg, 1974, pp. 513–527

LAWLER, E. L., A Note on the Complexity of the Chromatic Number Problem, *Inform. Process. Lett.* 5(1976), 66–67

MITCHEM, J., On Various Algorithms for Estimating the Chromatic Number of a Graph, *Comput. J.* 19(1976), 182–183

The independent set approach is described in all details in the following book:

CHRISTOFIDES, N., *Graph Theory: An Algorithmic Approach*, Academic Press, London, 1975, Chapter 4

and some improvement to it have been presented in

ROSCHKE, S. I., and A. L. FURTADO, An Algorithm for Obtaining the Chromatic Number and an Optimal Coloring of a Graph, *Inform. Process. Lett.* 2(1973), 34–38

WANG, C. C., An Algorithm for the Chromatic Number of a Graph, *J. ACM* 21(1974), 385–391

The recursive-largest first (RLF) algorithm which combines the strategy of sequential coloring with independent set approach appeared in

LEIGHTON, F. T., A Graph Coloring Algorithm for Large Scheduling Problems, *J. Res. Nat. Bur. Standards* 84(1979), 489–506

The LFS algorithm has been proposed in

WELSH, D. J. A., and M. B. POWELL, An Upper Bound for the Chromatic Number of a Graph and its Application to Timetabling Problem, *Comput. J.* 10(1967), 85–86

and the sequential algorithms are reviewed in the paper

MATULA, D. W., G. MARBLE and J. D. ISAACSON, Graph Coloring Algorithms, in R. C. Read (ed.), *Graph Theory and Computing*, Academic Press, London, 1972, pp. 109–122

See also

DUNSTAN, F. D. J., Sequential Colorings of Graphs, in *Proc. 5th British Combinatorial Conference*, 1975, pp. 151–158

The backtracking sequential algorithm was first proposed by J. R. Brown in

BROWN, J. R., Chromatic Scheduling and the Chromatic Number Problem, *Management Sci.* 19(1972), 456–463

and several improvements to this method have been presented in

BRÉLAZ, D., New Methods to Color the Vertices of a Graph, *Comm. ACM* 22(1979), 251–256

KORMAN, S. M., The Graph-Colouring Problem, in N. Christofides, A. Mingozzi, P. Toth, and C. Sandi (eds.), *Combinatorial Optimization*, Wiley, New York, 1979, pp. 211–235

The SLF algorithm, called the *dynamic rearrangement algorithm*, has been also proposed, tested and evaluated in the papers

DÜRRE, K., Approximate Algorithms for Coloring Large Graphs, in J. R. Mühlbacher (ed.), *Datenstrukturen, Graphen, Algorithmen*, Carl Hanser Verlag, München-Vien, 1978, pp. 191–203, and

DÜRRE, K., J. HEUFT, and H. MÜLLER, Worst and Best Case Behaviour of an Approximate Graph Coloring Algorithm, in J. R. Mühlbacher (ed.), *Proc. the 7th Conf. Graphtheoretic Concepts in Computer Science (WG 81)*, Carl Hanser Verlag, München-Wien, 1982, pp. 339–348

The following paper presents a unified framework for the backtracking coloring algorithms:

JACKOWSKI, B. L., and M. KUBALE, A Generalized Implicit Enumeration Algorithm for Graph Coloring, Institute of Computer Science, Technical University, Gdańsk, 1982

The results of computational experiments performed with the coloring algorithms have been presented in the above-mentioned papers by Wang [1974], Leighton [1979], Matula et al. [1972], Brown [1972], Korman [1979], and Brélaz [1979]; see also

KUBALE, M., Empirical Comparison of Efficiency of Some Graph Colouring Algorithms, *Arch. Automat. Telemech.* 22(1978), 129–139

The algorithm which is the subject of Problem 4-26(a) has been proposed by A. A. Zykov. See the following papers for its improvements and evaluation:

CORNEIL, D. G., and B. GRAHAM, An Algorithm for Determining the Chromatic Number of a Graph, *SIAM J. Comput.* 2(1973), 311–318

McDIARIMID, C., Determining the Chromatic Number of a Graph, *SIAM J. Comput.* 8(1979), 1–14

The approximation algorithm of Problem 4-26(b) has been proposed in

DUTTON, R. D., and R. C. BRIGHAM, A New Graph Colouring Algorithm, *Comput. J.* 24(1981), 85–86

The edge-coloring problem is discussed extensively in

FIORINI, S., and R. J. WILSON, *Edge-Colourings of Graphs*, Research Notes in Mathematics 16, Pitman, London, 1977

Many classical results mentioned in this section can be found in almost every standard textbook on graph theory (see the References and Remarks in Section 3.3).

4.2. SCHEDULING PROBLEMS

This section deals with scheduling problems whose purpose is to find an optimal processing order of tasks (also called activities, jobs, or operations) on a set of machines subject to several constraints imposed on the tasks, machines, and their mutual relationships.

We begin our presentation (Section 4.2.1) with the classical network scheduling, known also as the critical path method (CPM). Then in the next section, the model is augmented with a set of unit resources (equivalent to machines). This extension, unlike the CPM model, has no known polynomial-time algorithm, so we present an enumerative method, which can be also used for solving the job shop problem considered in Section 4.2.3.

After discussing these three best known scheduling models, we introduce in Section 4.2.4 a general framework for studying scheduling problems. In the next two sections, we review the fundamental results in scheduling of a single and parallel machines. Finally, a special case of the parallel machine scheduling with precedence constraints is discussed, together with its efficient computer implementation.

Restricting the attention to very simple classical models, our main goal is to present basic ideas and approaches, and relatively simple scheduling rules and procedures. We also indicate and illustrate some algorithmic relations between scheduling problems and other topics presented in the book.

4.2.1.   NETWORK SCHEDULING

First network methods for project planning and scheduling appeared in the late 1950s under the acronyms CPM (critical path method) and PERT (program evaluation and review technique). They have been rapidly accepted as very powerful and practical techniques in a variety of areas such as research and development, construction, production, and maintenance. Originally, these two methods were designed to deal mainly with precedence constraints relating elements of a project and their duration times. Here the objective is to minimize the total time required for the completion of the entire project. The difference between CPM and PERT techniques lies in the interpretation and assumption about the duration of each project activity. In CPM models, all duration times are deterministically known, whereas PERT is a network model in which relations among different activities are still completely determined but duration times are assumed to carry some uncertainty of a known distribution.

PERT models, as well as other probabilistic scheduling techniques, are not discussed in this book, which is focused on deterministic discrete models and algorithms only.

In this section, we first discuss briefly some network construction methods and the standard critical path method. Then a critical path technique is applied to solve the network scheduling model under some resource constraints. The critical path technique is again used in Section 4.2.6 while discussing scheduling problems on parallel machines.

Network Construction

A *project* consists basically of two kinds of elements: *activities* and *precedence constraints* (relations) imposed on activities. Activities are characterized by their duration times and the precedence constraints form a binary relation between activities. If a and b are two activities and a precedes b, which we denote as $a \prec b$, it means that a must be completed before b can be started. In practically all real-world projects, the precedence constraints correspond to a transitive relation; that is, if $a \prec b$ and $b \prec c$, then $a \prec c$. However, for the sake of simplicity (but without any loss of generality) we may always remove from a project those constraints which result from the transitivity. Problem 4-32 suggests how it can be done very efficiently.

Precedence constraints have another property which follows from their origin. If all activity duration times are positive, a *consistent project* (i.e., correctly built) has no sequence of activities whose precedence constraints form a cycle.

There are two basic ways of assigning a network to a project and they differ in representing the activities. In the *activity-on-node* representation (AN or *activity network*), an activity is represented by a node and precedence constraints are represented by arcs. Therefore, an activity network is a digraph in which nodes are in one-to-one correspondence with the project activities and two nodes are joined by an arc if the corresponding activities are in precedence relation. Since a digraph can model any binary relation defined on its node set, every project can be easily represented by an activity network. The two properties of a project may now be formulated in terms of the network elements as follows. An activity network has no directed cycle and transitive arcs can be removed without any loss of generality.

Example 4-7

We shall use the following project to illustrate our considerations:

Activity, x	Immediate Predecessors of x, $P(x)$	Immediate Successors of x, $S(x)$	All Successors of x, $S'(x)$
b	a	d, e	d, e, g, h, i
c	a	e	e, g, h, i
d	b	g	g, i
e	b, c	g, h	g, h, i
f	a	g, h	g, h, i
g	d, e, f	i	i
h	e, f	i	i
Dummy source, a	$\varnothing$	b, c, f	b, c, d, e, f, g, h, i
Dummy sink, i	g, h	$\varnothing$	$\varnothing$

Figure 4-9 shows the corresponding activity network which has been augmented with two dummy activities, a *source a* and a *sink i* which, respectively, precede and follow all project activities. This convention will be used throughout the section. ∎

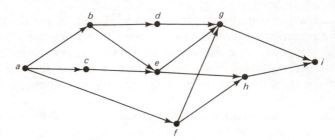

Figure 4-9 *Sample activity network.*

In the other network representation of a project, called the *activity-on-arc* network (AA or *event network*), project activities are assigned to arcs and the precedence constraints among the activities are preserved. The nodes of such a representation are called *events* since they correspond to time moments at which activities begin and end. Although it is not quite obvious how to construct an event network for a given project, this representation of projects has enjoyed greater popularity in network analysis than the activity network representation. This motivates our choice; we use the event network representation for the temporal analysis of networks in this section, and activity networks will be utilized for resource-constrained network scheduling in Section 4.2.2.

Let us first consider four activities *b*, *c*, *d*, and *e* of our example project and construct a network that preserves the precedence constraints among these activities represented by arcs. For instance, we may assign the activities to four separate arcs and then introduce some other arcs to maintain the precedence constraints. Figure 4-10(b) shows the resulting network in which dashed arcs (activities), called *dummies*, indicate only the precedence constraints. We can easily observe that shrinking dummy activities *bd* and *ce* [see Fig. 4-10(c)] does not alter the precedence constraints. However, activity *be* cannot be removed. We ask the reader to verify that the network in Fig. 4-10(d) is an event network corresponding to the activity network of Fig. 4-9, and that it has the minimum number of dummy activities among all possible event networks.

We should be satisfied with the event network of the sample project since it has less nodes and less arcs than its activity network, counting even the dummy ones. This is always the case, although there is no known efficient algorithm for constructing event networks with the minimum number of dummy activities. Moreover, in the light of the result due to Krishnamoorthy and Deo [1979], it is very unlikely that minimal event networks can be constructed in polynomial time since the problem is NP-complete.

In practice, we may use several rules which when applied to a project *D* (i.e., an activity-on-node network) attempt to construct its event network *E* with the minimum number of dummy activities. One may try first to minimize the number of nodes in the event network and then concentrate on minimizing the number of dummy activities. Although these two quantities cannot be minimized simulta-

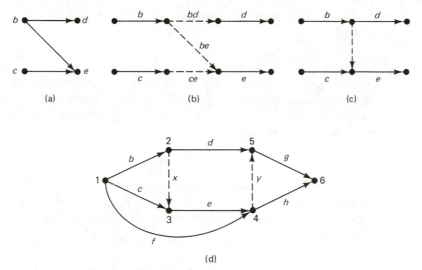

Figure 4-10   *Construction of an event network.*

neously (see Syslo [1981]), such an approach is very practical and leads quickly to quite good approximations of the optimal (minimal) event networks.

We now formally describe a heuristic method for constructing minimal event networks. Let us assume that, as an input, given is an activity network (digraph) $D = (V, A)$ which represents a project. Step 1 tests if the project is consistent and if so, finds a topological numbering of the nodes (activities). Step 2 removes transitive arcs (precedence relations) from D. Finally, steps 3 and 4 construct an event network.

Step 1.   Test if D is a circuit. An efficient algorithm for this task is offered in Problem 4-30. If D has no cycle, then the nodes of D can be numbered with integers in such a way that if (i, j) is an arc in D, then $i < j$. Such a numbering (ordering) of nodes of an acyclic digraph is called *topological*.

Step 2.   Find the *transitive reduction* of D. In the other words, remove from D all arcs (i, j) for which there exists in D another path from i to j. Problem 4-32 suggests an efficient algorithm for finding the transitive reduction of a digraph.

Step 3.   Let $P(i)$, $S(i)$, and $S'(i)$ denote, respectively, the set of all activities that immediately precede, immediately succeed, and succeed activity i in D (see Example 4-7 for an illustration). Initially, as an event network corresponding to D, we assume that $E = (W, B)$, where $W = \{w_i, w_i': i \in V\}$, and B consists of all arcs $\{(w_i, w_i'): i \in V\}$ corresponding to the project activities and dummy arcs $\{(w_i', w_j): (i, j) \in A\}$ corresponding to the precedence constraints. Now apply the following rules, which minimize the number of nodes in an event network (activities i and j in these rules are the project activities).

Rule 1

Activities i and j may share the same tail node if they have the same sets of immediate predecessors [i.e., when $P(i) = P(j)$]. Such i and j are called *branching* activities. Therefore, if $P(i) = P(j)$, then merge the tails of activities (w_i, w_i') and (w_j, w_j').

Rule 2

Activities i and j may share the same head node if they have the same sets of immediate successors [i.e., when $S(i) = S(j)$]. Such i and j are called *merging* activities. Therefore, if $S(i) = S(j)$, merge the heads of activities (w_i, w_i') and (w_j, w_j').

Rule 3

Activity i may share its head node with the tail node of activity j if the set of successors $S'(i)$ of i is contained in the set of successors of every immediate predecessor of j. In other words, in this case i and j must satisfy

$$\{S'(k): k \in P(j)\} = S'(i)$$

where $i \in P(j)$. Therefore, if this condition holds, shrink the dummy activity (w_i', w_j).

Comment: The first two rules are obvious, but rule 3 is somewhat complicated. Yet it can be viewed as a generalization of rule 2 (see Fig. 4-11).

Step 4. In this step we are allowed to use all available tools to minimize further the number of dummy activities in E. Since it is very unlikely to come up with an efficient exact method, any set of fast heuristic rules can be applied. We leave the details to the reader, who could consult the literature (see a review paper by Sysło [1981] and also Problem 4-33).

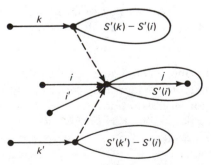

Figure 4-11 *An illustration of Rule 3.*

Once a network representation of a project is constructed we may formally assign duration times to the activities (0 to dummy activities) and proceed to temporal analysis of the network.

In the remainder of this section (4.2.1.) we assume that a project is represented by an event network, called simply a network. Therefore, the activities are on arcs and we will not make any distinction between the project activities and the dummy activities.

Temporal Analysis of a Network

Let $G_p = (W, B, p)$ be a network with a nonnegative function p defined on the arc set B. Therefore, $p((i, j))$, or simply p_{ij} for $(i, j) \in B$ is *the duration time* of activity (i, j) and if (i, j) is a dummy activity, then we have $p_{ij} = 0$.

Temporal analysis of a network refers to a variety of questions that are concerned with the timing of activities and events. The basic questions relate to how long the project will take, when a particular activity may be started, and how long an activity can be delayed without delaying the completion of the entire project.

Let us assume that the project starts at time 0. It is evident that to complete the project it will take at least as long as to complete every chain of activities leading from source to sink of the network. Therefore, the shortest project duration is equal to the length of a longest path from source to sink. Moreover, a particular activity may start when all the activities that precede it have been completed. Since our network is acyclic, we may assume that the nodes are in topological order. Therefore, to process event i we have to consider only events j such that $j < i$. Let π_i^e denote the *earliest* possible time at which event i could possibly occur (i.e., π_i^e is the length of a longest path from 1 to i). Hence we have

$$\pi_1^e = 0$$

$$\pi_i^e = \max\{\pi_j^e + p_{ji} : (j, i) \in B\} \qquad \text{for} \quad i = 2, 3, \ldots, n \qquad (4\text{-}9)$$

This recurrence relation is equivalent to an algorithm for finding a longest path in a sorted acyclic network and defines a forward-pass procedure whose complexity is $O(|B|)$.

The time of the last event π_n^e is the earliest realization of the entire project represented by the network G_p. A longest path in G_p is called *critical*; so are its activities, since the delay of any such activity would cause the delay of the whole project. A network can have several critical paths. On the other hand, activities that do not lie on a critical path have some freedom in their processing. Assuming the duration of the project equal to T (which is at least equal to π_n^e) we may now find π_i^l, the *latest* possible time at which event i could occur without delaying the completion of the project at time T. The latest event times satisfy the following recurrence relation:

$$\pi_n^l = T$$

$$\pi_i^l = \min\{\pi_j^l - p_{ij}: (i, j) \in B\} \qquad \text{for} \quad i = n - 1, n - 2, \ldots, 1 \qquad (4\text{-}10)$$

This relation defines a backward-pass procedure whose complexity is also $O(|B|)$. The difference $\pi_i^l - \pi_i^e = s_i$ is the *slack time* of event i, which represents the maximum possible delay of event i that causes no delay in π_n^l. If the project has no explicit due date, we may assume that $\pi_n^l = \pi_n^e$ in the backward pass. Once all the earliest and latest event times are known, we may turn our attention to activities. In particular, the following quantities can be calculated using activity durations p_{ij} and π_i^e and π_i^l times:

Earliest start time: $\text{ES}_{ij} = \pi_i^e$

Earliest finish time: $\text{EF}_{ij} = \text{ES}_{ij} + p_{ij} = \pi_i^e + p_{ij}$

Latest finish time: $\text{LF}_{ij} = \pi_j^l$

Latest start time: $\text{LS}_{ij} = \text{LF}_{ij} - p_{ij} = \pi_j^l - p_{ij}$

Activity slack time: $s_{ij}^1 = \text{LF}_{ij} - \text{EF}_{ij} = \text{LS}_{ij} - \text{ES}_{ij} = \pi_j^l - \pi_i^e - p_{ij}$

for every activity $(i, j) \in B$. Events and activities with zero slack times belong to critical paths of the network, and any delay in their occurrence or performance results in some delay of the project completion time.

Example 4-8

We illustrate the foregoing calculations with the network shown in Fig. 4-10(d). The activity duration times and the other quantities are listed in Table 4-5.   ∎

The activity slack time s_{ij}^1 assumes that all activities that precede (i, j) are accomplished as early as possible (i.e., they are started at their earliest start times) while all activities that follow it are started as late as possible (i.e., at their latest start times). Therefore, s_{ij}^1 is a measure of the maximum possible flexibility in scheduling activity (i, j). In our example, activity d may be started no earlier than at time 5 and to avoid delaying the project, it must be completed by time 10. Since it requires 3 units of time only, there are 2 extra units which can be absorbed before or after activity d is completed or during its processing. These 2 units form the slack time of activity d. However, if activity g has been scheduled to start at 9, the slack time of activity d would have been only 1. This kind of freedom in scheduling activities is called *float*. The activity slack time is called the *total float*, and the other measures of

Table 4-5. Event and Activity Times for the Network of Fig. 4-10(d)

Events

i	Critical	Earliest Time, π_i^e	Latest Time, π_i^l	Slack Time, $s_i = \pi_i^l - \pi_i^e$
1	*	0	0	0
2	*	5	5	
3	*	5	5	0
4	*	9	9	0
5		9	10	1
6	*	13	13	0

Activities

(i,j)	Critical	p_{ij}	Start Times ES_{ij}	Start Times LS_{ij}	Finish Times EF_{ij}	Finish Times LF_{ij}	Float Times Total, s_{ij}^1	Float Times Safety, s_{ij}^2	Float Times Free, s_{ij}^3	Float Times Independent, s_{ij}^4
$b = (1,2)$	*	5	0	0	5	5	0	0	0	0
$c = (1,3)$		3	0	2	3	5	2	2	2	2
$f = (1,4)$		5	0	4	5	9	4	4	4	4
$x = (2,3)$	*	0	5	5	5	5	0	0	0	0
$d = (2,5)$		3	5	7	8	10	2	2	1	1
$e = (3,4)$	*	4	5	5	9	9	0	0	0	0
$y = (4,5)$		0	9	10	9	10	1	1	0	0
$h = (4,6)$	*	4	9	9	13	13	0	0	0	0
$g = (5,6)$		3	9	10	12	13	1	0	1	0

float are defined as follows:

$$\textit{Safety float}: \qquad s_{ij}^2 = \pi_j^l - \pi_i^l - p_{ij}$$

$$\textit{Free float}: \qquad s_{ij}^3 = \pi_j^e - \pi_i^e - p_{ij}$$

$$\textit{Independent float}: \qquad s_{ij}^4 = \max\left\{0, \pi_j^e - \pi_i^l - p_{ij}\right\}$$

The *total float* of an activity x represents the delay in the start time of x, which will not cause the delay in the project completion time provided that all activities that precede x are completed as early as possible and all activities that succeed x are accomplished as late as possible. The *safety float* measures the maximum delay of an activity when preceding activities are completed as late as possible. The *free float* is the maximum time an activity may be delayed without effecting the start time of any activity that succeeds it. Finally, the *independent float* is the most individual scheduling freedom of an activity and measures the maximum delay in start time of the activity that is independent of any other scheduling decision in the network. Table 4-5 illustrates the calculation of the floats for our sample network.

In summary, the temporal analysis can be applied to a deterministic network, that is, to networks with constant and known in advance duration times of all activities. The duration of the project is equal to the length of a critical path of the network, and any delay in a critical activity can lead to the delay of the entire project. Noncritical activities can be scheduled with some flexibility represented by various measures of activity float.

There are three main extensions of the deterministic network models considered so far. The first extension, known as PERT model, accepts probabilistic activity durations and determines event and activity times and durations in probabilistic terms. We do not discuss this model here. In the second possible extension, activity durations are treated as variables from a given range and there is a cost involved in processing every activity which is a nonincreasing function of the activity duration. Therefore, there is a time/cost trade-off for each activity and for the entire project. This extension is the subject of Problem 4-37.

Finally, the most practical extension of the simple CPM model involves some resource constraints. In the CPM model, we schedule activities assuming that there are no limits on resources (such as machines, row materials, personnel, etc.) and as a consequence, many activities utilizing the same resource could take place simultaneously. The resource-constrained network model in a simplest form is discussed in the next section, where a Pascal implementation of a related algorithm is also provided.

4.2.2. RESOURCE-CONSTRAINED NETWORK SCHEDULING

The critical path calculations presented in the preceding section assume that the resources necessary to carry out activities (some of them over the same period of time) are available in sufficient amount. We now assume that some resource requirements are added to the basic CPM model. There can be several different types of resources of different amount and the activities may require all or only some of them. We will, however, consider a simplified problem. Our resource-constrained network model is assumed to satisfy the following conditions:

1.   There is a set of resources, one unit of each type.
2.   Each activity requires a unique resource unit during its processing, and upon the activity completion this resource unit is available again.

It follows that no two activities may use the same resources simultaneously. Hence the resources in our model can be considered as a shop consisting of machines, and each activity needs exclusively one machine for the time period it is performed. Therefore, no two activities may be processed on any machine at the same time and no activity needs more than one machine for its processing. Thus the set of all activities can be partitioned into mutually disjoint subsets containing those activities which are assigned to the same machines. In the standard CPM model with no resources, some activities from the same set of this partition could be performed simultaneously, but now we have to impose additional precedence constraints on all activities in each of the sets of the partition. In other words, we have to find an order in which machines will be used to process the activities. These additional precedence constraints among activities assigned to the same machine, together with the original constraints of the project, form an augmented network which should remain feasible (i.e., with no cycles). A critical path in the augmented network determines the duration of the project for a particular ordering of activities on the machines. Hence our problem is to minimize the critical path in an augmented network over all feasible augmentations.

We now formulate this problem more precisely in a form which will be convenient for calculations.

In this section (4.2.2.) we utilize activity networks (i.e., project activities are on nodes of a network). Let $G = (V, A)$ be such a network. Consequently:

1. G is acyclic.
2. G has exactly one source and one sink: the node without predecessors and the node without successors, respectively. Nodes other than source and sink lie on a path from source to sink. In what follows, we number the source with 1 and the sink with n, where n is the total number of nodes in G.
3.   There is assigned the duration time (called now the *processing time*) p_i to each node (activity) i in V.

Without loss of generality we may also assume that:

4. G contains no transitive arc. The existence of such arcs causes no problem in our consideration; however, their removal may speed up the calculations significantly.

The reader should have no major difficulty with switching from event networks used in the preceding section to activity networks utilized in this section. It could be helpful to rewrite the CPM calculations for activity networks (see Problem 4-36). Although the activities are in nodes, it is still possible to operate the processing times assigned to arcs of G. To this end, the processing time of activity i should be moved from node i on every arc emanating from i. If the sink activity is a real one (i.e., with a nonzero processing time), then an additional dummy arc should be added from the original sink to a new dummy sink.

Disjunctive Networks and Their Properties

An activity network $G = (V, A)$ of a project is called *conjunctive*, and its arcs, *conjunctive arcs*, since they belong to every augmentation. Let us now assume that the set of activities V is partitioned into nonempty and mutually disjoint subsets $V_1, V_2, \ldots, V_m$ such that:

5.   Every activity in V_k ($k = 1, 2, \ldots, m$) has to be processed by using exclusively a unique machine available at one unit, and upon the completion of its processing, the machine is again fully available for other activities from the same set V_k.

Condition 5 can be expressed in graph-theoretic terms by introducing a *disjunctive* pair of arcs $[(i, j), (j, i)]$ for each pair $i, j \in V_k$ of different activities ($i \neq j$) and such that there is no path connecting i with j or j with i in G. The last assumption should be clear since if i precedes j in G and both have to be processed by using the same machine, there is no freedom in choosing the order of i and j (i.e., i must be finished before j can be started).

Example 4-9

Consider a conjunctive network shown in Fig. 4-12(a) and assume the following machine partition of activities (nodes): $V_1 = \{5, 9, 7, 4\}$, $V_2 = \{3, 8\}$, and $V_3 = \{2, 6, 10\}$. Figure 4-12(b) shows the corresponding disjunctive network.   ∎

Let D denote the set of all disjunctive arcs of G generated by a given partition $\{V_k\}$ of V. The processing time function can be defined on disjunctive arcs in the same way that it has been defined on conjunctive arcs. Therefore, we set $p_{ij} \leftarrow p_i$ and $p_{ji} \leftarrow p_j$ for the disjunctive pair $[(i, j), (j, i)]$. A pair of disjunctive arcs is called *settled* if one of the arcs has been added to a subset $C \subset D$ of *chosen* arcs and the other has been *rejected*. By choosing a disjunctive arc (i, j) we introduce precedence of activity i over activity j on their common machine. A set C of settled disjunctive

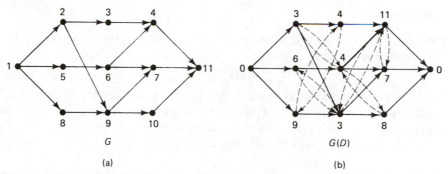

Figure 4-12 *Sample pair of conjunctive and disjunctive networks. In (a) activity numbers are in nodes and in (b) activity processing times are in nodes.*

arcs is *feasible* if it satisfies:

1′. $(i, j) \in C$ if and only if $(j, i) \in D - C$.
2′.   The network $G(C) = (V, A \cup C)$ is acyclic.

A feasible set of disjunctive arcs C, also called a *selection*, resolves all machine conflicts by imposing an additional order on those activities that must use the same machine. The project duration for a selection C corresponds to the length of a critical path in $G(C)$, and therefore the problem of minimizing the project duration time now reduces to finding a minimaximal path in $G(D)$ that is a critical path in a feasible network $G(C)$ that is minimal over all selections C of D.

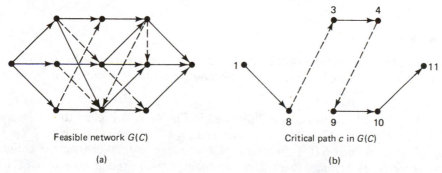

Feasible network $G(C)$ Critical path c in $G(C)$

(a)   (b)

Figure 4-13 (a) *Feasible network of the network of Fig. 4-12(b) and* (b) *its critical path.*

Example 4-9 (cont.)

Figure 4-13(a) shows a feasible network of $G(D)$ in Fig. 4-12(b), and its critical path is extracted in Fig. 4-13(b).   ∎

We will now describe an implicit enumeration method for finding a mini-maximal path in a disjunctive network $G(D)$. The method generates a sequence of selections $C_1, C_2, \ldots$ such that every C_l in the sequence is obtained from one of the previous selections C_k $(k < l)$ by *switching* a disjunctive arc of C_k for its partner in the pair.

First, we review basic properties of feasible networks $G(C)$ and introduce some assumptions simplifying and making possible further consideration. By assumption 2 about G and by the construction of $G(C)$ we have:

3′.   Every disjunctive arc in a feasible network $G(C)$ lies on a path from source to sink.

Regarding the activity processing times, we assume that they satisfy the strict triangle inequality; that is:

4′.   In every feasible network $G(C)$, an arc $(i, j) \in A \cup C$ is the unique shortest path from i to j. In other words, if there is a path $(i, h_1, h_2, \ldots, h_g, j)$ $(g \geqslant 1)$ from i to j in $G(C)$, then

$$p_{ij} < p_{ih_1} + p_{h_1 h_2} + \ldots + p_{h_{g-1} h_g} + p_{h_g j}$$

From the meaning of numbers p_{ij} and the interpretation of the network model, it follows that this assumption is realistic.

The following two theorems justify our approach.

Theorem 4-4

Let $G(C)$ be a feasible network and let c denote a critical path in $G(C)$. Any network $G(C')$ obtained from $G(C)$ by switching a disjunctive arc of c is feasible.   ∎

The proof of this theorem is illustrated in Fig. 4-14. It is clear that if switching a disjunctive arc $(i, j) \in c$ for (j, i) results in a cycle in $G(C')$, where $C' = C - \{(i, j)\} \cup \{(j, i)\}$, then $G(C)$ contains a path b from i to j, other than the arc (i, j). Now, since (i, j) belongs to c, it is the longest path from i to j in $G(C)$ and according to assumption 4′, the arc (i, j) is also the unique shortest path from i to j in $G(C)$. Therefore, the arc (i, j) is the only path from i to j in $G(C)$ and its switching for (j, i) cannot create a cycle in $G(C')$.

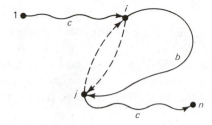

Figure 4-14 *An illustration for the proof of Theorem* 4-4.

The next theorem shows not only that switching of disjunctive arcs along a critical path in a feasible network results in a feasible network but also that it is necessary to switch one such arc if we are still searching for a minimaximal path.

Theorem 4-5

Let $G(C)$ be a feasible network and let c denote a critical path in $G(C)$. If there exists a feasible network $G(C')$ with a critical path c' shorter than c, at least one of the disjunctive arcs of c is switched for its partner in $G(C')$. ∎

This is quite clear, since if no disjunctive arc of c is switched in $G(C')$, then $G(C')$ contains the path c and a critical path c' of $G(C')$ cannot be shorter than c.

Figure 4-13(b) shows that a critical path in a feasible network can have more than one disjunctive arc. We will introduce some measures that will help us to decide in which order critical disjunctive arcs should be chosen for switching. But first we show how a topological ordering in $G(C)$ can be modified to yield that in $G(C')$ in a network that has been obtained from $G(C)$ by switching a critical disjunctive arc.

Example 4-9 (cont.)

Assume that $G(C')$ is obtained from $G(C)$ in Fig. 4-13(a) by switching arc $(8, 3)$. Figure 4-15 shows the subnetwork of $G(C)$ generated by the first (in topological order) six nodes. It is easily seen that the simple exchange of nodes 8 and 3 in their positions does not preserve the ordering, due to the existence of arc $(2, 3)$. This can be resolved by applying the next theorem. ∎

Figure 4-15 *Subnetwork of $G(C)$ in Fig. 4-13 generated by the first six nodes.*

Theorem 4-6

Let (i, j) be a critical disjunctive arc of a feasible network $G(C)$ and c_{ij} denote the set of nodes that lie between i and j in a topological ordering of the nodes in $G(C)$. Then c_{ij} contains no activity (node) which must be processed on the same machine as i and j. Moreover, the nodes in c_{ij} can be partitioned into mutually disjoint subsets $c_{ij} = c_i \cup c_j$ such that in $G(C)$ there is no path from any node in c_i to j and from i to any node in c_j. ∎

A proof of this theorem is left to the reader (see Problem 4-41). The importance of this theorem lies in that when a feasible network $G(C)$ is transformed to another feasible network $G(C')$ by switching a critical disjunctive arc of $G(C)$, a topological ordering in $G(C')$ can be constructed from that of $G(C)$ by considering only some nodes in $G(C)$ (see also Problem 4-40 and the REVERSE procedure in the Pascal implementation of the algorithm).

Balas's Algorithm

We now discuss enumeration process of feasible networks. Let $G(C) = (V, A \cup C)$ be a feasible network and let c denote its critical path. If c contains no disjunctive arc of $G(C)$, then according to Theorem 4-5, no feasible network, which has a critical path shorter than c, can be generated from $G(C)$ by switching a disjunctive arc. If c contains a disjunctive arc (i, j), let $G(C')$ denote the feasible network obtained from $G(C)$ by switching (i, j); therefore, $C' = C \cup \{(j, i)\} - \{(i, j)\}$. Network $G(C')$ is acyclic by Theorem 4-4. The path c can be partitioned as $c = (c_1, (i, j), c_2)$, where c_1 and c_2 are, respectively, a longest path from 1 to i and a longest path from j to n in $G(C)$ (see Fig. 4-16). If c' denotes a critical path in $G(C')$, then evidently $p(c') > p(c)$ only if c' contains (j, i), the sole arc of $G(C')$ which is not in $G(C)$. Let us consider a longest path c'' in $G(C')$ among all paths containing the arc (j, i). It can be partitioned as $c'' = (c_1'', (j, i), c_2'')$, where c_1'' and c_2'' are, respectively, a longest path from 1 to j and a longest path from i to n in $G(C')$ [or equivalently, in $G(C'')$, where $C'' = C - \{(i, j)\}$, since c_1'' and c_2'' cannot

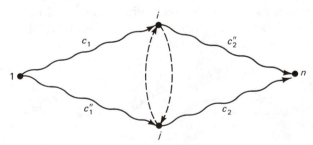

Figure 4-16 *Switching a disjunctive arc (i, j).*

contain (i, j) (see Fig. 4-16)]. Let us denote

$$\delta(i, j) = p(c'') - p(c) = p(c_1'') + p_{ji} + p(c_2'') - p(c_1) - p_{ij} - p(c_2)$$

Therefore, if $\delta(i, j) > 0$, then c'' is a critical path in $G(C')$; otherwise, $p(c'') = p(c) + \delta(i, j)$ is only a lower bound for $p(c')$, the length of a critical path in $G(C')$.

In a lower bound to $p(c')$ we may consider also other paths of $G(C')$, for instance, (c_1, c_2'') and (c_1'', c_2) of length

$$p(c_1, c_2'') = p(c) - p_{ij} - p(c_2) + p(c_2'')$$
$$p(c_1'', c_2) = p(c_1'') + p(c) - p(c_1) - p_{ij}$$

Hence we can define the following lower bound to $p(c')$:

$$p(c') \geqslant \max\{ p(c''), p(c_1, c_2''), p(c_1'', c_2)\} = p(c) + \Delta(i, j)$$

where we assumed that

$$\Delta(i, j) = -p_{ij} + \max\{q_1 + q_2 + p_{ji}, q_1, q_2\}$$
$$q_1 = p(c_1'') - p(c_1) \quad \text{and} \quad q_2 = p(c_2'') - p(c_2)$$

Note that all the quantities in $\Delta(i, j)$ are well defined, since by assumption 3', i and j lie on some paths from source to sink. Values $\Delta(i, j)$, calculated for all disjunctive arcs of a critical path in $G(C)$, can be used as a criterion for choosing an arc to be switched for its partner. Since we are searching for a minimaximal path in $G(D)$, a preference should be given to arcs with smaller $\Delta(i, j)$. Let π_i denote the length of a longest path from 1 to i in $G(C)$.

Table 4-6. Longest Paths in $G(C)$ and in Its Subnetworks

	Node i										
	1	2	3	4	5	6	7	8	9	10	11
π_i in $G(C)$ of Fig. 4-13(a)	0	0	9	13	0	6	24	0	24	27	35
π_i' in $G(C - \{(8,3)\})$	0	0	3	10	0	6	24	0	21	24	32
π_i'' in $G(C - \{(4,9)\})$	0	0	9	13	0	6	24	0	9	12	31

Example 4-9 (cont.)

We now find $\Delta(8, 3)$ and $\Delta(4, 9)$, for the two disjunctive arcs of the critical path exhibit in Fig. 4-13(b). Table 4-6 contains the length of longest paths from source in $G(C)$, and in $G(C)$ after removing either $(8, 3)$ or $(4, 9)$. Hence we have

$$\Delta(8, 3) = -p_{83} + \max\{q_1 + q_2 + p_{38}, q_1, q_2\}$$
$$= -9 + \max\{3 + 6 + 4, 3, 6\} = 4$$

where $q_1 = \pi'_3 - \pi_8 = 3$, $q_2 = (\pi'_{11} - \pi'_8) - (\pi_{11} - \pi_3) = 32 - 26 = 6$; and $\Delta(4, 9)$ $= -p_{49} + \max\{q_1 + q_2 + p_{94}, q_1, q_2\} = -11 + \max\{-4 + 7 + 3, -4, 7\} = -4$, where $q_1 = \pi''_9 - \pi_4 = -4$, $q_2 = (\pi''_{11} - \pi''_4) - (\pi_{11} - \pi_9) = 18 - 11 = 7$. Therefore, a critical path in the network obtained by switching arc $(8, 3)$ passes its complementary arc $(3, 8)$ and has length $\pi_{11} + \Delta(8, 3) = 35 + 4 = 39$. While switching arc $(4, 9)$ we have $\Delta(4, 9) = -4$. Note that no critical path in the latter switched network contains the introduced arc $(9, 4)$. ∎

  Now we are ready to describe an implicit enumeration algorithm for finding an optimal selection of disjunctive arcs that minimizes the length of a critical path. The algorithm starts with a feasible selection, whose arcs are called *normal*, and their disjunctive partners, *reverse*, and generates a sequence of feasible networks. Problem 4-43 provides a heuristic method for finding reasonable starting solutions. Every network $G(C')$ in the sequence is obtained from some preceding network $G(C)$ in a *forward step* which consists of switching one normal arc of $G(C)$. Whenever we are sure that further generation of feasible networks from $G(C')$ cannot improve the current best selection, we abandon $G(C')$ and *backtrack* to $G(C)$. Let us assume that $G(C')$ was obtained from $G(C)$ by switching arc (i, j). To avoid producing the same network more than once, some disjunctive arcs are temporarily *fixed* in $G(C')$, in the sense that they cannot be switched in any descendant network of $G(C')$. Therefore, the reverse arc (j, i) of (i, j) is fixed in $G(C')$ and in every descendant network of $G(C')$ in the search tree. On the other hand, when we backtrack from $G(C')$ to $G(C)$, we fix in $G(C)$ the normal arc (i, j). Thus each feasible network $G(C)$ generated by the algorithm contains a set of fixed disjunctive arcs C_f. The normal arcs in C_f correspond to abandoned networks generated directly from $G(C)$ or from its ancestor networks, whereas the reverse arcs in C_f correspond to the sequence of networks that lead from the initial network to $G(C)$. The disjunctive arcs of $G(C)$ which are currently not fixed, that is, those in $C - C_f$, are called *free*. A free arc of $G(C)$ that belongs to a critical path is called a *candidate*. At each stage of the algorithm, only candidates are used to generate new feasible networks from $G(C)$.

  One of the drawbacks of the algorithm is its weak dominance test, whose role is to abandon as many as possible unnecessary forward steps. Let π^* denote the value of the currently best solution. If

$$\pi\big(G(C_f)\big) \geqslant \pi^*$$

then all potential descendant networks of $G(C)$ can be abandoned, where $\pi(D(C_f))$ is the length of a critical path in the network obtained from the conjunctive network G by augmenting it with only fixed arcs of $G(C)$. Computational results with the algorithm reported in many papers show that it usually takes a considerable amount of time to prove that the current best solution is optimal indeed.

The algorithm described above was proposed by E. Balas in 1969. We now present it in a compact form as Algorithm 4-7. The arcs of a current feasible network $G(C)$ are identified in the algorithm by special labels, which mark all arcs of $G(D)$ in accordance to their status in $G(C)$. Let l_{ij} denote the label of arc (i, j). We assume the following values:

If (i, j) is a conjunctive arc in $G(D)$, then $l_{ij} = 3$.

If (i, j) is a normal disjunctive arc in $G(D)$, then

$$l_{ij} = \begin{cases} 2, & \text{if } (i, j) \text{ is free in } G(C) \\ 4, & \text{if } (i, j) \text{ is fixed in } G(C) \\ 0, & \text{if the partner } (j, i) \text{ of } (i, j) \text{ is fixed} \end{cases}$$

If (i, j) is a reverse arc in $G(D)$, then

$$l_{ij} = \begin{cases} 3, & \text{if } (i, j) \text{ is fixed in } G(C) \\ 1, & \text{if the partner } (j, i) \text{ of } (i, j) \text{ is present in } G(C) \\ & \text{either as a free or a fixed arc} \end{cases}$$

It can easily be seen that the current feasible network $G(C)$ consists of those arcs which are labeled with at least 2 and the conjunctive arcs of $G(C)$ [i.e., the conjunctive arcs of $G(D)$ and the fixed arcs of $G(C)$] have labels equal to at least 3.

Algorithm 4-7 uses three procedures: DMATES, CRITICALPATH(lab, π), and REORDER(i, j). Procedure DMATES matches the disjunctive arcs which belong to the same pair. This information about the location of disjunctive mates in the network representation is utilized in the algorithm whenever it deals with a disjunctive arc and has to perform some actions on its partner. Procedure CRITI-CALPATH(lab, π) finds the lengths π_i of longest paths from source 1 to all other nodes in the network formed by the arcs currently marked with a label equal to at least lab. Procedure REORDER(i, j) adjusts a topological ordering of a current feasible network after switching disjunctive arc (i, j) for its partner.

Algorithm 4-7:   Balas's Implicit Enumeration Algorithm for Network Scheduling

```
begin   (* given a disjunctive network G(D) and its initial selection *)
    DMATES;                 (* initialization *)
    for every arc (i, j) of G(D) do
```

```
            if (i, j) is a conjunctive arc then l_ij ← 3
            else if (i, j) is a normal arc then l_ij ← 2
                    else l_ij ← 1;
    π* ← inf;                    {* inf is a large number *}
    r ← 1;                       {* r is the level number of the search tree *}
    repeat                       {* until search is exhausted *}
            CRITICALPATH(3, d);  {* finding upper bound, d_n = π(G(C_f)) *}
            if d_n ≥ π* then back ← true
            else
            begin
                    back ← false;
                    CRITICALPATH(2, π);  {* finding a critical path in current network *}
                    if π_n < π* then update the best current solution and π*;
                    for every disjunctive arc (i, j) in the critical path of the current network do
                            add (i, j) to the list of disjunctive critical arcs, criticalarc;
                    {* We assume that the arcs added to criticalarc on the rth level of the search
                        are in a nondecreasing order of their Δ's and they lie between ll_r and ul_r *}
                    b ← ll_r > ul_r
            end;
            if back or b then
                    while b and (r > 0) do
                    begin    {* backtracking *}
                            {* let (j, i) be the most recently fixed reverse arc *}
                            remove (j, i) from the current network (i.e., l_ji ← 1);
                            fix arc (i, j) (i.e., l_ij ← 4);
                            REORDER(j, i);
                            r ← r − 1;
                            b ← ll_r > ul_r;
                            if b then
                            begin {* there is no unscanned critical disjunctive arc on level r *}
                                    r ← r − 1;
                                    if r > 0 then
                                        for all critical disjunctive arcs (i, j) on level r do
                                                l_ij ← 2
                            end
                    end;
            if r > 0 then
            begin {* forward step *}
                    {* let (i, j) be the next-to-be-considered critical disjunctive arc of the current
                        network *}
                    remove arc (i, j) from the current network (i.e., l_ij ← 0);
                    fix arc (j, i) (i.e., l_ji ← 3);
                    REORDER(i, j);
                    r ← r + 1
            end
    until r = 0
end
```

Example 4-9 (cont.)

We continue to complete the solution of our sample problem. As shown in Fig. 4-13, the initial feasible network has a critical path that contains two disjunctive arcs $(4, 9)$ and $(8, 3)$. We found also that $\Delta(4, 9) = -4$ and $\Delta(8, 3) = 4$; hence first we switch $(4, 9)$ and then $(8, 3)$. Figure 4-17 shows the search tree of the algorithm applied to our example network. Nodes of the tree are labeled with the sets of fixed arcs in the

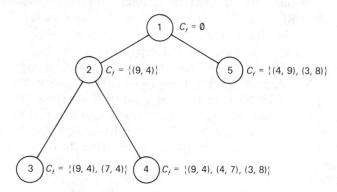

Figure 4-17 *Search tree of the sample problem.*

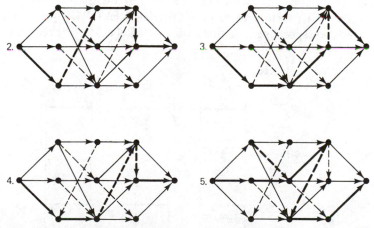

Figure 4-18 *Feasible networks generated in the course of Balas's algorithm applied to the sample problem. Numbers refer to the nodes of the search tree (Fig. 4-17). The initial network (number 1) is shown in Fig. 4-13. The critical paths are marked in heavy lines.*

corresponding feasible networks which are listed in Fig. 4-18. We encourage the reader to complete these illustrations with all necessary algorithm calculations. ∎

Computer Implementation of Balas's Algorithm

The algorithm, as presented in Algorithm 4-7, is implemented in procedure NETWORKSCHEDULING. The basic data structures employed in the procedure are illustrated in Fig. 4-19.

A backward-star form is used to represent the entire disjunctive network $G(D)$. Arrays INARC, LABELS, and MATE are of dimension 1..M, where M is the number of all arcs in $G(D)$. Array NR[1..N1], where N1 = N + 1 and N is the number of all nodes in $G(D)$, contains the pointers to the first neighbor of each node. Hence all starting nodes of the arcs going to node I are stored in array INARC from NR[I] through NR[I + 1] − 1, where the conjunctive arcs of $G(D)$ are only from NR[I] through NRC[I]. Processing times of activities are stored in array TIME[1..N]; therefore, the length of arc (I, J), is equal to TIME[I] for every arc (I, J) emanating from node I. Array MATE is constructed in the beginning of the procedure (by calling procedure DMATES) and matches the disjunctive arcs belonging to the same pair. If (I, J) is a disjunctive arc, then it is stored as INARC[P] = I, where NRC[J] < P < NR[J + 1]. The partner of (I, J), arc (J, I) is stored as INARC[Q] = J, where NRC[I] < Q < NR[I + 1]. The DMATES procedure assigns in this case MATE[P] = Q and MATE[Q] = P. The algorithm employed in the DMATES procedure is identical to that used in its undirected version MATES applied in the coloring procedure INTERSEQCOLORING (see Section 4.1.3).

Array ORDER[1..N] specifies a topological ordering of the nodes in a current feasible network. On input, it corresponds to an initial feasible network. Array REVORD[1..N] is conjugate with ORDER in a sense that it identifies the node positions in ORDER; therefore, we have ORDER[REVORD[I]] = I. Arc labels are stored in the corresponding positions of LABELS[1..M]. Procedures DMATES,

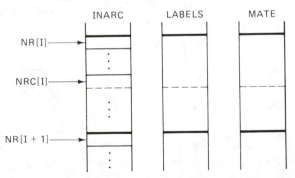

Figure 4-19 *Data structures used in the* NETWORKSCHEDULING *procedure.*

CRITICALPATH(LAB,D,E), and REORDER(I,J) are implementations of the corresponding procedures used in Algorithm 4-7.

Global Constants

N   number of activities (nodes) in the network

N1   N + 1

M   number of arcs in the network

MAX   sufficiently large integer number, an upper bound on the total number of candidates at any level of the search tree. This number is never greater than M^2

Data Types

```
TYPE   ARRN   = ARRAY[1..N] OF INTEGER;
       ARRN1  = ARRAY[1..N1] OF INTEGER;
       ARRM   = ARRAY[1..M] OF INTEGER;
       ARRMAX = ARRAY[1..MAX] OF INTEGER;
```

Procedure Parameters

```
PROCEDURE NETWORKSCHEDULING(
    N,INF                 :INTEGER;
    VAR NRC,ORDER,TIME :ARRN;
    VAR NR                :ARRN1;
    VAR INARC             :ARRM;
    VAR COUNT             :INTEGER;
    VAR LONGESTPATH    :ARRN);
```

Input

N   number of nodes in the network

INF   maximal integer number available in the system that is used

NR[1..N1]   array of pointers in a backward-star form

INARC[1..M]   array of arc initial nodes

NRC[1..N]   array of pointers to the last conjunctive arcs (see Fig. 4-19)

ORDER[1..N]   array of an initial topological ordering of nodes

TIME[1..N]   array of processing times

Output

ORDER[1..N]   array containing a topological ordering of the nodes
                       in the solution

LONGESTPATH[1..N]   array of the earliest possible starting times of activi-
                       ties in the solution

COUNT   number of feasible networks generated by the proce-
                       dure

Comment: Variable COUNT can be used to terminate calculations after a specified number (say 1000) of feasible networks has been generated. To do this, one should: declare label 1, insert the statement IF COUNT = 1000 THEN GOTO 1 before the statement COUNT: = COUNT + 1, and insert label 1 before the first FOR statement in the output of the best solution.

Example

For the network of Example 4-9 the input is

$$N = 11, \quad INF = 1000,$$

and

$$NR[1..12] = [1, 1, 3, 5, 9, 11, 14, 17, 19, 23, 25, 28]$$
$$NRC[1..11] = [0, 1, 3, 6, 9, 11, 15, 17, 20, 23, 27]$$
$$INARC[1..27] = [1, 6, 2, 8, 3, 6, 7, 9, 1, 9, 5, 2, 10, 6, 9, 4, 1, 3, 2, 8, 5, 4, 9, 6, 4, 7, 10]$$

If we assume an initial network to be that shown in Fig. 4-13(a), we may take

$$ORDER[1..11] = [1, 8, 5, 2, 3, 6, 4, 9, 10, 7, 11]$$

and the solution obtained is

$$COUNT = 4$$
$$ORDER[1..11] = [1, 8, 5, 2, 3, 6, 9, 7, 4, 10, 11]$$
$$LONGESTPATH[1..11] = [0, 0, 9, 19, 0, 6, 12, 0, 9, 12, 30]$$

Pascal Procedure NETWORKSCHEDULING

```
PROCEDURE NETWORKSCHEDULING(
    N,INF               :INTEGER;
    VAR NRC,ORDER,TIME  :ARRN;
```

```
    VAR NR                  :ARRN1;
    VAR INARC               :ARRM;
    VAR COUNT               :INTEGER;
    VAR LONGESTPATH         :ARRN);


VAR DELTA,I,I1,J,K,L,P,Q,R,S,T,T1,T2:INTEGER;
    B,BACK                              :BOOLEAN;
    D,DELARR,E,F,OPTORD,PI,REVORD    :ARRN;
    LABELS,LL,MATE,TREE,UL              :ARRM;
    CRITARC                             :ARRMAX;


PROCEDURE DMATES;
    (*  THIS PROCEDURE MATCHES DISJUNCTIVE ARCS
        OF THE SAME PAIR  *)
VAR I,J,K,L:INTEGER;
BEGIN
    FOR I:=1 TO N DO
    BEGIN
        PI[I]:=NRC[I]+1;
        D[I]:=PI[I];
        E[I]:=NR[I+1]-1
    END;
    FOR I:=1 TO N DO
        FOR K:=PI[I] TO E[I] DO
        BEGIN
            J:=INARC[K];
            L:=D[J];
            LABELS[L]:=I;
            CRITARC[L]:=K;
            D[J]:=L+1
        END;
    FOR I:=1 TO N DO
    BEGIN
        FOR K:=PI[I] TO E[I] DO
            D[LABELS[K]]:=CRITARC[K];
        FOR K:=PI[I] TO E[I] DO
            MATE[K]:=D[INARC[K]]
    END
END;  (*  DMATES  *)


PROCEDURE CRITICALPATH(LAB:2..3;VAR D,E:ARRN);
    (*  THIS PROCEDURE FINDS THE LENGTH OF LONGEST PATHS
        FROM SOURCE TO ALL OTHER NODES IN THE NETWORK
        FORMED BY ARCS WITH LABELS AT LEAST LAB  *)
VAR H,I,J,K,L,R,S:INTEGER;
BEGIN
    D[1]:=0;
    FOR K:=2 TO N DO
```

```
   BEGIN
      J:= ORDER[K];
      R:=-1;
      FOR L:=NR[J] TO NR[J+1]-1 DO
          IF LABELS[L] >= LAB THEN
          BEGIN
             I:=INARC[L];
             S:=D[I]+TIME[I];
             IF S > R THEN
             BEGIN R:=S;    H:=L  END
          END;
      D[J]:=R;   E[J]:=H
   END
END;   (*  CRITICALPATH  *)


PROCEDURE REORDER(I,J:INTEGER);
   (* THIS PROCEDURE MODIFIES TOPOLOGICAL ORDERING AFTER
      SWITCHING DISJUNCTIVE ARC (I,J) FOR ITS REVERSE  *)

VAR K,L,P,Q,R,S,U,V:INTEGER;
    NODORD          :ARRN;

PROCEDURE MOVE(VAR G,H:INTEGER);
   (*  THIS PROCEDURE MOVES NODE G TO POSITION H IN ARRAY ORDER *)
BEGIN
   ORDER[H]:=G;
   REVORD[G]:=H;
   H:=H-1
END;   (*  MOVE  *)

BEGIN
   K:=REVORD[I];   L:=REVORD[J];
   ORDER[L]:=-J;
   Q:=L;
   WHILE Q > K+1 DO
   BEGIN
      R:=ORDER[Q];
      IF R < 0 THEN
         FOR U:=NR[-R] TO NR[-R+1]-1 DO
         BEGIN
            S:=INARC[U];    V:=REVORD[S];
            IF (V > K) AND (LABELS[U] >= 2) THEN ORDER[V]:=-S
         END;
      Q:=Q-1
   END;
   P:=0;
   FOR U:=K+1 TO L DO
      IF ORDER[U] < 0 THEN
      BEGIN
         P:=P+1;
         NODORD[P]:=-ORDER[U]
      END;
```

```
      Q:=L;
      FOR U:=L-1 DOWNTO K DO
          IF ORDER[U] > 0 THEN
              MOVE(ORDER[U],Q);
      FOR U:=P DOWNTO 1 DO
          MOVE(NODORD[U],Q)
  END;  (*  REORDER  *)

BEGIN  (*  MAIN BODY  *)
   (*  INITIALIZATION  *)
    DMATES;
    FOR I:=1 TO N DO
        REVORD[ORDER[I]]:=I;
    FOR I:=1 TO N DO       (*  SETTING LABELS  *)
    BEGIN
        FOR J:=NR[I] TO NRC[I] DO
            LABELS[J]:=3;
        P:=REVORD[I];
        FOR J:=NRC[I]+1 TO NR[I+1]-1 DO
        BEGIN
            Q:=REVORD[INARC[J]];
            IF P > Q THEN LABELS[J]:=2
            ELSE LABELS[J]:=1
        END
    END;

    R:=0;    (* R IS THE LEVEL NUMBER OF THE SEARCH TREE  *)
    K:=0;
    T:=INF;
    COUNT:=0;
    REPEAT    (*  UNTIL THE SEARCH IS EXCHAUSTED  *)
        CRITICALPATH(3,D,E);
        BACK:=D[N] >= T;
        B:=TRUE;
        IF NOT BACK THEN
        BEGIN
            CRITICALPATH(2,PI,E);
            IF PI[N] < T THEN
            BEGIN  (*  UPDATING THE BEST CURRENT SOLUTION  *)
                FOR I:=1 TO N DO
                    OPTORD[I]:=ORDER[I];
                T:=PI[N]
            END;
            J:=N;   L:=K+1;
            WHILE J <> 1 DO
            BEGIN
                I1:=E[J];  I:=INARC[I1];
                IF LABELS[I1] = 2 THEN
            BEGIN  (* ARC (I,J) IS DISJUNCTIVE AND FREE *)
                LABELS[I1]:=1;         (* FINDING DELTA OF (I,J) *)
                CRITICALPATH(2,D,F); (* LONGEST PATH WITH NO (I,J) *)
                LABELS[I1]:=2;
                T1:=D[J]-PI[I];
```

```
            T2:=D[N]-D[I]-PI[N]+PI[J];
            DELTA:=T1+T2+TIME[J];
            IF T1 > DELTA THEN DELTA:=T1;
            IF T2 > DELTA THEN DELTA:=T2;
            DELTA:=DELTA-TIME[I];
            P:=L;   Q:=1;
            WHILE (P <=K) AND (DELARR[Q] <= DELTA) DO
            BEGIN P:=P+1;   Q:=Q+1 END;
            K:=K+1;    Q:=K-L+1;
            FOR S:=K DOWNTO P+1 DO
            BEGIN
                CRITARC[S]:=CRITARC[S-1];
                DELARR[Q]:=DELARR[Q-1];
                Q:=Q-1
                    END;
                    CRITARC[P]:=I1;
                    DELARR[Q]:=DELTA
                END; (* LABELS[I1]=2 - FREE ARC (I,J)  *)
                J:=I
            END; (* WHILE J <> 1 - TRAVERSING CRITICAL PATH *)
            B:=L > K;
            IF NOT B THEN
            BEGIN
                R:=R+1;
                LL[R]:=L;
                UL[R]:=K
            END
    END; (*  NOT BACK  *)
    IF BACK OR B THEN
        WHILE B AND (R > 0) DO
        BEGIN                   (* BACKTRACKING STEP *)
            P:=TREE[R];
            Q:=MATE[P];
            LABELS[P]:=1;       (* REMOVE REVERSE ARC (J,I) *)
            LABELS[Q]:=4;      (* FIX ARC (I,J) *)
            REORDER(INARC[P],INARC[Q]);
            B:=LL[R] > UL[R];
            IF B THEN
            BEGIN
                R:=R-1;
                IF R > 0 THEN
                    FOR S:=UL[R]+1 TO UL[R+1] DO
                        LABELS[CRITARC[S]]:=2
            END
        END;
    IF R > 0 THEN
        BEGIN                   (* FORWARD MOVE *)
            K:=UL[R];
            COUNT:=COUNT+1;
            I1:=LL[R];
            P:=CRITARC[I1];
            LL[R]:=I1+1;
            Q:=MATE[P];
```

```
        LABELS[Q]:=3;        (* FIX ARC (J,I) *)
        LABELS[P]:=0;
        REORDER(INARC[P],INARC[Q]);
        TREE[R]:=Q
    END
UNTIL R = 0;
                        (* OUTPUT THE BEST SOLUTION *)
FOR I:=1 TO N DO
BEGIN
    ORDER[I]:=OPTORD[I];
    REVORD[ORDER[I]]:=I
END;
FOR I:=1 TO N DO
BEGIN
    P:=REVORD[I];
    FOR J:=NRC[I]+1 TO NR[I+1]-1 DO
    BEGIN
        Q:=REVORD[INARC[J]];
        IF P > Q THEN LABELS[J]:=2
        ELSE LABELS[J]:=1
    END
END;
CRITICALPATH(2,LONGESTPATH,E)

END; (* NETWORKSCHEDULING *)
```

Computational Results

The procedure NETWORKSCHEDULING has been tested on a number of literature instances of the network scheduling problem and the job shop problem. In
particular, we used three large-size job shop problems (4–6 in Table 4-7) given by H.
Fischer and G. L. Thompson in Muth and Thompson [1963, pp. 236–237], two of
which have not yet been solved completely by any existing method. The networks
corresponding to problems 4, 5, and 6 have respectively 38, 102, and 102 nodes and
222, 1010, and 2020 arcs. Table 4-7 presents our computational results, which in the
case of the last three problems support our prediction that Balas's algorithm
generates very large number of feasible networks to prove optimality.

4.2.3. FLOW SHOP AND JOB SHOP
SCHEDULING

Here we consider certain scheduling problems which can be modeled and solved by
using the network scheduling methods described in the preceding section. However,
the problems have special structure of the precedence constraints which in many
cases allow the use of specialized methods. We briefly discuss the basic properties,

Table 4-7. Computational Results from the Application of the Procedure
NETWORKSCHEDULING to Some Network
Scheduling Problems[a]

Problem Description	Solution			Time[d]	COUNT
	Obtained	Best Known	Optimal		
1. Example 4-9	30	30	30	4 msec	4
2. 2×3 job shop Balas [1969]	31	31	31	5 msec	10
3. 4×5 job shop Balas [1967]	13	13	13	100 msec	132
4. 6×6 job shop	55	55	55	24.3 sec	10,293
	58			1.3 sec	500
	55			3.7 sec	1,500
5. 10×10 job shop		935[b]	?		
	1180			10 sec	500
	1175			19 sec	1,500
	1154			195 sec	20K
	1096			935 sec	100K
	1091			1857 sec	200K
6. 5×20 job shop		1165[c]	?		
	1614			14 sec	150
	1468			63 sec	1,500
	1468			307 sec	15K
	1465			1830 sec	100K

[a]An $m \times n$ job shop instance consists of m machines and n jobs, each requiring m operations, with one operation on each of the machines. K denotes 1000.
[b]Lageweg, B. J., *Combinatorial Planning Models*, Mathematisch Centrum, Amsterdam, 1982.
[c]McMahon, G., and M. Florian, On Scheduling with Ready Times and Due Dates to Minimize Maximum Lateness, *Oper. Res.* 23(1975), 475–482.
[d]On Amdahl 470 V/8.

algorithms, complexity, and some important easy solvable special cases of the problems.

In this section, activities, now called *operations*, are combined into larger units called *jobs*. Each operation corresponds to the processing of a job on a certain machine. We assume that there are no precedence constraints among jobs and among operations of different jobs, except those of disjunctive character imposed on the operations which are to be processed by the same machine. We make the usual assumptions about operations and machines:

1. Individual operations are not splittable and no operation can be performed on more than one machine at a time.

2 No machine can handle more than one operation at a time.

Let $\mathcal{J} = \{J_1, J_2, \ldots, J_n\}$ be a set of jobs and $\mathcal{M} = \{M_1, M_2, \ldots, M_m\}$ be a set of machines. Sometimes we will use "job j" instead of J_j and "machine i" instead of M_i. Each job J_j consists of m_j operations $O_{1j}, O_{2j}, \ldots, O_{m_j j}$; where m_j may vary per job. Some operations of a job may be processed by the same machine and in general m_j may be even greater than m. The time of an uninterrupted processing of operation O_{ij} is denoted by p_{ij}. The vector of processing times of J_j will be denoted by $\mathbf{p}_j$. The matching of operations with machines is done through numbers m_{ij} $(i = 1, 2, \ldots, m_j, \ j = 1, 2, \ldots, n)$ such that operation O_{ij} is to be processed on machine m_{ij}. The objective is to find a processing order of operations on each machine such that the maximum completion time is minimized.

If the operations of any job are not related by precedence constraints, the problem is referred to as an *open shop*. Therefore, in an open shop, the order in which a job is processed by machines is immaterial. We do not discuss this case here, and refer the reader to Gonzales and Sahni [1976] for some efficient algorithms and complexity results.

We restrict our attention to the problems in which machine orderings are specified for all jobs. In a *flow shop* each job passes all machines in the same order, hereafter assumed to be $(M_1, M_2, \ldots, M_m)$. An obvious example of such a situation is an assembly line, where the workers and work stations represent the machines. Therefore, a flow shop is characterized by $m_j = m$ and $m_{ij} = i$ for $i = 1, 2, \ldots, m$ and $j = 1, 2, \ldots, n$. In a *job shop* different machine orderings may be specified for the jobs. A job in a flow and job shop scheduling problem is a sequence of operations specified by a machine ordering in which it is to be processed. A job shop problem can be considered as a special instance of the network scheduling problem, but the networks are of little use in flow shop scheduling (see Problem 4-44).

Informally, a schedule can be represented by the *Gantt chart*, in which machines are shown along the vertical axis and a time scale is along the horizontal axis.

Example 4-10

Figure 4-20 illustrates the use of the Gantt charts in representing schedules of job shop and flow shop problems, respectively. An instance of a job shop consists of three jobs which are split into two, three and three operations, respectively. Machine orderings and the processing time vectors are, respectively, $(m_{11}, m_{21}) = (2, 1)$, $\mathbf{p}_1 = (2, 1)$ for job J_1; $(m_{12}, m_{22}, m_{32}) = (1, 3, 2)$ and $\mathbf{p}_2 = (1, 3, 2)$ for job J_2; and $(m_{13}, m_{23}, m_{33}) = (1, 2, 3)$ and $\mathbf{p}_3 = (3, 2, 2)$ for job J_3. A flow shop instance also

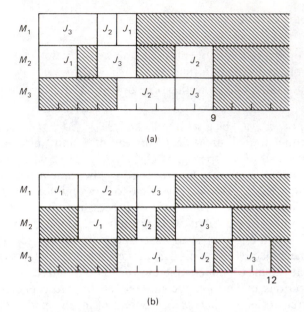

Figure 4-20   *Representation of schedules by the Gantt charts.*

consists of three jobs with the following processing times on three machines: $\mathbf{p}_1 = (2, 2, 4)$, $\mathbf{p}_2 = (3, 1, 1)$, and $\mathbf{p}_3 = (2, 3, 2)$, respectively.   ∎

In the sequel, we adopt the following notation, which is consistent with a more general convention introduced in Section 4.2.4. The job shop and flow shop problems are denoted by $J\|C_{\max}$ and $F\|C_{\max}$, respectively. Notice that neither the number of machines m nor the number of jobs n are specified, for they are considered as free parameters. If we restrict our attention to a particular number of machines, then we write $Jm\|C_{\max}$ and $Fm\|C_{\max}$. Therefore, $F2\|C_{\max}$ is the flow shop problem of scheduling n jobs on two machines. The number of jobs is always a free parameter.

Flow Shop Scheduling

We present here a simple argument showing that flow shop scheduling on at least three machines is computationally very hard. However, two-machine problems can be easily solved by Johnson's algorithm. Finally, a special case of the problem is shown to be reducible to the traveling salesman problem.

Although the order of machines is the same for all jobs in a flow shop problem, it is not clear how this assumption influences the orderings of jobs on the machines. However, it is easy to show that if the processing times are positive, every flow shop problem has an optimal solution with the same processing order of jobs on the first two machines and with the same processing order on the last two machines (see Problem 4-45). It follows that in a three-machine flow shop we may restrict ourselves to the same processing orders on all the machines. In problems with at least four machines, the machine processing orderings in the optimal solutions can vary per machine, as it can be illustrated by the example with four machines and two jobs with the processing times $\mathbf{p}_1 = (4, 1, 1, 4)$ and $\mathbf{p}_2 = (1, 4, 4, 1)$.

We further restrict the flow shop problem by assuming an identical processing order on all the machines. Such a problem is called a *permutation flow shop* problem. Its schedule can be completely specified by one permutation of jobs and hence is called a *permutation schedule*. A permutation flow shop is denoted in our convention by $PF\|C_{\max}$. As we already know, when processing times are positive, then $F2\|C_{\max}$ is equivalent to $PF2\|C_{\max}$ and $F3\|C_{\max}$ is equivalent to $PF3\|C_{\max}$.

Complexity of Flow Shop Scheduling

Before presenting a positive result which will lead to efficient scheduling of two-machine and some of the three-machine flow shops, we show that the existence of a polynomial-time algorithm for solving the general flow shop problem on at least three machines is very unlikely. To this end, we shall introduce a special version of the knapsack problem (see Section 2.1). The knapsack problem and its variations are the most popular problems used in proving that some scheduling problems are intractable. For the purpose of this section, we define the *decision knapsack* (DK) problem (DKP) as follows:

Given positive integers $a_1, a_2, \ldots, a_q$, and b.
Is there a subset $B \subset \{1, 2, \ldots, q\}$ such that $\displaystyle\sum_{j \in B} a_j = b$?

The DK problem is NP-complete; therefore, it represents a large family of problems which have no known polynomial solution methods and the existence of such a method for any of these problems would imply the polynomial solvability of all NP-complete problems.

For an instance $a_1, a_2, \ldots, a_q, b$ of the DK problem, we construct an instance of the $PF3\|C_{\max}$ problem by defining $n = q + 1$ jobs with the processing time vectors $(0, a_1, 0), (0, a_2, 0), \ldots, (0, a_q, 0)$ and $(b, 1, \sum_{j=1}^{q} a_j - b)$. The problem now is to find if there exists a permutation of jobs of length $\sum_{j=1}^{q} a_j + 1$.

Figure 4-21 shows an example schedule. Each schedule has length at least $\sum_{j=1}^{q} a_j + 1$, due to the processing times of job $q + 1$, and there exists a schedule of exactly such length if there is a subset B of the a_j's such that $\sum_{j \in B} a_j = b$. If this is the case, the jobs with processing times in B can be processed on the second machine before job $q + 1$ and the resulting schedule has the shortest possible length. Hence the DK problem can be solved by solving the corresponding instance of the

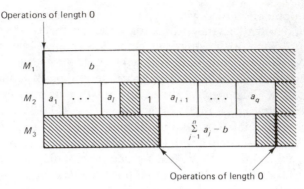

Operations of length 0

Figure 4-21 *Flow shop schedule corresponding to a* DK *problem instance.*

three-machine flow shop problem. It follows that we can expect the latter problem to be at least as hard as the former one.

Johnson's Algorithm

The $F2\|C_{\max}$ problem can be solved efficiently by applying Johnson's algorithm, which is considered as the first major result in scheduling theory. S. M. Johnson proved the following theorem, also known as *Johnson's rule* which gives rise to a very fast solution method.

Theorem 4-7 (Johnson [1954])

In a $F2\|C_{\max}$ problem, if

$$\min\{ p_{1i}, p_{2j}\} \leqslant \min\{ p_{2i}, p_{1j}\}$$

then job J_i precedes job J_j in an optimal schedule. ∎

Different proofs of this theorem can be found in almost every general text on scheduling. The theorem has a very useful corollary which helps to construct a solution iteratively from both its ends. Let us consider the value $x = \min_{1 \leqslant i \leqslant n}\{ p_{1i}, p_{2i}\}$. We can have two cases, either $x = p_{1k}$ for some k or $x = p_{2l}$ for some l. In the former case, we have $\min\{ p_{1k}, p_{2j}\} \leqslant \min\{ p_{2k}, p_{1j}\}$ for every job J_j ($j \neq k$). Therefore, according to the theorem, job J_k precedes every other job J_j in an optimal solution. By the same argument, in the latter case, job J_l follows every other job J_i ($i \neq l$). Algorithm 4-8 generates an optimal permutation $\pi = (\pi_1, \pi_2, \ldots, \pi_n)$ for a two-machine flow shop problem following these observations. It can be easily implemented in $O(n \log n)$ time.

Algorithm 4-8:   Johnson's Algorithm
for Solving Two-Machine Flow Shop Problem

```
begin
    N₁ ← {j: p₁ⱼ < p₂ⱼ}; (* Initial jobs on machine 1 *)
    N₂ ← {j: p₁ⱼ ≥ p₂ⱼ}; (* Final jobs on machine 2 *)
    order the jobs in N₁ according to nondecreasing p₁ⱼ;
    order the jobs in N₂ according to nonincreasing p₂ⱼ;
    π is formed by the ordered set N₁ followed by the ordered set N₂
end
```

Example 4-11

To illustrate the algorithm, let us consider a five-job problem with the following processing time vectors $\mathbf{p}_1 = (4, 2)$, $\mathbf{p}_2 = (1, 3)$, $\mathbf{p}_3 = (4, 4)$, $\mathbf{p}_4 = (5, 6)$, $\mathbf{p}_5 = (3, 2)$. Algorithm 4-8 first creates the sets $N_1 = \{2, 4\}$ and $N_2 = \{1, 3, 5\}$, then orders them $N_1 = (2, 4)$ and $N_2 = (3, 1, 5)$, and finally generates a schedule $\pi = (2, 4, 3, 1, 5)$, shown in Fig. 4-22.   ■

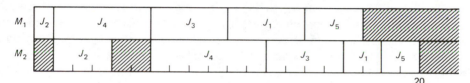

Figure 4-22   *Optimal schedule for the problem of Example 4-11.*

There are several special cases of the general permutation flow shop problem with more than two machines which can be efficiently solved by special-purpose algorithms. A large class of the cases can be handled by Johnson's algorithm or its modifications. One such case appeared in Johnson's original presentation. He proved that if the second machine in the $F3\|C_{\max}$ problem is dominated by one of the others, then an optimal permutation of jobs can be found by applying his algorithm to a two-machine problem. More precisely, machine 2 is *dominated* by machine i ($i = 1$ or 3) if

$$\min_j p_{ij} \geq \max_j p_{2j} \tag{4-11}$$

This assumption makes it possible to prove a theorem similar to Theorem 4-7 which gives rise to the two-machine flow shop problem with the processing times $p_{1j} + p_{2j}$ on the first machine and $p_{2j} + p_{3j}$ on the second machine.

It has been computationally tested that even if the second machine is not dominated by any other, the transformation of the three-machine problem to two machines yields reasonably good approximate solutions to the original problem. Johnson's algorithm can also be used to generate approximate schedules for an arbitrary number of machines (see Problem 4-48).

As pointed out by Coffman and Denning [1973], the three-machine model with the second machine time dominated by one of the others can have some applications in computer systems. Consider, for instance, a situation in which the second machine represents CPU activities while the first and the third apply to input and output processing, respectively. Assume also that input, execute, and output phases are disjoint and appear in a sequence. In several applications, the time spent on processing input and output dominates the computation time; therefore, condition (4-11) might very well be met and the jobs could be optimally ordered by Johnson's algorithm.

No-Wait Flow Shop

One of the assumptions used very often in scheduling without formulating it explicitly is that in the job processing there is always enough intermediate storage space to hold partially processed jobs which cannot be processed further because other machines are busy. In many practical situations, however, once the processing of a job begins on its first machine, processing on the subsequent machines must be carried out with no delays between any two machines. The flow shop problem with no intermediate storage is sometimes referred to as the *no-wait flow shop problem*. Possible applications of this problem can be found in the steel industry, where very high temperature of processed materials have to be maintained during the production process and any delay between operations of the same job may result in cooling that makes the process difficult or impossible to continue. Figure 4-23 shows a no-wait version of the schedule shown in Fig. 4-20(b). The no-wait flow-shop problem can be transformed to the traveling salesman problem and then solved by one of the algorithms designed for the latter problem (see Section 3.8).

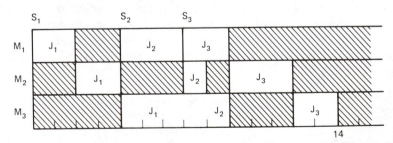

Figure 4-23   *No-wait schedule.*

Let us consider an example of a no-wait schedule of Fig. 4-23. It is clear that the length of such a schedule is equal to the total length of the last job plus the sum of intervals $S_{j+1} - S_j$, where S_j is the starting time of J_j. In general, if $\pi = (\pi_1, \pi_2, \ldots, \pi_n)$ is a permutation schedule, then

$$C_{\max}^{\text{no-wait}}(\pi) = \sum_{j=1}^{n-1} c_{\pi_j \pi_{j+1}} + \sum_{k=1}^{m} p_{k\pi_n} \qquad (4\text{-}12)$$

where c_{ij} denotes $S_j - S_i$. Now the problem is to minimize $C_{\max}^{\text{no-wait}}(\pi)$ over all permutation schedules π. First, we have to determine the coefficients c_{ij}. One can easily see that a particular c_{ij} does not depend on any other job that precedes J_i or follows J_j. Time c_{ij} reflects how close neighbor jobs can be pushed toward each other provided that no delay is introduced in the processing of each of these two jobs. The coefficient c_{ij} may also be interpreted as follows: Assume that both jobs arrive at the same time; then c_{ij} is the total delay (measured from the start of job J_i) incurred by job J_j when it follows J_i in a no-wait schedule. How long we have to delay job J_j depends on how far we have to push $k - 1$ nondelayed operations of J_j to match their end with the end of k nondelayed operations of J_i. Hence

$$c_{ij} = \max_{1 \leqslant k \leqslant m} \left\{ \sum_{l=1}^{k} p_{li} - \sum_{l=1}^{k-1} p_{lj} \right\} \qquad \text{for } i, j = 1, 2, \ldots, n \qquad (4\text{-}13)$$

In order to unify expression (4-12) we introduce a dummy job J_0 with zero processing times on all machines, $p_{i0} = 0$ $(i = 1, 2, \ldots, m)$, which represents the beginning and the end of a schedule. It is easy to verify that (4-13) gives $c_{0j} = 0$ for every starting job J_j of a schedule and $c_{i0} = \sum_{l=1}^{m} p_{li}$ for every job J_i which terminates a schedule. Hence (4-12) can be written as

$$C_{\max}^{\text{no-wait}}(\pi') = \sum_{j=0}^{n} c_{\pi_j' \pi_{j+1}'}$$

where $\pi' = (\pi_0', \pi_1', \ldots, \pi_n')$ is a permutation of $\{0, 1, 2, \ldots, n\}$, and we assume that $\pi_{n+1}' = \pi_0'$.

Minimization of $C_{\max}^{\text{no-wait}}(\pi')$ over all permutations π' corresponds to solving the traveling salesman problem on $n + 1$ "cities" $0, 1, 2, \ldots, n$ with the "distance matrix" $[c_{ij}]$ defined by (4-13) for $i, j = 0, 1, 2, \ldots, n$.

Job Shop Scheduling

Recall that in a job shop scheduling problem, a job may pass the machines in an arbitrary order which may vary per job, that a job may be processed by the same machine several times, and that there is no bound on the number of operations a job may consist of. A schedule of a job shop problem is usually illustrated by the Gantt chart [see Fig. 4-20(a)]; however, the problem itself is very often represented by a disjunctive network. This concept has been introduced in Section 4.2.1 and we refer

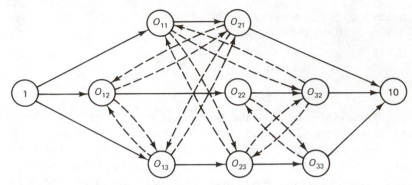

Figure 4-24   *Disjunctive network representing the job shop problem of Fig.* 4-20(*a*).

the reader there for a detailed description of its use. As an example, we show in Fig. 4-24 a disjunctive network corresponding to the job shop problem of Fig. 4-20(a).

    In general, operations of a particular job form a conjunctive path from source to sink, and the operations to be performed on the same machine are mutually connected by pairs of disjunctive arcs. Two operations of the same job which are to be processed by the same machine are, however, not joined by disjunctive arcs since their order on the machine is already determined by the conjunctive arcs. Hence, if in the example of Fig. 4-24, we add one more operation to the first job and assign it to the second machine, then the node corresponding to the new operation will not be connected by disjunctive arcs with the first operation of J_1, which is also assigned to machine 2.

    The disjunctive network model of the job shop problem is a special case of that used in network scheduling. Thus any method designed to solve the latter problem can also be applied to solving the job shop problem. Some computational results of the application of the NETWORKSCHEDULING procedure to the job shop problems are reported in the preceding section and for the results of more extensive experiments the reader is referred to the published literature.

Jackson's Algorithm

Finally, we mention a special case of the job shop problem $J2|m_j \leqslant 2|C_{max}$, which can be solved by a polynomial-time algorithm. It corresponds to flow shop problem $F2||C_{max}$ and may be considered as the latter one with relaxed machine requirements. Assumption $m_j \leqslant 2$ in the second field of the problem description means that every job consists of at most two operations, which, since this is a job shop, can be assigned to any or to both available machines. For the sake of simplicity we may assume that if a job consists of two operations, they have to be processed by different machines. Thus all jobs can be partitioned into the following four groups according to the number and the machine assignments of operations:

$$\mathcal{J}_{12} = \{J_j: m_j = 2, m_{1j} = 1, m_{2j} = 2\}, \quad \mathcal{J}_{21} = \{J_j: m_j = 2, m_{1j} = 2, m_{2j} = 1\}$$
$$\mathcal{J}_1 = \{J_j: m_j = 1, m_{1j} = 1\}, \qquad\qquad \mathcal{J}_2 = \{J_j: m_j = 1, m_{1j} = 2\}$$

The solution to the $J2|m_j \leqslant 2|C_{max}$ problem is presented in the following theorem proved originally by J. R. Jackson in 1956, who was inspired by Johnson's theorem (see Theorem 4-7).

Theorem 4-8 (Jackson [1956])

An optimal solution to a problem $J2|m_j \leqslant 2|C_{max}$ can be formed by the following processing order:

$$\text{On machine } M_1: \quad \mathcal{J}_{12} - \mathcal{J}_1 - \mathcal{J}_{21}$$
$$\text{On machine } M_2: \quad \mathcal{J}_{21} - \mathcal{J}_2 - \mathcal{J}_{12}$$

where jobs in $\mathcal{J}_{12}$ and $\mathcal{J}_{21}$ are ordered by Johnson's algorithm, and jobs in $\mathcal{J}_1$ and $\mathcal{J}_2$ are in arbitrary order. ∎

The theorem can be proved by applying an interchange argument, explained in the next section. The details of the proof are left to the reader.

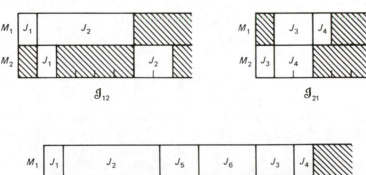

Figure 4-25 *Solution of the job shop problem of Example 4-12.*

Example 4-12

Let us consider a job shop problem with the following job partition:

$$\mathcal{J}_{12} = \{J_1 = (1,1), J_2 = (5,2)\}, \mathcal{J}_{21} = \{J_3 = (1,2), J_4 = (2,1)\}$$
$$\mathcal{J}_1 = \{J_5 = 2, J_6 = 3\}, \qquad \mathcal{J}_2 = \{J_7 = 1\}$$

where the jobs are represented by their processing times on corresponding machines. Figure 4-25 shows the schedules of $\mathcal{J}_{12}$, $\mathcal{J}_{21}$, and the schedule of all jobs constructed by applying Theorem 4-8. ■

Theorem 4-8 is the strongest possible result in a sense that both problems $J2|m_j \leqslant 3|C_{\max}$ and $J3|m_j \leqslant 2|C_{\max}$ are as hard as the (decision) knapsack problem (see Problem 4-54).

4.2.4. GENERAL SCHEDULING PROBLEM

The problems we have discussed so far indicate that there are three main types of objects that define a scheduling problem, namely *activities* (also called *jobs*, *tasks*, or *operations*), *machines* (or *processors*), and additional *resources*.

In a network scheduling problem (see Sections 4.2.1 and 4.2.2), given is a partially ordered set of activities which are to be completed in the shortest possible time period subject to some resource constraints. We discussed only problems in which the resources consist of m different units and each activity needs for its processing at most one specific unit. Such resources can be interpreted as machines, so our presentation was in terms of activities to be performed on machines.

In a flow or job shop problem (see Section 4.2.3), the set of all activities (called there operations) is partitioned into jobs which are to be processed on a set of machines subject to some precedence constraints among operations. Again, machines are the only resources.

In what follows, we use the terms *task* for a simplest activity to be performed and *machine* for a processing unit to which the tasks are to be assigned. Let $\mathcal{Z} = \{Z_1, Z_2, \ldots, Z_n\}$ denote the set of tasks and $\mathcal{M} = \{M_1, M_2, \ldots, M_m\}$ denote the set of machines. Where it causes no confusion, we simply use "task j" instead of Z_j and "machine i" instead of M_i. Informally, a *schedule* is an assignment of tasks to machines and to the time periods satisfying the restrictions imposed on both types of objects and on their relations. Therefore, a schedule is a synonym of a feasible solution to a scheduling problem.

Machines are distinguished among other resources by their special properties and relations to tasks. We assume throughout this section that:

Each machine can process at most one task at a time.

Each task can be processed on at most one machine at a time.

Each task Z_j can be characterized by a set of data containing:

(t1) A *processing time* (also called *duration time* in network scheduling) p_j. The processing time usually includes both directed processing time and facility setup time.

(t2) A *release date* (*ready* or *arrival* time) r_j, at which task j becomes available for processing. When all tasks arrive simultaneously at the starting time of the project, we assume r_j to be 0. Release dates should not be confused with the earliest starting times of activities calculated in network scheduling, for the latter result from the precedence constraints.

(t3) A *due date* d_j, the point in time by which task j should be completed. The due dates that must be respected are called *deadlines*.

(t4) A *weight* w_j, which indicates the relative importance of task j.

(t5) A nondecreasing (real) *cost function* f_j, where $f_j(t)$ is the cost of completing task j at time t.

In a schedule, it is natural to assume that tasks start as early as possible (a motivation behind this assumption is discussed at the end of this section). Therefore, given a schedule, we can compute the *starting time* S_j and the *completion* (or *finishing*) time C_j of each task Z_j.

The measures that evaluate schedules, called simply the optimality criteria, are usually some functions of task completion times and other quantities related to task parameters. The most often used are:

(t7) The *lateness* L_j of task j, which is the amount of time by which the completion time exceeds the due date, that is, $L_j = C_j - d_j$.

(t6) The *flow time* F_j, which is equal to the amount of time task j spends in the system, that is, $F_j = C_j - r_j$.

(t8) The *tardiness* T_j, which is the lateness of task j if it fails to meet its due date or zero otherwise, that is, $T_j = \max\{0, L_j\}$.

The flow time of a task measures the task time spent in the system between the task arrival and its departure. The lateness reflects the relation between the completion time and the due date and can be negative whenever a task is completed early. Negative lateness corresponds to better service and positive to poorer one. When the cost of processing all tasks depends only on positive lateness, and there is no benefit from early finished tasks, the tardiness can be used as a proper task performance measure.

Tasks can be mutually related by *precedence constraints* represented by an acyclic digraph (network) G. In the network scheduling, the task system may form an arbitrary acyclic network, whereas in job and flow shop problems tasks are combined into unrelated chains.

The relations between tasks and machines are usually expressed by the collection of integers $m_j(j = 1, 2, \ldots, n)$, where $1 \leqslant m_j \leqslant m$, specifying that task Z_j must be processed on machine m_j. Relaxing this very rigid constraint we may be given for each task a subset $\mathfrak{M}_j \subset \mathfrak{M}$ of machines and values p_{ij} such that task j can be processed on one of the machines $M_i \in \mathfrak{M}_j$ in time p_{ij}. Very often, we have $\mathfrak{M}_j = \mathfrak{M}$ and $p_{ij} = p_j$ for every $M_i \in \mathfrak{M}$ and for every task j; therefore, each task can be processed on one of the *identical* (*parallel*) machines in $\mathfrak{M}$.

Apart from machines, there may be some other resources required by tasks during their execution. We will, however, not consider other resources than machines.

In some applications, for instance in the case of computer processors as machines, the processing of a task may be interrupted and resumed later. Schedules that allow interruptions of tasks are called *preemptive*.

Finally, we have to describe how schedules are evaluated. The objective in any scheduling problem is generally to minimize the cost related to the scheduling decisions. The cost functions, also called *performance measures*, are defined in terms of characteristics and some other quantities related to schedules. The objective in the problems discussed in Sections 4.2.1 through 4.2.3 was to minimize the maximum completion time (also called *schedule time* or *make-span*). We express this measure as $C_{\max} = \max\{C_j: j = 1, 2, \ldots, n\}$. It is a special case of a more general optimality criterion involving cost functions and is called the *maximum cost*

$$f_{\max} = \max\{f_j(C_j): j = 1, 2, \ldots, n\} \tag{4-14}$$

Hence $C_{\max}$ is $f_{\max}$ for $f_j(t) = t$. The maximum cost may also involve other task quantities; we can have, for instance:

Maximum flow time:	$F_{\max} = \max\{F_j: j = 1, 2, \ldots, n\}$
Maximum lateness:	$L_{\max} = \max\{L_j: j = 1, 2, \ldots, n\}$
Maximum tardiness:	$T_{\max} = \max\{T_j: j = 1, 2, \ldots, n\}$

Another collection of performance measures is obtained by considering the *total weighted cost* (or *the weighted sum of costs*)

$$\sum w_j f_j = \sum_{j=1}^{n} w_j f_j(C_j) \tag{4-15}$$

which is also to be minimized. Thus we have the weighted sum of completion times $\sum w_j C_j$, flow times $\sum w_j F_j$, latenesses $\sum w_j L_j$, and tardinesses $\sum w_j T_j$. If $w_j = 1$ ($j = 1, 2, \ldots, n$), then (4-15) becomes the *sum of costs* (denoted by $\sum f_j$) and we have, respectively, ΣC_j, ΣF_j, ΣL_j, and ΣT_j. If $w_j = 1/n$ $(j = 1, 2, \ldots, n)$, then (4-15) becomes the *average cost* and again we may consider $\bar{f} = \Sigma f_j/n$, $\bar{F} = \Sigma F_j/n$, $\bar{L} = \Sigma L_j/n$, and $\bar{T} = \Sigma T_j/n$.

Some of the performance measures are pairwise equivalent; that is, they have the same sets of optimal schedules. Every average criterion is equivalent to the corresponding unweighted sum of costs. We also have

$$\sum w_j C_j = \sum_{j=1}^{n} w_j F_j + \sum_{j=1}^{n} w_j r_j = \sum_{j=1}^{n} w_j L_j + \sum_{j=1}^{n} w_j d_j$$

Therefore, $\Sigma w_j C_j$, $\Sigma w_j F_j$, and $\Sigma w_j L_j$ are equivalent, and similarly, ΣC_j, ΣF_j, and ΣL_j are equivalent. It can also be shown that every schedule that minimizes $L_{\max}$ also minimizes $T_{\max}$ (see Problem 4-56).

Related to lateness and tardiness measures there are also measures which involve only the number of *tardy* (late) *tasks*. For a given schedule, let U_j be defined as 1 if $C_j > d_j$, and 0 otherwise. Thus U_j is the unit penalty, ΣU_j is the number of tardy tasks, and $\Sigma w_j U_j$ is its weighted version. Note that there is no need to introduce $U_{\max}$. Yet some other performance measures are defined in Problems 4-58, 4-59, and 4-60.

For convenience and simplicity, the scheduling problems are usually defined by using a three-field notation $\alpha|\beta|\gamma$. The full details of this convention can be found in Graham et al. [1979]. We restrict our attention to those values of the fields that correspond to the problems considered in this chapter. We make a general assumption that there are n tasks to be processed on m machines.

The first field α specifies the machine environment, and α is a positive integer if m is constant and equal to α. Otherwise, the first field is empty means that m is assumed to be variable. To distinguish the shop scheduling problems we assumed $\alpha = O$ for an open shop, $\alpha = F$ for a flow shop, $\alpha = PF$ for a permutation flow shop, and $\alpha = J$ for a job shop.

The second field describes task characteristics. If β is empty, we assume that no preemptions are allowed, no resources (except machines) are required by tasks, tasks are independent (i.e., no precedence constraints among tasks are present), the release dates are the same (we assume that $r_j = 0$, $j = 1, 2, \ldots, n$), and the due dates and processing times are arbitrary. Otherwise, in the second field we can use for instance:

pmtn	if preemptions are allowed
res	when tasks require some additional resources
prec	if the precedence constraints between tasks are present
tree (*forest*)	if the precedence constraints form a tree (or a forest)
r_j	if the release dates may differ and are specified
$p_j = 1$	if each task has unit processing time (UPT)

Notice that due dates d_j are not assumed to appear in β, since their significance to the problem depends entirely on the performance measure which is specified in the next field.

The last field γ refers to the optimality criterion. Therefore, γ is either the maximum cost $f_{\max}$ or the (weighted) total cost $\sum w_j f_j$. As pointed out earlier, we may assume that

$$f_{\max} \in \{C_{\max}, F_{\max}, L_{\max}\} \quad \text{and} \quad \sum w_j f_j \in \{\sum w_j C_j, \sum w_j T_j, \sum w_j U_j, \sum C_j, \sum T_j, \sum U_j\}.$$

Using the foregoing convention, some of the scheduling problems discussed so far in this section can be described as follows:

$m = 0| prec|C_{\max}$ minimize the maximum completion time of n related tasks. This is the network scheduling problem with no resource (machine) constraints.

$F2|no\text{-}wait|C_{\max}$ the flow shop problem on two machines and with no delays between the operations of the same job.

Let us again consider the performance measures. Each of the measures we have introduced is a function of the task completion times $C_1, C_2, \ldots, C_n$ so that they can be expressed as $\Phi(C_1, C_2, \ldots, C_n)$. Furthermore, these functions are *regular* measures in the sense that

$$\Phi(C_1, C_2, \ldots, C_n) < \Phi(C_1', C_2', \ldots, C_n')$$

if $C_j < C_j'$ for at least one task Z_j. A regular measure is therefore a nondecreasing function in every variable, and hence there is no advantage in delaying the completion of any task.

In the class of nonpreemptive scheduling problems, regular measures make it possible to consider task schedules and task sequences as equivalent notions. A task schedule that can be completely determined by a permutation of tasks is called a *permutation schedule*. We have already encountered such schedules in the network scheduling and in the permutation flow shop. It is clear that every schedule can uniquely determine a sequence of tasks. For instance, we may order the tasks lexicographically with respect to 2-tuples (S_j, m_j), where S_j is the starting time of task j and m_j is the machine to which j is assigned. The resulting order is linear, since if $S_j = S_{j'}$ for some $j = j'$, then $m_j \neq m_{j'}$ for two tasks cannot be started at the same time on the same machine. On the other hand, for a given sequence of tasks on each machine, the earliest possible starting times of tasks can be found by applying the forward-pass calculations similar to that used in the network scheduling; every task is performed as soon as possible, consistent with the imposed restrictions. This justifies our early informal definition of a schedule.

In the sequel, schedules are represented and illustrated with the Gantt charts introduced in Section 4.2.3 (see Fig. 4-20).

4.2.5. SINGLE MACHINE SCHEDULING

Our goal is to present the fundamental results on scheduling a single machine together with problems that can be solved by the related methods. We also show that even slight generalizations (relaxations of the problems restrictions) may turn these problems into the class of intractable problems.

Maximum Cost Measures

We consider the maximum-cost measure (4-14), and assume first that the release dates are equal, that is, $r_j = 0$ ($j = 1, 2, \ldots, n$). When no additional restrictions are imposed on the tasks, then the maximum completion time satisfies $C_{max} = \sum_{j=1}^{n} p_j$ and is clearly sequence-independent. Therefore, the $1\|C_{max}$ problem vanishes in this case.

Before proceeding further, we shall prove a general property of single-machine schedules for equal task release dates and regular measures. The machine *idle time* is defined as $I = C_{max} - \sum_{j=1}^{n} p_j$. We have

Theorem 4-9 (Conway et al. [1967])

For a single-machine problem with equal release dates, there exists an optimal schedule with respect to any regular measure that has no machine idle time and no task preemptions. ■

The proof of this theorem is illustrated in Fig. 4-26. If the idle time I is positive and for two neighbor tasks k and l we have $S_l > C_k$, then shifting all the tasks that follow k by $S_l - C_k$ would only decrease their completion times. On the other hand, if a task k is split into several parts, then moving all the parts to the right most one would again only decrease the completion times of the tasks, which were originally scheduled between the first and the last parts of task k. Since a performance measure is assumed to be regular, none of these two transformations of a schedule can

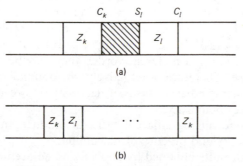

Figure 4-26 *An illustration for the proof of Theorem 4-9.*

increase the measure value. It follows that in a single-machine scheduling problem with a regular measure, when the release dates are not involved, it is sufficient to consider only schedules without idle time and with no preemptions.

If tasks are released at different dates, then clearly it is always advantageous for a regular maximum cost measure to schedule an available task. Therefore, the $1|r_j|C_{max}$ can be easily solved by ordering the tasks according to nondecreasing r_j.

Assume now that additionally some precedence constraints are imposed on the tasks; therefore, our problem is $1|prec, r_j|C_{max}$. Let an acyclic digraph G represent the precedence constraints and assume that the nodes of G are in a topological order. Note that if i and j are two tasks such that i precedes j in G, then $S_j \geqslant C_i \geqslant r_i + p_i$ in every schedule. Therefore, we may further assume that $r_j \geqslant r_i + p_i \geqslant r_i$ whenever i precedes j. Hence for a given set of release dates and task precedence constraints, we may set new release dates:

$$r_j' \leftarrow \max\{r_j, \max\{r_i + p_i : i \text{ precedes } j \text{ in } G\}\}, \qquad j = 1, 2, \ldots, n$$

We can now ignore the precedence constraints. Thus the problem $1|prec, r_j|C_{max}$ can be reduced to $1|r_j'|C_{max}$ for the modified release dates but with no precedence constraints. Then it can be solved by ordering the tasks according to nondecreasing r_j'. Therefore, problems $1|r_j|C_{max}$ and $1|prec, r_j|C_{max}$ can be solved in $O(n \log n)$ and $O(e + n \log n)$ time, respectively, where e is the number of arcs in the precedence graph. Summarizing, we have

Theorem 4-10

The maximum completion time in the $1|prec, r_j|C_{max}$ is maximized by ordering the tasks according to nondecreasing modified released dates.   ■

Adjacent Pairwise Interchange Method

Ordering the tasks according to a certain quantity can be viewed as an iterative modification of a given schedule of tasks by using local transpositions of tasks. Let us consider, for instance, three independent tasks Z_1, Z_2, Z_3 and assume that $r_1 = 0$, $p_1 = 1$, $r_2 = 2$, $p_2 = 2$, and $r_3 = 3$, $p_3 = 1$. Figure 4-27(a) shows the schedule (Z_1, Z_3, Z_2) in which Z_3 and Z_2 violate the optimal ordering. The effect of interchanging these two tasks is shown in Fig. 4-27(b). It is obvious that whenever two adjacent tasks are not in a preference order, they may be interchanged with no resultant cost increase. Therefore, alternatively, an optimal task ordering can be produced by interchanging adjacent tasks, if necessary. However, this method could have $O(n^2)$ worst-time complexity.

The idea of interchanging adjacent tasks to produce an optimal ordering is widely used in the theory and practice of scheduling.

One of the first results obtained by applying an adjacent pairwise interchange argument was the solution of the $1\|L_{max}$ problem given by J. R. Jackson in 1955. He

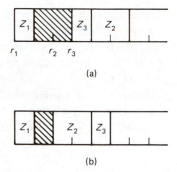

Figure 4-27 *Interchange argument.*

proved that the maximum lateness on a single machine is minimized by ordering the tasks according to nondecreasing due dates d_j. Let us assume that two adjacent tasks Z_i and Z_j satisfy $d_j < d_i$. When we interchange these two tasks, their completion times C_i' and C_j' and latenesses L_i' and L_j' in the resulting schedule satisfy

$$L_j' = C_j' - d_j \leqslant C_j - d_j = L_j$$
$$L_i' = C_i' - d_i = C_j - d_i \leqslant C_j - d_j = L_j$$

Hence

$$L_{\max}' = \max\{L_l': l = 1, 2, \ldots, n\} \leqslant L_{\max}, \text{ where } L_l' = L_l(l = 1, 2, \ldots, n, l \neq i, j)$$

and we may therefore conclude that scheduling the tasks by earliest due dates (EDD) minimizes $L_{\max}$. It follows from Problem 4-56 that the maximum tardiness is also minimized by scheduling according to EDD. Hence

Theorem 4-11 (Jackson [1955])

The maximum task lateness in the $1\|L_{\max}$ problem and the maximum task tardiness in the $1\|T_{\max}$ problem are minimized by scheduling the tasks according to earliest due dates. ∎

In a similar way we may solve the problems $1\|L_{\min}$ and $1\|T_{\min}$, where $L_{\min} = \min\{L_j: j = 1, 2, \ldots, n\}$ and $T_{\min} = \min\{T_j: j = 1, 2, \ldots, n\}$ and they are to be maximized (see Problem 4-60). Problem 4-61 presents an answer to the question when a set of tasks with precedence constraints can be scheduled to meet *all* due dates (called *deadlines*).

When we allow nonzero release dates, the $1|r_j|L_{\max}$ problem turns out to be very difficult and no polynomial-time algorithm is known for its solution (see

Problem 4-62). However, when additionally the precedence constraints are introduced and the tasks can be preempted, then the $1|pmtn, prec, r_j|L_{max}$ can still be solved by a polynomial-time algorithm (see Problem 4-63).

We shall now consider the single-machine scheduling problem with the general maximum cost measure $f_{max} = \max\{f_j(C_j): j = 1, 2, \ldots, n\}$, where f_j are monotone nondecreasing cost functions. We assume that the release dates are equal, and tasks may be related by precedence constraints which are given in the form of an acyclic digraph G with topologically ordered nodes. Therefore, the problem can be denoted as $1|prec|f_{max}$. Again using an interchange argument, we will show the following theorem.

Theorem 4-12 (Lawler [1973])

Let W be the set of nodes of G with no successors. If k is such that $f_k(P) = \min\{f_j(P): j \in W\}$, where $P = \sum_{j=1}^{n} p_j$, there exists an optimal schedule for $1|prec|f_{max}$ in which task k is last. ∎

Consider a schedule S' in which k is not last [see Fig. 4-28(a)], and modify S' to obtain S as shown in Fig. 4-28(b). Notice that if the precedence constraints are satisfied in S', they are also satisfied in S, since both k and k' must belong to W. Except for task k, no other task is completed later in S than in S'. Because the cost functions are monotone and nondecreasing, no cost can be greater in S than in S', except possibly that for k. By the theorem assumption, however, task k is such that $f_k(P) \leqslant f_{k'}(P)$. Therefore, the maximum cost for S is not greater than the cost of S'.

The scheduling algorithm that results from the last theorem can be implemented in $O(n^2)$ time (see Problem 4-64).

Although problem $1|r_j|C_{max}$ can be solved in polynomial time (see Theorem 4-10), the existence of a polynomial algorithm for its counterpart with the general maximum cost function, $1|r_j|f_{max}$, is very unlikely. The last claim is based on the status of $1|r_j|L_{max}$ (see Problem 4-62). When preemptions are allowed, problems

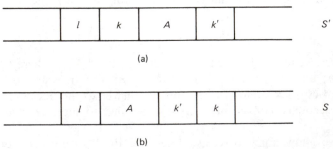

(a)

(b)

Figure 4-28 *An illustration for the proof of Theorem 4-12.*

$1|pmtn, r_j|f_{max}$ and $1|pmtn, prec, r_j|f_{max}$ can be solved efficiently (Baker et al. [1980]).

Total Cost Measures

Our goal now is to minimize the total cost $\sum w_j f_j = \sum_{j=1}^{n} w_j f_j(C_j)$ related to a particular scheduling situation. One of the classical results in scheduling theory concerns the total completion time and its proof is again based on an interchange argument.

Theorem 4-13 (Smith [1956])

The total weighted completion time in the $1||\sum w_j C_j$ problem is minimized by ordering the tasks according to nondecreasing ratio p_j/w_j. ∎

Assume that S is a schedule that contains a pair of adjacent tasks i and j such that $p_i/w_i > p_j/w_j$ or equivalently, $w_j p_i > w_i p_j$. Let S' denote the schedule obtained from S by interchanging i and j. The completion times in S' are $C_i' = C_i + p_j$, $C_j' = C_j - p_i$, and $C_k' = C_k$ for $k \neq i, j$. Hence

$$\sum_{j=1}^{n} w_j C_j' = \sum_{j=1}^{n} w_j C_j + w_i p_j - w_j p_i > \sum_{j=1}^{n} w_j C_j$$

Therefore, the interchange of tasks i and j has reduced the value of $\sum w_j C_j$, and any schedule that contains such tasks can be improved.

Scheduling tasks according to nondecreasing ratio p_j/w_j is called *Smith's rule* or a *weighted shortest processing time* (WSPT) scheduling. If $w_j = 1$ ($j = 1, 2, \ldots, n$), then WSPT becomes SPT.

As a corollary to the theorem above, we have that the mean weighted latenesses are also minimized by WSPT schedules. This result is surprising because the scheduling rule which ignores due dates turns out to be optimal for a due-date-oriented performance measure.

Smith [1956] also gave an efficient algorithm to solve the $1|C_j \leq d_j|\sum C_j$ problem, that is, to find a schedule in which each task is completed by its given due date and the total unweighted completion time is minimized (see Problem 4-68). The reader may be struck by the fact that the weighted version of this problem (i.e., $1|C_j \leq d_j|\sum w_j C_j$) belongs to the class of hard combinatorial problems which have no known polynomial algorithms.

If for a given set of independent tasks and their due dates, there is no schedule which meets all the due dates, we may want to optimize another performance measure involving late tasks and their latenesses. Usually, some penalties are assessed on late tasks, but no benefit is gained from completing tasks early. Thus the tardiness measures should be considered. Unfortunately, the most general problems involving total tardiness, $1\|\sum w_j T_j$ and $1\|\sum T_j$ have no polynomial algorithms, although the status of the latter problem is uncertain at this moment. We refer the reader to Rinnooy Kan et al. [1975] and to Baker and Schrage [1978] for some enumerative and dynamic programming algorithms, respectively, for solving $1\|\sum w_j T_j$.

Minimizing the Number of Late Tasks

Another performance measure that takes into account due dates involves the unit penalty U_j. As happens to many pairs of scheduling problems, $1\|\sum w_j U_j$ is very hard, but $1\|\sum U_j$ can be solved efficiently by a polynomial-time algorithm. The former claim can be shown by transforming the DK problem to the $1\|\sum w_j U_j$ problem. This justifies the use of dynamic programming approach (see Lawler and Moore [1969] and Problem 4-71) and polynomial-time approximation algorithms (see, e.g., Sahni [1976]) for solving $1\|\sum w_j U_j$.

We now present an algorithm for solving the unweighted problem $1\|\sum U_j$. The algorithm is based on the following property of a set of tasks that can be completed by their due dates. This property has been obtained by Jackson as a by-product of scheduling tasks to minimize $L_{\max}$ (see Theorem 4-11).

Theorem 4-14

A set of tasks can be scheduled with all the tasks completed by their due dates if and only if such a schedule is obtained by ordering the tasks according to nondecreasing due dates. ∎

We leave the proof to the reader (see Problem 4-61, where precedence constraints are also included). It follows from the theorem that there exists a schedule minimizing the number of late tasks which has the following property: All tasks to be completed on time are scheduled first by nondecreasing due dates followed by an arbitrary schedule of late tasks. Therefore, it is sufficient to find an optimal set of on-time tasks.

Example 4-13

Let us consider a set of six tasks with the following parameters:

	z_j					
	1	2	3	4	5	6
p_j	2	4	1	2	3	1
d_j	3	5	6	6	7	8

Figure 4-29(a) shows the ordering of tasks according to nondecreasing due dates,

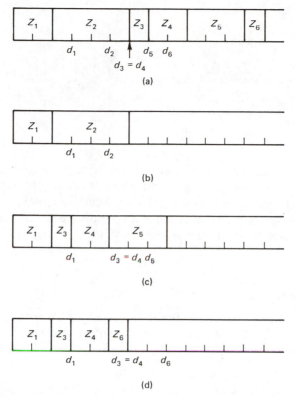

Figure 4-29 *An illustration of Moore–Hodgson's algorithm.*

which minimizes the maximum lateness L_{max} (see Theorem 4-11). Only one task in this schedule is completed on time. ∎

An efficient algorithm for finding the maximum number of on-time tasks has been proposed by J. M. Moore and then improved by T. J. Hodgson, and it is referred to as Moore–Hodgson's algorithm (see Moore [1968]). First the tasks are ordered according to nondecreasing due dates $d_1 \leqslant d_2 \leqslant \ldots \leqslant d_n$, and then an optimal set of on-time tasks is constructed as described in Algorithm 4-9.

Algorithm 4-9: Moore–Hodgson's Algorithm

begin
  $S_0 \leftarrow \varnothing$;
  for every task $j = 1, 2, \ldots, n$ **do**

```
begin
    S_j ← S_{j-1} ∪ {j};
    if ∑ p_j > d_j then
       j∈S_j
    begin
        find l such that p_l = max{p_k: k ∈ S_j};
        S_j ← S_j − {l}
    end
  end
end
```

Set S_n contains the solution. The algorithm starts with the empty set, augments the partial solution with a subsequent task, and if its due date is not met, a task with the longest processing time in the current partial solution is removed. In other words, the algorithm searches a list of tasks ordered according to nondecreasing due dates and if a tasks violates the due date, a longest task among partially accepted is removed.

Example 4-13 (cont.)

Task 2 is the first late task in EDD schedule (see Fig. 4-29(a)) and it is longer than the first one; therefore, it is removed. Next steps of the algorithm are shown in Fig. 4-29(b)–(d).   ∎

The correctness proof of the algorithm can be found in Lenstra et al. [1983].

Concluding the last topic of this section, we know that the general problem $1\|\sum w_j U_j$ with arbitrary weights is hard, but it can be solved by a dynamic programming algorithm in $O(n\sum_{j=1}^n p_j)$ steps (see Problem 4-71). If all weights w_j are equal, the problem becomes polynomially solvable.

4.2.6.   PARALLEL MACHINE SCHEDULING

In the general problem of scheduling n tasks on parallel machines we are given a set of n tasks $\mathcal{Z} = \{Z_1, Z_2, \ldots, Z_n\}$ which are to be processed on machines from a set $\mathcal{M} = \{M_1, M_2, \ldots, M_m\}$. Let p_{ij} denote the processing time of task Z_j on machine M_i. The usual restrictions imposed on tasks and machines also hold in this case; that is, no machine can process more than one task at a time and every task can be performed on at most one machine at a time. It follows from these assumptions that if a task can be split into several parts to be completed on different machines, the processing periods of the parts cannot overlap in time. If preemptions are not allowed, each task must be assigned to exactly one machine.

In general, the tasks may be allowed to be performed only on certain machines from $\mathcal{M}$ and if the machines are of different types, the processing time usually varies from machine to machine. For the sake of simplicity, we assume, however, that every task can be processed by each machine. The parallel machine problems can be classified according to machine speed. The machines are *identical* if $p_{ij} = p_j$ ($i = 1, 2, \ldots, m$), *uniform* if $p_{ij} = s_i p_j$ for a given *speed factor* s_i on machine M_i ($i = 1, 2, \ldots, m$), and *unrelated* if p_{ij} are arbitrary. We will use symbols P, Q and R, respectively, in the first field of our three-field notation to distinguish these three cases. For instance, $P\|C_{\max}$ denotes the problem of scheduling n independent tasks on m identical machines to minimize maximum completion time, and $P2\|C_{\max}$ is its special version when only two machines are available.

In the remainder of this section we restrict our consideration to identical machines and review some classical results on scheduling independent tasks. Then in Section 4.2.7, the problem of scheduling dependent tasks is discussed together with a solution algorithm and its Pascal implementation.

Nonpreemptive Scheduling

Recall from the preceding section that the $1\|C_{\max}$ problem has a trivial solution and the $1\|\Sigma w_j C_j$ problem can be efficiently solved by applying Smith's rule, that is, by ordering tasks according to nondecreasing ratio p_j/w_j.

We will show first that introducing one more machine, that is, allowing the tasks some freedom in choosing a machine, both problems change their complexity status and become as hard as NP-complete problems. To this end, we will show how these problems can be used for solving yet another decision version of the knapsack problem, the partition problem.

In PARTITION we are given positive integers $a_1, a_2, \ldots, a_q, b$ such that $\sum_{i=1}^{q} a_i = 2b$ and question is if there exists a subset $B \subset \{1, 2, \ldots, q\}$ such that $\sum_{i \in B} a_i = b$.

For a given instance of PARTITION, defined by positive integers $a_1, a_2, \ldots, a_q, b$, an instance of $P2\|C_{\max}$ is defined as follows: $n = q$ and $p_j = a_j$ ($j = 1, 2, \ldots, n$). Clearly, there exists a subset B such that $\sum_{i \in B} a_i = b$ if and only if there exists a schedule with $C_{\max} \leqslant b$. The machines in $P2\|C_{\max}$ play a role of subsets of elements and maximum completion time corresponds to the subset weight. Therefore, PARTITION is reducible to $P2\|C_{\max}$, and since the former problem is NP-complete, the latter one is not easier. This fact implies also that all generalizations of $P2\|C_{\max}$ with a modified first field are also very hard computationally, for instance, $P3\|C_{\max}, \ldots, P\|C_{\max}, Q2\|C_{\max}, \ldots, Q\|C_{\max}, \ldots, R\|C_{\max}$.

In the transformation of PARTITION to $P2\|\Sigma w_j C_j$, we assume in the latter problem that $n = q$, $p_j = w_j = a_j$ ($j = 1, 2, \ldots, n$) and the threshold value of the total cost $y = \sum_{1 \leqslant i \leqslant j \leqslant q} a_i a_j - b^2$. The reader is asked to show that PARTITION has a solution if and only if this $P2\|\Sigma w_j C_j$ problem has a schedule with cost $\leqslant y$ (see Problem 4-73).

It is interesting that if all weights are equal, the $P\|\Sigma C_j$ problem can be solved by a polynomial-time algorithm which is a generalization of the algorithm for solving one machine problem $1\|\Sigma C_j$. Such an algorithm has been proposed by Conway, Maxwell, and Miller (CMM; see Conway et al. [1967]). Recall that the solution to $1\|\Sigma C_j$ is obtained by applying the shortest processing time (SPT) rule which orders the tasks according to nondecreasing processing times p_j. To solve $P\|\Sigma C_j$ with more than one machine, we also order the tasks by applying SPT rule and then assign Z_1 to machine M_1, Z_2 to machine $M_2,\ldots,$ Z_m to machine M_m, Z_{m+1} to machine $M_1,\ldots.$ Therefore, task Z_j is assigned to machine M_l, where $l = j(\bmod m)$.

An optimality proof is based on the fact formulated in Problem 4-65. Without loss of generality, we may assume that the number of tasks satisfies $n = km$ (if not, we may add some dummy tasks with zero processing times). Therefore, each machine will be assigned k tasks. Notice that the processing time of every lth task on a machine is counted $k + 1 - l$ times in the criterion value ΣC_j. Hence

$$\sum_{j=1}^{n} C_j = \sum_{j=1}^{n} v_j p_j, \quad \text{where } (v_1, v_2,\ldots, v_n) = (l,\ldots, l, l - 1,\ldots, l - 1,\ldots, 1,\ldots, 1)$$

Since v_j's are in nonincreasing order, ΣC_j is minimized when the processing times are in nondecreasing order and, therefore, when the SPT rule of scheduling is applied.

Example 4-14

Let us consider the set of six tasks with the processing times $p_j = j$, which are to be scheduled on two machines. Figure 4-30(a) shows the SPT schedule. It follows from the criterion studing that any permutation of the tasks that are scheduled in the same positions does not alter the criterion value. Figure 4-30(b) shows another optimal schedule with tasks 3 and 4 interchanged in their SPT schedule. Notice that these two schedules, which are optimal for the total cost, have different maximum completion time, 12 and 11, respectively. Therefore, there is a room for introducing a secondary measure into the $P\|\Sigma C_j$ problem.   ∎

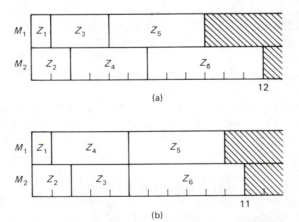

(a)

(b)

Figure 4-30   *An example of the $P2\|\Sigma C_j$ problem.*

A similar method can be used for scheduling uniform parallel machines in the $Q\|\Sigma C_j$ problem (see Problem 4-76). The total completion time problem for unrelated machines, $R\|\Sigma C_j$, can be also solved efficiently (see Problem 4-77).

List Scheduling

The complexity of the $P\|C_{\max}$ problem justifies the use of enumerative and approximation algorithms for its solution. The former ones find optimal solutions but may require prohibitive running time in the worst case, whereas the latter methods are designed to generate reasonably good feasible schedules and work in polynomial time. An example of a heuristic algorithm for solving $P\|C_{\max}$ (and several other problems) is *list scheduling* (*LS*) *strategy* (see Section 4.2.7 for an exact, list scheduling algorithm).

The *LS algorithm* schedules the tasks according to a given priority list of the tasks, and at each step the first available task on the list is assigned to the machine with the currently task earliest finishing time. In other words, whenever a machine becomes available, the list is searched from left to right and the first unexecuted ready task encountered in the search is scheduled on the machine.

Example 4-15

Let us apply the LS algorithm to solve the following instance of $P3\|C_{\max}$: $n = 5$, $p_1 = 2$, $p_2 = 2$, $p_3 = 1$, $p_4 = 1$, $p_5 = 3$ and the priority list $(1, 2, 3, 4, 5)$. Figures 4-31(a) and (b) show the solution produced by the LS algorithm and an optimal solution, respectively. ∎

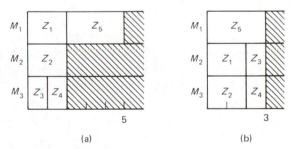

(a)   (b)

Figure 4-31   (*a*) *The list schedule and* (*b*) *An optimal schedule of Example* 4-15.

A natural question arises: How good are the solutions generated by the LS algorithm? Such a question is usually answered by assessing the worst case.

Let for a given instance of $Pm\|C_{\max}$, $C_{\max}(A)$ denote the maximum completion time of the schedule produced by algorithm A and $C_{\max}^*$ denote the value of an optimal solution. If $C_{\max}(A)$ is always equal to $C_{\max}^*$, A is called an optimization algorithm; otherwise, A is an approximation algorithm. To answer our question about the performance of the LS approach, we have to bound its worst-case behavior. Such a bound is called a *performance guarantee* and usually takes the form of the inequality $C_{max}(A) \leqslant r C^*_{max}$, where r is a constant.

The following theorem evaluates the LS algorithm.

Theorem 4-15 (Graham, see Coffman [1976], Chapter 5)

The list scheduling algorithm applied to an instance of the $Pm \| C_{max}$ problem generates a solution which satisfies

$$C_{max}(\text{LS}) \leqslant (2 - 1/m) C^*_{max} \qquad \blacksquare$$

Thus LS schedules can be at most two times worst than optimal. A comparatively easy proof of this theorem is left to the reader (see Problems 4-80 and 4-81 for some hints). Example 4-15 shows that Theorem 4-15 gives the best performance guarantee possible for the list scheduling and the $P \| C_{max}$ problem [see Problem 4-81(b)].

While discussing the $P \| \sum C_j$ problem, we presented the CMM algorithm and pointed out that it may generate many different schedules of the same total cost $\sum C_j$. Now one may ask how good these solutions are in terms of the maximum completion time C_{max}. Let $C_{max}(\text{CMM})$ denote the minimum C_{max} among all schedules generated by the CMM algorithm applied to a given instance of $Pm \| C_{max}$. It has been proved that the CMM algorithm can behave as badly as the LS algorithm, namely

$$C_{max}(\text{CMM}) \leqslant (2 - 1/m) C^*_{max}$$

It is interesting and somewhat unexpected that when precedence constraints are added to $Pm \| C_{max}$, converting it to the $Pm | prec | C_{max}$ problem, a natural generalization of the LS algorithm still has the same performance ratio as in Theorem 4-15.

One may wonder if the LS algorithm can produce better solutions (in the worst case) for restricted and easy generated priority list. Looking at Example 4-15, we can conclude that the poor performance of the LS algorithm might be attributed to delaying the longest tasks. A natural way of avoiding this is to form a priority list according to nonincreasing order of processing times p_j. The list scheduling algorithm applied to such ordering is called the *largest processing time* (LPT) algorithm. Figure 4-31(b) shows the solution of Example 4-15 obtained by the LPT algorithm.

Example 4-16

Let us apply the LPT algorithm to the following instance of $P3 \| C_{max}$: $n = 7$, $p_1 = p_2 = 5$, $p_3 = p_4 = 4$, $p_5 = p_6 = p_7 = 3$. Figure 4-32(a) shows the solution obtained by LPT algorithm and an optimal schedule is shown in Fig. 4-32(b). $\blacksquare$

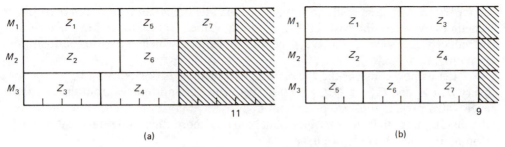

Figure 4-32 *An example of* LPT *scheduling.*

A performance guarantee for the LPT algorithm is given in the following theorem.

Theorem 4-16 (Graham, see Coffman [1976], Chapter 5)

The LPT algorithm applied to an instance of the $Pm\|C_{\max}$ problem generates a solution which satisfies

$$C_{\max}(\text{LPT}) \leqslant \left(\frac{4}{3} - \frac{1}{3m} \right) C_{\max}^* \qquad \blacksquare$$

This means that the length of an LPT schedule is at most 33% greater than optimal. The proof of this theorem is much more complex than that of the preceding one. Example 4-16 illustrates that the bound given in Theorem 4-16 is the best possible (see Problem 4-82).

The problem of scheduling dependent tasks on parallel machines is discussed in the next section and a surprising behavior of the LS algorithm on such problems is illustrated in Problem 4-85.

Parallel machine scheduling with deadlines is known as the *bin packing* problem (see Problems 4-86 and 4-87 for some approximation algorithms).

Preemptive Scheduling

We present only basic results on scheduling preemptive tasks on identical parallel machines to minimize the maximum and total completion time. First, recall Theorem 4-9, saying that preemptions are never advantageous in scheduling a single machine under a regular performance measure. We shall demonstrate that this result generally cannot be extended to parallel machines. In a special case, however, when the total completion time is to be minimized, it still holds. McNaughton [1959] proved that for $P\,|\,pmtn|\Sigma\,C_j$, no schedule with a finite number of task preemptions has a smaller criterion value than an optimal nonpreemptive schedule (in contrast to Theorem 4-9, the original proof of this result is quite complex). Therefore, in dealing with $P\,|\,pmtn|\Sigma\,C_j$ we may restrict our attention to schedules with no interruptions in task performance. It is easily seen that machine idle times can be also removed from schedules with no increase in the criterion value. Hence the CMM algorithm, introduced for solving $P\|\sum C_j$, generates schedules which are also optimal for $P|pmtn|\sum C_j$.

Let us now consider the $P|pmtn|C_{max}$ problem. It is obvious that $C_{max}^* \geq p_j$ for every task Z_j and $C_{max}^* \geq (\sum_{j=1}^n p_j)/m$. The former bound follows from the fact that even when preemptions are permitted, $C_j \geq p_j$ for every task j. The latter holds by the assumption that no machine can process more than one task at a time. Combining these bounds, we have

$$C_{max}^* \geq \max\left\{ \max_{1 \leq j \leq n} p_j, \left(\sum_{j=1}^n p_j \right)/m \right\} \tag{4-16}$$

We will prove that there is always a preemptive schedule that meets this bound.

Theorem 4-17 (McNaughton [1959])

There is an optimal schedule for a $P|pmtn|C_{max}$ problem which meets the bound (4-16) and has at most $m - 1$ preemptions.   ∎

We prove this theorem by providing a suitable scheduling rule, called *McNaughton's rule*: For the tasks in an arbitrary order, fill the machines successively splitting a task whenever it exceeds the bound. It is clear that only $m - 1$ preemptions can be introduced (on the first $m - 1$ machines), no task is split more than once, and no parts of a task are overlapped. Since the bound (4-16) can be evaluated in $O(n)$ time, the algorithm works in $O(n)$ time.

Example 4-17

Let us consider the following instance of $P3|pmtn|C_{max}$: $n = 6$, $p_1 = 4$ $p_2 = 5$, $p_3 = 2$, $p_4 = 1$, $p_5 = 2$, $p_6 = 1$. The bound in (4-16) is equal to 5 and Fig. 4-33(a) shows a preemptive schedule of length 5 with two preemptions found by applying McNaughton's rule to ordering $(1, 2, 3, 4, 5, 6)$. Another optimal schedule to this problem with no split tasks is shown in Fig. 4-33(b).   ∎

It is natural to ask how big an improvement can be obtained by preemptive scheduling as compared to nonpreemptive scheduling. An answer can be formulated for the more general problem of scheduling n dependent tasks on m identical machines. Let C_{max}^N and C_{max}^P denote the lengths of optimal solutions for two particular instances of $Pm|prec|C_{max}$ and $Pm|prec, pmtn|C_{max}$, respectively. Then

$$C_{max}^N \leq (2 - 1/m)C_{max}^P$$

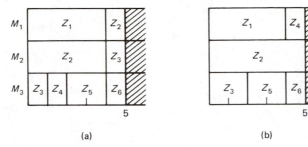

Figure 4-33   *Preemptive scheduling of parallel machines.*

Thus a nonpreemptive optimal solution to a $P|prec|C_{max}$ can be at most two times worse than a preemptive one.

4.2.7. PARALLEL MACHINE SCHEDULING WITH PRECEDENCE CONSTRAINTS

As we know from the preceding section, the $P\|C_{max}$ problem belongs to a class of most difficult combinatorial problems. Nevertheless, augmenting it with some additional restrictions might convert it to an easy problem. Clearly, the problem becomes trivial if all tasks have the same processing time. By introducing simultaneously the precedence constraints among the tasks, nontrivial computational problems are introduced.

We shall focus on the $P|tree, p_j = 1|C_{max}$ problem, which is to minimize the maximum completion time of n unit-time (equal-length, in general) tasks on m identical machines subject to precedence constraints which form a tree (a forest, in general). We assume that in the rest of this section a *tree* (called also an *in-tree*) is a directed rooted tree with exactly one arc outgoing from every node except the root. A *forest* is a collection of trees. Therefore, trees and forests form a subfamily of acyclic digraphs.

We present an almost linear (in n) algorithm for solving $P|tree, p_j = 1|C_{max}$ originally proposed by T. C. Hu in 1961 and then efficiently implemented by R. Sethi in 1976. The algorithm illustrates the use of two common scheduling techniques: list scheduling and critical path scheduling. List scheduling in this case is sometimes referred to as level scheduling, since the priority list is constructed according to the node level numbers. The *level* of a node u in a tree is defined as the number of nodes (including u) on the path from u to the root. Therefore, the root is at level 1. The node level can be defined for any acyclic digraph as the maximum number of nodes on any path to the terminal node. A task is said to be *ready* (for processing) when all its predecessors have been executed. The *level scheduling* rule is defined similarly to the list scheduling:

> Whenever a machine becomes available, schedule on it an unexecuted ready task at the highest level.

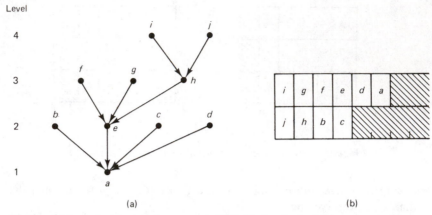

Figure 4-34 (*a*) *A tree task system and* (*b*) *Its level schedule on two machines.*

Example 4-18

Let us consider the tree-structured task system shown in Fig. 4-34 and apply the level rule to schedule the tasks on two machines. Tasks *i* and *j* at level 4 are assigned first to the first time unit. During the second time unit we assign any two of the three ready tasks at level 3, say *g* and *h*. During the next time unit, first we assign *f* to one of the machines. At this stage of scheduling, all ready tasks are at level 2, but *e* cannot be assigned during the third time unit, for its predecessor *f* would be executed at the same time. We choose instead task *b*. Then we can assign *e* and one of *c* or *d*, say *c*. Finally, *d* is assigned to the fifth time unit and the terminal task *a* must be assigned to the next time unit. ∎

The schedule in Fig. 4-34(b) illustrates an important property of optimal schedules for $P|tree, p_j = 1|C_{max}$ problems. Task *a*, as the only terminal, can be scheduled after all other tasks are processed. It takes at least five time units to process nine tasks on two machines; therefore, all feasible schedules are at least of length 6.

Once the levels are assigned to the tree nodes, a list of tasks in order of nonincreasing level can be created. It is clear that the list scheduling applied to this list is equivalent to the level rule. The main result of this section is formulated in the following theorem.

Theorem 4-18 (Hu [1961])

The level rule constructs an optimal solution to the $P|tree, p_j = 1|C_{max}$ problem; that is, it finds a minimal-length nonpreemptive schedule for a tree-structured system of unit-processing time tasks and identical machines. ∎

Although the result is quite simple and appealing, its proof is involved and we do not discuss it here.

Generalizations

Before presenting an efficient implementation of the level strategy, we mention some possible generalizations and applications of this very simple scheduling model.

First, we relax the restriction put on the structure of the precedence constraints. We observe that the level rule can be applied directly to a forest. This follows from the fact that if the level rule does not construct an optimal schedule for forest-structured tasks, the rule cannot guarantee an optimal schedule for the tree-structured task system formed by adding a new task as an immediate successor of all terminals of the forest. Notice that node levels of different trees of a forest can be calculated separately, but a priority list is one for the entire forest. It can easily be seen that priority lists constructed according to levels are not unique, even for trees.

Another family of precedence graphs to which the level rule can be applied is formed by *antiforests*. An antiforest can be obtained from a forest by reversing all its arcs; therefore, except for initial nodes, every other node has exactly one arc coming in. An optimal schedule for an antiforest is the reverse of an optimal schedule obtained for the corresponding forest.

It is not difficult to devise an example which will show that the level rule may fail to produce an optimal schedule for a union of trees and antitrees. For instance, let us consider the task system shown in Fig. 4-35(a), which is to be scheduled on three machines. Assume that the nodes form a level priority list in order of their labels. The list scheduling applied to this list generates a solution of length 4 [Fig. 4-35(b)], whereas an optimal schedule has length 3 [Fig. 4-35(c)].

The $P|prec, p_j = 1|C_{max}$ problem, where *prec* denotes an arbitrary precedence constraint, is computationally very hard. When restricted to only two machines, it can be solved by a polynomial-time algorithm. One such algorithm, presented by Coffman and Graham [1972], leads to list scheduling. Nevertheless, finding an appropriate priority list in this case is much more complex than in Hu's algorithm.

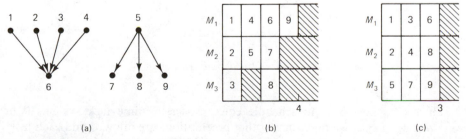

(a)   (b)   (c)

Figure 4-35 *Level scheduling of a mixed task system: (a) System with labels assigned; (b) Level list schedule; (c) Optimal schedule.*

However, once such a priority list is found, Hu's and Coffman–Graham's algorithms can use the same list scheduling procedure.

We now relax the restriction put on the other parameter of the second field, namely on the processing time. But first we will interpret the level scheduling rule in yet another way. Recall that the tasks are arranged by levels, starting from the highest. The level of a node u is exactly the length of the path from u to the terminal node. Therefore, one can say that the task to be scheduled next is the one that heads the current longest (critical) path in the precedence graph. Scheduling according to this rule is called *critical path* (CP) *scheduling*. Hence critical path scheduling is list scheduling applied to a list of tasks arranged by nonincreasing longest path. Now, when arbitrary processing times are allowed, we have to recalculate the node levels as the lengths of longest paths to the terminal.

Example 4-19

Let us consider the instance of $P2|tree|C_{\max}$ shown in Fig. 4-36(a). Figure 4-36(b) shows a critical path schedule and another list schedule is shown in Fig. 4-36(c). The CP schedule is not optimal in this case.   ∎

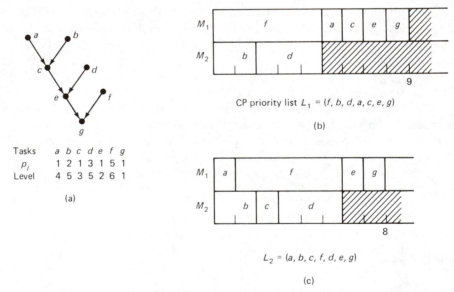

CP priority list $L_1 = (f, b, d, a, c, e, g)$

(b)

Tasks	a	b	c	d	e	f	g
p_j	1	2	1	3	1	5	1
Level	4	5	3	5	2	6	1

(a)

$L_2 = (a, b, c, f, d, e, g)$

(c)

Figure 4-36   *Critical path scheduling.*

Since we know how to schedule equal-processing-time task systems (when *prec = tree* or $m = 2$) we may suggest that preemptions are allowed and each task is split into a chain of one-unit tasks. Figure 4-37(a) illustrates the effect of such transformation applied to the instance of Example 4-19. The tree with each task split

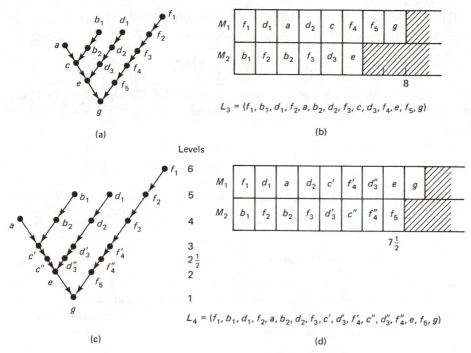

$$L_3 = (f_1, b_1, d_1, f_2, a, b_2, d_2, f_3, c, d_3, f_4, e, f_5, g)$$

(a)   (b)

$$L_4 = (f_1, b_1, d_1, f_2, a, b_2, d_2, f_3, c', d_3', f_4', c'', d_3'', f_4'', e, f_5, g)$$

(c)   (d)

Figure 4-37 *Critical path scheduling with preemptions.*

into a chain of unit tasks is shown in Fig. 4-37(a) and Fig. 4-37(b) shows a level schedule obtained for list L_3. Therefore, at least in this example, splitting the tasks into unit-time subtasks leads to a better level schedule [compare with Fig. 4-36(b)]. Such splitting of tasks (i.e., into unit-time tasks) is equivalent to allowing task preemptions only at integral time points. This may not guarantee an optimal solution when preemptions are allowed at any time instant (i.e., when considering problem $P|tree, pmtn|C_{max}$). Let us consider again the task system of Example 4-19 and split the tasks as is shown in Fig. 4-37(c). A level schedule obtained for list L_4 turns out to be an optimal preemptive solution. The $P|pmtn, tree|C_{max}$ has a polynomial-time algorithm, which, however, is not a simple refinement of the level rule. The $P|pmtn, prec|C_{max}$ problem with arbitrary precedence constraints is computationally hard. The effect of introducing preemptions into parallel machine scheduling was discussed at the end of the preceding section.

The scope and usefulness of CP scheduling have been illustrated many times in previous sections. The temporal analysis in Section 4.2.1 can now be interpreted as the CP scheduling of n arbitrarily dependent tasks on an unlimited number of machines. Then in the next section, we introduced some limits on the number and the use of machines. Thus the problem became very hard, so we applied an implicit enumeration method in which critical path scheduling is performed in every step.

The pure critical path scheduling can be also applied to $Pm|prec|C_{max}$ problems yielding only approximate solutions which satisfy

$$C_{max}(\text{CP}) \leqslant (2 - 1/m)C_{max}^*$$

where $C_{max}(\text{CP})$ denotes the solution value found by CP scheduling. If no precedence constraints are present among tasks, the worst-case performance ratio becomes

$$C_{max}(\text{CP}) \leqslant \left(\frac{4}{3} - \frac{1}{3m}\right)C_{max}^*$$

Note that in this case, CP scheduling is equivalent to LPT scheduling (compare Theorem 4-16). In spite of these worst-case bounds, CP scheduling behaves very well on random problems (see Kohler [1975]).

Implementation of Hu's Algorithm

We now present an efficient implementation of the level rule for solving the $Pm|tree, p_j = 1|C_{max}$ problem, that is, for scheduling tree-structured unit-time tasks on m machines to minimize the maximum completion time. The algorithm based on Hu's theorem (4-18) consists of two main steps:

Step 1.   Construct a task list L according to nonincreasing levels.

Step 2.   Apply to L the list scheduling rule.

The list L can be easily constructed by applying a longest-path algorithm to a precedence tree. Starting from the terminal, L is created backward, listing nodes by their level numbers. The computational time spent in this step in bounded by a constant times the number of arcs in the precedence graph, that is, $O(n)$, since every tree has $n - 1$ arcs.

Once the tasks (tree nodes) are arranged in list L, the list scheduling can be applied. Recall that according to the level rule, whenever a machine becomes available, the first unexecuted ready task on the list is assigned to the machine. It may, however, be very costly to find such a task by simple scanning of list L from one of its ends.

Example 4-20

As an example, let us consider the tree-structured task system shown in Fig. 4-38, which is to be scheduled on two machines. Assume that list L [shown also in Fig. 4-38(a)] is scanned from left to right. Task $2k + 1$ is assigned to the first machine during the first time unit. Then we have to spent $k - 1$ steps on finding the first ready task, 2. Task $2k$ is then assigned to the first machine during the second time period and again we spend $k - 2$ steps to find the next ready task, now 3. Therefore, it will take $O(k^2)$ steps to construct an optimal schedule for this task system.   ∎

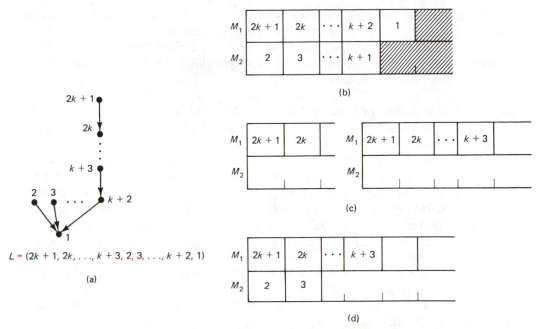

Figure 4-38 *Worst case of naive level scheduling.*

It is possible, however, to implement Hu's algorithm to run in almost linear time. We shall describe such an implementation proposed in Sethi [1976]. Two properties of the problem and level lists are crucial in Sethi's implementation: (1) all tasks have the same processing time, and (2) all the predecessors of task Z_j appear before it in every priority list L. The algorithm constructs a schedule by constructing schedules for successively augmented sublists (Z_1), (Z_1, Z_2), (Z_1, Z_2, Z_3),.... In-stead of searching a list for the first ready task, it assigns the first unscheduled task. Figure 4-38(c) shows the schedules corresponding to some partial lists for the task system of Fig. 4-38(a). The schedules for sublists can easily be constructed by making use of the following modification of level scheduling:

Modified Level (ML) Scheduling Rule

Let U_{i-1} be the list schedule for $(Z_1, Z_2, \ldots, Z_{i-1})$ and let t be the first time unit by which all predecessors of Z_i have been processed. The list schedule U_i for $(Z_1, Z_2, \ldots, Z_i)$ can be constructed from U_{i-1} by assigning Z_i to any machine during the first time unit u, $u \geqslant t$, at which a machine is available. ∎

We now show that both level rules produce the same schedule. The proof is by induction on the list length. Let us assume that first $i - 1$ tasks are scheduled by both rules during the same time units and consider task Z_i. It cannot be scheduled before time unit t due to precedence constraints and the ML rule assumption. After time unit t, in level scheduling, task Z_i will be reached when there is a machine free and no task Z_j ($j \leqslant i$) can be assigned. However, this situation corresponds to the first idle time after t in schedule U_{i-1}.

A basic difference between ML scheduling and level scheduling is that the former assigns consecutive tasks to machines, whereas the latter fills in consecutive time units with tasks.

An efficient implementation of the ML rule makes use of a fast algorithm for merging disjoint sets. In the beginning we have n one-element sets $set(t) = \{t\}$ ($t = 1, 2, \ldots, n$) corresponding to n time units, which can be necessary for scheduling n tasks. During the course of the algorithm, if j is a member of $set(u)$, then u is the first time unit at which a machine is available for processing j. In an implementation of the ML rule we need to be able to find the set that contains a given task and to determine the first time unit at which a task becomes ready. To this end, in the latter case we maintain variables $ready(j)$, which hold such information for each task j. In the beginning, we set $ready(j) = 1$ for all initial tasks j (i.e., for tasks with no predecessors). For the other tasks l, we have $ready(l) = 1 + \max\{t_k$: task k immediately precedes task $l\}$, where t_k denotes the time unit to which task k has been assigned. Since in a level list all predecessors of l appear before l, value $ready(l)$ is well defined when task l is encountered.

We have also to keep track of the number of machines available at each time unit. Let $time(i)$ store such information. In the beginning, $time(i) = m$ for every $i = 1, 2, \ldots, n$, where m is a given number of machines that can be used. When we schedule task j such that $ready(j) = t$ and t belongs to $set(u)$, then j is assigned to a machine during time unit u and $time(u) \leftarrow time(u) - 1$. If $time(u)$ becomes 0, no machine is available at u any more and we have to find the first time unit after u at which there is a free machine. It is clear that this can be accomplished by considering the set $set(v)$, which contains the next-to-u time unit, $u + 1$. By the definition of the sets, v is the first time unit at or after $u + 1$ at which a machine is available. Thus, when $time(u)$ becomes 0, we have to merge $set(u)$ and $set(v)$ and call the new set $set(v)$.

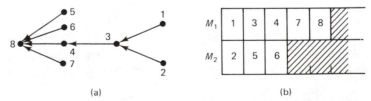

(a)   (b)

Figure 4-39   *An example of* ML *scheduling.*

Example 4-21

Let us illustrate the algorithm with the task system shown in Fig. 4-39(a), to be scheduled on two machines. In the beginning we set $time(i) = 2$ and $set(i) = \{i\}$ $(i = 1, 2, \ldots, 8)$ and $ready(1..8) = (1, 1, -, -, 1, 1, 1, -)$. Table 4-8 shows the algorithm actions when applied to list $(1, 2, \ldots, 8)$ and the solution obtained is shown in Fig. 4-39(b). ∎

Table 4-8. Results of the Application of the ML Scheduling Algorithm to the Task System of Fig. 4-39

Task,	Schedule Time,	Changes in:		
j	t_j	$time(\cdot)$	$ready(\cdot)$	$set(\cdot)$
1	1	$time(1) \leftarrow 1$		
2	1	$time(1) \leftarrow 0$	$ready(3) \leftarrow 2$	$set(2) \leftarrow \{1, 2\}$
3	2	$time(2) \leftarrow 1$	$ready(4) \leftarrow 3$	
4	3	$time(3) \leftarrow 1$		
5	2	$time(2) \leftarrow 0$		$set(3) \leftarrow \{1, 2, 3\}$
6	3	$time(3) \leftarrow 0$		$set(4) \leftarrow \{1, 2, 3, 4\}$
7	4	$time(4) \leftarrow 1$	$ready(8) \leftarrow 5$	
8	5	$time(5) \leftarrow 1$		

There are two set operations used in the algorithm: FIND(t) that determines the set which contains element (task) t, and MERGE(u, v) that merges sets $set(u)$ and $set(v)$ and calls the new set $set(v)$. These operations are implemented by using tree representation of sets. Each set is represented as a single rooted tree. Each node (including roots) in the tree represents an element in the set, and the root of the tree represents additionally the entire set. Operation FIND is implemented with the collapsing step which makes all nodes reached during the search for the root, sons of the root. In operation MERGE we make use of the weighted union rule, which balances the height of trees. More details on efficient implementations of the set operations can be found in Aho et al. [1974].

The algorithm performs n operations FIND and at most $n - 1$ operations MERGE. It has been proved by R. E. Tarjan that all these operations can be performed in time $O(n\alpha(n))$, where $\alpha(n)$ is a very, very slowly growing function of n known as a functional inverse of Ackermann's function. For instance, we have $\alpha(n) \leqslant 3$ for all $n \leqslant 2^{65536}$! Therefore, Sethi's implementation of Hu's algorithm solves the $Pm|tree, p_j = 1|C_{\max}$ problem in time which is almost linear in the number of tasks. A compact description of the implementation is presented in Algorithm 4-10.

Algorithm 4-10:   Sethi's Implementation
of Hu's Algorithm

```
begin
    determine the node levels and form a priority list of nodes (tasks)
        in order of nonincreasing levels;
    for j ← 1 to n do
    begin                      ⟨* initialization *⟩
        time(j) ← m;
        set(j) ← ⟨j⟩;
        ready(j) ← 1      ⟨* ready(j) for noninitial nodes are updated later *⟩
    end;
    for j ← 1 to n do
    begin
        t ← ready(j);
        u ← FIND(t);
        tᵢ ← u;
        time(u) ← time(u) − 1;
        if j < n then
        begin
            if time(u) = 0 then
            begin
                v ← FIND(u + 1);
                merge set(u) and set(v) and call the new set set(v)
            end;
            for every node i adjacent out of node j do
                if ready(i) < u + 1 then
                    ready(i) ← u + 1
        end ⟨* j < n *⟩
    end ⟨* task j *⟩
end
```

Computer Implementation of Hu's Algorithm

The LEVELSCHEDULING procedure finds an optimal schedule of N tasks on M machines subject to precedence constraints. The tasks are assumed to have unit processing times and the precedence constraints are restricted to those that form trees. Some possible modifications and extensions of the procedure are discussed in the text and are also suggested in Problems 4-93, 4-94, and 4-95. The precedence constraints are assumed to be given in a backward-star form (see Section 3.1), in which arcs are grouped according to their terminal nodes.

Procedure LEVELSCHEDULING uses one function and three other procedures. Procedure OUTARCREP constructs a forward-star representation of the precedence constraints which is used for reducing indegrees of nodes and for updating ready times. Procedure LEVELLIST evaluates the node levels and orders the nodes according to nonincreasing level. Function FIND(T) returns the set which

contains task T, and procedure MERGE merges two sets. Variable SETLAB[U] is
the name of set U. Hence SETLAB[FIND(T)] is the first time unit at which a
machine is available for processing task T.

Although the structure of precedence constraints makes it possible to use a
representation which is simpler and typical for trees, we decided to use a general
digraph representation which is suitable for other types of precedence constraints
and may be employed in some procedure modifications suggested in the problems.

Global Constants

M   number of machines which can be used
N   number of tasks in the system
N1   N + 1

Data Types

```
TYPE  ARRN  = ARRAY[1..N] OF INTEGER;
      ARRN1 = ARRAY[1..N1] OF INTEGER;
      ARRM  = ARRAY[1..M] OF INTEGER;
```

Procedure Parameters

```
PROCEDURE LEVELSCHEDULING(
    M,N              :INTEGER;
    VAR NRI          :ARRN1;
    VAR INARC        :ARRM;
    VAR SCHEDULE :ARRN);
```

Input

M   number of machines
N   number of tasks
NRI[1..N1]   array of pointers in backward star form
INARC[1..M]   array of arcs in backward star form

Output

SCHEDULE[1..N]   array of schedule times; SCHEDULE[J] is the time unit
                 to which task J has been assigned in the optimal solution
                 found by the procedure

Example

For the task system of Example 4-21, the data representation in backward star form is as follows:

$$M = 2, \qquad N = 8$$
$$NRI[1..9] = [1, 1, 1, 3, 4, 4, 4, 4, 8]$$
$$INARC[1..7] = [1, 2, 3, 5, 6, 4, 7]$$

The solution obtained is

$$SCHEDULE[1..8] = [1, 1, 2, 3, 2, 3, 4, 5]$$

Pascal Procedure LEVELSCHEDULING

```
PROCEDURE LEVELSCHEDULING(
    M,N            :INTEGER;
    VAR NRI        :ARRN1;
    VAR INARC      :ARRM;
    VAR SCHEDULE   :ARRN);

VAR I,J,L,P,R,S,T,U,V            :INTEGER;
    FATHER,INDEG,LEVEL,LIST,
    OUTDEG,READY,TIME,SETLAB     :ARRN;
    NRO                          :ARRN1;
    OUTARC                       :ARRM;

PROCEDURE OUTARCREP(VAR OUTDEG:ARRN;VAR NRO:ARRN1;VAR OUTARC:ARRM);
    (*  THIS PROCEDURE CONSTRUCTS FORWARD-STAR REPRESENTATION  *)

VAR I,J,K,L:INTEGER;
    AUX     :ARRN;
BEGIN
    FOR I:=1 TO N DO NRO[I]:=0;
    FOR J:=1 TO N DO
        FOR L:=NRI[J] TO NRI[J+1]-1 DO
        BEGIN
            I:=INARC[L];
            NRO[I]:=NRO[I]+1
        END;
    J:=1;
    FOR I:=1 TO N DO
    BEGIN
        L:=NRO[I];
        OUTDEG[I]:=L;
        NRO[I]:=J;
        AUX[I]:=J;
        J:=J+L
    END;
    NRO[N+1]:=J;
    FOR J:=1 TO N DO
        FOR L:=NRI[J] TO NRI[J+1]-1 DO
```

```
      BEGIN
         I:=INARC[L];
         K:=AUX[I];
         OUTARC[K]:=J;
         AUX[I]:=K+1
      END
END;   (*  FORWARD-STAR REPRESENTATION  *)


PROCEDURE LEVELLIST(VAR LEVEL,LIST:ARRN);
   (*  THIS PROCEDURE EVALUATES NODE LEVELS AND ORDERS
       NODES ACCORDING TO NONINCREASING LEVEL  *)

VAR I,J,K,L,LEV,P,Q:INTEGER;
    AUX            :ARRN;
BEGIN
   FOR I:=1 TO N DO
      AUX[I]:=OUTDEG[I];
   P:=N;
   FOR I:=1 TO N DO
      IF AUX[I] = O THEN
      BEGIN
         LEVEL[I]:=1;
         LIST[P]:=I;
         P:=P-1
      END;
   Q:=N;
   WHILE (P > O) AND (P < Q) DO
   BEGIN
      J:=LIST[Q];
      Q:=Q-1;
      LEV:=LEVEL[J]+1;
      FOR L:=NRI[J] TO NRI[J+1]-1 DO
      BEGIN
         I:=INARC[L];
         K:=AUX[I]-1;
         IF K <> O THEN AUX[I]:=K
         ELSE
         BEGIN
            LEVEL[I]:=LEV;
            LIST[P]:=I;
            P:=P-1
         END
      END
   END
      (*  IF P = Q THEN PRECEDENCE GRAPH CONTAINS A CYCLE  *)
END;  (*  LEVELLIST  *)

FUNCTION FIND(I:INTEGER):INTEGER;
   (*  THIS FUNCTION FINDS THE SET CONTAINING I  *)

VAR PTR,X,Y:INTEGER;
```

```
BEGIN
   PTR:=I;
   WHILE FATHER[PTR] > 0 DO
      PTR:=FATHER[PTR];
   X:=I;
   WHILE FATHER[X] > 0 DO
   BEGIN
      Y:=FATHER[X];
      FATHER[X]:=PTR;
      X:=Y
   END;
   FIND:=PTR
END;  (*  FIND  *)

PROCEDURE MERGE(U,V:INTEGER);
   (*  THIS PROCEDURE MERGES TWO SETS U AND V  *)

VAR X:INTEGER;
BEGIN
   X:=FATHER[U]+FATHER[V];
   IF FATHER[U] > FATHER[V] THEN
   BEGIN
      FATHER[U]:=V;
      FATHER[V]:=X
   END
   ELSE
   BEGIN
      FATHER[V]:=U;
      FATHER[U]:=X;
      SETLAB[U]:=SETLAB[V]
   END
END;  (*  MERGE  *)

BEGIN  (*  MAIN BODY  *)
   OUTARCREP(OUTDEG,NRO,OUTARC);
   LEVELLIST(LEVEL,LIST);
   FOR I:=1 TO N DO
   BEGIN
      INDEG[I]:=NRI[I+1]-NRI[I];
      FATHER[I]:=-1;
      SETLAB[I]:=I;
      TIME[I]:=M
   END;
   FOR I:=1 TO N DO
      READY[I]:=1;
   P:=1;
   WHILE P <= N DO
```

```
BEGIN
    I:=LIST[P];       (*  PROCESSING NODE I  *)
    P:=P+1;
    T:=READY[I];
    U:=FIND(T);
    R:=SETLAB[U];
    SCHEDULE[I]:=R;
    S:=TIME[R]-1;
    IF S > 0 THEN TIME[R]:=S
    ELSE
    BEGIN
        V:=FIND(R+1);
        MERGE(U,V)
    END;
    FOR L:=NRO[I] TO NRO[I+1]-1 DO
    BEGIN
        J:=OUTARC[L];
        INDEG[J]:=INDEG[J]-1;
        IF READY[J] < R+1 THEN
            READY[J]:=R+1
    END
  END
END;  (*  LEVEL SCHEDULING  *)
```

PROBLEMS

4-29. Show that if every node of a digraph G has an arc coming in and an arc going out, then G has a cycle.

4-30. Let G be a digraph given in a forward star form. Design and write a Pascal procedure which tests if G is acyclic and if so, it finds a topological ordering of its nodes. Be sure that your algorithm works in $O(n + m)$ time, where n is the number of nodes and m is the number of arcs of G. [*Hints:* Show first that (1) every subdigraph of an acyclic digraph is acyclic, and (2) every acyclic digraph has a source and a sink.]

4-31. Given is the following set of precedence relations among six activities: $a \prec e, a \prec f, b \prec d, b \prec e, c \prec f$, and $c \prec d$.

(a) Identify the error in the following event network:

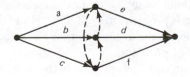

      (b) Construct a correct activity network for this project.

      (c) Construct a correct event network for this project.

4-32.   (First read Problem 3-17)   The *transitive reduction* of a digraph G is a subdigraph G^- of G which has the same transitive closure as G, and G^- has a minimal number of arcs among all such subdigraphs. Show how one can modify the algorithms for finding the transitive closure of a digraph proposed in Problem 3-17 to obtain efficient algorithms for the transitive reduction of an acyclic digraph.

4-33.   Construct for the following set of activities and precedence constraints the corresponding activity network D and an event network E with the minimum number of nodes (by applying steps 1 to 3 to D). Then try to reduce further the number of dummy activities in E.

Activity x	Immediate Successors of x
a	e, f, g
b	f, g
c	f
d	g

If you have succeeded and obtained a minimum-event network, state formally a construction rule you have applied.

4-34.   Given is an event network G_p which has no cycle and the nodes are already in a topological order. Write a Pascal procedure which for G_p represented in a forward (backward) star form calculates all event and activity times (earliest, latest, slack, and float).

4-35.   **The linear programming formulation of the critical path model.**   The problem of minimizing the duration of a project represented by its event network $G_p = (W, B, p)$ can be formulated as the following linear programming problem:

minimize
$$\pi_n - \pi_1$$

subject to $\pi_j - \pi_i \geqslant p_{ij}$ for every $(i, j) \in B$ (4-17)

    π_i's are not restricted in sign $(i = 1, 2, \ldots, n)$

where π_i denotes the starting time of event i.

    (a) Show the correctness of this formulation. How can one solve this problem without using a standard LP method?

    (b) Write a dual problem for (4-17). How can we solve it? Give an interpretation of the complementary slackness theorem (Theorem 1-4) for this pair of problems.

    (c) Illustrate your consideration with the help of the network of Example 4-8.

4-36.   Discuss and design a CPM algorithm for a project given as an activity network (that is an activity-on-node network), and implement it in Pascal. Apply your algorithm to the network of Fig. 4-9 with the activity duration times given in Table 4-5. Compare the results with those in Table 4-5.

4-37.   **The time / cost network scheduling model.**   Let $G_p = (W, B, p)$ be an event network and assume that the activity duration times are functions of invested costs (e.g., labor, capital, etc.). In the simplest form of the time/cost model, activity duration p_{ij} and activity costs satisfy the following conditions:

1. p_{ij} is bounded by a minimum m_{ij} and a maximum M_{ij} feasible duration.
2. The cost c_{ij} incurred in performing an activity $(i, j) \in B$ is a linear function of the activity duration p_{ij}, that is,

$$c_{ij}(p_{ij}) = u_{ij} - v_{ij}p_{ij} \qquad \text{for} \quad m_{ij} \leqslant p_{ij} \leqslant M_{ij}$$

where u_{ij} is the normal cost and v_{ij} is the cost per unit time.

(a) Extend the LP formulation of the CPM model (Problem 4-35) to the problem of minimizing (over feasible p_{ij}) the total cost $C(T) = \Sigma_{(i, j) \in B} c_{ij}(p_{ij})$ of carrying out the project in time T (i.e., $\pi_n - \pi_1$ is assumed to be equal to T).
(b) Formulate the dual problem and interpret the complementary slackness theorem.
(c) Show that the time/cost relationship between the project duration T and the cost $C(T)$ of its realization is a piecewise linear function determined for T that are in the interval between the shortest and the longest duration of the project.

4-38.   Show how the time/cost network model with concave cost functions can be transformed to the linear case approximating the cost functions by piecewise linear functions. Formulate the corresponding LP problem.

4-39.   Given is a set of n activities each of which is to be performed on one of two machines. All activities need the same processing time and R_{ij} is the requirement of activity i for resource j. Assume that not more than r_j units of resource j can be used at a time. The problem is to minimize the maximum completion time. Show how this problem can be transformed to the maximum matching problem (see Section 3.7).

4-40.   Let $G(C)$ be a feasible disjunctive network and let V_k denote a subset of activities (nodes) which are to be performed on the same machine. Prove that every topological ordering in $G(C)$ has the following property: If (i, j) $(i, j \in V_k)$ is a critical disjunctive arc in $G(C)$, then the nodes i and j are not separated in the ordering by any other node of V_k.

4-41. Prove Theorem 4-6.

4-42. Perform all calculations of Balas's algorithm on the network of Example 4-9 and compare the results obtained with Figs. 4-17 and 4-18.

4-43. (Giffler and Thompson [1960]—see Baker [1974, pp. 189, 196]) A heuristic algorithm for constructing a feasible network of a given disjunctive network $G(D)$. The algorithm generates a feasible network $G(C)$ and a topological ordering of its nodes *order*. An activity is said to be *schedulable* if all its predecessors are scheduled. Let R and S denote the set of scheduled activities and the set of schedulable activities, respectively. In the beginning we set $R \leftarrow \{1\}$, *order*$(1) \leftarrow 1$, $r \leftarrow 1$, and S consists of all activities with predecessor 1. The set of disjunctive arcs C is empty. Let S_j and C_j denote respectively the earliest starting time and the earliest finishing time of activity j. These quantities are well defined for schedulable activities by the CPM method. The algorithm iterates the following steps until all activities (nodes) are in R.

Step 1. Let $C_{j*} = \min\{C_j: j \in S\}$ and k^* denote the machine on which j^* is to be executed.

Step 2. Choose an arbitrary $i \in S$ which is to be performed on k^* and such that $S_i < C_{j*}$.

Step 3. Set $r \leftarrow r + 1$, *order*$(r) \leftarrow i$, $R \leftarrow R \cup \{i\}$. Add to C disjunctive arcs from i to all other unscheduled activities which are to be performed on k^*. Modify S.

Implement this algorithm in Pascal and apply it to the problem discussed in Example 4-9.

Also implement the following modification of step 2 and compare the computational results: Instead of choosing an arbitrary activity, always select an activity that satisfies the conditions of this step and has a minimum processing time.

4-44. Show how a flow shop scheduling model could be interpreted as a network scheduling problem. What are the drawbacks of such a representation?

4-45. Consider the $F\|C_{\max}$ problem with positive processing times. Use an interchange argument to show that it has an optimal solution with the same processing order of the jobs on the first two machines and with the same processing order on the last two machines.

4-46. Prove that the permutation flow shop problem is equivalent to the following problem: Find a permutation $\pi = (\pi_1, \pi_2, \ldots, \pi_n)$ of columns $\{1, 2, \ldots, n\}$ of a given $m \times n$ matrix $[p_{ij}]$, $p_{ij} > 0$, that minimizes

$$\max_{l_1, l_2, \ldots, l_{n-1}} \left\{ \sum_{k=1}^{l_1} p_{k\pi_1} + \sum_{k=l_1}^{l_2} p_{k\pi_2} + \ldots + \sum_{k=l_{n-1}}^{m} p_{k\pi_n} \right\} \quad (4\text{-}18)$$

where integers $l_1, l_2, \ldots, l_{n-1}$ satisfy the relations $1 \leqslant l_1 \leqslant l_2 \leqslant \ldots \leqslant$

$l_{n-1} \leqslant m$. Illustrate your considerations with examples and interpret function (4-18) on the Gantt chart. Find the sequence $(l_1, l_2, \ldots, l_{n-1})$ which minimizes (4-18) for the optimal solution of Example 4-11.

4-47. Implement Johnson's algorithm as a Pascal procedure.

4-48. The following heuristic algorithm for solving the $m \times n$ permutation flow shop problem has been proposed in Cambel et al. [1970] (see Baker [1974]). The algorithm reduces the problem to a sequence of two-machine flow shops with modified processing times.

```
begin
    π* ← (1, 2, ..., n);
    z* ← z(π*);      {* z(π) denotes the length of schedule π *}
    for i ← 1 to m − 1 do
    begin
               i                         i
        p'₁ⱼ ←  Σ  pₖⱼ and p'₂ⱼ ←  Σ  pₘ₋ₖ₊₁,ⱼ for j = 1, 2, ..., n;
              k=1                      k=1
        apply Johnson's algorithm to find a permutation π, a solution of the
            (2 × n) PF shop problem with processing times p'ₖⱼ (k = 1, 2; j =
            1, 2, ..., n);
        if z(π) < z* then
        begin z* ← z(π);   π* ← π end
    end
end
```

Implement this algorithm in Pascal and test on a set of random data.

4-49. Show that in the no-wait shop problem with positive processing times every feasible solution is a permutation schedule.

4-50. (a) Construct a no-wait schedule for the job shop problem of Fig. 4-20(a).
(b) Find an optimal no-wait schedule for the problem of Fig. 4-20(b) by constructing and solving the corresponding instance of the TSP problem.

4-51. The length of a no-wait flow shop schedule can be also expressed in terms of the delays on the last machine. Derive a corresponding formula and a transformation to the TSP problem.

4-52. Prove Theorem 4-8 by using an interchange argument.

4-53. Implement Jackson's theorem 4-8 in the form of a Pascal procedure.

4-54. Show that the DK problem can be reduced to the decision version of $J2|m_j \leqslant 3|C_{\max}$.

4-55. Formulate the open shop, job shop, and flow shop problems by specifying precisely the set of single tasks, the set of machines, precedence constraints, task and machine characteristics, and the relations between tasks and machines.

4-56.   Prove that every schedule that is optimal with respect to L_{max} is also optimal with respect to T_{max}, but that these two performance measures are not equivalent.

4-57.   Show that the maximum and total cost performance measures involving completion times, flow times, latenesses, and tardiness are regular.

4-58.   The *earliness* of task Z_j in a schedule is defined as $E_j = \max\{0, -L_j\}$, where L_j is the lateness of task Z_j. Consider the maximum earliness E_{max} and the average earliness $\bar{E}$. Are these two performance measures regular?

4-59.   Consider a schedule of n tasks on a single machine and define the *waiting time* of task Z_j as $W_j = C_j - p_j$. Show that the weighted waiting time in the $1\|\Sigma\, w_j W_j$ problem is minimized by WSPT scheduling and SPT scheduling minimizes the maximum waiting time in the $1\|W_{max}$ problem. Which schedules maximize the mean waiting time?

4-60.   (Conway et al. [1967])   Consider the problem of scheduling n tasks on a single machine and define the *slack time* of task j at a time t to be $Q_j = d_j - p_j - t$. Prove that a sequence of tasks formed according to nondecreasing slack time (at any given moment, assume that $t = 0$) maximizes the minimum lateness and the minimum tardiness.

4-61.   Consider the $1\|prec\|C_{max}$ problem and assume that the tasks are in a topological order. Define the modified due dates as follows:

$$d'_j = \min\{d_j, \min\{d_l - p_l : j < l\}\}, \qquad j = 1, 2, \ldots, n$$

Hence d'_j is either the original due date or the earliest due date imposed by the successors of task j. Prove that all tasks can be completed on time (consistent with the precedence constraints) if and only if this can be accomplished by scheduling the tasks according to nondecreasing modified due dates.

4-62.   Consider $1|r_j|L_{max}$, the problem of minimizing the maximum lateness in scheduling on one machine n independent tasks with release dates. Show that a DK problem can be efficiently transformed to a decision version of $1|r_j|L_{max}$ which asks if the problem has a schedule satisfying $L_{max} \leqslant y$ for a given threshold y.

4-63.   Show how Jackson's rule can be applied to solve the $1|pmtn, prec, r_j|L_{max}$ problem. (*Hint:* First modify the release and due dates so that they reflect the precedence constraints.)

4-64.   Based on Theorem 4-12, design an algorithm that generates an optimal schedule for a $1|prec|f_{max}$ problem. Assume that the cost functions $f_j(t)$ can be evaluated in constant time for every j and t, and that the graph which represents *prec* is given in a star representation (Section 3.1).

4-65. **(a)** Let $u_1 \geqslant u_2 \geqslant \ldots \geqslant u_n$ and $v_1, v_2, \ldots, v_n$ be two sequences of integers. Prove that a permutation $\pi = (\pi_1, \pi_2, \ldots, \pi_n)$ minimizes the sum

$$\sum_{i=1}^{n} u_i v_{\pi_i}$$

if and only if $v_{\pi_1} \leqslant v_{\pi_2} \leqslant \ldots \leqslant v_{\pi_n}$.

(b) Determine the order of $x_1, x_2, \ldots, x_n$ which minimizes $\sum_{j=1}^{n} \sum_{i=1}^{j} x_i$.

(c) Consider the sum $\sum_{i=1}^{m} \sum_{j=1}^{n_i} (n_i - j + 1) x_j$, where $n_1 + n_2 + \ldots + n_m = n$. Prove that if this sum is minimal, then the n_i values satisfy the inequalities $|n_i - n_k| \leqslant 1$ for every pair $i, k = 1, 2, \ldots, m$.

(d) Provide some applications of parts (a) through (c) to scheduling problems.

4-66. There are n programs, each of length l_i, which are to be stored on a computer magnetic tape. Assume that all programs are retrieved equally often and that the tape reading head is positioned in front of the first program before every search. Find an ordering of programs that minimizes the mean retrieval time.

4-67. Extend Smith's rule to solving $1 \| pmtn, r_j | C_j$.

4-68. (Smith [1956]) Consider the $1|C_j \leqslant d_j|\Sigma C_j$ problem; that is, all tasks have to be completed by their due dates and the total completion time is to be minimized. Prove that there exists an optimal schedule with task Z_k last if and only if it satisfies the following conditions:

$$d_k \geqslant \sum_{j=1}^{n} p_j, \quad \text{and } p_k \geqslant p_i \quad \text{for all } i \text{ for which } d_i \geqslant \sum_{j=1}^{n} p_j$$

Using this fact, design a polynomial-time algorithm for solving the $1|C_j \leqslant d_j|\Sigma C_j$ problem.

4-69. Show that a longest processing time (LPT) scheduling maximizes the average cost in $1 \| \overline{F}$.

4-70. Construct an instance of $1 \| \overline{T}$ to illustrate that SPT scheduling does not always minimize the mean tardiness.

4-71. Dynamic programming approach to solving $1 \| \Sigma w_j U_j$. Let $f(j, t)$ denote the minimum total cost incurred in completing j tasks subject to the constraint that task j is completed by time t. Prove that $f(j, t)$ satisfies the following relations:

$$f(0, t) = 0 \ (t \geqslant 0), \quad f(j, t) = +\infty \quad (j = 1, 2, \ldots, n; t < 0)$$

$$f(j, t) = \min\{f(j, t-1), f(j-1, t-p_j), w_j + f(j-1, t)\}$$

$$(j = 1, 2, \ldots, n; t \geqslant 0) \qquad (4\text{-}19)$$

and show that $f(n, T)$ determines the optimal solution, where T is sufficiently large. Design an algorithm based on relation (4-19).

4-72.   Implement Moore–Hodgson's algorithm. What is the complexity of your procedure?

4-73.   Complete details of the proof that PARTITION can be transformed to the $P2\|\sum w_j C_j$ problem.

4-74.   Count the number of optimal solutions for the $Pm\|\sum C_j$ problem.

4-75.   Formulate the $P\|C_{max}$ problem as a 0–1 linear programming problem with variables x_{ij} $(i = 1, 2, \ldots, m; j = 1, 2, \ldots, n)$ such that $x_{ij} = 1$ if task Z_j is assigned to machine M_i and $x_{ij} = 0$, otherwise.

4-76.   Design a polynomial-time algorithm for solving the $Q\|\sum C_j$ problem. (*Hint:* First notice that if Z_j is the kth last task on machine M_i, then it contributes $kp_{ij} = ks_i p_j$ to the total cost. Consider ks_i as a weight and apply the results of Problem 4-65.)

4-77.   Formulate the $R\|\sum C_j$ problem as a linear transportation problem with variables x_{ijk} $(i = 1, 2, \ldots, m; j, k = 1, 2, \ldots, n)$ such that $x_{ijk} = 1$ if task Z_j is in the kth last position on machine M_i, and 0 otherwise.

4-78.   Is the algorithm CMM for solving $P\|\sum C_j$ an example of list scheduling?

4-79.   Implement the LS and LPT scheduling rules for solving the $P\|C_{max}$ problem and determine the complexity of your procedures in terms of m and n.

Comment: Possibly there is an $O(n \log m)$ implementation of the LS algorithm and an $O(n \log mn)$ implementation of the LPT algorithm.

4-80.   Show that the value of an optimal solution C_{max}^* to $Pm\|C_{max}$ satisfies the inequality

$$C_{max}^* \geqslant \max\left\{ \max_{1 \leqslant j \leqslant n} p_j, \left(\sum_{j=1}^{n} p_j \right) / m \right\}$$

4-81.   Performance guarantee for the list scheduling algorithm.

  (a) Prove Theorem 4-15. [*Hint:* Let Z_k be a task finished at time $C_{max}(LS)$. Note that by the algorithm, the finishing times of processors are at least equal to $C_{max}(LS) - p_k$. Combine these relations and apply the bound of Problem 4-80.]

  (b) Generalize the instance of Example 4-15 to an infinite family of instances which yield the equality in the bound of Theorem 4-15.

4-82.   Generalize the instance of Example 4-16 to an infinite family of instances which yield the equality in the bound of Theorem 4-16.

4-83.   Construct an infinite family of instances of the $Pm\|pmtn\|C_{max}$ problem which requires $m - 1$ preemptions in optimal solutions.

4-84.   Formulate the $R\|pmtn\|C_{max}$ problem as a linear programming problem with variables x_{ij} being the time spent by task Z_j on machine M_i.

Comment: In view of the results presented in Section 1.1.5, $R\|pmtn\|C_{max}$ can be solved by a polynomial-time algorithm.

4-85. **The anomalies of list scheduling** (see Coffman [1976, pp. 165–227]). Consider the problem of scheduling dependent tasks on parallel machines and assume that list scheduling is used as a scheduling rule. Clearly, the solution depends on the number of machines, the task processing times, the precedence constraints, and the priority list. A *list scheduling anomaly* refers to the situations when decreases in processing times, increases in the number of processors, the use of different priority lists, and the removal of certain precedence constraints result in an increase of schedule lengths. Given the following instance of the problem: $m = 3$, $n = 9$, $p_1 = 3$, $p_2 = 2$, $p_3 = 2$, $p_4 = 2$, $p_5 = 4$, $p_6 = 4$, $p_7 = 4$, $p_8 = 4$, $p_9 = 9$, priority list $L = (Z_1, Z_2, \ldots, Z_9)$, and the following precedence constraints:

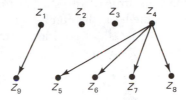

Find the list schedules for the original problem and for the following modifications:

(a) Instead of L, use $L' = (Z_1, Z_2, Z_4, Z_5, Z_6, Z_3, Z_9, Z_7, Z_8)$.
(b) Increase the number of machines to $m' = 4$.
(c) Decrease the processing times by 1, $p'_j = p_j - 1$ ($j = 1, 2, \ldots, 9$).
(d) Weaken the precedence constraints by removing arcs (Z_4, Z_5) and (Z_4, Z_6).

4-86. (Coffman [1976, p. 11]) Consider the problem of scheduling n independent tasks on parallel machines in which the due dates (called deadlines) must be respected and the goal is to minimize the number of machines required to meet the common deadline d. Show that this problem is equivalent to the $Pm|p_j = 1, res(d)|C_{max}$ problem, where $m \geqslant n$, which is to minimize the maximum completion time of n unit-time tasks on m identical machines ($m \geqslant n$) subject to the resource constraints. There is only one resource of quantity d at each moment of time.

4-87. **Bin packing problem.** The *bin packing problem* refers to the scheduling problem formulated in the beginning of Problem 4-86 and it calls for the minimum number of bins of size d for packing into them nonsplittable weights a_j. Given an arbitrary list of weights $L = (a_1, a_2, \ldots, a_n)$. There are four basic bin-packing algorithms. In each of them, a_1 is packed first and a_k is packed before a_{k+1} for all $k \geqslant 1$.

1. *First-fit* (FF). Each a_k is placed into the lowest indexed bin into which it fits.

2. *Best-fit* (BF). Each a_k is placed into a bin for which the resulting unused capacity is minimal.

3. *First-fit decreasing* (FFD). The list L is ordered according to nonincreasing weight and FF is applied to the resulting list.

4. *Best-fit decreasing* (BFD). The same as algorithm 3, with FF replaced by BF.

(a) Assume that the bins are of capacity 14 and apply the algorithms above to the following list of weights (6, 10, 12, 3, 2, 5, 2, 2).

(b) Provide suitable examples showing that none of the algorithms is superior to another.

4-88. Consider an arbitrarily structured system of unit-time tasks to be scheduled on m machines. Denote by v_j the number of tasks with the level number at least j and let l be the length of a critical path in the precedence graph G.

(a) Show that the length $C^*_{\max}$ of an optimal schedule satisfies the inequality

$$C^*_{\max} \geqslant \max_{0 \leqslant j \leqslant l} \left\{ j + \lceil v_{j+1}/m \rceil \right\} \qquad (4\text{-}20)$$

(b) Show that the minimum number of machines m^*_t required to execute all the tasks in t units of time satisfies

$$m^*_t \geqslant \max_{0 \leqslant j \leqslant l} \left\{ \lceil v_{j+1}/(t-j) \rceil \right\}, \qquad \text{where } t \geqslant l \qquad (4\text{-}21)$$

(c) Provide suitable instances of the problem for which (4-20) and (4-21) hold as equalities. Consider also trees as G and compare the bounds with the results of Hu's algorithm.

4-89. Apply the CP scheduling to the following task system to be scheduled on three machines:

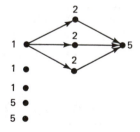

where the numbers in the nodes denote the task processing times. Find an optimal schedule and generalize this example to an arbitrary (but fixed) number of machines.

4-90. Consider a feasible schedule for a $Pm|tree$, $p_j = 1|C_{\max}$ problem and show that if no idle time occurs before the second-to-last time unit, the schedule is optimal.

4-91. Find an optimal schedule for the $P3|forest, p_j = 1|C_{max}$ problem, where the task system is formed by the union of the two tree-structured systems shown in Figs. 4-34(c) and 4-39(a).

4-92. Apply the level scheduling rule to the task systems defined in Examples 4-18 through 4-21 and Problem 4-91 modified by taking the antitrees as the precedence constraints.

4-93. Adapt the LEVELSCHEDULING procedure for precedence constraints that form a forest.

4-94. Modify the LEVELSCHEDULING procedure so that it can handle precedence constraints that form an antiforest.

4-95. Modify the LEVELSCHEDULING procedure to handle arbitrary precedence constraints using LIST as an input. Run your modification on some literature examples of $P2|prec, p_j = 1|C_{max}$.

4-96. Implement the list scheduling rule as a Pascal procedure for solving $Pm|prec|C_{max}$. Assume that a priority list is given as an input.

4-97. Implement the critical path scheduling for solving $Pm|prec|C_{max}$.

REFERENCES AND REMARKS

Informal methods of scheduling based on intuition and experience have been used for a long time. Formal scheduling models appeared during World War I and were handled in the form of the Gantt charts. Then in the mid-1950s, critical path methods came to light and replaced Gantt charts. The latter are still widely used but mainly as graphic illustrations.

First formal mathematical models of scheduling problems began to appear in the mid-1950s, when S. M. Johnson, J. R. Jackson, and W. E. Smith published their fundamental results. Since that time numerous articles have appeared in journals on operations research, management science, industrial engineering, combinatorics, and computer science. For an exhaustive presentation of the most important problems we refer to the following books (listed in historical order):

CONWAY, R. W., W. L. MAXWELL, and L. W. MILLER, *Theory of Scheduling*, Addison-Wesley, Reading, Mass., 1967

ASHOUR, S., *Sequencing Theory*, Springer-Verlag, Berlin, 1972

COFFMAN, E. G., JR., and P. J. DENNING, *Operating Systems Theory*, Prentice-Hall, Englewood Cliffs, N.J., 1973

BAKER, K., *Introduction to Sequencing and Scheduling*, Wiley, New York, 1974

LENSTRA, J. K., *Sequencing by Enumerative Methods*, Mathematisch Centrum, Amsterdam, 1976

RINNOOY KAN, A. H. G., *Machine Scheduling Problems*: *Classification*, *Complexity and Computations*, Nijhoff, The Hague, The Netherlands, 1976

ELMAGHRABY, S. E., *Activity Networks*: *Project Planning and Control by Network Models*, Wiley, New York, 1977

LENSTRA, J. K., E. L. LAWLER, and A. H. G. RINNOOY KAN, *Theory of Sequencing and Scheduling*, Wiley, New York, 1983

to the collections of papers:

MUTH, J. F., and G. L. THOMPSON (eds.), *Industrial Scheduling*, Prentice-Hall, Englewood Cliffs, N.J., 1963

ELMAGHRABY, S. E. (ed.), *Symposium on the Theory of Scheduling*, Springer-Verlag, Berlin, 1973

COFFMAN, E. G., JR. (ed.), *Computer and Job Shop Scheduling Theory*, Wiley, New York, 1976

Operations Research 26(1978), No. 1

and to a recent review paper

GRAHAM, R. L., E. L. LAWLER, J. K. LENSTRA, A. H. G. RINNOOY KAN, Optimization and Approximation in Deterministic Sequencing and Scheduling: A Survey, *Ann. Discrete Math.* 5(1979), 287–326

The classification scheme for scheduling problems discussed in Section 4.2.4 first appeared in Conway et al. [1967]. Its present form is described in full detail in Graham et al. [1979], which is also a good source of complexity results and performance bounds.

The classical network scheduling models and techniques such as CPM, PERT, and time/cost models were introduced in late 1950s and early 1960s, and now their description can be found in nearly every book on operations research, network analysis, and applied graph theory. We especially recommend Chapters 9 and 10 of Baker [1974], Elmaghraby [1977], and

PHILLIPS, D. T., and A. GARCIA - DIAZ, *Fundamentals of Network Analysis*, Prentice-Hall, Englewood Cliffs, N.J., 1981, Chapter 4

A primal–dual algorithm for the determination of the complete time/cost curve (Problem 4-37) has been proposed by D. R. Fulkerson; see

FORD, L. R., JR., and D. R. FULKERSON, *Flows in Networks*, Princeton University Press, Princeton, N.J., 1962

and its implementation in ALGOL 60 is given in

KUCHARCZYK, J., and M. M. SYSŁO, *Optimization Algorithms in ALGOL 60*, Państwowe Wydawnictwo Naukowe, Warsaw, 1975 (in Polish)

The time/cost network scheduling problem with nonlinear cost–time functions is discussed in

FALK, J. E., and J. L. HOROWITZ, Critical Path Problems with Concave Cost–Time Curves, *Management Sci.* 19(1972), 446–455

Almost every description of network analysis also provides an illustration on how to construct activity and event networks for a given project. For a formal treatment of the network construction methods, we refer the reader to a review paper,

SYSŁO, M. M., Optimal Constructions of Event-Node Networks, *RAIRO Rech. Opér.* 15(1981), 241–260

where it is also shown that dummy activities in event networks are enforced by precedence constraints which contain the subgraph of Fig. 4-10(a) as an induced subgraph.

The NP-completeness of the minimum-dummy-activities problem has been established in

KRISHNAMOORTHY, M. S., and N. DEO, Complexity of the Minimum-Dummy-Activities Problem in a PERT Network, *Networks* 9(1979), 189–194

Disjunctive graphs were introduced to scheduling theory in the paper

ROY, B., and B. SUSSMAN, Les Problèmes d'ordonnancement avec constraintes disjonctives, SEMA, Note D.S. No. 9, Dec. 1964

and since then they have been rediscovered many times. Our presentation is based on Balas's contribution; see

BALAS, E., Machine Sequencing via Disjunctive Graphs: An Implicit Enumeration Algorithm, *Oper. Res.* 17(1969), 941–957

Some improvements to Balas's algorithm appeared in

GRABOWSKI, J., Formulation and Solution of the Sequencing Problem with Parallel Machines, *Zastos. Mat.* 16(1978), 215–232

A critical review of the existing (by 1973) methods for resource-constrained project scheduling appeared as

BENNINGTON, G. E., and L. F. McGINNIS, A Critique of Project Planning with Constrained Resources, in Elmaghraby [1973, pp. 1–28]

See also

DAVIS, E. W., Project Scheduling under Resources Constraints—Historical Review and Categorization of Procedures, *AIIE Trans.* 5(1973), 297–313

Another approach to resource allocation in project networks has been proposed and elaborated on in the context of the job shop problem formulated as a network model. Instead of constructing a new feasible solution at every stage as Balas's algorithm does, it constructs a solution iteratively. We illustrate such an algorithm in Problem 4-43, which presents a heuristic, one-pass iterative method. Exact methods are all of enumerative type, since it is unlikely that a polynomial-time algorithm can generate an exact solution. Implicit enumeration methods for job shop scheduling via disjunctive graphs are discussed in

LAGEWEG, B. J., J. K. LENSTRA, and A. H. G. RINNOOY KAN, Job-Shop Scheduling by Implicit Enumeration, *Management Sci.* 24(1977), 441–450

Some efficient algorithms (for $O2\|C_{max}$ and $O|pmtn|C_{max}$) and complexity results for open shop scheduling are presented in

GONZALES, T., and S. SAHNI, Open Shop Scheduling to Minimize Finish Time, *J. ACM* 23(1976), 665–779

The complexity of job shop and flow shop problems is discussed in

GAREY, M. R., D. S. JOHNSON, and R. SETHI, The Complexity of Flowshop and Jobshop Scheduling, *Math. Oper. Res.* 1(1976), 117–129

GONZALES, T., and S. SAHNI, Flowshop and Jobshop Schedules: Complexity and Approximation, *Oper. Res.* 26(1978), 36–52

The fundamental results on the two-machine flow shop problem (Theorem 4-7) and its generalizations appeared in

JOHNSON, S. M., Optimal Two- and Three-Stage Production Schedules with Setup Times Included, *Naval Res. Logist. Quart.* 1(1954), 61–68

Special, polynomially solvable cases of the flow shop problem are reviewed in

SZWARC, W., Permutation Flow-Shop Theory Revised, *Naval Res. Logist. Quart.* 25(1978), 557–570

See also

SZWARC, W., The Critical Path Approach in the Flow-Shop Problem, *Opsearch* 16(1979), 98–102

In the last two papers, the flow shop problem is studied by using critical path concepts, suggested already by S. M. Johnson in his fundamental paper of 1954 (see also Problem 4-46).

Branch-and-bound methods for solving the permutation flow shop problem are classified and reviewed in

LAGEWEG, B. J., J. K. LENSTRA, and A. H. G. RINNOOY KAN, A General Bounding Scheme for the Permutation Flow-Shop Problem, *Oper. Res.* 26(1978), 53–67

The no-wait flow shop problem and its generalizations to groups of nonde-layed operations have been discussed in

GRABOWSKI, J., and M. M. SYSŁO, On Some Machine Sequencing Problems I, *Zastos. Mat.* 13(1973), 339–345
SYSŁO, M. M., On Some Machine Sequencing Problems II, *Zastos. Mat.* 14(1974), 93–97

A no-wait version of the job shop problem can be also transformed to the TSP problem; see

LENSTRA, J. K., and A. H. G. RINNOOY KAN, Some Simple Applications of the Travelling Salesman Problem, *Operational Res. Quart.* 26(1975), 717–733

The classical result on the two-machine job shop problem (Theorem 4-8) appeared in

JACKSON, J. R., An Extension of Johnson's Results on Job Lot Scheduling, *Naval. Res. Logist. Quart.* 3(1956), 201–203

The fundamental results on scheduling a single machine have been presented in

JACKSON, J. R., Scheduling a Production Line to Minimize Maximum Tardiness, RR 43, Management Science Research Project, University of California, Los Angeles, 1955
SMITH, W. E., Various Optimizers for Single-Stage Production, *Naval Res. Logist. Quart.* 3(1956), 59–66

The single-machine scheduling with arbitrary cost functions has been solved (Theorem 4-14) in

LAWLER, E. L., Optimal Sequencing of a Single Machine Subject to Precedence Constraints, *Management Sci.* 19(1973), 544–546

and some of its polynomially solvable extensions are presented in

BAKER, K. R., E. L. LAWLER, J. K. LENSTRA, and A. H. G. RINNOOY KAN, Preemptive Scheduling of a Single Machine to Minimize Maximum Cost Subject to Release Dates and

Precedence Constraints, TR 8028/0, Econometric Institute, Erasmus University, Rotterdam, 1980

The complexity of $1\|\sum w_j T_j$ can be established by using the DK problem (see Lenstra [1976]). Dynamic programming and enumerative algorithms for this problem are presented, respectively, in

BAKER K. R., and L. E. SCHRAGE, Finding an Optimal Sequence by Dynamic Programming: An Extension to Precedence-Related Tasks, *Oper. Res.* 26(1978), 111–120

RINNOOY KAN, A. H. G., B. J. LAGEWEG, and J. K. LENSTRA, Minimizing Total Costs in One-Machine Scheduling, *Oper. Res.* 23(1975), 908–927

The hardness of $1\|\sum w_j U_j$ has been established in the fundamental paper on computational complexity

KARP, R. M., Reducibility among Combinatorial Problems, in R. E. Miller and J. W. Thatcher (eds.), *Complexity of Computer Computations*, Plenum Press, New York, 1972, pp. 85–103

A dynamic programming algorithm for this problem working in $O(n\sum p_j)$ time has been presented in

LAWLER, E. L., and J. M. MOORE, A Functional Equation and Its Application to Resource Allocation and Sequencing Problems, *Management Sci.* 16(1969), 77–84

and approximation algorithms are discussed in

SAHNI, S., Algorithms for Scheduling Independent Tasks, *J. ACM* 23(1976), 116–127

Moore–Hodgson's algorithm appeared in

MOORE, J. M., An *n* Job, One Machine Sequencing Algorithm for Minimizing the Number of Late Jobs, *Management Sci.* 15(1968), 102–109

and has been extended to the case when certain tasks have to be performed on time, in

SIDNEY, J. B., An Extension of Moore's Due Date Algorithm, in Elmaghraby [1973, pp. 393–398]

Fundamental results on scheduling parallel machines appeared in Conway et al. [1967] and in

McNaughton, R., Scheduling with Deadlines and Loss Functions, *Management Sci*. 6(1959), 1–12

The performance guarantees of scheduling algorithms such as LS, LPT, CMM, and CP as well as the bounds on the LS anomalies (Problem 4-85) are discussed by R. L. Graham in Chapter 5 of Coffman [1976] and in Graham et al. [1979]; see also an easy introduction to the subject in

Garey, M. R., R. L. Graham, and D. S. Johnson, Performance Guarantees for Scheduling Algorithms, *Oper. Res.* 26(1978), 3–21

The LPT algorithm has been recently improved to an algorithm with an asymptotic performance ratio of 5/4; see

Langston, M. A., Improved LPT Scheduling for Identical Processor Systems, *RAIRO Tech. Sci. Inform*. 1(1982), 69–75

Approximation algorithms for the bin-packing problem (see Problem 4-87) are reviewed in

Garey, M. R., and D. S. Johnson, Approximation Algorithms for the Bin Packing Problems: A Survey, in G. Ausiello and M. Lucertini (eds.), *Analysis and Design of Algorithms in Combinatorial Optimization*, Springer-Verlag, New York, 1981, pp. 147–172

Parallel machine scheduling with additional resource constraints is discussed in

Błażewicz, J., J. K. Lenstra, and A. H. G. Rinnooy Kan, Scheduling Subject to Resource Constraints: Classification and Complexity, *Discrete Appl. Math.* 5(1983), 11–24

The algorithm implemented in the LEVELSCHEDULING procedure is due to T. C. Hu and appeared originally in

Hu, T. C., Parallel Sequencing and Assembly Line Problems, *Oper. Res.* 9(1961), 841–848

Its almost linear implementation has been proposed in

Sethi, R., Scheduling Graphs on Two Processors, *SIAM J. Comput.* 5(1976), 73–82

and is also described in Coffman [1976, pp. 51–99]. A polynomial-time algorithm for solving the $P2|\,prec,\ p_j = 1|C_{\max}$ problem has been presented in

Coffman, E. G. Jr., and R. L. Graham, Optimal Scheduling for Two Processor System, *Acta Inform*. 1(1972), 200–213

The critical path scheduling of dependent tasks on parallel machines has been experimentally found as a very promising approximation method; see

KOHLER, W. H., A Preliminary Evaluation of the Critical Path Method for Scheduling Tasks on Multiprocessor Systems, *IEEE Trans. Comput.* 24(1975), 1235–1238

Set operations and their efficient implementations used in the LEVELSCHED-ULING procedure are discussed in

AHO, A. V., J. E. HOPCROFT, and J. D. ULLMAN, *The Design and Analysis of Computer Algorithms*, Addison-Wesley, Reading, Mass., 1974, Chapter 4

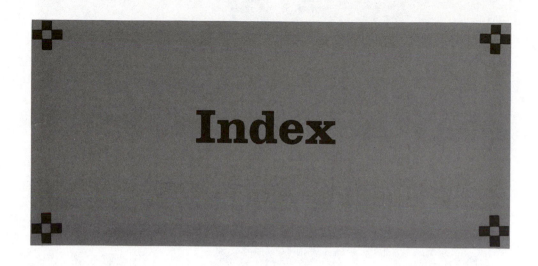

Index

AA (*see* Event network)
Absolute ∈–approximate algorithm,
 135
Activity, 446, 484
 branching, 450
 critical, 451
 dummy, 448
 merging, 450
 schedulable, 520
Activity network (*or* AN) (*or*
 activity–on–node network), 447
Activity–on–arc network (*see* Event
 network)
Activity–on–node network (*see*
 Activity network)
Acyclic graph (network), 226
Adjacency matrix, 233
Adrabiński, A., 362, 369–70, 388
Aggregation method for integer
 linear constraints, 121–23, 172
Aho, A. V., 511, 534
Airline crew scheduling problem,
 178
Alavi, Y., 443
Algorithm:
 all–integer dual (*or* Gomory's),
  86–87
 all–integer primal, 97–99
 approximately maximum
  independent set (*or* AMmIS)
  coloring, 403–4
 backtracking sequential coloring,
  425
 Balas's (for scheduling problem),
  460–66
 Balas 0–1 additive, 104–6
 Bellman–Moore's, 235–37
 best–fit decreasing (*or* BFD), 526
 best–fit (*or* BF), 526
 Busacker–Gowen's, 303
 Christofides's (for coloring), 401
 Christofides's (for TSP), 381
 Chvátal's, 215
 CMM (*see*
  Conway–Maxwell–Miller
  algorithm)
 composite simplex, 53
 cutting plane, 80–81
 Dijkstra's (*or* label–setting), 228
 Dinic's, 270, 299

Algorithm (cont.):
 dual simplex, 42–44
 ellipsoid (*or* Khachian), 23–26
 first–fit decreasing (*or* FFD),
  526
 first–fit (*or* FF), 525
 Floyd's, 243, 249
 Ford–Fulkerson (for max–flow
  problem), 57, 270, 299
 Ford–Fulkerson (for transportation
  problem), 65–66
 Gomory's (*or* all–integer dual),
  86–87
 greedy, 126, 268
 Hu's, 504–12
 insertion, 361–62
 Jackson's, 482
 Johnson's, 478
 Khachian (*or* ellipsoid), 23–26
 Kruskal's, 253–55
 label–correcting (*or*
  Bellman–Moore's), 235–37
 labeling (for max–flow problem),
  57
 label–setting (*or* Dijkstra's), 228,
  236
 largest–first (*or* LF) sequential
  coloring, 405
 lexicografic dual simplex, 82–84
 max–flow, 270
 maximum independent set (*or*
  MmIS) coloring, 402–4
 Moore–Hodgson's, 495
 nearest–neighbor (*or* Prim's), 259,
  261
 primal–dual simplex, 54
 Prim's (*or* nearest–neighbor), 259,
  261
 recursive–largest–first (*or* RLF)
  coloring, 404
 reduction (for covering problems),
  183
 reduction (for knapsack problem),
  128
 revised simplex, 5–6, 9–13
 r–optimal (*or* r–opt), 368
 saturation–largest–first (*or* SLF)
  coloring, 407, 441
 sequential with interchange (*or* SI)
  coloring, 414

Algorithm (cont.):
 simple sequential (*or* S) coloring,
  404
 simplex, 4
 smallest–last sequential (*or* SLS)
  coloring, 407
 Sollin's, 265, 268
 tableau simplex, 30–31
 3–optimal approximate for TSP,
  374–76
 2–optimal approximate for TSP,
  370–71
 update, 439
All–integer dual (*or* Gomory's)
 algorithm, 86–87
All–integer (*or* Gomory) cut, 85–86
All–integer primal algorithm, 97–99
Alternating path, 323
Alternating tree, 324
AMmIS (*see* Approximately
 maximum independent set coloring
 algorithm)
AN (*see* Activity network)
Antiforest, 505
Appel, K., 442
Approximately maximum
 independent set, 403
Approximately maximum
 independent set (*or* AMmIS)
 coloring algorithm, 403–4
Approximation algorithm:
 absolute ∈–approximate, 135
 ∈–approximate, 136
 fully polynomial time, 174
 insertion, 361–62
 for knapsack problem, 135–42
 r–optimal (*or* r–opt) for TSP,
  368–76
 for set–covering problem, 215–17
Arbitrary insertion, 362
Arc (*or* directed edge), 222
 candidate, 462
 chosen, 456
 conjunctive, 456
 disjunctive, 456
 free, 462
 normal, 462
 rejected, 456
 reverse, 462
 settled, 456

Arrival time (*see* Release date)
Artificial variable, 6
Ashour, S., 527
Assignment problem:
 bottleneck, 76
 linear, 74, 322, 339
Asymmetric traveling salesman
 problem, 345
Augmenting path:
 for matching problem, 323
 for max–flow problem, 57, 270
Ausiello, G., 533
Average cost, 486

Back–plane wiring problem, 381
Backtracking sequential coloring
 algorithm, 425
Backward recurrence relation, 144
Backward star, 225
Baker, K. R., 493–94, 520–21,
 527–28, 531–32
Balas, E., 2, 104, 112–13, 115,
 175, 218, 460, 463, 469, 529
Balas's algorithm (for scheduling
 problem), 460–66
Balas 0–1 additive algorithm, 104–6
Balinski, M., 218–19
Bandwidth of a graph, 439
Bartels, R. H., 18, 35
Bartz, A. E., 300
Bar–Yehuda, R., 215, 220
Basic (feasible) solution, 4
Basic variable, 4
Basis, 4
Bayer, G., 299
Bazewicz, J., 533
Beale, E. M. L., 36
Bell, E. J., 36
Bellman, R. E., 143, 174, 235–37,
 251
Bellman–Moore's (*or*
 label–correcting) algorithm,
 235–37
Bellmore, M., 386–87, 391
Bennington, G. E., 529
Berge, C., 268, 321, 323, 340
Berge's theorem, 323
Bernhart, F. R., 442
Best–fit decreasing (*or* BFD)
 algorithm, 526
Best–fit (*or* BF) algorithm, 526
BF (*see* Best–fit algorithm)
BFD (*see* Best–fit decreasing
 algorithm)
BFS (*see* Breadth–first search)
Biggs, N. L., 397, 442
Bin packing problem, 340, 502,
 525–26
Bipartite graph (*or* network), 55,
 322, 397
Bipartite matching problem, 322
Birkhoff–von Neumann theorem, 339
Bloniarz, P., 252
Blossom, 326
BNP (*see* Bounded knapsack
 problem)
Bock, F., 390
Bodin, L., 389
Bottleneck assignment problem, 76
Bounded knapsack problem (*or*
 BNP), 119
Bradley, G. H., 76
Branch–and–bound algorithm for
 TSP, 346
Branching, 346

Branching activities, 450
Bray, T. A., 319
Breadth–first search (*or* BFS), 247
Brélaz, D., 407, 434, 441, 444–45
Brigham, R. C., 441, 445
Brocklehurst, E. R., 33, 35
Brooks, R., 399
Brown, G. G., 76
Brown, J. R., 427, 444–45
Brucker, P., 168–69, 174
Burkard, R. E., 77, 339, 342–43,
 386
Busacker, R. G., 303, 318
Busacker–Gowen's algorithm, 303

Cambel, H. G., 521
Candidate arc, 462
Capacity:
 of a cut, 59, 298
 of an edge, 56, 269
 of a node, 270, 296
Capital budgeting problem, 172
Cardinality matching problem, 321
Caterer problem, 316
Chang, S. K., 172
Change–making problem, 119, 166,
 172
Charnes, A., 32, 35
Chartrand, G., 443
Cheapest insertion, 362
Cheapest path, 227, 302
Cheriton, D., 268
Cheung, T. Y., 300
Chiba, N., 443
Chosen arc, 456
Christofides, N., 99, 116, 171, 219,
 374, 381, 385–90, 401, 443–44
Christofides's algorithm (for
 coloring), 401
Christofides's algorithm (for TSP),
 381
Chromatic index, 441
Chromatic number, 395
Chvátal, V., 215, 220
Chvátal's algorithm, 215
Circuit (*see* Cycle)
Clausen, J., 22, 32, 35
Clique, 398
 separation, 438
Clique number, 398
CMM (*see* Conway–Maxwell–Miller
 algorithm)
Coffman, E. G., Jr., 480, 500–501,
 505, 525, 527–28, 533
Color class, 396
Coloring, 394
 complete, 395
 equivalent, 396
 partial, 395
 redundant, 396
 sequential, 404–33
Coloring algorithm:
 approximately maximum
  independent set (*or* AMmIS),
  403–4
 backtracking sequential, 425
 largest–first (*or* LF) sequential,
  405
 maximum independent set (*or*
  MmIS), 402–4
 recursive–largest–first (*or* RLF),
  404
 saturation–largest–first, 407, 441
 sequential with interchange (*or*
  SI), 414

Coloring algorithm (cont.):
 simple sequential (*or* S), 404
 smallest–last sequential (*or* SLS),
  407
Coloring function, 396
Coloring partition, 396, 399
Complementary slackness theorem,
 39
Complete coloring, 395
Complete (*or* perfect) matching, 212,
 338
Completion (*or* finishing) time (of a
 schedule), 485
Composite simplex algorithm, 53
Conjunctive arc, 456
Connected component, 266
Connected graph, 226
Conradt, D., 327, 342–43
Consistent project, 447
Convex hull, 27
Conway, R. W., 489, 497, 522,
 527–28, 532
Conway–Maxwell–Miller (*or* CMM)
 algorithm, 497
Cooper, W. W., 35
Corneil, D. G., 441, 445
Cornuéjols, G., 381, 389
Cost function, 485–87
Covering problems, 176
CP (*see* Critical path scheduling)
CPM (*see* Critical path method)
Crashing, 13
Critical activity, 451
Critically *k*–chromatic graph, 439
Critical (*or* longest) path, 248, 451
Critical path (*or* CP) scheduling, 506
Critical path method (*or* CPM), 446
Croes, A., 389
Crowder, H., 391
Cullen, F. H., 218
Cut, in a network, 59
Cut (*or* cutting plane), 80
 all–integer (*or* Gomory), 85–86
 Gomory (*or* all–integer), 85–86
Cutting plane algorithm, 80–81
Cutting plane (*or* cut), 80
Cutting stock problem, 172
Cycle (*or* circuit), 226
 fundamental, 267
 Hamiltonian, 345
 negative–weight, 238, 243

Daellenbach, H. G., 36
Dantzig, G. B., 3, 21, 26, 35, 46,
 54, 80, 99, 124, 167, 173, 248,
 252, 387
Davis, E. W., 530
Deadline, 485, 491
Decision knapsack problem (*or*
 DKP), 477
Degenerate basic feasible solution,
 21
Degenerate LP problem, 21
Delivery problem, 179
Dembo, R. S., 173
Denardo, E. V., 239, 251
Denning, P. J., 480, 527
Dennis, K., 33, 35
Deo, N., 249–50, 387, 391, 448,
 529
Depth–first search, 146
Derigs, U., 77, 334, 339, 341–3
D'Esopo, 237–37, 251
Dial, R. B., 239, 251
Diet problem, 31

Digraph (*or* directed graph), 222
Dijkstra, E. W., 228, 251, 259, 267, 307
Dijkstra's shortest–path (*or* label–setting) algorithm, 228
Dinic, E. A., 270, 285, 299
Dinic's algorithm, 270, 299
Directed edge (*see* Arc)
Directed graph (*see* Digraph)
Disjunctive arcs, 456
Distinct representatives problem, 323, 338
DKP (*see* Decision knapsack problem)
Dobkin, D. P., 28, 35
Dominated machine, 479
Dörfler, W., 341
Dorhout, B., 77
Doyle, T., 389
Dreyfus, S. E., 174, 250
Dual algorithm:
    all–integer (*or* Gomory's), 86–87
    for LP (*see* Dual simplex algorithm)
    for min–cost flow problem (*see* Busacker–Gowen's algorithm)
Dual feasible solution, 40
Duality, 37
Duality theorem, 38
Dual linear program, 37–38
Dual simplex algorithm, 42–44
Dual simplex tableau, 83
Due date, 485
Duffuaa, S., 35
Dummy activity, 448
Duration (*or* processing) time, 455, 485
Dürre, K., 444
Dutton, R. D., 441, 445
Dynamic programming, 143–46, 170
Dynamic search tree, 170
Dzikiewicz, J., 252

E-approximate algorithm, 136
Earliest due dates (*or* EDD) scheduling, 451
Earliest finish time, 452
Earliest start time, 451–52
Earliness, 522
EDD (*see* Earliest due dates scheduling)
Edge, 222
    matched, 323
    parallel, 222, 316
    saturated, 271
    usable, 278
Edge capacity, 56, 269
Edge coloring, 441–42
Edge cover, 213, 339
Edge–covering problem, 213
Edge packing (*or* matching), 212, 320
Edmonds, J., 59, 76, 270, 299, 307, 320, 324, 326, 341
Eight Queens problem, 114
Eilon, S., 171, 374, 390
Elias, P., 298
Ellipsoid, 24
Ellipsoid (*or* Khachian) algorithm, 23–26
Elmaghraby, S. E., 172, 528
Equivalent coloring, 396
Euler, L., 442
Even, S., 215, 220, 299, 342
Event, 448

Event network (*or* AA) (*or* activity–on–arc network), 448
Examination scheduling, 394–95, 441
Exposed (*or* free) node, 323
Extreme point, 27

Facility location problem, 101, 179
Falk, J. E., 529
Farthest insertion, 362
Farthest–insertion approximate TSP (*or* FITSP) algorithm, 362
Fault testing problem, 179
Fayard, D., 116, 168, 173, 176
FCP (*see* Four–color problem)
Feasible network, 457
Feasible solution, 4
    basic, 4
    dual, 40
    primal, 4
Feinstein, A., 298
FF (*see* First–fit algorithm)
FFD (*see* First–fit decreasing algorithm)
Finishing (*or* completion) time (of a schedule), 485
Fiorini, S., 441, 445
First–fit decreasing (*or* FFD) algorithm, 526
First–fit (*or* FF) algorithm, 525
Fisher, H., 469
FITSP (*see* Farthest–insertion approximate, TSP algorithm)
Fixed–charge problem, 79
Float, 452–54
    free, 454
    independent, 454
    safety, 454
    total, 454
Flood, M. M., 386
Florian, M., 469
Flow conservation equation, 56, 269–70
Flow (*or* flow pattern), 56, 269
    maximum, 56, 269
    minimum cost (*or* min–cost), 301
    multi commodity, 297, 317
    saturating, 271
Flow pattern (*see* Flow)
Flow shop, 475–81
    no–wait, 480
    permutation, 477
Flow time, 485
Flow value, 56, 269
Floyd, R. W., 242, 251
Floyd's algorithm, 243, 249
Floyd–Warshall algorithm (*see* Floyd's algorithm)
Ford, L. R., Jr., 54, 56, 59, 67, 76, 251, 270, 296, 298, 316, 528
Ford–Bellman–Moore algorithm (*see* Bellman–Moore's algorithm)
Ford–Fulkerson max–flow algorithm, 57, 270, 299
Ford–Fulkerson transportation algorithm, 65–66
Forest, 226, 503
    anti–, 505
    minimum spanning, 253
Forward recurrence relation, 144
Forward star, 224
Four–color problem (*or* FCP), 442
Fox, B. L., 76, 239, 251
Frederickson, G. N., 391
Fredman, M. L., 252

Free arc, 462
Free float, 454
Free (*or* exposed) node, 323
Frieze, A. M., 174, 391
Fujii, M., 340, 343
Fulkerson, D. R., 54, 56, 59, 67, 76, 270, 296, 298, 316, 318, 387, 528
Fully polynomial time approximation scheme, 174
Fundamental cycle, 267
Furtado, A. L., 443

Gabow, H., 327, 341
Gács, P., 25, 35
Galbiati, G., 391
Galil, Z., 300
Gallai, T., 213, 219, 439
Gallai's theorem, 213, 219
Gantt chart, 475
Garcia-Diaz, A., 319, 387, 528
Garey, M. R., 173, 388, 443, 530, 533
Garfinkel, R. S., 94, 99, 171, 217–19
Gass, S. I., 9, 15, 35
General knapsack problem (*or* GNP), 119
Geoffrion, A. M., 112, 115
Ghouila–Houri, A., 268
Giffler, B., 520
Gill, A., 172
Gill, P. E., 20, 35, 36
Gilmore, P. C., 172, 381
GLF ordering (*see* Greedy LF ordering)
Glover, F., 114, 116, 173, 251, 300, 320
GNP (*see* General knapsack problem)
Golden, B., 362, 369–70, 389
Goldfarb, D., 37
Goldsmith, D. L., 443
Golub, G. H., 18, 35
Gomory, R. E., 81–82, 99, 172, 381, 388
Gomory (*or* all–integer) cut, 85–86
Gomory's (*or* all–integer dual) algorithm, 86–87
Gonzales, T., 380, 389, 475, 530
Goodman, S. E., 268
Goodness function, 397
Gotlieb, C. C., 250
Gotlieb, L. R., 250
Gowen, P. J., 303, 318
Grabowski, J., 529, 531
Graham, B., 441, 445
Graham, R. L., 487, 500–501, 505, 528, 533
Graph (*or* network), 222
    acyclic, 226
    bipartite, 322, 397
    connected, 226
    critically *k*–chromatic, 439
    directed (*or* digraph), 222
    k–chromatic, 395
    k–colorable, 395
    k–partite, 397, 440
    Mycielski's, 398, 433, 439
    undirected, 222
    weighted (*see* Network)
Graves, G. W., 76
Graves, R., 100
Greedy algorithm:
    for knapsack problem, 126
    for MST problem, 268

Greedy LF (*or* GLF) ordering, 427
Greenberg, H. 170, 175
Greenberg, H. J., 36, 100
Guha, D. K., 219

Haken, W., 442
Hakimi, S. L., 391
Haldi, J., 91, 94, 99, 113, 116
Hall's theorem, 338
Hamacher, H., 300
Hamilton, W. R. Sir, 345
Hamiltonian cycle, 345
Hamiltonian cycle problem, 345
Hamiltonian path (*or* tree), 345, 381
Hamiltonian path problem, 345
Hamiltonian tree (*or* path), 345, 381
Hammer, P. L., 77, 112, 116, 173, 219
Hansen, P., 188
Harary, F., 219, 249
Heap, 254
Hecht, M. S., 391
Hedetniemi, S. T., 268
Hegerich, R. L., 170, 175
Held, M., 386, 388
Helgason, R. V., 319
Heske, A., 341–42
Heuft, J., 444
Hitchcock, F. L., 317
Hochbaum, D. S., 220
Hodgson, T. J., 495
Hoffman, A. J., 386
Holt, R. C., 249, 253
Hong, S., 391
Hopcroft, J. E., 342, 534
Horowitz, E., 139, 146, 149, 167, 171, 173, 175, 250, 268, 387
Horowitz, J. L., 529
Hu, T. C., 317, 503–4, 533
Hudson, P. D., 167
Hu's algorithm, 504–12
Hyperplane, 27

Ibarra, O. H., 174
Identical (*or* parallel) machines, 486, 497
Idle time, 489
ILP (*see* Integer linear programming)
Immediate predecessor, 226
Immediate successor, 226
Implicit enumeration, 102–4
  for knapsack problem, 146–62
  for set–partitioning problem, 194–210
  for 0–1 ILP, 102–4
Incremental (*or* modified) network, 302
Independent float, 454
  approximately maximum, 403
  maximal (*or* MIS), 396
  maximum (*or* MmIS), 396
Independent (node) set (*or* node packing), 212, 394
Independent set approach (to coloring), 399
Ingargiola, G. P., 127, 173
Initial matching, 327–28
Inner node, 325
Insertion:
  arbitrary, 362
  cheapest, 362
  farthest, 362
  nearest, 362
Insertion algorithm, 361–62

Integer linear programming (*or* ILP), 77
Interchange argument, 490–93
In–tree, 503
Iri, M., 307, 320
Isaacson, J. D., 444
Isotonic path, 22

Jackowski, B. L., 445
Jackson, J. R., 483, 490–91, 494, 527, 531
Jackson's algorithm, 482
Ja'Ja', J., 392
Jarvis, J. J., 218
Jenkyns, T. A., 391
Job, 474, 484
Job shop, 475
Johnson, D. S., 173, 220, 388, 443, 530, 533
Johnson, E. L., 100, 386
Johnson, S. M., 387, 478, 527, 530–31
Johnson's algorithm (*or* rule), 478

Kainen, P., 443
Kameda, T., 341
Kanellakis, P., 390
Kantorovitch, L., 3, 317
Karel, C., 387
Kariv, O., 342
Karney, D., 251, 320
Karp, R. M., 59, 76, 270, 299, 307, 320, 342, 388, 532
Karzanov, A. V., 284, 300
Kasami, T., 343
k–chromatic graph, 395
k–colorable graph, 395
k–coloring, 395
Kendall, K. E., 166, 173
Kennington, J. L., 319
Kernighan, B. W., 369, 374, 390
Kerschenbaum, A., 268
Khachian, L. G., 23, 35
Khachian (*or* ellipsoid) algorithm, 23–26
Kim, C. E., 174, 391
Kindorf, J. R., 112, 116
Klee, V., 22, 31, 35
Klein, M., 317–18
Klingman, D., 76, 251, 300, 320
Knapsack problem, 124
  bounded (*or* BNP), 119
  decision (*or* DKP), 477
  general (*or* GNP), 119
  one-dimensional, 119
  shortest–path formulation, 169
  0–1 multi–dimensional, 120
  0–1 multiple, 120
  0–1 (*or* 0–1 KP), 118
Knödel, W., 387
Koch, J. A., 442
Kohler, W. H., 508, 534
Kolesar, P. J., 175
Koncol, R. D., 219
König, D., 317, 439, 442
König–Egerváry theorem, 297
Konig's theorem, 397, 439
Koopmans, T. C., 317
Kornam, S. M., 116, 219, 427, 444–45
Korsh, J. F., 127, 173
k–partite graph, 397, 440
k–person TSP, 385
Krarup, J., 35
Kress, M., 35

Krishnamoorthy, M. S., 448, 529
Kruskal, J. B., 253, 267
Kruskal's algorithm, 253–55
Kubale, M., 445
Kucharczyk, J., 14, 35, 44, 53, 66, 76, 89, 99, 107, 116, 528
Kuester, J. L., 100
Kuhn, H. U., 54
Kumar, M. P., 270, 285, 300
Kumar, V. K. P., 60, 392

Label, 57, 228
  permanent, 228
  temporary, 228
Label–correcting (*or* Bellman–Moore's) algorithm, 235–37
Labeling algorithm (for max–flow problem), 57
Label–setting (*or* Dijkstra's) algorithm, 228, 236
Lageweg, B. J., 469, 530–31
Land, A., 100, 112, 116
Langston, M. A., 533
Largest–first (*or* LF) ordering, 405
Largest–first (*or* LF) sequential coloring algorithm, 405
Largest processing time (*or* LPT) scheduling, 500
Lasdon, L. S., 21, 35, 99
Lasky, J. S., 217, 219
Lateness, 485
Latest finish time, 451
Latest start time, 451
Lauriere, M., 168, 173
Lawler, E. L., 77, 174, 220, 268, 313, 318, 320, 327, 341–43, 381, 387, 440, 443, 492, 494, 528, 531–32
Layer, 271
Layered network, 271
Leighton, F. T., 404, 434, 444–45
Lemke, C. E., 42, 53, 210, 219
Lenstra, J. K., 218, 381, 387, 392, 496, 527–28, 530–33
Lesniak–Foster, L., 443
Level of a node, 503
Level scheduling, 503
Lewis, P. M., 389
Lexicographically negative, 84
Lexicographically positive, 84
Lexicographic dual simplex algorithm, 82–84
LF (*see* Largest–first ordering)
Lick, D. R., 443
Lin, S., 369, 374, 390
Linear assignment problem, 74, 322, 339
Linear inequalities, 27
Linear program (in standard form), 2
  dual, 37–38
  (non)degenerate, 21
  primal, 37
  simplified, 27
Linear programming (*or* LP), 2
Linear strict inequalities, 23, 25–26
Linked adjacency list, 224
List of edges, 224
List scheduling anomaly, 525
List scheduling (*or* LS), 499–501
Little, J. D. C., 387
Lloyd, E. K., 442
Loading problem, 120
Local search (or neighborhood search), 368

Longest (*or* critical) path, 248, 451
Look–ahead procedure, 427
Lossy network, 297, 317
Lovász, L., 25, 35, 220
LP (*see* Linear programming)
LP equivalent problem, 26–28
LPT (*see* Largest processing time
    scheduling)
LQ (*see* Orthogonal matrix
    factorization)
LS (*see* List scheduling)
LU (*see* Triangular matrix
    factorization)
Lucertini, M., 533
Luenberger, D. G., 4, 30, 35, 38, 53

Machine (*or* processor), 484
    dominated, 479
Machine–assignment problem, 75
Machines:
    parallel (*or* identical), 486, 497
    uniform, 497
    unrelated, 497
Maffioli, F., 391
Maheshwari, N., 60, 270, 285, 300
Mahr, B., 252
Make–span (*see* Schedule time)
Malhotra, V. M., 60, 270, 273,
    285, 300
Map coloring, 394
Marble, G., 444
Marginal cost (*or* shadow cost), 168
Marsten, R. E., 218
Martello, S., 125, 146, 150, 157,
    167, 171–76
Martin, G., 100
Matched edge, 323
Matched (*or* saturated) node, 323
Matching (*or* edge packing), 212,
    320
    complete (*or* perfect), 212, 338
    initial, 327–28
    maximum–cardinality, 321
    perfect (*or* complete), 212, 338
Matching problem, 212, 321
    bipartite, 322
    cardinality, 321
    weighted, 322
Mates, 323
Mathews, G. B., 172
Matrix:
    adjacency, 233
    optimal policy (*or* path), 243
    path (*or* optimal policy), 243
    reachability (*or* transitive closure),
        249, 267
    transitive closure (*or* reachability),
        249, 267
Matrix factorization, 18–20
Matula, D. W., 406, 415, 433, 440,
    444–45
Max–flow algorithm, 57, 270, 299
Max–flow min–cut theorem, 59, 298
Max–flow problem (*see* Maximum
    flow problem)
Maximal independent set (*or* MIS),
    396
Maximum–cardinality (*or* maximum)
    matching, 321
Maximum cost, 486
Maximum flow, 56, 270
Maximum flow (*or* max–flow)
    problem, 56, 269
Maximum independent set (*or*
    MmIS), 396

Maximum independent set (*or*
    MmIS) coloring algorithm, 402–4
Maximum matching (*see*
    Maximum–cardinality matching)
Maxwell, W. L., 497, 527
McDiarmid, C., 445
McGinnis, L. F., 529
McMahon, G., 469
McMillan, C., Jr., 94, 99, 113, 116
McNaughton, R., 501–2, 533
McNaughton's rule, 502
Meeks, H. D., 36
Mendelsohn–Dulmage theorem, 340
Menger, K., 297, 386
Menger's theorem, 297
Merging activities, 450
Micali, S., 343
Miller, L. W., 497, 527
Miller, R. E., 532
Min–cost flow problem (*see*
    Minimum–cost flow problem)
Mingozzi, A., 99, 171, 386, 444
Minieka, E., 249, 318, 343
Minimum–cost flow, 301
Minimum–cost (*or* min–cost) flow
    problem, 301
Minimum spanning forest, 253
Minimum spanning tree (*or* MST),
    227, 253
Minimum spanning tree (*or* MST)
    problem, 253
Minty, G. J., 22, 31, 35
MIS (*see* Maximal independent set)
Mitchem, J., 443
Mize, J. H., 100
ML (*see* Modified level scheduling)
MmIS (*see* Maximum independent
    set)
Modified level (*or* ML) scheduling,
    509
Modified (*or* incremental) network,
    302
Moore, E. F., 235–37, 251
Moore, J. M., 494–95, 532
Moore–Hodgson's algorithm, 495
Mote, J., 300
MST (*see* Minimum spanning tree)
Mühlbacher, J. R., 341, 444
Müller, H., 444
Müller–Merbach, H., 167, 173
Multi–commodity flow, 297, 317
Munro, I., 341
Murray, W., 20, 35, 36
Murty, K. G., 51, 53, 68, 76, 387
Muth, J. F., 469, 528
Mycielski, J., 398, 433, 439
Mycielski's graph, 398, 433, 439

Narula, S. C., 112, 116
Nearest insertion, 362
Nearest–insertion approximate TSP
    algorithm, 362, 381, 385
Nearest–neighbor algorithm (*see*
    Prim's algorithm)
Negative–weight cycle, 238, 243
Neighborhood search (*or* local
    search), 368
Nemhauser, G. L., 94, 99, 171,
    217–19, 381, 386–87, 389
Network (*or* weighted graph), 222
    activity–on–node (*see* Activity
        network)
    activity–on–node (*see* Event
        network)
    activity (*or* AN), 447

Network (cont.):
    acyclic, 226
    bipartite, 55
    event (*or* AA), 448
    feasible, 457
    incremental (*or* modified), 302
    layered, 271
    lossy, 297, 317
    modified (*or* incremental), 302
    transportation, 55
Network representation, 223–25
    backward star, 225
    forward star, 224
    linked adjacency list, 224
    list of edges, 224
    weight matrix, 223
Network scheduling, 446–74
Nievergelt, J., 250, 387
Nijenhuis, A., 300
Ninamiya, K., 343
Nishizeki, T., 443
Node capacity, 270, 296
Node cover, 213, 339
Node (*or* vertex), 222–23
    demand, 55
    exposed (*or* free), 323
    free (*or* exposed), 323
    inner, 325
    matched (*or* saturated), 323
    outer, 325
    pseudonode, 326
    reference, 272
    saturated (*or* matched), 323
    sink (*or* target), 56, 228, 269,
        301, 447
    source (*or* origin), 56, 228, 269,
        301, 447
    supernode, 265, 317
    supply, 55
Node packing (*or* independent node
    set), 212, 394
Node potential, 272
Node–covering problem, 213
Node–packing problem, 212
Noltemeier, H., 252
Nondegenerate linear program, 21
Nonpreemptive scheduling, 497–99
Normal arc, 462
No–wait flow shop, 480

Oliver, R. M., 320
One–dimentional knapsack problem,
    119
One–tree, 266
Open shop scheduling problem, 475
Operation, 474, 484
Opsut, R., 443
Optimal policy (*or* path) matrix, 243
Orchard–Hays, W., 54
Ordering:
    greedy LF (*or* GLF), 427
    largest first (*or* LF), 405
    saturation–largest–first (*or* SLF),
        407, 441
    smallest–last (*or* SL), 406
    topological, 449
Ore, O., 398–99, 439, 442
Origin (*see* Source node)
Orthogonal matrix factorization (*or*
    LQ), 18–20
Outer node, 325
Out–of–kilter algorithm, 318–19

Padberg, M. W., 218, 391
Pang, C. Y., 250

Papadimitriou, C. H., 37, 100, 313, 320, 343, 390
Pape, U., 237, 251, 327, 342
Parallel edges, 222
Parallel (*or* identical) machines, 486, 497
Partial coloring, 395
Partition problem, 497
Path (*or* simple path), 226
  alternating, 323
  augmenting, 57, 270, 323
  cheapest, 227, 302
  critical (*or* longest), 248, 451
  Hamiltonian, 345, 381
  isotonic, 22
  longest (*or* critical), 248, 451
  shortest, 226
Path (*or* optimal policy) matrix, 243
Perfect (*or* complete) matching, 212, 338
Performance guarantee, 500
Performance measure, 486
  regular, 488
Permanent label, 228
Permutation flow shop, 477
Permutation schedule, 477, 488
PERT (*see* Program evaluation and review technique)
Peterson, C. C., 112, 116
PFI (*see* Product form of the inverse)
Phillips, D. T., 319, 387, 528
Pierce, A. R., 250
Pierce, J. F., 210, 217, 219
Pivot (operation), 7, 28
Plane, D. R., 94, 99, 113, 116
Plateau, G., 116, 168, 173, 176
Potential of a node, 272
Potts, R. B., 320
Powell, M. B., 405, 444
Powell, S., 100, 112, 116
(*p*,*q*)-interchange, 414
Precedence constraints, 446, 485
Preemptive scheduling, 486, 501–3
Prim, R. C., 259, 267
Primal algorithm, 5
  all–integer, 97–99
  for LP (*see* Simplex algorithm)
  for min–cost flow problem, 317–19
Primal–dual algorithm:
  for min–cost flow problem, 318–19
  simplex, 54
  for transportation problem, 65
Primal linear program, 37
Primal simplex tableau, 30
Prim's algorithm (*or* nearest–neighbor algorithm), 259, 261
Principle of optimality, 143
Processing (*or* duration) time, 455, 485
Processor (*or* machine), 484
Product form of the inverse (*or* PFI), 7
Program evaluation and review technique (*or* PERT), 446
Project, 446
  consistent, 447
Pseudonode, 326

Rabinowitz, P., 36
Randolph, P. H., 36
Ratliff, H. D., 218

Reachability (*or* transitive closure) matrix, 249, 267
Read, R. C., 444
Ready time (*see* release date)
Recursive–largest–first (*or* RLF) coloring algorithm, 404
Reduction algorithm:
  for knapsack problem, 128
  for set covering problems, 183
Redundant coloring, 396
Redundant constraint, 27
Reference node, 272
Regular performance measure, 488
Reid, J. K., 20, 35–37
Reingold, E. M., 249–50, 253, 347, 387
Reiss, S. P., 28, 35
Rejected arc, 456
Relative cost, 5
Release date (*or* ready, *or* arrival time), 485
Resource, 484
Reverse arc, 462
Revised simplex algorithm, 5–6, 9–13
Rinnooy Kan, A. H. G., 218, 387, 392, 494, 528, 530–33
Rivest, R. L., 253
RLF (*see* Recursive–largest–first coloring algorithm)
Roberts, F. S., 443
r–opt (*see* r–optimal solution)
r–optimal (*or* r–opt) algorithm, 368
r–optimal (*or* r– opt) solution of TSP, 368
Roschke, S. I., 443
Rosenkrantz, D. J., 381, 384–85, 389
Rosenstiehl, P., 252
Roy, B., 443, 529
Rudeanu, S., 112, 116, 219

S (*see* Simple sequential coloring algorithm)
Saaty, T., 443
Safety float, 454
Sahni, S., 138–39, 146, 149, 167, 171–75, 250, 268, 380, 387, 389, 475, 494, 530, 532
Saito, N., 443
Salkin, H. M., 80, 82, 86, 99, 100, 113, 116, 171–73, 218–19
Sandi, C., 99, 171, 386, 444
Saturated edge, 271
Saturated (*or* matched) node, 323
Saturating flow, 271
Saturation degree, 441
Saturation–largest–first (*or* SLF) coloring algorithm, 407, 441
Saturation–largest–first (*or* SLF) ordering, 407, 441
Saunders M. A., 20, 36
SC (*see* set–covering problem)
Schedulable activity, 520
Schedule, 484
  permutation, 477, 488
  preemptive, 486
Schedule time (*or* make–span), 486
Scheduling:
  critical path (*or* CP), 506
  earliest due dates (*or* EDD), 491
  examination, 394–95, 441
  largest processing time (*or* LPT), 500

Scheduling (cont.):
  level, 503
  list (*or* LS), 499–501
  modified level (*or* ML), 509
  network, 446–74
  nonpreemptive, 497–99
  preemptive, 486, 501–3
  shortest processing time (*or* SPT), 493
  time/cost network, 519
  weighted shortest processing time (*or* WSPT), 493
Schrage, L. E., 494, 532
Schrijver, A., 218
Search tree:
  branch–and–bound, 102, 146, 170, 346
  dynamic, 170
  static, 146
Selection, 457
Separable point sets, 28
Separation clique, 438
Sequential coloring, 404–33
Sequential with interchange (*or* SI) coloring algorithm, 414
Set covering, 176
Set–covering problem (*or* SC), 96, 176
Sethi, R., 503, 509, 530, 533
Set packing, 176
Set–packing problem (*or* SP), 176
Set partitioning, 176
Set–partitioning problem (*or* SPP), 176
Settled arc, 456
Shadow cost (*or* marginal cost), 168
Shannon, C. E., 298
Shapiro, J. F., 77, 169, 174
Shepardson, F., 218
Shisha, O., 35
Shor, N. Z., 23
Shortest path, 226
Shortest–path algorithm:
  Bellman–Moore's (*or* label–correcting), 235–37
  Dijkstra's (*or* label–setting), 228
  Floyd's, 243, 249
Shortest–path problem, 227
  between all pairs of nodes, 242
  single source and arbitrary weights, 228
  single source and nonnegative weight, 235
  between two nodes, 228
Shortest–path tree (*or* skim tree), 232
Shortest processing time (*or* SPT) scheduling, 493
SI (*see* Sequential with interchange coloring algorithm)
Sidney, J. B., 532
Simonnard, M., 54
Simple path (*see* Path)
Simple sequential (*or* S) coloring algorithm, 404
Simplex algorithm, 4
  composite, 53
  dual, 42–44
  lexicographic dual, 82–84
  primal–dual, 54
  revised, 5–6, 9–13
  tableau, 30–31
Simplex multipliers, 5

Simplex tableau:
  dual, 83
  primal, 30
Simplified linear programming, 27
Sink node (or target), 56, 228, 269,
  301, 447
Skim tree (or shortest–path tree),
  232
SL (see Smallest–last ordering)
Slack time, 452, 522
Slack variable, 3
Sleator, D., 300
SLF (see Saturation–largest–first
  ordering)
SLS (see Smallest–last sequential
  coloring algorithm)
Smale, S., 37
Smallest–last (or SL) ordering, 406
Smallest–last sequential (or SLS)
  coloring algorithm, 407
Smith, D. R., 388
Smith, W. E., 493, 523, 527, 531
Smith's rule, 493
Sollin, M., 265, 268
Sollin's algorithm, 265, 268
Sommer, D., 114, 116
Source node (or origin), 56, 228,
  269, 447
SP (see Set–packing problem)
Spanning tree, 226
  minimum (or MST), 227, 253
Spath, H., 342
Speed factor, 497
Spielberg, K., 219
Spira, P. M., 252
SPP (see Set–partitioning problem)
SPT (see Shortest processing time
  scheduling)
Stable marriage problem, 323, 339,
  343
Starting time (of a schedule), 485
Static search tree, 146
Steiglitz, K., 37, 313, 320, 343, 390
Sterns, R. E., 389
Stewart, W., Jr., 389
Subgraph, 226
Supernode, 265, 317
Sussman, B., 529
Swamy, M. N. S., 249, 268
Sweeney, D. W., 387
Switching, 458
Syslo, M. M., 14, 35, 44, 53, 66,
  76, 89, 99, 107, 116, 252, 362,
  369–70, 381, 388, 449–50,
  528–29, 531
Szekeres–Wilf inequality, 406
Szwarc, W., 77, 530

Tableau simplex algorithm, 30–31
Tardiness, 485
Tardy task, 487
Target (see Sink node)
Target flow, 301
Tarjan, R. E., 268, 511

Task, 484
Temporary label, 228
Thatcher, J. W., 532
Thompson, G. L., 469, 520, 527
3–exchange, 368–67, 374–76
3–optimal (or 3–opt) approximate
  TSP algorithm, 374–76
Thulsiraman, K., 249, 268
Time/cost network scheduling, 519
Time transportation problem, 76
Tomizawa, N., 77
Topological ordering, 449
Total cost, 486
Total float, 452
Toth, P., 99, 125, 146, 150, 157,
  167, 171–76, 386, 444
Transitive closure (or reachability)
  matrix, 249, 267
Transitive reduction, 446, 449, 518
Transportation network, 55
Transportation problem, 54–55
  maximum flow formulation, 60–61
  time, 76
Traub, J. F., 173
Trauth, C. A., 94, 99
Traveling salesman curve, 381
Traveling salesman problem (or
  TSP), 344
Tree, 226
  alternating, 324
  (binary) search, 102, 146, 170,
    346
  Hamiltonian, 381
  in–, 503
  minimum–weight spanning (or
    MST), 227, 253
  one–, 266
  shortest–path (or skim), 232
  skim (or shortest–path), 232
  spanning, 226
Triangle inequality, 345, 380, 384,
  458
Triangular matrix factorization (or
  LU), 18–20
TSP (see Traveling salesman
  problem)
Tucker, A. W., 3, 54, 386
2–exchange, 368–70
2–optimal (or 2–opt) approximate
  TSP algorithm, 370–71
Two–processor scheduling problem,
  340

Ullman, J. D., 534
Undirected graph, 222
Uniform machines, 497
Unrelated machines, 497
Update algorithm, 439
Usable edge, 278

Van Slyke, R., 268
Van Valkenburg, M. E., 391
Van Vliet, D., 239, 251

Variable:
  artificial, 6
  basic, 4
  slack, 3
Vazirani, V. V., 343
Vertex, (see Node)
Vertex independent number, 440
Vizing, V. G., 441
von Neumann, J., 3
Vuillemin, J., 253

Wagner, H. M., 174
Waiting time, 522
Walukiewicz, S., 35
Wang, C. C., 444–45
Warshall, S., 251
Weight, 485
Weighted graph (see Network)
Weighted matching problem, 322
Weighted shortest processing time
  (or WSPT) scheduling, 493
Weight matrix, 223
Weight of a path, 226
Weight of a spanning tree, 226
Weingartner, H., 172
Welsh, D. J. A., 405, 444
Werra, D. de, 443
Whitman, D., 300
Whitney, H., 386, 439
Wilf, H. S., 300
Wilson, R. J., 441–42, 445
Wing, M., 172
Wirth, N., 163, 201, 343
Witzgall, C., 319, 341
Wolfe, P., 21, 26, 36, 100, 386
Woolsey, R. E. D., 9, 99, 100
Wright, J. W., 172
WSPT (see Weighted shortest
  processing time scheduling)

Yao, A. C. C., 268
Young, R., 98–99

Zadeh, N., 318
Zahn, C. T., Jr., 341
Zemel, E., 175
Zielinski, S., 113, 116
Zionts, S., 100, 112–13, 116, 166,
  173
Zykov, A. A., 441, 445
0–1 knapsack problem (or 0–1 KP),
  118
0–1 linear programming problem (or
  0–1 ILP), 100–1
0–1 many–knapsack problem (or 0–1
  MKP), 120
0–1 multi–dimensional knapsack
  problem, 120
0–1 multiple knapsack problem, 120
0–1 ILP (see 0–1 integer linear
  programming)
0–1 KP (see 0–1 knapsack problem)
0–1 MKP (see 0–1 many–knapsack
  problem)

INDEX OF CODES

ALLINTEGER, 92
BABTSP, 358
BACKTRACKSEQCOLORING, 430
BALAS, 109
BUSACKER, 310
DIJKSTRA, 234
DSIMPLEX, 47
FITSP, 367
FLOYD, 235
INTERSEQCOLORING, 419
KNAPAPPROX, 140
KNAPBACKTRACK, 158
KNAPRED, 131
KRUSKAL, 257

LEVELSCHEDULING, 514
MATCH, 335
MAXFLOW, 289
NETWORKSCHEDULING, 470
ORDERING, 410
POM, 241
PRIM, 263
PSIMPLEX, 16
SEQCOLORING, 413
SETPARTBACKTRACK, 204
SETPARTRED, 188
THREEOPT, 377
TRANSPORT, 68
TWOOPT, 372